D0853806

NATIONAL
GEOGRAPHIC
COMPLETE
BIRDS
OF THE WORLD

NATIONAL GEOGRAPHIC

COMPLETE BIRDS OF THE WORLD

EDITED BY TIM HARRIS

NATIONAL GEOGRAPHIC
WASHINGTON, D.C.

CONTENTS

FOREWORD

When archaeologist Howard Carter first entered Tutankhamen's tomb in 1922, his impatient benefactor, Lord Carnarvon, called down, 'Can you see anything?' An astonished Carter could only reply, 'Yes, wonderful things.' As birders and bird lovers we have extraordinary opportunities to see 'wonderful things' almost anywhere in the world, often in great abundance and variety, even in our own backyards. The diversity of the world's birds—almost 9,700 living species—is astonishing. As a way to organize this diverse array, ornithologists have grouped birds into families of related species. *National Geographic Complete Birds of the World* will take you on a worldwide tour of all 193 bird families.

A fascination with birds usually begins at home, in backyards and nearby parks, among friends with similar interests, and the focus is usually on learning the local birds—perhaps a few hundred species. Rightly so; these home-based experiences will give you confidence in your skills. If you venture farther afield, the experience can be a revelation. While many bird families, such as waterfowl, sandpipers, or thrushes, have representative species distributed around the globe, other families have very proscribed distributions and are highly adapted to specific habitats—think kiwis or penguins.

Personal travel can be expensive and many now worry about their 'carbon footprint,' but travel has some overarching benefits beyond a quest to see new birds. As a group, birders spend hundreds of millions of dollars every year on ecotourism. When you travel, don't be bashful about letting people know that you are looking for birds. Communities that host special birds and special habitats benefit economically from your visits. Good stewardship of natural areas does not have to run counter to local business interests. Wear your binoculars proudly wherever your search leads you. We are fortunate to live at a time when birding field guides are available for almost all the countries and regions of the world. *National Geographic Complete Birds of the World* offers something else, a wide-angle view with a worldwide scope. Each family account gives specific details about individual species and an overview of the family's plumage, biology, and distribution, including a map of the family's worldwide range. Armed with this information and a regional field guide, your observations of birds will be placed in a whole-earth context—highlighting the similarities between unrelated families, as well as the unique adaptations of others. Some of you will go on to travel the world and perhaps see and identify thousands of birds; others will stay closer to home and keep an eye on the pulse of your local marsh or woodlot. Wherever you are, an appreciation of the complete spectrum of the world's birds will enhance your observations.

And what a spectrum it is. Consider the Ostrich, the Shoebill, and the Plains Wanderer, species so different from all others that they warrant their own exclusive families. They are not alone; 16 other families also have only a single member—Magpie-Goose, Hammerkop, Kagu, Sunbittern, Limpkin, Crab Plover, Ibisbill, Hoatzin, Oilbird, Cuckoo-Roller, Hoopoe, Sapayoa, Bristlehead, Palmchat, Olive Warbler, and Bananaquit. The largest family, the tyrant-flycatchers of the New World, boasts 400 species. About 12,000 additional species are known from the fossil record of birds. One informed researcher has estimated that 1,634,000 species—a number hard to imagine—have existed over the 150 million years of avian evolution. Our living species represent less than one percent of the species that have lived and gone extinct during the evolution of birds. Even the exact number of living bird species is never constant. Taxonomists differ on what constitutes a species, keeping the exact number in flux. With a greater understanding of vocalizations, behavior, and molecular data, researchers continue to recognize new species teased apart ('split') from those previously

An **Osprey** (*Pandion haliaetus*) snatches a fish. In the 20th century, Ospreys suffered severe decline in North America and Europe due to hunting and pesticide contamination, but with greater protection their numbers have strongly recovered. Finland.

recognized. When the 'taxonomic dust' settles in a few decades, the current world list of about 9,700 species could grow to 15,000—or more! Out in the field, discoveries of new species have mostly occurred in remote mountain valleys and jungle habitats in South America and Asia, but there have also been remarkable finds much closer to home. For example, in the United States the Gunnison Sage-Grouse was first recognized in the 1990s and formally named in 2000. This large grouse closely resembles the Greater Sage-Grouse, but dedicated researchers noticed subtle differences in its vocalizations, morphology, and plumage. The Gunnison Sage-Grouse has a small home range in parts of Colorado and Utah, where it is declining in numbers. From the moment it was first described, it became a candidate for listing as an endangered species.

When you consider planet Earth's avifauna, consider these facts as well: BirdLife International, an alliance of conservation organizations, estimates that about 1,200 species, or 12 percent of the world's bird species, are threatened with extinction. In that light, putting names to birds is not a trivial pursuit; it can be the first step in preserving and protecting them. Without names, birds are generic and often ignored, but once you attach a name to a species, both it and you are transformed. For then you can consider its nesting requirements, its feeding niche, its migratory pathways—the suite of singular qualities that set it apart from all other species. And you begin to care about its welfare, not in an abstract way but with knowledge and understanding. Because of our connection with birds, it's natural for many of us to become active conservationists. As a birder, your grassroots awareness of the local environment will place you among the best informed about conservation issues in your community. It's my hope that this book will help to inspire you to participate in preserving Earth's diversity of birds far into the future.

JONATHAN ALDERFER
Chief Consultant, National Geographic Birding Program

HOW TO USE THIS BOOK

This book is a concise but authoritative guide to the bird families of the world. It is not intended to be a field guide, and neither does it go into the finer points of birds' anatomy, functions, or evolution. However, anyone with an interest in birds will find it a useful addition to their library. In essence, *National Geographic Complete Birds of the World* provides an overview of the huge variety of avian life on our planet, both visually stunning and packed with information. It illustrates the features that distinguish one bird family from another and describes the range of variation within families. The text describes the birds' appearance, activities, and environment to give a complete picture of each family's distinct characteristics. The text below guides you through all the elements of a typical article, showing how each one contributes to the overall portrait of a bird family.

Following an introductory paragraph, the presentation of the main text for each family is divided into eight sections—Structure, Plumage, Behavior, Voice, Habitat, Movements, Diet, and Breeding. An optional section, Taxonomy, is included wherever relevant.

Fact Panel

Each family account has a fact panel that includes a map and at-a-glance information on the number of genera and species, their distribution, and any species that are considered to be under threat.

Map The map summarizes each family's distribution across the world. These maps do not differentiate seasonal (winter/summer) ranges for families with migratory species, but information on migration is found in the Movements section of the main text.

Number of genera and number of species These numbers are, in all but a few cases, based on the information given in the 2003 *The Howard and Moore Complete Checklist of the Birds of the World*, 3rd edition, one of the most authoritative bird classification systems. In the few family accounts where the numbers in this book differ (such as monarch flycatchers), it is because certain genera (singular: genus) classified as *incertae sedis* ('of uncertain placement') by Howard and Moore have been grouped within existing families, in accordance with other taxonomic authorities.

Conservation status BirdLife International estimates that around 1,200 species, or 12 percent, of the world's bird species, are threatened with extinction. The threats come in many forms, from deforestation to long-line fishing, global warming, and hunting. Never before has it been so vital to monitor the status of our planet's birdlife. The status information in this panel shows at a glance how many, if any, species are facing threats that could lead to population decline or even extinction. The grades of possible threat follow the categories used by the global environmental network IUCN, the International Union for Conservation of Nature.

- **Critically Endangered** species are those facing an extremely high risk of extinction in the wild, usually due to the loss of 80–90 percent of the population, and/or because of continuing threats such as loss of habitat. Critically Endangered species are estimated to face a 50 percent probability of extinction within 10 years or three generations.
- **Endangered** species are those facing a very high risk of extinction in the wild, for reasons similar to those given above. Endangered species are thought to face at least a 20 percent probability of extinction within 20 years or five generations.
- **Vulnerable** species are those facing similar threats to Critically Endangered or Endangered species, but at less risk. Vulnerable species are considered to face at least a 10 percent probability of extinction within 100 years.
- **Near Threatened** species are not currently at risk, but face threats (such as habitat loss) that could endanger the population in the future. Least Concern species are

Knowing the basic parts of a bird—bird topography—will help you to understand the plumage and morphology descriptions presented in the text. In general, all birds have a similar arrangement of feathers, so specific names of feather groups carry over to most families. We present two labeled illustrations; more detailed information can be found in most field guides.

ear coverts
crown
eye ring
upper mandible
nape
secondaries
tertials
lower mandible
rump
throat
uppertail coverts
tail feathers
breast
greater coverts
flank
undertail coverts
belly
tarsus
primaries

Wood Thrush (*Hylocichla mustelina*)

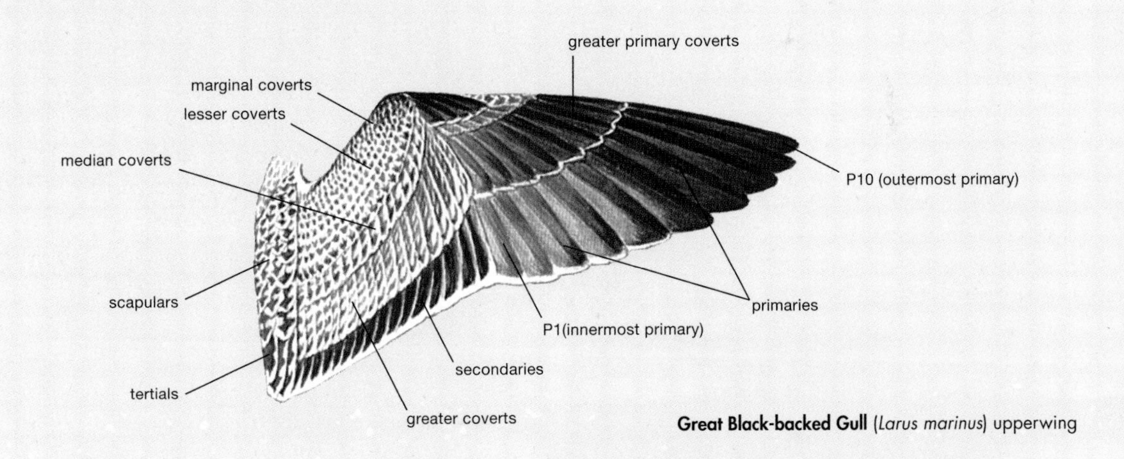

Great Black-backed Gull (*Larus marinus*) upperwing

abundant, widespread, and/or not facing any threats to population numbers or distribution.

Distribution This section adds to the information shown on the map by listing the main regions of the world where the family occurs.

Structure

This section covers the physical features that make each bird family distinctive. It includes length, measured from the tip of the bill to the end of the tail, and—in large or highly aerial species—wingspan (measured from one outstretched wingtip to the other). It also includes major anatomical points such as the size of the head, the size and shape of the bill, the shape of the body, wings, and tail, the length of the legs, and the structure of the feet. Body structure helps one to identify particular species, and can also give clues to the birds' lifestyle: for example, birds that migrate long dis-

Despite its relatively small size, the male North American **Painted Bunting** (*Passerina ciris*) is one of the most brightly plumaged of all birds, with a dazzling array of colors. Florida, U.S.

tances generally have longer and more pointed wings than those that have a sedentary lifestyle.

Plumage

Feathers do more than just keep a bird warm—their colors and shapes convey essential information to other birds and to animals, including us. This section describes the range of colors and markings seen on adult males and females; breeding plumage, and features visible only in flight or during displays; juvenal plumage, where this differs from that of adults; and any morphs (variant colorations seen in one particular species). It includes any unusual feather formations, such as crests or tail streamers. The text also mentions any distinctive molt patterns. Coloration of bare parts (bill, legs, and areas of bare skin) is also covered in this section.

Behavior

In this section, the birds' typical activities are described, including information on whether they are usually solitary or more often seen in groups; methods by which they look for food; and the way in which they move around in the air, on the ground, or in water. There may also be information on distinctive behaviors such as threat displays, courtship displays, and social interaction, although courtship is covered in greater depth in the section on breeding.

Voice

Different species are often distinctive for the sounds they make: calls, which are single notes or short series of notes; songs, which tend to be longer, more complex vocalizations used for specific purposes, such as to advertise territory; and alarm calls or contact calls. This section also mentions any nonvocal sounds that birds make, such as clapping the wings together or clicking the bill.

Habitat

Many species have evolved to fit into a specific habitat. In a few cases, unrelated species occupying similar ecological niches have even come to look alike, such as hummingbirds in the Americas and sunbirds in Africa; this process is called 'convergent evolution.' This section outlines where in the world each family is found; their typical habitat, such as forests, marshes, mountains, or deserts; and where they live—for example, in the tops of trees, or by rivers. It describes breeding territories and wintering areas, where these differ. In addition, it may mention possible threats to particular species, such as loss or degradation of their habitat, being hunted, or attacks from introduced predators (such as cats or rats).

Movements

Some species spend all year in one territory, while others may relocate to warmer areas or lower ground in winter, or may move around to find the most abundant supplies of food. This section describes whether birds are sedentary (staying in one place year-round), nomadic (moving around within a certain area), or migratory (moving to a different habitat, or even to a different part of the world). It may also mention any other types of movement that birds may make, such as the dispersal of young birds away from the area where they were raised.

Diet

The lifestyle of many birds is closely related to their food. Diet obviously influences the ways in which they forage. The energy that the food provides can also relate to the birds' level of activity; for example, small birds that do a lot of flying, such as swifts and hummingbirds, eat high-energy foods such as insects and nectar, respectively. Diet can even influence body structure, as in flamingos, which feed on tiny aquatic creatures, and have bills with 'filters' (lamellae) that enable them to sift their food out of the water. This section describes the main constituents of each family's diet.

Breeding

This section covers everything from attracting a mate to building a nest and rearing young. It includes details of courtship behavior, whether birds breed singly or in colonies, and how the young are raised: whether by the mother alone, by both parents, or with help from other individuals as well. The Breeding section also covers the type and placement of nests, the typical size of a clutch, and, in some cases, whether the birds have more than one brood in a season. Lastly, there is a brief description of the chicks' lives from hatching to fledging. Information about breeding habits is not always easy to come by—for example, in the case of little-known or very secretive birds—but this book includes the most recent data generally available.

A male **Malleefowl** (*Leipoa ocellata*) constructs his nest mound. In this unique incubation method, the male buries his mate's eggs in the mound to keep them warm until hatching. Australia.

Sunbirds, like this **Yellow-bellied Sunbird** (*Nectarinia jugularis*), resemble hummingbirds but are in fact unrelated; the two families are examples of convergent evolution. Queensland, Australia.

Taxonomy

The sequence of families in this book is based on taxonomy, reflecting their evolutionary relatedness and age; the oldest families come first. However, the composition and affinities of bird families are in an almost constant state of flux, and developments in genetic analysis have challenged many traditional views on relationships. *National Geographic Complete Birds of the World* generally follows the taxonomy of Howard and Moore, but common areas of divergence from this system are highlighted in this section.

Images

The photographs for each bird family have been chosen to show a range of representative species, performing typical activities, and set in their natural habitat. Many of the family accounts also feature artworks, some of which have been specially commissioned for this book, showing the structure and plumage of particular species in greater detail.

Additional Reading

General and family-specific references are listed, along with useful web sites.

Glossary

Unfamiliar terms used in the text are explained.

TINAMOUS *Tinamidae*

A unique family, the tinamous probably represent a very early branch of the evolutionary 'family tree' of birds. Although at first glance they resemble other terrestrial birds such as quails and partridges, their nearest relatives are thought to be ratites (ostriches, emus, and relatives), although their lineage is perhaps even older. Tinamous retain very primitive characters, including reptilian-like blood proteins and palate structure. However, unlike the flightless ratites, they have a well-developed keel and pectoral muscles, and they are capable of brief but strong flights.

The cryptically patterned plumage of this **Red-winged Tinamou** (*Rhynchotus rufescens*) helps camouflage the bird in long grass. Argentina.

Structure The birds range from 6 to 19 in. (14.5–49 cm.) in length, with females larger than males. They have a small, rounded head on a slender neck; a plump body; short, rounded wings; a very short tail; and fairly short legs. In body shape, tinamous are reminiscent of Old World quails or short-tailed francolins (family Phasianidae) but have various structural differences, including a longer, thinner bill (which is markedly decurved in some species) and a raised or absent hind toe. Males possess a penis; females have a smaller phallic organ in the cloaca. Tinamous have the smallest heart, relative to body size, of any bird, and small lungs; this combination reportedly causes them to become exhausted rapidly when pressed to run or fly.

Andean Tinamou (*Nothoprocta pentlandii*), showing the decurved bill. Peru.

Plumage Tinamous have a unique feather structure, in which the barbules of each feather are securely joined rather than hooked (as in other birds). The back and rump feathers are loosely attached, like those of doves and pigeons; it may be that when a predator tries to capture the bird, these loose feathers come away, enabling the bird to escape and leaving the predator with only a mouthful of feathers.

In most species, the sexes are alike, but females of some species in genus *Crypturellus* are brighter or more heavily patterned. Typically, the plumage is cryptically colored in various shades of gray, brown, or buff, enabling the bird to blend in well with its surroundings. There may be contrasting colors on the head, chest, or undertail coverts. The back and wings are usually patterned with vermiculated, barred, or streaked markings; the plumage of some species is especially quail-like. Two species (genus *Eudromia*) have wispy, elongated crest feathers.

The bill may be blackish, gray, or horn- to straw-colored, often with a paler lower mandible; in a few species, it is red or reddish. The iris

Number of genera 9
Number of species 47
Conservation Status 5 species Vulnerable
Distribution Central and South America

is dark brown, red, or pale yellow. Feet and leg colors may be flesh, yellow, red, brown, or gray.

Chicks hatch with a dense coat of cryptically marked down. They then develop juvenal plumage, which is generally similar to that of adults; this is replaced relatively quickly by a complete postjuvenal molt.

Behavior Tinamous are terrestrial and secretive. Most forage by picking food from the ground or vegetation. When threatened, a tinamou will freeze, walk or run away, or explode quail-like from the ground with whistling wings. At the nest, the male freezes even when intruders come very close. As a last resort, if the nest or chicks are threatened, he uses a distraction display to lure intruders away. Forest-dwelling species are generally more solitary; those on open terrain can often be found in small groups, as in Puna Tinamou (*Tinamotis pentlandii*), or in large flocks, as in Elegant Crested Tinamou (*Eudromia elegans*), which can number more than 100 individuals.

Voice Songs include a variety of clear, haunting, quavering, or flute-like whistles or *whoo* sounds, often rising or falling; also piping notes. Many species sing primarily around dawn and dusk, but some have a brief daytime song.

Habitat Tinamous occur in a wide variety of tropical and subtropical habitats, from lowland forests, cloud forest, and dry or seasonally flooded savannas, to *puna* grassland, and at elevations from sea level to more than 16,000 ft. (5,000 m.).

Movements Generally sedentary, but montane species may make limited altitudinal movements in response to weather conditions, while others undergo limited local movements to track seasonal food resources.

Diet Mainly arthropods, especially insects. Also snails, slugs, and worms; small amphibians and reptiles, and sometimes small mammals; and plant matter such as fruit, seeds, tender shoots, flowers, and tubers. Food is eaten whole or pecked into small pieces. Chicks are reported to eat a higher proportion of arthropods.

Elegant Crested Tinamou inhabits dry grassland and woodlands. Argentina.

Breeding In the tropics, tinamous may breed year-round, but generally breeding is synchronized with periods of seasonal food abundance. The birds are territorial. The breeding strategies are extremely complex and poorly known. Some species appear to be monogamous. Others may be monogamous only in early breeding attempts, and then shift to serial polyandry: females may lay a single clutch or distribute their eggs among the nests of several males, each male incubating and rearing one clutch. Simultaneously, males are polygynous, calling to attract females to their territory or nest; they may fertilize multiple females, and initiate new clutches once the first chicks are independent.

The nest is usually a simple pile of vegetation on the ground, although Ornate Tinamou (*Nothoprocta ornata*) builds a more cuplike structure, while Highland Tinamou (*Nothocercus bonapartei*) may nest in a cavity or under an overhang. Clutches comprise one to 16 eggs. Incubation takes 16–20 days, and hatching is synchronous. The chicks are precocial, able to feed themselves within a few days, and are independent in 10–20 days.

Taxonomy There are two subfamilies, mainly differing by the location of the nostrils on the bill: about midway down in the Tinaminae, and at the base in the Rhynchotinae.

A male **Highland Tinamou** at the nest. Tinamous are notable for their large, shiny eggs, in colors including blue, green, purple, red, and brown. Costa Rica.

OSTRICH *Struthionidae*

The Struthionidae is an Old World family comprising just one species, one of the world's most unmistakable birds: Ostrich (*Struthio camelus*).

Structure Ostriches are the largest living birds. Adult males stand up to 8 ft. (2.4 m.) tall and weigh up to 280 lb. (127 kg.). The females are slightly smaller. Ostrich is a flightless species. The head is small, with a fairly short, rounded bill, and the neck is long. The birds have very long, powerful legs. Their feet have only two toes, which are adapted for fast running and used in defense against predators.

Plumage Ostrich plumage is downy, lacking strong, stiff flight and tail feathers. In males, the body and wings are mostly black and white; females are duller and browner. The head and legs are mostly bare, with pinkish skin, and the neck is covered with fine, downy feathers. The subspecies that lives in far eastern Africa, Somali Ostrich (*S. c. molybdophanes*), is blacker, and its head and neck are more blue-gray, than the other forms.

Behavior Ostriches occur mostly in pairs or family groups of up to 40.

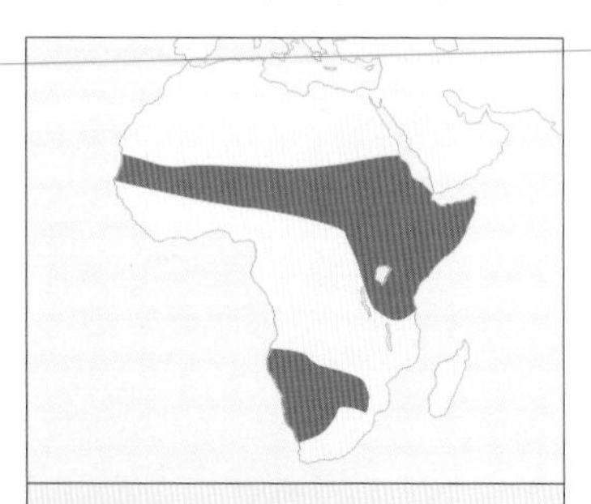

Number of genera 1
Number of species 1
Conservation Status Least Concern
Distribution Africa; formerly more widespread into Arabia

Male **Ostriches** fiercely defend their territory and their mates; the bird to the right is chasing a rival away. Namibia.

They are always alert: periods of foraging alternate with standing, head raised, looking for predators. The birds are keen-sighted and often detect approaching danger earlier than many antelopes. Their normal walking pace is a sedate 2–2.5 miles an hour (3.5–4 km/h.), but they trot at up to 18.5 miles an hour (30 km/h.). When escaping from predators such as lions or leopards, they can reach 37–43.5 miles an hour (60–70 km/h.), usually in a jerking side-to-side motion.

Voice Normally silent. Adult males become vocal at the onset of the breeding season, with a deep, booming note like a distant lion roaring and audible for up to 1.8 miles (3 km.). The call is given either to attract females or to alert other males.

Habitat Short-grassed plains and semi-desert savannas, occasionally in open woodland; Somali Ostriches also occur in dry or semi-desert scrub to 10 ft. (3 m.) high.

Movements Mostly sedentary, but in particularly arid areas becomes nomadic when not breeding, dispersing more widely in the rainy season.

Diet Ostriches feed on short, ground-level plants or the flowers and fruit of acacias; they occasionally eat grasshoppers and locusts.

Breeding Displaying males flap the wings, raise the tail, and rub necks with the female, while hissing or grunting. The nest is a scrape on the ground, usually with an all-round view; to make it, the male lies down and scratches a hollow with his feet. Eggs laid by a paired female may be added to by other females in the family group, forming clutches of up to 25. High predation, mostly by Egyptian Vulture (*Neophron percnopterus*), jackals, and hyenas, means few chicks survive. The young stay with the parents for up to 9 months. The average lifespan is 30 years.

Taxonomy Some authorities view Somali Ostrich as a separate species.

RHEAS *Rheidae*

The largest New World birds, the rheas are the evolutionary equivalents of Ostriches.

Structure Both species are flightless. They stand 5 ft. (1.6 m.) high, and weigh about 50 lb. (23 kg.); females are slightly smaller than males. Rheas have a long, slender neck, an elongated, pointed bill, powerful legs with three toes, and no tail. The wings, which are used mainly for balance while running and in breeding displays, have a strong and usually concealed claw at the elbow, which is used in defense.

Plumage Soft and mostly gray-brown. The head, neck, and thighs have short, downy feathers. Lesser Rhea (*Pterocnemia pennata*) has white spots on the upperparts, which help camouflage the bird while it is foraging in vegetation.

Behavior Rheas forage in family groups of between 10 and 30 birds, and are constantly on the move. They often associate with herds of deer, guanacos, and vicunas, a beneficial arrangement in which the birds' keen eyesight and the mammals' ability to pick up scent are combined to detect predators.

Voice Usually silent, uttering only hoarse cries and hissing threats when alarmed. Breeding males, however, give a deep, resounding, booming call when displaying.

A **Lesser Rhea** accompanies a group of chicks. Males are solely responsible for parenting the young, especially for the first six months. Chile.

Both **Greater Rhea**, shown here, and Lesser Rhea have long, soft, shaggy plumage.

Habitat Rheas live on open plains, especially pampas. Greater Rhea (*Rhea americana*) also occurs on grasslands with tall scrub, and at the margins of forests. Lesser Rhea lives on the open, arid *puna* of the Andean plateaux, at altitudes up to 11,000 ft. (3,500 m.), and on the lower Patagonian flatlands.

Movements Mostly sedentary, but flocks of rheas may cover a large area when feeding.

Diet Mainly a variety of plants, including the sharp, spiny casings of thistle burrs. Also grasshoppers and other insects, frogs, small lizards, snakes, and small birds.

Breeding In the breeding season, from July to January, males dispute territories by intertwining their necks, biting, kicking, and chasing each other in circles. A successful male attracts mates by herding groups of females into the territory with outstretched wings, while giving a booming call. The male builds the nest, a platform of vegetation concealed in bushes, and leads the females to it. Each female in the male's harem lays her eggs in turn before leaving to mate with another male and repeat the process. Incubation is carried out by the male only, and lasts for up to six weeks. The chicks are reared by the male only, and young birds remain within the group until they become adult at 2–3 years old.

Number of genera 2
Number of species 2
Conservation Status Least Concern
Distribution Much of South America

CASSOWARIES *Casuariidae*

The purpose of the cassowary's casque (as seen on this **Southern Cassowary**) is unknown. It has been suggested that the casque amplifies the sounds that the birds make, or protects the head as the birds move through thick vegetation. Australia.

Cassowaries are flightless birds related to Emu (*Dromaius novaehollandiae*). There are just three species: Dwarf Cassowary (*Casuarius bennetti*), Northern Cassowary (*Casuarius unappendiculatus*), and Southern Cassowary (*Casuarius casuarius*). All are uncommon and difficult to observe in the wild, so comparatively little is known about them. They are vulnerable to habitat loss and hunting; Dwarf Cassowary, in particular, is classified as Near Threatened, and is now absent from many areas due to hunting pressure.

Structure A large bird: Southern Cassowary is the biggest member of the family, standing at 4.9–5.9 ft. (1.5–1.8 m.). In Australia, it is the second tallest native bird after Emu. Cassowaries have a long neck and thick, powerful legs with three toes. Their skull is topped by a bony helmet called a casque, which is covered in horny material; Southern Cassowary's casque stands 6.6 in. (17 cm.) high, but those of other species are smaller.

Plumage Coarse, shaggy black feathers cover the body. Their hairlike structure provides insulation and protection as the bird forages in thorny vegetation. The bare skin of the neck and face is a striking blue, with bright red wattles or warts. Juveniles are a much plainer brown and lack a casque.

Behavior Cassowaries walk slowly through the forest, foraging on the ground or from low vegetation. They are shy birds, but can kick with great force when threatened.

Voice The birds produce a variety of sounds, particularly in the breeding season. Male Southern Cassowaries make a loud booming noise, while their threat call, usually given to warn intruders, is a rumbling growl. The male also makes a guttural coughing sound to keep in contact with the young when foraging.

Habitat Tropical forests. Dwarf Cassowary is confined to highland regions, while Northern Cassowary is a bird of coastal swamps and lowland riverine forest. Where the species' distributions overlap, they avoid competition by occupying different altitudinal ranges.

Movements All species are sedentary.

Diet Cassowaries mainly eat a wide variety of fruits. They also take seeds, berries, and other plant matter, but may occasionally prey on invertebrates and small animals. The Southern Cassowary has also been known to feed on carrion.

Breeding Southern Cassowary breeds in the dry season, June to November, usually nesting close to a tree trunk or in grasses at the forest edge. The nest is a rudimentary collection of leaves and plant litter, in which three to five large, coarse-shelled, pale green eggs are laid. The male carries out most of the parental duties, and remains with the young for around nine months.

Number of genera 1
Number of species 3
Conservation Status 2 species Vulnerable, 1 species Near Threatened
Distribution Northeast Australia, New Guinea and surrounding islands

EMU *Dromaiidae*

Emu's long body feathers hang loosely in a distinctive, shaggy mat that looks almost like hair. Australia.

Australia's largest native bird, Emu (*Dromaius novaehollandiae*) is found through much of the continent except parts of the extreme north, east, and southeastern coastal regions. A monotypic relative of cassowaries and ostriches, it is a familiar and unmistakable bird of the outback.

Structure A long-necked, flightless bird, Emu stands up to 6 ft. 6 in. (2 m.) tall and typically weighs around 77 lb. (35 kg.). Females are, on average, larger than males, and can weigh up to 120 lb. (55 kg.). The legs are long and powerful. The hard-soled feet have three toes, all of which point forward.

Plumage Dull gray-brown to dark brown. The long body feathers provide excellent protection against elemental extremes, from hot sun to snow and sleet. The bare parts of the neck and head are whitish to pale blue. Chicks have a spotted crown and black to brown body stripes; these markings are lost as the birds grow older. Birds in southeastern Australia show a pale white neck ruff during the breeding season, while birds in the north are paler overall.

Behavior Although solitary birds are often seen, Emu is gregarious, found in loose groups of four to nine birds and occasionally in larger flocks when on the move. Wary, but also inquisitive, the bird is known to approach and follow humans, simply to investigate what they are doing. It is a strong runner, with a bouncy gait, and can reach speeds of up to 30 miles an hour (50 km/h.). It can also swim well.

Voice One typical sound is a single- or double-note drumming, which is made mainly by the female. The male emits low, rolling grunts. Young birds have a higher-pitched peeping call.

Habitat Emu ranges through a wide variety of habitats, from tropical woodland to farmland and dry plains. The bird avoids only the most arid deserts and proximity to urban areas.

Emu's large eggs are dark green in color.

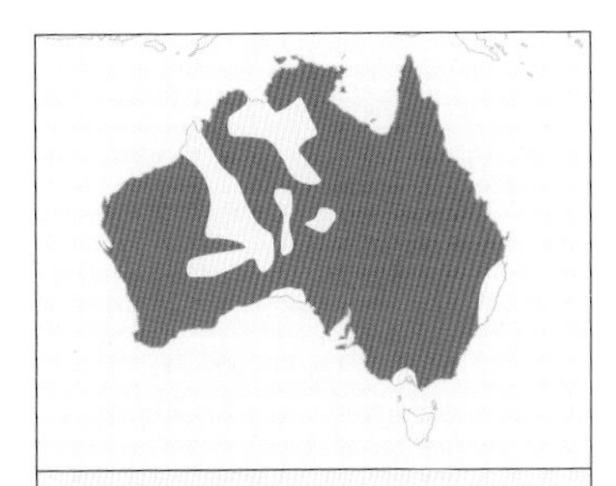

Number of genera 1
Number of species 1
Conservation Status Least Concern
Distribution Australia

Movements Mainly sedentary or nomadic. Post-breeding birds in the southern part of Western Australia migrate north to south toward the coast in spring. Elsewhere, Emus tend to be more sedentary, but range widely within their territory depending on rainfall and the availability of food.

Diet Emu is omnivorous, with seasonal variations: for example, birds often eat grasshoppers and other insects in summer.

Breeding Pairing occurs from December to January, and egg-laying between February and July, peaking in April and May. The nest is little more than a scrappy collection of bark, grasses, and other vegetation in a shallow depression. Clutches comprise 5–11 eggs. The male incubates the eggs for eight weeks and raises the young alone, remaining with them for up to 18 months.

Taxonomy A race occurring on Tasmania was hunted to extinction by European settlers in the mid-19th century, along with two isolated endemic Emu species found on King Island and Kangaroo Island.

KIWIS *Apterygidae*

The primitive-looking kiwis are flightless birds with tiny, vestigial wings and long bills. Kiwis are ratites—relatives of Emus, cassowaries, rheas, and ostriches—yet they have a very different morphology from these other families.

The birds live mainly in forests. All five species have declined since New Zealand was colonized by Europeans, due to habitat loss and predation, particularly by introduced mammals. North Island Kiwi (*Apteryx mantelli*) is classified as Endangered, and the other species listed as Vulnerable.

South Island Kiwi (*A. australis*) feeds at night, with invertebrates making up most of its diet.

Structure All kiwis are tailless and generally similar in appearance. The sexes are alike except that females have a longer bill than males. Kiwis have no keel on their sternum and therefore lack an anchor for wing muscles. The most pronounced external feature is the long, slightly decurved bill, which can grow to 8 in. (20 cm.) in some species. Unusually for a bird, the bill has nasal openings located near the tip; these have very large olfactory lobes that help the bird detect prey above or below the ground. A valve near the base allows the bird to eject water and detritus through the nostrils while foraging. The legs are widely spaced and relatively muscular. Their strong feet have four clawed toes, used for burrowing and swimming.

Plumage Kiwis have a distinctive, shaggy, hairlike covering of brown feathers, which are rough and waterproof at the tips and fleecelike at the base. Unlike the feathers of birds that fly, kiwi feathers lack an aftershaft and hooks on the barbs. Young birds have softer feathering, but look similar to adults.

Behavior The birds are almost entirely nocturnal, emerging from their burrow systems to feed after sunset. They prefer to keep to cover when leaving their roosting burrows, although they may emerge into open areas such as shoreline to feed after dark. Kiwis forage in soil and leaf litter. They have poor eyesight, so locate their food by means of their highly developed sense of smell and sensitive, probing bill. Larger insects are also detected by sound. Kiwis are highly territorial and, unusually for birds, mark their home range and their burrows with pungent droppings.

Voice All five species have a similar vocal range. Their atmospheric, far-carrying calls are heard mainly at night, when males make a high-pitched, repetitious whistling *ah-eeel*, or, in both spotted species, a more warbling whistle. Female kiwis produce shorter, more guttural notes; pairs often duet. A variety of hisses, grunts, and purrs are made during courtship and mating, and are used as contact calls between paired birds. Snorting, growling, and bill-snapping may be heard during territorial encounters.

Habitat Kiwis are birds of swampy, dense evergreen forest. They occupy a wide altitudinal range, from coastal areas to the alpine zone at around 3,900 ft. (1,200 m.). Loss of their traditional rain forest habitat has forced the birds to adapt; the largest population of North Island Kiwis now occurs in a pine plantation.

Movement Highly sedentary; often remain on the same territory for life.

Diet The birds are omnivorous, but most of their food comprises invertebrates. Earthworms are a favorite prey item, but kiwis also eat a wide variety of beetles, grubs, caterpillars, and spiders, with opportunistic feeding occurring to take advantage

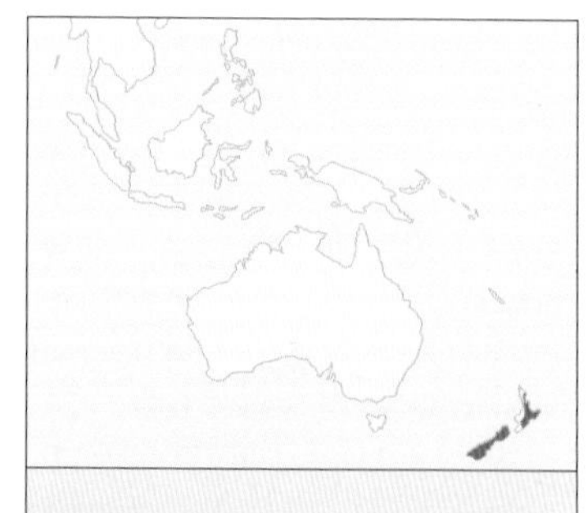

Number of genera 1
Number of species 5
Conservation Status 1 species Endangered, 4 species Vulnerable
Distribution New Zealand

A six-day-old **North Island Kiwi** chick forages among ferns. The chicks emerge from the egg fully feathered, looking like miniature versions of the adults. They can walk and feed themselves only a few days after hatching. New Zealand.

of seasonal abundance. Their body can accumulate large deposits of fat—up to 30 percent in North Island Kiwi and South Island Kiwi (*A. australis*)—which they need for the rigors of the long breeding season.

Breeding Uniquely among ratites, kiwis are monogamous for life: some Little Spotted Kiwis (*A. owenii*) have been recorded as remaining paired for 10 years. The courtship display involves the birds crossing bills while circling each other slowly. Breeding takes place in spring, around August, but the season may extend as late as February.

Kiwis lay the largest egg, in proportion to their own body weight, of any living bird. The yolk is also proportionally 50 percent larger than those of other birds. The female lays between one and three eggs in the nesting burrow, with a gap of 14–30 days between them (depending on species). The eggs are incubated by the male for 62–93 days. Despite this lengthy process, North Island Kiwi, South Island Kiwi, and Okarito Kiwi (*A. rowi*) sometimes manage to rear two broods in a season. Kiwi chicks lack an egg tooth, so break out of the shell using their feet; they are independent by 14–20 days after hatching.

Taxonomy Kiwis all belong to a single genus, *Apteryx*. Some authorities recognize just three species: Little Spotted Kiwi, Great Spotted Kiwi (*A. hasstii*), and Brown Kiwis, with the last comprising North Island, South Island, and Okarito Kiwis. Evidence of their ancestors in the fossil record has been found as far back as the Eocene epoch. Recent studies suggest that kiwis do not have a common ancestral origin with the extinct moas of New Zealand, as was once thought.

MEGAPODES *Megapodiidae*

Megapodes (whose name means 'large-feet') are medium to large ground-dwelling birds. They are the most primitive members of the order Galliformes, which also includes cracids, grouse, turkeys, pheasants, partridges, guineafowl, and New World quail. The species are usually divided into three groups: brush turkeys, scrubfowl, and the monotypic Malleefowl (*Leipoa ocellata*).

Some megapode species, like this
, build huge mounds of soil or sand with plant matter mixed into it, to incubate their eggs.

Structure The megapodes range from 12 to 27.5 in. (30–70 cm.) in length. The sexes are similar, although males tend to be slightly larger. They have the plump body, small head, and stout bill typical of Galliformes. Megapodes have large, rounded wings, which are more suitable for a quick take-off to escape predators than for covering any great distance. Their legs and feet are remarkably strong and well-developed, owing to the family's distinctive ground-nesting habits. The hind toe is placed on the same level as the front-facing toes—a feature not found in any other Galliformes except for New World cracids. Most species have bare skin on the head and neck; in the brush turkeys, both males and females have combs and wattles (those of the males being larger), which become more prominent during the breeding season. Maleo (*Macrocephalon maleo*) is unique among the family, having a curious 'skullcap' that rises to a bony knob at the back of the skull. Some scientists believe this knob may act as a shock absorber when the bird is cracking open hard-shelled food with its bill.

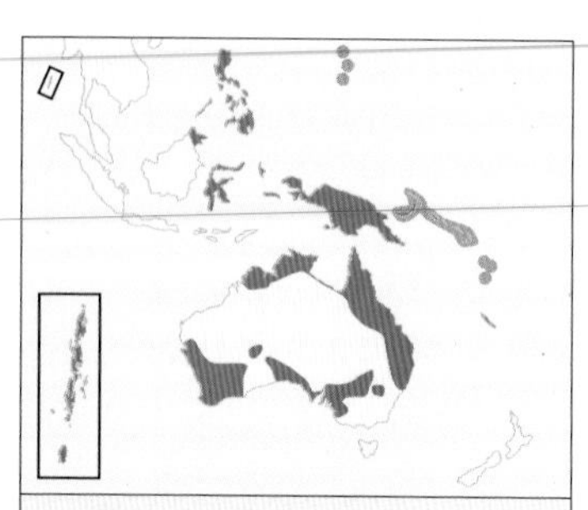

Number of genera 7
Number of species 22
Conservation Status 3 species Endangered, 6 species Vulnerable, 2 species Near Threatened
Distribution Australia, islands of eastern Indian Ocean and southwest Pacific Ocean

Plumage Most megapodes have drab body plumage that helps them avoid detection by predators. Males are often more brightly colored; male brush turkeys have more vivid yellowish, red, or blue facial skin, particularly in the breeding season. Malleefowl has the most striking plumage, with barred gray, black, chestnut, and ginger upperparts.

Behavior Away from their breeding sites, megapodes are hard to observe in the wild, so comparatively little is known about their behavior. They are generally thought to occur in pairs or alone, but some species, such as Brush Turkey (*Alectura lathami*) of Australia, form communal roosts of up to 30 birds at dusk. Malleefowl usually remain solitary for up to nine months outside the breeding season. When threatened, most species will run rather than fly; their powerful legs make them adept runners. The Black-billed Brush Turkey (*Talegalla fuscirostris*) will further outwit a predator by running in a rapid zig-zag across the forest floor, making it harder to pursue.

Voice Most megapodes are relatively vocal, which is something of a necessity in dense rain forest. They use a variety of calls to maintain contact, defend their territorial boundaries, and sound the alarm. Some scrubfowl call to each other constantly during periods of peak activity, early and late in the day. During the breeding season, paired Orange-footed Scrubfowl (*Megapodius reinwardt*) proclaim their territory with loud, whistling calls that may be heard up to 3.75 miles (6 km.) away. Male Malleefowl and Brush Turkeys produce loud, booming territorial calls; the latter has an inflatable throat sac, which gives its voice greater resonance.

Habitat Most species inhabit tropical rainforests and other types of wooded habitat, with the greatest diversity occurring in New Guinea. Malleefowl inhabits the semi-arid

Megapodes, like this **Orange-footed Scrubfowl**, use their strong claws to rake over leaf litter, soil, and vegetation while foraging; the disturbed ground is often exploited by other birds looking for an easy meal. Komodo Island, Indonesia.

zone (known as 'mallee' country) of southern Australia.

Movements Most seem to be sedentary, but very little definitive information is available for the most secretive species. Some scrubfowl have been observed 'island-hopping' across several miles of open sea.

Diet The birds are generally omnivorous, but there are wide variations. Some scrubfowl mainly eat invertebrates. Malleefowl is largely vegetarian, although it occasionally takes fungi and small animals. For most species, seeds and fruit are major foods when in season, although chicks are fed extensively on small, nutrient-rich invertebrates.

Breeding Adults play no part in incubation; instead, the birds dig burrows or build mounds in which they lay the eggs. Mound-builders make a huge heap of soil and vegetation. As the vegetation rots, it generates heat, keeping the eggs at the correct temperature. The male Malleefowl actively monitors his mound. He can detect the internal temperature to within a few degrees, by inserting his bill into the structure. The bird regulates the temperature daily by removing or replacing part of the mound covering; this involves moving huge quantities of sand and soil. Burrow-nesters such as Polynesian Megapode (*M. pritchardii*) exploit local geothermal heating to maintain the correct temperatures. Several of these species rely entirely on warmth generated by decaying tree roots. Some megapodes are communal nesters. Two pairs of Orange-footed Scrubfowl may use the same mound at once. More extremely, an estimated 4,000–5,000 pairs of Moluccan Megapode (*Eulipoa wallacei*) occupy burrows on the same beach on Haruku Island in a single year. Chicks use their strong legs to kick open the eggshell, then burrow upward for 12–48 in. (30–120 cm.), taking up to several days to reach the soil surface. The young are totally independent, and their parents play no part in rearing them. They hatch almost fully feathered, and are able to fly soon afterward.

A male **Brush Turkey**. This Australian species is distinctive for its red head and yellow wattle. Queensland, Australia.

CURASSOWS AND GUANS *Cracidae*

Chachalacas, like this **Plain Chachalaca** (*Ortalis vetula*), are the smallest birds in the cracid family. Mexico.

This Neotropical family of medium to large, chicken-like birds, sometimes known collectively as cracids, is the most arboreal of the Galliformes (pheasants and related birds). Two subfamilies are typically recognized: the Penelopinae (chachalacas and guans, comprising 36 species) and the Cracinae (curassows, 14 species).

Structure The birds range from 16.5 to 36 in. (42–92 cm.) in length. The different species show slight to pronounced dimorphism in size: in most cases, males are larger than females, but the female Highland Guan (*Penelopina nigra*) is larger than the male. Curassows are proportionately stouter than the other members of the family.

The birds have a small head on a long, relatively slender neck. The head or face are festooned with a horn or knobs, a fleshy cere, dewlap, or wattles; these features are often less developed in females or immatures. Curassows in genera *Mitu* and *Pauxi* have a highly modified bill, with an exaggerated casque in Razor-billed Curassow (*M. tuberosum*) and in the helmeted curassows (genus *Pauxi*). Ornamentation increases with age, older males often being more conspicuously adorned.

Many species have an elongated and looped trachea, which enables them to produce booming or trumpeting calls. Guans and chachalacas have a distensible esophagus (gullet) to store food prior to digestion, while curassows have a true crop. All have a gizzard, but the extent of development varies between species.

Cracids have broad, rounded wings and a long, rounded tail. Long, powerful legs allow them to walk and run well. The feet have a long hind toe at the same level as the front toes; this helps the birds perch and move through vegetation.

Plumage Generally cryptically colored, with brighter bare parts and bill. The sexes are alike in chachalacas and guans (except Highland Guan). The curassows are dichromatic, or have dichromatic forms (except for Nocturnal Curassow, *Nothocrax urumutum*).

The plumage shows various shades of black, gray, brown, and olive, and is glossy in some species. It may be uniform in color, or patterned with paler scalloping, stripes, or bars; many species have differently colored tail tips, bands, or wing patches, or white on the underparts. Morphs, such as barred or rufous morphs, exist in several species; in a few, some individuals show white mottling. Many cracids have a crest, which is most elaborate in some curassows. Most guans have modified quills, on the outermost primaries, that they use to make sounds.

Bare-faced Curassow (*Crax fasciolata*) is highly dichromatic. This bird is a female; males are all black, apart from a white belly and a yellow bill and cere. Brazil.

Chicks hatch with dense, cryptic down, which is replaced by juvenal plumage. Adult plumage is usually acquired at about one year (two years in Highland Guan). The molt is poorly studied; adults probably have one long overlapping molt (for example, three generations of flight feathers may be present at once).

Number of genera 11
Number of species 50
Conservation Status 3 species Critically Endangered, 6 species Endangered, 7 species Vulnerable
Distribution Southern North America and South America

Behavior Cracids are gregarious, occurring in pairs or family groups; chachalacas are commonly found in larger flocks. The birds are primarily diurnal, but foraging has been reported before sunrise, after sunset, or during a full moon. During the day, they spend long periods maintaining their feathers by preening, sunning, and dustbathing. They roost in trees; some species, especially chachalacas, roost communally. Cracids fly well but rarely for long distances. Instead, they use their wings and tail to maneuver to a treetop, then launch themselves off and glide to their destination.

Voice Both sexes are highly vocal and very loud. Chachalacas and guans have a variety of calls and songs, which include cacophonous cackling, grunts, growls, screams, piping notes, and clear whistles, as well as the chachalacas' *cha-cha-lac-a* call, for which they are named. Most species duet. Male curassows have a song that comprises booming or hooting notes or whistles, and both sexes make an assortment of whistles or popping notes. Chicks produce a soft peeping sound.

Habitat Cracids prefer wooded habitats: humid to dry lowland and montane forests, woodland, mangroves, and savanna, from sea level to 12,795 ft. (3,900 m.). Loss and degradation of habitat, and subsistence hunting, pose the greatest threats to the birds.

Movements Generally sedentary, but some species undergo local seasonal movements, often following food sources, or due to changing water levels in the dry and wet seasons.

Diet Primarily seeds, fruit, leaves, and flowers. Cracids may also take insects and larvae, and other invertebrates including spiders, centipedes, and mollusks; rarely, they eat vertebrates including amphibians and small rodents. Grit is consumed to aid gizzard function; any bulky indigestible material is regurgitated.

Breeding The breeding season occurs at times of peak food availability. The birds perform their courtship displays during or before twilight; Nocturnal Curassow sings on moonless nights. Displays are performed by males only or by both sexes. In chachalacas, they include strutting, mutual preening, chasing, and courtship feeding. Guans (except for Horned Guan, *Oreophasis derbianus*) use their modified wing feathers to make 'wing-drumming' sounds in flight. Curassows perform 'dances' on the ground, involving various postures and movements such as wing-flapping and mutual feeding.

Of the species that have been studied, most are monogamous with a strong pair bond, and pairs defend their breeding territory. Some curassows are believed to be polygynous; males display at a lek and pair with several females.

The male builds the nest, sometimes assisted by the female. The nest, a platform of sticks or coarse vegetation and leaves, is usually placed in a tree or shrub; Highland Guan and Rufous-vented Chachalaca (*Ortalis ruficauda*) build theirs on the ground. Clutches comprise one or two eggs in curassows, or three or four in guans and chachalacas. They are incubated by the female for 24–36 days (longest in Northern Helmeted Curassow, *Pauxi pauxi*). Chicks leave the nest within hours or days of hatching, and are brooded by both sexes. They are fed for the first few days, occasionally much longer, by both parents or by the female only. They remain with their parents for several weeks; Red-billed Curassow (*Crax blumenbachii*) attends its young for four months.

White-winged Guan (*Penelope albipennis*), distinctive for its red wattle and purplish facial skin, is Critically Endangered, with possibly fewer than 100 individuals left. Peru.

GUINEAFOWL *Numididae*

A small Old World family of terrestrial birds, the guineafowl are most closely related to pheasants (family Phasianidae). They are endemic to sub-Saharan Africa, but one species, Helmeted Guineafowl (*Numida meleagris*), has been domesticated and introduced to countries around the world.

Structure Guineafowl are plump, large-bodied birds ranging from 16 to 22 in. (40–56 cm.) in length. The sexes are alike, but males are slightly larger or heavier than females. Most species are of similar structure, with slight variations in head and crest shape. They have a small, bare head adorned with wattles, a casque, or a crest, on a fairly long or slender neck. The bill is rather short, slightly decurved, and generally strong; it is largest in Vulturine Guineafowl (*Acryllium vulturinum*). The wings are short and rounded; these short wings, together with the large body, make the birds generally weak fliers. The tail is also fairly short and mostly obscured by the long uppertail coverts. Guineafowl have short legs (longest in Vulturine Guineafowl), but the legs and feet are strong. The two species in genus *Agelastes* have well-developed spurs on the legs, while Vulturine Guineafowl has vestigial spurs.

Plumage Mostly black or bluish-black. Four species are finely spotted with white, often in the form of streaks on the wings and tail. White-breasted Guineafowl (*Agelastes meleagrides*) has an entirely white breast. Vulturine Guineafowl has a broad tuft of chestnut feathers across the nape and the neck, and long, pointed hackles of black, bright blue, and white feathers.

All species have a bare head, which may be colored pink, red, blue, gray, or yellow. Black Guineafowl (*A. niger*) has a pale pink head and neck with a short, bristle-like crest. Helmeted Guineafowl has an erect, bone-colored casque and red and blue wattles. Plumed Guineafowl (*Guttera plumifera*) and Crested Guineafowl (*G. pucherani*) have prominent crests: long and straight in the former, and shorter and more curly in the latter.

Behavior Almost entirely terrestrial. All species are highly social, most forming groups of up to 30 birds;

Helmeted Guineafowl has become the best known and most widely domesticated of all the guineafowl species. Kruger National Park, South Africa.

Vulturine Guineafowl is the largest and perhaps most striking of the guineafowl: it is easily recognized by the bright blue breast and long, glossy, striped hackles. Kenya.

however, in the two *Agelastes* species, flocks generally number fewer than 10, consisting of males with harems or large family groups. Helmeted Guineafowl are the most gregarious, with flocks of more than 2,000 birds found outside the breeding season, usually close to a foraging area or water-hole. Dominant males often act as scouts, adopting prominent perches to scan the surrounding area for approaching predators such as jackals or baboons. The presence of a water hole appears to be more critical to some species than others; Vulturine Guineafowl, which occurs at the edges of desert areas, has rarely been recorded at water.

The birds forage in leaf litter and loose soil, scratching like chickens with their strong feet, and on the ground below fruiting trees. Open-country and grassland species often feed on swarming insects, such as locusts, by advancing in a line abreast and picking up the insects along the way. Most foraging takes place in the early morning, and the birds spend the hottest part of the day in the shade. Several species roost in trees and make noisy pre-roosting gatherings late in the day.

Voice All species are fairly noisy, especially in the early morning and evening. Calls are usually loud or raucous, and include rhythmic, harsh, grating and clucking phrases interspersed with softer or more musical notes; they often lessen in speed and volume before beginning again. The alarm calls of most species are particularly loud, often including high-pitched, squealing notes, Helmeted Guineafowl has a nasal, rising, staccato, almost braying alarm: *kek kek kek, ka ka ka kaaaaa ka ka.*

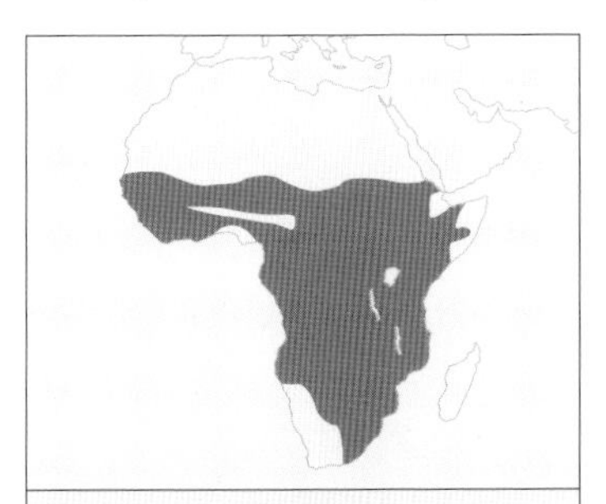

Number of genera 4
Number of species 6
Conservation Status 1 species Vulnerable
Distribution Sub-Saharan Africa

Crested Guineafowl is most commonly found in scrub or dense thickets, where it forages for food by scratching in leaf litter. South Africa.

Habitats Open scrub and grasslands or the edges of dense rain forest. The natural range of guineafowl is entirely within Africa, south of the Sahara. However, the birds have been introduced elsewhere and have become widely domesticated; in some cases, feral populations have become established from these domestic birds. White-breasted Guineafowl is considered vulnerable due to the continuing destruction and fragmentation of primary forests in its natural West African habitat, although it has been introduced elsewhere in the world.

Movements All guineafowl species are entirely sedentary.

Diet Mainly insects, including ants, termites, grasshoppers, and locusts, as well as small mollusks. Also seeds, leaves, bulbs, some roots (especially tubers), fallen fruit, and grain including millet, maize, and sorghum.

Breeding Those species that are well-known are monogamous, but the breeding behavior of the *Agelastes* species is almost unknown. At the onset of the breeding season, males become aggressive toward one another, and ritualized posturing, chases, and flights often ensue; the more successful males court, feed, and then pair with attendant females. Nests are simple scrapes or hollows in the ground, occasionally sparsely lined with leaves, grass, and feathers. Clutches comprise 4–20 eggs, ranging in color from white to pale rufous-brown and unmarked or finely speckled. The female incubates the eggs alone, for 23–28 days. The downy young (known as keets) leave the nest and forage with the parents within one or two days of hatching; they can fly short distances by two to three weeks of age.

NEW WORLD QUAIL *Odontophoridae*

Male **Northern Bobwhite** are easily recognizable due to the white throat and eye line. Females have similar plumage, but the throat and eye line are buff-colored. Texas, U.S.

The odontophorids comprise the New World quail, wood partridges, bobwhites, and wood quail. The birds resemble Old World quail in shape and plumage, but they belong to a different family.

Structure All New World quail are relatively chunky birds 7–15 in. (17–37 cm.) in length, with a 'potbelly.' They have a short, stout bill, similar to those of grouse and pheasants, but with unique serrations near the tip, probably used for cutting vegetation. The wings are stubby and rounded. The legs are short and stocky, with chickenlike feet that have three toes pointing forward; the legs and feet are well adapted for the birds' mostly terrestrial existence, both for running along the ground and for scraping the soil in search of food.

Plumage Some New World quail are sexually dimorphic. The Neotropical wood quail species are less so, and the sexes are essentially identical in the wood partridges. In most of the quails, the plumage is black, brown, and gray, well-suited for blending into their environment, but on close inspection, many have elaborate patterns of black and white streaks or spots. The male Montezuma Quail (*Cyrtonyx montezumae*) has an ornate black-and-white 'harlequin' face pattern, with black and white spotting along the sides of the body, yet the streaked back renders it nearly invisible. The widespread Northern Bobwhite (*Colinus virginianus*) has a distinctive white throat and extensive white streaking and spotting on the underparts. Several of the wood quail have bright rufous on the underparts or head. The wood partridges have a much longer tail than other genera. Several of the New World quail species have crests, or have modified crown feathers that form a 'topknot.' This feature is most notable in Mountain Quail (*Oreortyx pictus*), which has two elongated feathers sticking straight up 3–4 in. (7.5–10 cm.) from the top of the head.

Behavior The birds are usually found in groups from small family groups to large 'coveys,' which are probably made up of multiple families. These groups break up during the breeding season, when males and females pair off to nest. The birds typically forage on the ground by digging or scratching the surface in search of insects or seeds, and they have been known to tear leaves or buds from trees or shrubs. Males often sit up on exposed branches or rocks and vocalize. Flights are usually short and low, with an explosive take-off.

Voice All species have a variety of calls. Males give an advertising call during the breeding season, and both sexes give contact calls to locate each other. Songs are usually short and explosive in nature, as with Scaled Quail (*Callipepla squamata*), which gives a loud, repeated *kuck-yur*. In contrast, Singing Quail (*Dactylortyx thoracicus*), endemic to Mexico, has a longer, more complex song beginning with introductory whistles, and followed by a series of rollicking notes. Wood quails and wood partridges are very vocal, giving loud, emphatic, rollicking calls at dawn and dusk, often as duets. Most species have soft clucks as contact notes when in coveys.

Habitats Different species inhabit a wide variety of environments, in-

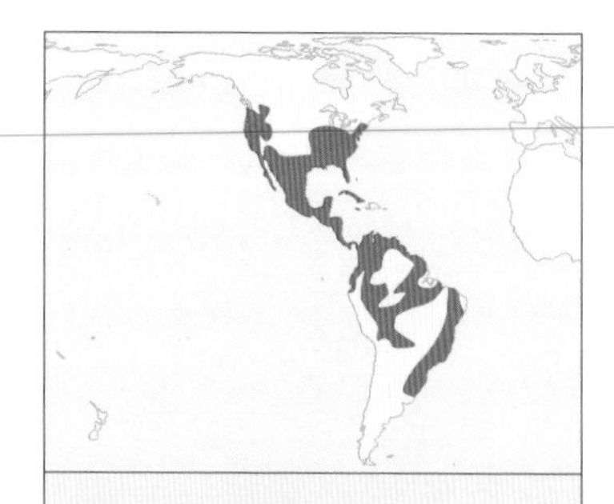

Number of genera 9
Number of species 32
Conservation Status 1 species Critically Endangered, 4 species Vulnerable
Distribution North, Central, and South America

A male **Gambel's Quail**. These birds are notable for their prominent head plume, found in both sexes; males also have a pale, buff-colored chest. Sonoran Desert, Arizona, U.S.

cluding desert, pine and oak forests, cloud forest, and rain forest. Several quail species are found in North America, mainly in the western states of the U.S., such as Gambel's Quail (*Callipepla gambelii*), which lives mainly in the Sonoran Desert, and California Quail (*Callipepla californica*), found primarily in the oak-dominated chaparral in the mountains of coastal California. The only quail species occurring in the eastern U.S. is the widespread Northern Bobwhite. There is a higher diversity of species in the Neotropics, including three species of wood partridges in the genus *Dendrortyx*, found mainly in the highlands of Mexico, and 15 species of wood quail in the genus *Odontophorus*, found mainly in both lowland and montane humid forests in Central and South America.

Movements New World quail are generally sedentary. However, some species will move around locally; one example is Northern Bobwhite, which occupies one area during the breeding season and another during the winter.

Diet Mostly seeds, buds, or bulbs that the birds find on the ground or just under the soil surface. They have also been known to take seeds or fruit from trees and shrubs. Most species feed exclusively on plant matter, while some, such as Montezuma Quail, supplement their diet with insects and grubs, particularly in the summer monsoon months.

Breeding Males are not thought to defend specific breeding territories, although some have been known to share incubation. The New World quail are generally monogamous, but some polygamy has been reported. Nests typically consist of a scrape on the ground, which is often lined with vegetation, and usually situated within dense cover. Few wood partridge nests have been found, but those that have been seen have generally been covered by a canopy or dome. Both males and females participate in nest-building, and both provide food for the young. The incubation period varies from 16 to 30 days, depending on species, and the eggs hatch synchronously. The young are precocial: they can walk and feed within a few hours, and fly within a couple of days. Some species have multiple broods in a year.

Taxonomy The Odontophoridae were formerly classified as a subfamily of the Phasianidae, but they are now considered by most authorities to be a distinct family.

Montezuma Quail is found on hillsides with grass, scrub, or oak or pine woods. Arizona, U.S.

GROUSE AND PHEASANTS *Phasianidae*

A large family, the Phasianidae include grouse, turkeys, capercaillies, partridges, Old World quail, francolins, tragopans, monals, pheasants, junglefowl, and peafowl. The junglefowl in the genus *Gallus* are the ancestors of all domestic chickens. Within genera most species are very similar in shape and structure, but there are wide differences in plumage, and sexual dimorphism of plumage and size is common.

Structure Small to very large or long-tailed; the smallest are the Old World quail at around 5 in. (12–13 cm.) long, whereas the long-tailed Reeve's Pheasant (*Syrmaticus reevesii*) is about 82 in. (210 cm.) and Green Peafowl (*Pavo muticus*) is about 98 in. (250 cm.), with the tail at least half the length of the bird. In most of the grouse the tail is rounded (or lyre-shaped in two species), and in the quails and francolins it is short and square-ended. All, apart from the quails and partridges, are heavily built birds, the heaviest being Green Peafowl at about 11 lb. (5 kg.) and some male Western Capercaillie (*Tetrao urogallus*) at around 14 lb. (6.5 kg.).

The bill in most species is short, relatively thick, and pointed, with the tip of the upper mandible overlapping the lower. All have short, rounded wings. The legs and feet are stout and very well adapted for ground-living, with thick soles and four toes. Some of the pheasants and the peafowl are long-legged, Several of the francolins, junglefowl, pheasants, and peafowl have pointed spurs on the rear of the leg, but spurs are absent in all of the grouse, which have feathered legs.

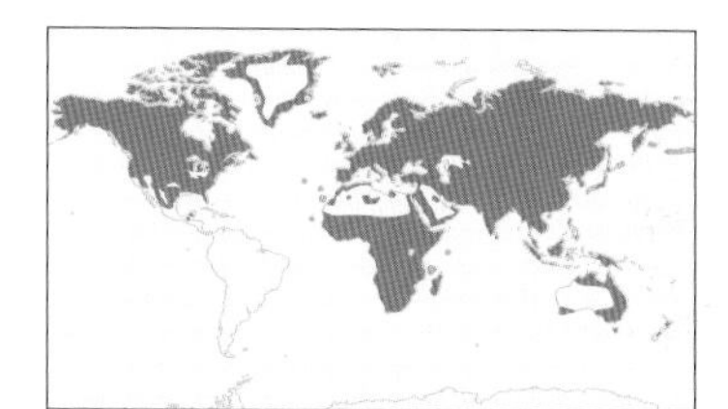

Number of genera 45
Number of species 172
Conservation Status 1 species Extinct, 2 species Critically Endangered, 9 species Endangered, 33 species Vulnerable
Distribution Widely distributed across North America and throughout the Old World

Plumage Highly variable, from drab browns and blacks to bright or deep iridescent blue to reds, golds, and silvery-white. The grouse are mainly brown or black and cryptically

A male **Blue Grouse** (*Dendragapus obscurus*) displaying. During displays, the bird's magenta neck sacs are inflated, thus exposing the white-based feathers around them, and the combs over the eyes stand out. U.S.

marked with fine barring, broad streaks, or large spots. Several of the grouse species have prominent red combs over the eyes. The females are usually duller or more cryptically marked. Several of the smaller grouse and the two species of ptarmigan undergo a body molt in late summer and become partly or entirely white. The quail, francolins, and partridges are mostly warm brown and buff, with long, two-tone flank stripes in the partridges, and richer or more intricate patterns in the bush quail and hill partridges. Several of these species have distinctive face patterns, often with patches of bare red skin around the eye in the francolins.

A male **Indian Peafowl** displays his fanned tail. The 'tail' actually comprises greatly elongated, modified uppertail coverts rather than actual tail feathers (rectrices). Sri Lanka.

Male pheasants are undoubtedly among the most brightly plumaged birds, with almost every color represented, often in a single species. The males of several species have elaborate, brightly colored face patches, crests, or ruffs, and a long, multicolored tail. Male tragopans have a bright blue or red face, which becomes enlarged during display; in male Temminck's Tragopan (*Tragopan temminckii*), two pale blue 'horns' extend back from the eyes, and a large electric-blue and red lappet inflates to cover the throat and upper breast. Junglefowl are also brightly colored: male Red Junglefowl (*Gallus gallus*) has glossy, deep brown and green plumage, with an orange-red ruff, a red comb and wattles, and a greenish-black, plumed tail. The peafowl have spectacular plumage, highly glossed in bottle-green and deep blue with long, flowing, filamentous tails displaying orange and black eye patterns.

Behavior Social behavior varies between genera and species, some species being entirely solitary, meeting only to mate with females, and others living usually in pairs or large family groups throughout the year. The more social birds often display fairly elaborate defense of females or territories. Most species forage on the ground, many scratching in the surface soil or litter, and several of the larger pheasants, monals, and tragopans dig in soft earth. Most species rarely fly far, but some pheasants roost in trees and often take food from within bushes and the lower levels of trees.

Voice A variety of dry or harsh notes or repeated crowing phrases, including cackles, hisses, grunts, and whistles. Displaying males also have a series of hooting, bubbling, clucking, or purring notes. Common Quail (*Coturnix coturnix*) has a whiplike *wet-my-lips* call. Territorial male junglefowl have the familiar cock-crowing call of domestic fowl.

Habitat Very varied, including lowland open fields and grasslands; desert margins; temperate, subtropical, and tropical forests; and montane forests and pastures. Several grouse species occur on open tundra well north of the Arctic Circle.

Movements Most species are largely sedentary, but some may make irregular or nomadic movements in search of food when not breeding; tundra-breeding populations of Willow Grouse (*Lagopus lagopus*) move up to 190 miles (300 km.) south to the edge of forested areas. Three species of quail are truly migratory, and of these the longest-distance migrant is Common Quail. This species breeds across Europe to central Asia, and moves south to winter in Africa along the Nile Valley and south of the Sahara; eastern birds winter in India south of the Himalayas.

A male **Temminck's Tragopan** engages in a display, extending the colorful lappets of skin under his throat. Vietnam.

Diet Grouse eat mostly seeds, buds, flowers, and shoots, including those largely shunned by other species, such as heather, sage, and pine needles. They may also take some insect food in spring or summer. In some areas, grouse may be entirely dependent on a single species of plant: for example, Sage Grouse (*Centrocercus urophasianus*) feeds entirely on sagebrush, and Willow Grouse on heather. Most pheasants and partridges take similar amounts of vegetable food (leaves, bulbs, roots, and tubers) but also eat earthworms, snails, and slugs, and may feed on fruit. Some species forage on spilled grain and grass seeds.

A male **Reeve's Pheasant**. This species has a boldly marked head and an extremely long, silvery, barred tail—the longest tail of any bird species. China.

Breeding The smaller species (the quails, francolins, and partridges) are mostly monogamous, but the grouse and pheasants are highly polygamous. Several of the grouse have well known and fairly elaborate displays, or leks, at which several males compete against each other in vocal or drumming displays, with fluffed-up plumage accompanied by wing- or tail-rattling dances or strutting postures, and possibly by songflights. Several species, including Sage Grouse, Sharp-tailed Grouse (*Tympanuchus phasianellus*), and the prairie-chickens, develop brightly colored combs and inflatable sacs on the sides of the neck. Male Capercaillies, normally shy and retiring, become bold and often aggressive to intruders, including humans, when defending their breeding territory. Pheasants have extremely noisy displays, which include repeated beating of their wings together with crowing, hissing, clucking, and cackling calls. Monogamous species lack the elaborate social displays of the polygamous species. Nests are almost always built on the ground, but several tragopans nest low down in trees. The nest is usually a simple, shallow scrape, sometimes lined with grass, twigs, and leaves. Clutch size varies from two eggs in the two argus pheasants to 5–10 in most grouse and up to 15 in some of the partridges and quail. Incubation periods vary from 17 days in quail to about 28 days in the larger pheasants. Chicks of most species leave the nest and forage with the adult female or the rest of the social group within a few days of hatching.

Two male **Black Grouse** (*Tetrao tetrix*) compete for females at a lek. During this communal display, males crouch and move in circles, while inflating the combs above their eyes and spreading their wings and tails to show the striking white feathers. Finland.

SCREAMERS *Anhimidae*

A pair of **Southern Screamers** with their chick. Both members of the pair advertise their territory by calling loudly, often in unison. Argentina.

The screamers are an ancient South American family. They are related to waterfowl (family Anatidae) and gallinaceous birds (family Phasianidae). The best-known species is Southern Screamer (*Chauna torquata*), which is common on the plains of southern South America. Horned Screamer (*Anhima cornuta*) is less common, and Northern Screamer (*C. chavaria*) is listed as Near Threatened.

Structure Screamers are big-bodied birds, 28–38 in. (70–95 cm.) in length. The largest species weighs up to 10 lb. (4.5 kg.). They have the most pneumatized skeleton of any birds: even the outermost toe bones are hollow. In addition, their skin contains many tiny air sacs.

The anatomy shows clear similarities to waterfowl, yet in body structure, bill shape, and feet, the birds resemble chickens or turkeys. The head is small, and the bill is short and thick at the base, like that of a chicken. The wings are large, broad, and armed with spurs up to 2 in. (5 cm.) long, and the tail is short. The legs are long, thick, and strong, with unwebbed toes. Horned Screamer has a long, thin 'horn' on the head, formed by a modified feather shaft.

Plumage The feathering is evenly distributed over the skin, rather than grouped in tracts as on most modern birds. Southern Screamer and Northern Screamer are gray with a black neck band, white cheeks, and a gray crest. They also have reddish bare skin around the eyes, and pink legs. Horned Screamer is black with a white belly and a barred black-and-white neck. Its soft parts are blackish, but the eye is red and the horn is gleaming white.

Behavior Screamers are most often seen in pairs, with males sitting on an elevated position watching for danger as the female or family group forages. The birds forage in shallow water or while walking across water vegetation. After breeding, adjacent nesting pairs may gather in small flocks for the winter. Commonly, one or both of the pair take flight in the late morning, and soar high over their breeding area, calling loudly.

Voice Screamers, as their name implies, have loud voices. However, rather than screams, the vocalizations are loud bugling or honking sounds, sometimes performed in a duet. All screamers also make low-pitched booming or rumbling sounds, only audible at close range.

Habitat The birds inhabit wetlands in lowland tropical areas. These habitats range from forested swamps with floating vegetation for the tropical species, to open marsh adjacent to grassland for Southern Screamer.

Movements All species are sedentary.

Diet Varied; includes insects and vegetable matter such as seeds.

Breeding Screamers appear to pair for life. They build a large nest of sticks and softer vegetation, placed just above the water's surface. Four to six eggs are laid, and are incubated by both parents for 40–45 days. The young are precocial; they fledge at 60–75 days, but the parents care for them for a further few weeks.

Number of genera 2
Number of species 3
Conservation Status 1 species Near Threatened
Distribution South America

MAGPIE-GOOSE *Anseranatidae*

The monotypic Magpie-Goose (*Anseranas semipalmata*) is considered by some authorities to be one of the oldest and most primitive of the order Anseriformes (ducks and geese). Its structure and behavior suggest that it may be an ancestral link between wildfowl and the screamers of South America.

One distinctive feature of **Magpie-Goose** is the knob on its crown. The large size of this bird's crown suggests that it is a male. Note also the bare, reddish face. Australia.

Structure The bird is 30–35 in. (75–90 cm.) in length. It has a bare face, a large bill with a hooked tip, and a long neck. The wings are large and rounded, and the legs are long, with partially webbed feet. The crown is peaked, this feature usually being more marked in males.

Plumage Black neck, breast, flight feathers, and tail; white elsewhere. Unlike other water birds, it does not lose all its wing feathers at once when molting, so has no flightless period.

Behavior Like all geese, Magpie-Goose is social. It is often found in flocks of several hundred to the low thousands, and occasionally up to 80,000. If a predator approaches, every bird in the flock raises its neck and may take flight almost vertically, honking loudly. The birds roost on dry earth banks or in mangroves and trees in wet woodlands. They feed on land, or forage from underwater mud by upending in the shallows.

Voice A series of noisy honks, which are louder and higher-pitched in males than in females.

Habitat Magpie-Goose occurs mostly at the edges of shallow wetlands with areas of sedges and rushes, or in swamps and flooded grasslands. In some areas, wetlands are only temporary or seasonal following rainfall, and birds move on once these dry out. The range formerly extended into southeast Australia, but the drainage of swamps and their replacement with agricultural land has reduced the extent of suitable habitat and may threaten further reductions in future.

Movements The bird's movements are mainly determined by the availability of food, and in dry years flocks may travel some distance in search of foraging areas. Some of the birds that breed in northern Queensland move north to southern New Guinea.

Diet Grazes on grasses, sedges, seeds, rush bulbs, and sedge rhizomes dug from the ground with the bill.

Breeding In the south of its range, Magpie-Goose breeds between August and October; in the north, between February and April. It is polygamous; following a noisy display, males usually pair with two females. Nests are built in small colonies, mostly in swampland, and made of rushes woven together with other vegetation. The clutch is usually around 20, occasionally more, as up to four females lay in the same nest. The downy young leave the nest within a day of hatching and are guarded and fed by the parental group for up to four months.

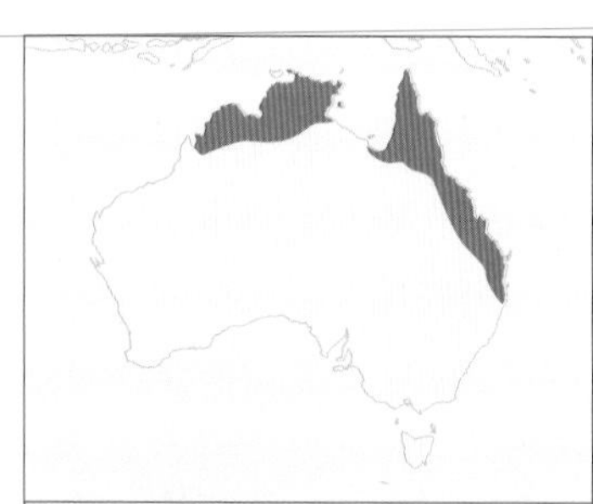

Number of genera 1
Number of species 1
Conservation Status Least Concern
Distribution Northern and eastern Australia, and southern New Guinea

DUCKS, GEESE, AND SWANS *Anatidae*

A large, cosmopolitan family, collectively known as wildfowl, and all more or less bound to bodies of water. The family comprises two major groups: the swans and geese (subfamily Anserinae) and the ducks (Anatinae). There are also two other subfamilies: the whistling ducks (Dendrocygninae) and, in Australia, Freckled Duck, *Stictonetta naevosa* (Stictonettinae), which may be the sole survivor of a primitive group of wildfowl.

The largest and most majestic of the wildfowl species are the swans. The four species found in the northern hemisphere are white, but Australian Black Swan (*Cygnus atratus*) and South American Black-necked Swan (*C. melancoryphus*) are not.

The 'true geese,' found only in the northern hemisphere, fall into two genera: 'gray geese' (*Anser*) and 'black geese' (*Branta*). Most species are highly migratory, many forming large, impressive flocks in winter.

A pair of **Torrent Ducks**. The male is the bird with the boldly patterned black and white head. Venezuela.

A male **Wood Duck** (*Aix sponsa*). This species declined dramatically in the 19th century, but has since recovered. U.S.

Ducks are the most abundant and widespread wildfowl and comprise the largest and most varied subfamily, showing a high degree of adaptive evolution. Less terrestrial than swans and geese, they inhabit all kinds of fresh water and marine environments. The most familiar group is the dabbling ducks (Anatini), which includes Mallard (*Anas platyrhynchos*), the ancestor of the 'farmyard duck.' Other groups include the sheldgeese and shelducks, steamer ducks, perching ducks, torrent ducks, pochards, eiders, sea ducks, and stifftails.

Structure Swans are very large aquatic birds, ranging from 43–67 in. (110–170 cm.) in length, with wingspans of 63–95 in. (160–240 cm.). The larger species are among the heaviest of all flying birds: for example, male Mute Swans (*C. olor*) may weigh around 27 lb. (12 kg.). Swans have a long neck, a heavy body, and large webbed feet. Structurally, geese are similar but smaller, ranging from 22–43 in. (56–110 cm.) in length, and more mobile. Swans and geese have a bill adapted for grazing on aquatic as well as ground vegetation: in both groups, the bill has lamellae in the form of serrated, horny 'teeth' along the edge of the maxilla.

Ducks can, broadly speaking, be divided into dabbling ducks and diving ducks. The former tend to be larger and bulkier, with a broad, flat, spatulate bill that also has lamellae. This feature is most highly developed in the shovelers, which feed on fine substances such as planktonic crustaceans. Diving ducks tend to be smaller and more compact, with a smaller, narrower bill adapted for eating both plant material and animal prey. The bill tends to be stronger and deeper in eiders and scoters, which feed chiefly on mollusks and crustaceans; in these species, the bill also has a swollen base housing salt-secreting nasal glands. In the sawbills or mergansers, which are mainly fish-eating ducks, the bill is long and narrow, again with serrated edges and even with a hooked tip adapted for holding onto prey.

Some diving species, such the goldeneyes (genus *Bucephala*) and, more particularly, the stifftails and Torrent Ducks, have a long, beaver-like tail that may serve as a rudder. The perching ducks (Cairinini) are a highly diversified group that has a well-developed hind toe and strong sharp claws to assist climbing.

Most ducks have relatively short, pointed wings, and some of the

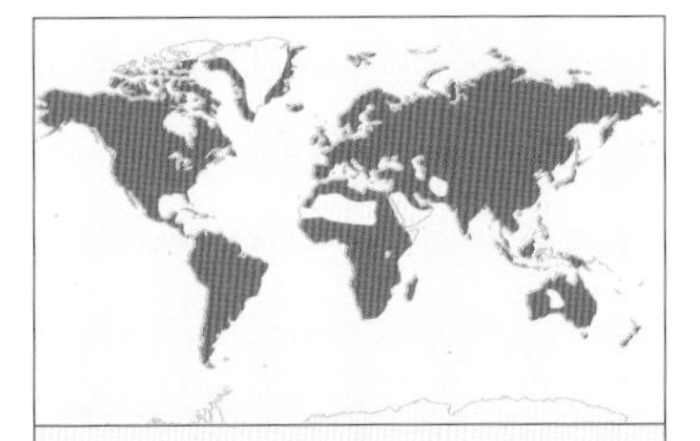

Number of genera 49
Number of species 158
Conservation Status 4 species Critically Endangered
Distribution Worldwide

Greater White-fronted Goose (*Anser albifrons*) belongs to the group of 'gray' geese, notable for their gray-brown plumage and brightly colored feet. Canada.

smaller dabbling ducks, such as the teals, are highly maneuverable when they are flying.

Of the less conventional species, steamer ducks are large, bulky, short-winged species, many of which are unable (or unwilling) to fly. Whistling ducks have a peculiar shape, with broad wings and long legs, and in flight look long-necked and slightly hump-backed. Torrent Duck (*Merganetta armata*) is a long-bodied, short-legged South American species whose streamlined shape is well adapted for swimming in fast-flowing Andean torrents. The peculiar Blue Duck (*Hymenolaimus malacorhynchos*) of New Zealand inhabits similar environments, and is also a strong swimmer, with tough claws for clinging to rocks.

Plumage Most swans are completely white in adult plumage, but juveniles are gray or brown. 'Gray geese' are usually brownish and grayish, with white ventral areas. However, Snow Goose (*Anser caerulescens*) and Ross's Goose (*A. rossii*) are white with black primaries (although both species have a dark 'blue morph'). The bill and legs may be more colorful pink or orange. The 'black geese' are much more strongly patterned with black, gray, brown, and white; Red-breasted Goose (*Branta ruficollis*) also has a beautiful chestnut-red breast. Juvenile geese are usually plainer than adults.

The greatest diversity in plumage occurs in the ducks. Males are much more colorful than their mates: some, such as Mandarin (*Aix galericulata*), Harlequin Duck (*Histrionicus histrionicus*), and Smew (*Mergellus albellus*), are among the most ornate of all birds. Male dabbling ducks are often subtly colored and intricately patterned, and many species have iridescent, shiny head feathers. A common feature of dabbling ducks is an iridescent speculum (colored patch on the wings), while many other species have either a white patch on the secondaries or a long white wing bar across the primaries and secondaries. Some species, such as Mandarin and Falcated Duck (*Anas falcata*), have spectacular plumes on the head.

For a short period in late summer, waterfowl molt their wing feathers and are temporarily flightless. During this time, male ducks molt into a subdued 'eclipse' plumage, which is either a much duller version of their 'full' plumage or is very similar to that of females. Eclipse plumage provides protective camouflage while the flightless males are vulnerable. Many species retain their male bill colors and, when regrown, their male wing patterns.

Female ducks have a much more functional, cryptic plumage that helps keep them hidden when incubating and tending young. Juveniles tend to resemble summer females except that their fresh plumage appears neat and immaculate. Many species have a profusely and delicately spotted belly (plainer and whiter in adult females, although this is reversed in scoters).

Some species of diving ducks have a colorful bill: for example, in Ruddy Duck (*Oxyura jamaicensis*) and White-headed Duck (*O. leucocephala*) the bill is bright sky blue. Eiders and scoters have a strikingly patterned bill, most notably in King Eider (*Somateria spectabilis*) and Surf Scoter (*Melanitta perspicillata*).

Less typical species such as the shelducks and sheldgeese may be boldly patterned, one of the most striking species being Common Shelduck (*Tadorna tadorna*), whose green head and black and white body plumage are offset by a thick orange ring around its upper body.

Behavior All species are essentially aquatic in summer, but some (swans, geese, and wigeon) are much more terrestrial in winter, often grazing on agricultural land. At this time of year, geese in particular may be found in spectacular flocks sometimes numbering tens or even hundreds of thousands, usually roosting at night on large lakes or in sheltered coastal areas. Most other species are also gregarious outside the breeding season, when large flocks may gather in areas with abundant food.

There is great divergence and flexibility in feeding methods, and all species are highly opportunistic. Many ducks, such as Common Shelduck, Common Teal (*Anas crecca*

Two **Red-breasted Mergansers** (*Mergus serrator*). These individuals are juvenile males or males in eclipse plumage. Nonbreeding males resemble females, but have larger white wing patches than the females. Sweden.

crecca) and Green-winged Teal (*A. c. carolinensis*), often siphon the mud for small plankton, while Ruddy Ducks and other stifftails forage in this way for insects and plant matter. Northern Shoveler (*A. clypeata*) siphons plankton from the water's surface with its large, spatulate bill, often feeding in large groups that mill around in circles. Others, such as Mallard and Northern Pintail (*A. acuta*), feed in shallow water by immersing their head and upending their body. Some species, such as Gadwall (*A. strepera*), simply pull up submerged weeds; they may even parasitize coots by stealing vegetation from them. Diving ducks feed on submerged vegetation or on animals such as insect larvae, mollusks, and crustaceans. The mergansers largely eat fish. Perching ducks, such as Mandarin and Wood Duck, often feed on the forest floor, eating seeds and nuts, such as acorns, chestnuts, and beechmast.

A pair of **Whooper Swans** (*C. cygnus*). Swans remain in pairs or family groups year-round. Location unknown.

Voice Most swans have deep, trumpeting calls, although Mute Swans grunt and hiss. Geese have various honking and yelping calls. Ducks have a huge variety of calls, particularly the males in display. Female ducks, however, tend to have simpler quacking, croaking, or grunting calls. Some species, such as Mute Swan and Common Goldeneye, make throbbing or whistling sounds with their wings as they fly.

Habitat Wetlands and coastal areas on every continent apart from Antarctica. Some species inhabit arid or semi-arid areas.

Movements In spring in the northern hemisphere, wildfowl disperse or migrate, many to the tundra, others into the taiga zone or into mid-continental lakes and marshes that were frozen during the winter. In fall, many species migrate south from these regions to winter in the closest icefree areas. Many species of swans and geese migrate to traditional stop-over and wintering sites. Their migrations are not simply from north to south, but also to the southwest, southeast and even east or west. Milder coastal regions in temperate areas receive huge influxes, many of which start surprisingly early in mid- to late summer. Several species of duck, such as Northern Pintail, Northern Shoveler, and Ferruginous Duck (*Aythya nyroca*), regularly penetrate into tropical regions, but perhaps only Garganey (*Anas querquedula*) can be thought of as a true trans-equatorial long-range migrant.

Since wildfowl are flightless in late summer, many species undertake molt migrations to areas with a reliable and abundant food source. One well-known example is provided by western European Common Shelducks, virtually all of which migrate in late summer to the Waddenzee, on the coast of Germany. Therefore, many species have not two but three basic migrations during the course of the year.

Within these primary movements, individual birds are highly opportunistic, especially at night, homing in on temporarily abundant food supplies. Migration itself is largely and sometimes almost exclusively nocturnal: it is often undertaken around the period of the full moon, when bodies of water may be more easily located. However, diurnal migration can be seen in many coastal areas. Birds may travel at considerable altitudes—the most famous example being Bar-headed Goose (*Anser indicus*), which routinely migrates over the Himalayas. Wildfowl may also fly extremely fast: a satellite-tracked Barnacle Goose (*Branta leucopsis*) flew from southwest Scotland to Norway in eight hours, at an average speed of 60 miles an hour (96 km/h.).

Some species that inhabit arid or semi-arid areas are nomadic and sometimes even irruptive, performing erratic movements dependent on drought and rainfall. Two examples are Ruddy Shelduck (*Tadorna ferruginea*) and Marbled Duck (*Marmaronetta angustirostris*). The former famously irrupted into northwestern Europe in 1892 and 2004, with birds in the earlier irruption reaching Greenland.

Diet A variety of food including aquatic vegetation, grass, insect larvae, mollusks and crustaceans, plankton, seeds, and nuts.

Breeding Pair formation often occurs on the wintering grounds, with males usually following females back to their natal areas. Whereas swans and geese tend to mate for life and maintain strong family bonds year-round, ducks usually maintain only seasonal pair bonds.

Swans and geese have relatively simple courtship displays that involve a lot of wing-flapping, honking, or trumpeting. Ducks' displays are more varied. Conventional displays include pursuit flights, communal courtship, and pair courtship. The last is often highly ritualized and includes head-flicking, head-bobbing, head-throwing, and neck-kinking, often accompanied by quacks, grunts, burps, whistles, or wheezes, depending on the species. In dabbling ducks, displays and pairing often start as soon as the males molt out of eclipse plumage in late fall. When mating, female ducks lie flat on the water or with the head and neck bent forward: the male may appear almost to drown the female! Pochard (*Aythya ferina*) and Canvasback (*A. valisineria*), practice communal courtship, with several males concentrating on one female. Ruddy Duck has a particularly bizarre display: while swimming, the male rapidly thumps its bill onto the breast feathers to form bubbles on the surface of the water. Many male dabbling ducks, such as Mallards, display to various females and often forcibly mate with them.

Waterfowl usually build nests on the ground. Most species nest close to water. However, Common Shelduck nests in rabbit burrows and other holes in the ground, while several species, including Common Goldeneye (*Bucephala clangula*), Hooded Merganser (*Lophodytes cucullatus*), Mandarin, and Wood Duck nest in holes in trees, regularly adapting to nest boxes. On fledging, the ducklings often have to jump down from considerable heights. Some species of *Aythya* and *Oxyura* may dump their eggs in the nests of other ducks.

Young wildfowl are precocial (although swans may help their young feed by bringing weed to the surface). In ducks, the female tends the young until they fledge, but in swans and geese both parents guard them. Brood sizes vary considerably: up to five or six in swans and geese, but often higher (usually five to 12, but up to 17 or 18) in ducks. Common Shelduck and Common Eider (*Somateria mollissima*) routinely creche their young under the guard of a small number of females. Fledging takes about six weeks, or longer in larger species.

Wildfowl Populations and People

Man has had a long and complex relationship with wildfowl. The birds have always been hunted, and later farmed, but at the same time they have been admired for their beauty. Unfortunately, over-hunting and the drainage and degradation of wetland habitats have reduced many species to dangerously low levels, and a few have been driven to extinction. In historic times, Crested Shelduck (*Tadorna cristata*), Pink-headed Duck (*Rhodonessa caryophyllacea*), Labrador Duck (*Camptorhynchus labradorius*), and Aukland Islands Merganser (*Mergus australis*) have become extinct. Four further species—Madagascar Teal (*Anas bernieri*), Laysan Duck (*A. laysanensis*), Meller's Duck (*A. melleri*), and Madagascar Pochard (*Aythya inotata*)—are critically endangered. Others, such as Hawaiian Goose (*Branta sandvicensis*), White-winged Duck (*Asarcornis scutulata*), Brazilian Merganser (*Mergus octosetaceus*), and Scaly-sided Merganser (*M. squamatus*), are now worryingly rare. Despite this, many species, such as Snow Goose and Ross's Goose, have increased considerably as a result of conservation. Other species, such as Canada Goose (*Branta canadensis*) and Ruddy Duck, have benefited from their introduction into areas where they do not occur naturally. ■

A female and male **Ruddy Duck** (*Oxyura jamaicensis*). This North American species was introduced to Europe, but threatens to wipe out the rare White-headed Duck (*O. leucocephala*) through interbreeding.

PENGUINS *Spheniscidae*

A familiar family of flightless birds that are adapted for a marine lifestyle. All breed on land or ice and winter at sea. The family has also evolved to live in a wide variety of habitats with an extreme range of temperatures, from 59°–82°F (15°–28°C) in the case of warmer-water species such as Galápagos Penguin (*Spheniscus mendiculus*), which breeds on the equator, to as low as -76°F (-60°C) for Emperor Penguin (*Aptenodytes forsteri*).

Structure Most species are similar in structure, but there is considerable variation in size. The sexes are similar in all species, although males are slightly larger than females. Penguins range from 16 to 45 in. (40–115 cm.) in length. All species have a flat-topped or pointed head and a short to medium-sized, stout bill; several species also have prominent or loose, tapering eye tufts or crests. They have a short, thick neck; a stout but tapering body; vestigial, paddle-like wings; a short, rudimentary tail; short legs; and strong, thickly webbed feet, set at the rear of the body. Under the skin, all species have a layer of subcutaneous fat which helps trap warm air beneath the plumage.

Rockhopper Penguin (*E. chrysocome*) has golden-yellow crests sweeping back from each eye over the crown. Falkland Islands.

Gentoo Penguin (*Pygoscelis papua*) lives on stony, ice-free ground, beaches, or in tussocks of grassland. Antarctic Peninsula.

Plumage Penguins have a uniform covering of thick, short, curved, overlapping feathers, which provide insulation and waterproofing; the feathers are longer in southern than northern species. Body coloration is very similar in all species, with bluish-gray to black upperparts and white underparts. The species differ mostly in the markings on the head and face: the areas visible above water. Two species have large orange sides to the neck. Seven have bright yellow crests of varying length emerging over the eyes. Four have a black face with white surrounds. Bill and leg colors are variably pale or bright red to black.

Behavior Mostly diurnal, but one species is active at night. Penguins are expert swimmers. They move like porpoises, bouncing through the waves, using their wings as flippers for rapid underwater propulsion and their tail and feet as a rudder for direction. Most can swim at 5–6 miles an hour (8–10 km/h) and dive for between two and eight minutes. The two largest species, King Penguin (*A. patagonicus*) and Emperor Penguin, have been known to spend up to 18 minutes underwater, at depths to some 1,000 ft. (320 m.). However, penguins are less adept on land. When walking, they shuffle on flattened legs or make short, rather ungainly hops, using their wings to balance. Species that breed on ice floes also move by lying down and 'tobogganing,' using their legs and long claws for propulsion.

Penguins can regulate their body temperature as they move between feeding in icy seas and coming ashore, where they may be exposed to hot sunshine; unable to sweat, they pant with their bill open, releasing heat from the fleshy gape.

Voice Generally silent outside the breeding area, but colonies are extremely noisy. Calls include trumpeting, braying, barking, cawing, and cackling in adults, and whistled notes in young. Individuals' calls can be recognized by their mate, parents, or young, even in the midst of a huge breeding colony.

Habitat Coastal areas of the Antarctic and sub-Antarctic, ranging north to the tropical seas around the Galápagos Islands. The birds breed on

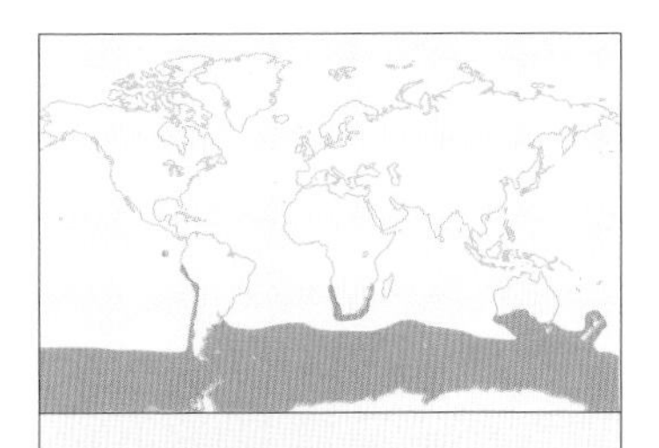

Number of genera 6
Number of species 17
Conservation Status 1 species Vulnerable, 2 species Near Threatened
Distribution South America, South Africa, Australia, New Zealand, Southern Ocean, Antarctica

A small group of **Chinstrap Penguins** (*P. antarcticus*) rests on an ice floe between fishing expeditions. The birds typically dive for krill around pack ice, close to the shore. Antarctica.

ice-floes, islands, and mainland areas with flat beaches and dunes or with rolling sand hills and little vegetation. Some species climb cliffs along regular pathways. In New Zealand, Fiordland Penguin (*Eudyptes pachyrhynchus*) breeds in coastal rain forests.

Movements Penguins disperse from their breeding areas to winter at sea, usually moving to preferred traditional feeding areas. The actual winter distribution of several species remains unknown.

Diet Principally fish; crustaceans, including krill; and cephalopods, mainly squid. The diet varies with the range of prey available: the more northerly species take a high proportion of fish (mainly anchovies, sardines, and small mullet and other schooling fish), while those around Antarctica take mostly crustaceans, especially krill. During the breeding season, some species may eat snow to obtain water. The feeding areas of smaller species are usually close to the breeding colonies, but those of the larger ones are more distant: for example, King Penguin may travel up to 560 miles (900 km.) to feed.

Breeding With the exception of Yellow-eyed Penguin (*Megadyptes antipodes*), which nests in isolated pairs or scattered groups, all species breed colonially, sometimes in concentrations of several hundred thousand pairs. All are monogamous, and pairs of some species bond for more than a single breeding season. Emperor Penguin breeds during the long, dark Antarctic winter. Nests vary from a loose pile of stones and driftwood to a platform of dry twigs and vegetation. Most species generally lay two eggs and take turns incubating and feeding at sea; incubation periods vary from 33–38 days in smaller species up to about 64 days in the larger species. The largest penguins make no nest, but hold the single egg, and later the chick, on the feet and in a fold of skin above the vent. Chicks are fed on regurgitated food. From the time they acquire their first covering of down feathers, they live in creches; however, the parents feed the young for a further two months.

Some penguin species, like this **Emperor Penguin**, find it faster and easier to travel across ice by 'tobogganing' on their belly, pushing with their feet and flippers. Antarctica.

LOONS *Gaviidae*

A small family of northern-hemisphere birds that spend virtually all their lives on water, swimming and diving.

Structure Loons range from 21 to 36 in. (53–91 cm.) in length and 3.1–11.8 lb. (1.4–5.4 kg.) in weight. Males are slightly larger than females. The birds have a dagger-shaped bill, which in Common Loon (*Gavia immer*) and Yellow-billed Loon (*G. adamsii*) is particularly large and strong. They also have a fairly long neck, a long body, relatively small wings, and a short tail. The legs are set far back on the body and have large webbed feet.

Plumage Males and females are alike. Breeding plumage is strikingly patterned. In all species, the underparts and underwings are white. Arctic Loon (*G. arctica*) and Pacific Loon (*G. pacifica*) have a pale gray head and hindneck, a black throat patch bordered by black-and-white stripes, and black-and-white upperparts. Red-throated Loon (*G. stellata*) has a gray head and a dark red throat patch but no bold upperpart markings. Common and Yellow-billed Loons have a greenish-black head and neck, the latter with a collar of fine vertical green and white stripes, and a black-and-white back. The feathers are dense and compact. Loons have 10 primaries, 22–24 secondaries, and 16–20 rectrices.

A **Red-throated Loon** taking off from water. This species has a distinctive style of flight, in which it holds its head and neck slightly below the level of the body. Finland.

Loons, like this **Common Loon**, brood their chicks by carrying them on their back. U.S.

Behavior Loons spend most of their time on water, where they hunt and roost. They dive frequently to find food, reaching depths of up to 20 ft. (6 m.) and remaining submerged for up to 2 minutes, although 45 seconds is the average for Arctic Loon. When alarmed, loons can submerge most of their body, keeping just the head and neck above the water level. The birds spend little time on land, where they have difficulty walking.

Voice Loons are highly vocal in the breeding season but silent during the rest of the year. The songs are loud and far-carrying, involving eerie wailing, moaning, yodel-like notes, cackling, and cooing. Common Loon's screaming has been likened to idiotic laughter. Pairs sometimes duet. Calls are a variety of croaking and barking sounds.

Habitat The birds breed on relatively undisturbed freshwater lakes in boreal forest or tundra, or along the littoral zone of rivers and estuaries. However, they spend much of the rest of the year mostly on sheltered inshore marine waters, or sometimes on unfrozen freshwater lakes.

Movements Migratory and dispersive. Loons leave their breeding lakes in fall and mostly winter farther south. For example, Common and Red-throated Loons, which breed as far north as northern Greenland, winter south to the Pacific coast of Mexico.

Diet Mostly smaller fish, up to 11 in. (28 cm.) long, caught after underwater pursuit. Also, mollusks, crustaceans, and amphibians.

Breeding Loons are very territorial. They are monogamous and probably form lifelong pair bonds. The nest, built by both partners, is rarely more than 3 ft. (1 m.) from the shore, a low mound of aquatic vegetation with a shallow cup. Nest sites are often reused. Usually, two eggs are laid, and are incubated by the pair for 24–30 days. Both parents also feed the young, which fledge after 38–77 days.

Number of genera 1
Number of species 5
Conservation Status Least Concern
Distribution Holarctic

ALBATROSSES *Diomedeidae*

Buller's Albatross (*Thalassarche bulleri*) is characterized by bold yellow stripes on the bill. Snares Islands, New Zealand.

At sea, the grace and elegance of the albatrosses is unrivaled, as they cruise for great distances with seemingly effortless flight. They are long-lived but their low breeding productivity means that populations are slow to rebound from losses.

Structure Albatrosses are 28 to 52 in. (71–135 cm.) in length, males being larger than females. They have very long, narrow wings, with a wingspan of 71–138 in. (180–351 cm.); some species have the largest wingspan of any flying bird.

The large, heavy bill is covered by horny plates, and has a hooked maxilla tip. The opening of each nostril has a tubular casing ('tubenose'). Large salt glands above the eyes process excess salt from food and water, and secrete it through the nostrils. Albatrosses have an extraordinary sense of smell, which is used to locate food, and is also important in recognizing other individuals. The tail is fairly short. The legs are strong, with webbed feet.

Plumage In most species, the sexes are alike or nearly so. Genera *Diomedea*, *Thalassarche*, and *Phoebastria* (except for one mostly blackish-brown species, *P. nigripes*) have blackish-brown flight feathers and are variably white or brown above and white below. The dark and light pattern of the underwing varies between species. Sooty-albatrosses (genus *Phoebetria*) are more uniformly dark and have a longer, wedge-shaped tail; juveniles are generally similar to adults. 'Giant' albatrosses (genus *Diomedea*) show complex variations by age and sex. Albatrosses undergo an annual molt at sea. In *Diomedea* species, the primaries are molted every two years, with the inner primaries and outer primaries molted in alternate years.

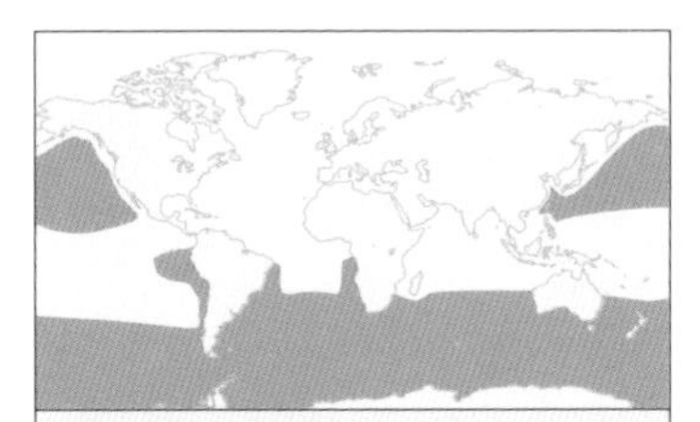

Number of genera 4
Number of species 13
Conservation Status 3 species Critically Endangered, 7 species Endangered, 9 species Vulnerable
Distribution Pelagic, coming to shore only to breed on oceanic islands

Behavior Albatrosses walk and swim well, but are most renowned for their flight. The birds' long wings enable them to cover huge distances with minimal effort. They glide on stiff wings for long periods, flapping only occasionally, taking advantage of the wind. Sometimes, they arc above the waves and glide through troughs, often skimming the water's surface with a wing tip. However, albatrosses are unable to make quick changes in flight. Take-offs are laborious unless it is very windy, requiring a 'run' on land or 'pattering' along the sea surface; in windless weather, they may not be able to get airborne. Landings are often awkward, and on the ground can be hazardous, because the birds do not maneuver well and sometimes land with a tumble.

Prey is plucked from the surface in flight or while swimming, or taken underwater by plunge- or surface-diving. Some species are known to dive to at least 16 ft. (5 m.). Black-browed Albatross (*T. melanophrys*), uses its wings to 'fly' under the water in pursuit of prey. At sea,

The courtship display of a **Waved Albatross** pair involves the male and female repeatedly snapping and fencing with their bills and pointing their heads skyward. Galápagos Islands, Ecuador.

Black-footed Albatrosses (*Phoebastria nigripes*) at the nest, where a single egg will be incubated by both parents for about 65 days. Most breed on Hawaii and Midway Atoll. This is the commonest albatross in the northeast Pacific. Hawaii.

the birds are generally solitary, or may be found feeding with other albatrosses or other species of seabirds. They often follow ships, taking advantage of the air current eddies created by the vessels, or waiting for offal or scraps.

Voice The birds make various sounds when foraging, including shrieks, grunts, and croaks. Breeding displays include bugling, whistling, or grunting notes; the birds also make clattering sounds with the bill.

Habitat Albatrosses are pelagic apart from during the breeding season, when they nest on oceanic islands from sea level to 2,100 ft. (640 m.).

Movements Most species are migratory or disperse from breeding sites in search of food. The movements of some species are not well known, but preliminary findings from telemetry data for some individuals of a few species indicate extraordinary distances traveled in a relatively short time period.

Diet Squid, a variety of fish and fish eggs, krill, and other invertebrates. Also offal, and even carrion, some obtained from refuse discarded by ships or fishing bycatch.

Breeding Albatrosses have a long maturation period: most species do not breed until eight to ten years of age. Most species are colonial, but a few are loosely colonial or solitary (including species in genus *Phoebetria*). They are monogamous, pairing for life. Most breed annually, but some (including all *Diomedea* and *Phoebetria*) breed biannually. Courtship involves stereotyped displays or dances. Most species build a raised nest of vegetation and mud on the ground; the nest may be reused. Nests of *Phoebastria* species are not as elaborate, and Waved Albatross (*P. irrorata*) nests on bare ground. The clutch comprises just one egg. Because the egg is proportionately very large, the female usually cannot lay another if it or the chick is lost. Incubation lasts 10 to 11 weeks, with another five or six weeks of brooding the chick. In large species, the length of the nesting period permits only one breeding attempt every other year. The parents take turns feeding the chick; the one who is brooding endures foodless periods of days or weeks. The chick is fed on partially digested food mixed with stomach oil; this high-calorie diet enables it to grow and build substantial fat reserves. Chicks are fed as long as they remain at the nest site and fledge when they are ready, leaving the colony without further parental assistance. They are independent at about five or six months.

Taxonomy Four genera are recognized: 'sooty' (*Phoebetria*); 'northern Pacific' (*Phoebastria*); 'great' (*Diomedea*); and 'mollymawks' (*Thalassarche*). However, the taxonomy, and the number of species and subspecies, are still matters of debate.

Black-browed Albatross (*T. melanophrys*) belongs to one of the smaller genera, often called mollymawks. Even so, its wingspan can be up to 98 in. (250 cm.). Antarctica.

PETRELS AND SHEARWATERS *Procellariidae*

The 'tubenosed' seabirds of this diverse family spend much of their time at sea. They have exceptional flight capabilities and are well adapted for ocean life. Because many species only come ashore to breed at remote sites, their life history is incomplete, and the family includes some of the least known of all bird species.

Structure The birds are 9–39 in. (23–99 cm.) in length, with a wingspan of 22–81 in. (56–205 cm.). Males are larger than females.

The size, shape, and structure of the bill varies greatly across the family, from delicate to heavy and stout, but in all cases the bill is covered by horny plates and has a hooked upper tip. The nostrils extend as a tube along the culmen ridge; excess salt from food and water is extracted by large glands above the eyes and secreted from the nostrils. The maxilla has a sharp cutting edge to cut squid or fish; in prions (genus *Pachyptila*), it has lamellae, which are used to filter plankton out of the water. The massive, hooked bill of the two giant petrel species (genus *Macronectes*) enables them to tear flesh from carcasses in a manner reminiscent of vultures.

All species have proportionately long, narrow wings, a relatively short tail, legs that are positioned relatively far back on the body, and webbed feet. Diving species have more laterally compressed tarsi

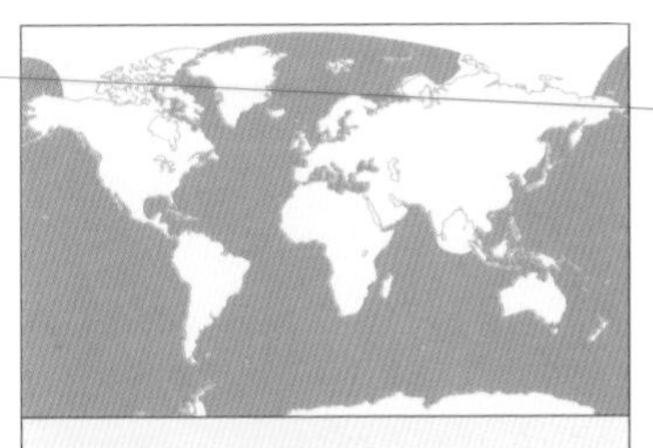

Number of genera 14
Number of species 74
Conservation Status 9 species Critically Endangered, 9 species Endangered, 17 species Vulnerable
Distribution Pelagic, found in oceans worldwide

Cape Petrel (*Daption capense*) is a fulmar-petrel of the southern oceans. Huge, noisy flocks often congregate behind fishing boats in winter. Ocean around Antarctica.

The tubular nostrils of this **Northern Fulmar** (*Fulmarus glacialis*) are visible as an open-ended ridge extending part of the way along the top of the bill. Scotland, U.K.

(shanks). Some species have long claws to facilitate digging burrows. All share an intense musty smell, originating from the production of stomach oil, which is used as a defense against predators and as high-energy food for nestlings.

Plumage All members of the family have dense plumage that gives excellent insulation. The sexes are alike, and juveniles are similar to adults. There is no seasonal variation. In most species, the coloration is blackish-brown to blue-gray above and paler or white below. A few are wholly dark brown, and one species, Snow Petrel (*Pagodroma nivea*), is entirely white. Many have a distinctive dorsal or ventral wing pattern (especially gadfly petrels and prions; also Antarctic and Cape petrels). A few species in different genera (such as *Macronectes, Fulmarus, Pterodroma,* and *Puffinus*) are polymorphic, with dark, light, and intermediate morphs. Chicks hatch with a thin coat of down. The molt is poorly known for many species: it is typically complete, and usually occurs at sea following breeding.

Behavior Gregarious. Most species occur in small to large flocks, or form aggregations in areas of plentiful food resources. Flight usually comprises several flaps followed by a glide, some species arcing high into the air; the birds can cover long distances with little effort. They are also good swimmers, but the leg position does not allow most to walk or stand for any period; only the albatross-like giant petrels can walk or stand well. Snow Petrel frequents ice floes; it does not swim or even rest on water as other species do.

The birds forage by diving from the surface or plunge-diving. Several species use their wings to 'fly' underwater in pursuit of prey. Petrels and shearwaters visually search for food or feeding birds, but also use their sense of smell. Once food is located, they use various methods to obtain it, including surface seizing and pursuit diving; some species dive to depths of more than 220 ft. (70 m.). *Pterodroma* petrels can catch flying fish on the wing. Fulmar-petrels and shearwaters will approach boats for offal or bycatch. Petrels and shearwaters also use their excellent sense of smell to locate nesting burrows and recognize other individuals.

Voice Quiet or relatively so away from breeding colonies, with some vocalization when aggregations form at feeding sites. At breeding colonies the birds are very vocal: individual species give a multitude of sounds during courtship and between pairs, including squeaks, whistles, coos, grunts, and croaks.

Habitat Petrels and shearwaters are mostly pelagic, but they nest in a wide range of habitats, from towering Arctic cliffs and rocky Antarctic islands to sandy tropical atolls and tropical forests, at altitudes from sea level to 7,200 ft. (2,200 m.).

Movements Most species are migratory, some with a transequatorial migration. Others are relatively sedentary, dispersing locally following breeding. Movements of many species during breeding and post-breeding are little known. Recent telemetry studies show a few individuals of certain species cover extraordinary distances to obtain food while breeding.

Diet Most petrels and shearwaters are highly opportunistic diurnal and nocturnal feeders, taking squid, fish and fish eggs, crustaceans, jellyfish, and other marine organisms. Prions consume a high percentage of krill (euphausiid crustaceans), captured as they swim or skim the surface, or in flight as they dip from near the surface. The diets of migratory species vary by season, geographic area, and prey abundance. Generally, prey size is correlated with body size, sometimes even within species that show sexual dimorphism. Many species also eat scraps dumped by ships or left by predatory fish or cetaceans. Carrion is also taken. The diet of giant petrels (genus *Macronectes*) comprises a large percentage of carrion (seals and birds) scavenged on land; these are the only members of the family that regularly feed onshore. Giant petrels are also formidable predators and kill by battering or drowning seal pups, penguin chicks, and other birds such as petrels and smaller albatross species.

Breeding The birds are colonial or loosely colonial.They are monogamous, breeding at 3–12 years of age, depending on species. Most probably mate for life. Many species perform communal courtship flights. Fulmar-petrels make a nest on the ground, in a crevice, or on a cliff ledge, and visit the colony by day. Most prions, gadfly-petrels, and

A flock of **Sooty Shearwaters** (*Puffinus griseus*) feeds with a humpback whale. The birds follow the whales and seize the fish that a whale disturbs as it forages. This species breeds on temperate and subantarctic islands in the southern-hemisphere summer, and undertakes a marathon trans-Equatorial migration to spend the northern-hemisphere summer in the northern Pacific and Atlantic. Alaska.

Zino's Petrel (*Pterodroma madeira*) is one of the world's rarest seabirds. Up to 80 pairs nest on mountain ledges on Madeira, Portugal. South of Madeira.

shearwaters nest in crevices or burrows dug by themselves or other animals and visit the colony only under the cover of darkness. Some nest on the ground under bushes, and a few species are active by day. A single egg is laid each season. Incubation lasts six to nine weeks and is shared by both parents, the birds alternating in foodless bouts of a few days to two weeks or more. The chick is fed an energy-rich mixture of partially digested prey and stomach oil. However, the parents abandon it once it is fully feathered. Following desertion, chicks exercise their wings outside the burrow at night for 6–15 days, then take their first flight to the ocean. Without further parental assistance they are (depending on the species) independent at 41–132 days.

Taxonomy This is still not well resolved. Four to five subfamilies are commonly recognized, primarily by morphology and breeding behavior: fulmar-petrels, prions, gadfly-petrels, petrels, and shearwaters. Some genera remain problematical (such as *Aphrodroma, Halobaena, Procellaria, Bulweria, Pseudobulweria*).

Morphologically, the fulmar-petrels (genera *Macronectes, Fulmarus, Thalassoica, Daption*, and *Pagodroma*) are the most diverse group, but they share similarities in diet and breeding biology. Four further genera are more compact: gull-like in coloration in the two fulmar species (genus *Fulmarus*); boldly patterned in Antarctic Petrel (*Thalassoica antarctica*) and Cape Petrel (*Daption capense*); and all-white in Snow Petrel (*Pagodroma nivea*).

In the other groups—gadfly-petrels (genera *Pterodroma* and *Aphrodroma*), petrels (genera *Pseudobulweria, Procellaria*, and *Bulweria*), prions (genera *Pachyptila* and *Halobaena*), and shearwaters (genera *Puffinus* and *Calonectris*)—most members are easier to classify as such by general appearance and structure, bill structure, and flight behavior.

Recent studies suggest that the diving petrels (family Pelecanoididae) should be included in the same family as petrels and shearwaters.

STORM PETRELS *Hydrobatidae*

Small seabirds, storm petrels are well adapted for flying and swimming. When foraging, they have a habit of fluttering just above the sea, with their feet dangling or pattering on the surface, so that they seem to be walking on the water. The life history of many species is poorly known; for example, the breeding grounds of Ringed Storm Petrel (*Oceanodroma hornbyi*) are still unknown, and those of Markham's Storm Petrel (*O. markhami*) were discovered only in 1987.

Structure These are small, delicate birds, 5–10 in. (13–26 cm.) in length, with a wingspan of 13–22 in. (32–56 cm.). They are classified in two subfamilies, distinguished by the relative proportions of the wings and legs: Oceanitinae, with shorter, more rounded wings, shorter tail, and long legs; and Hydrobatinae, with longer, pointed wings, long (often forked) tail, and shorter legs.

All species have a hooked bill and tubed nostrils on the culmen ridge. Well developed salt glands filter excess salt from food and water, and secrete it through this 'tube-nose.' The sense of smell is extraordinary and is used to locate food at long distances. The birds have webbed feet and are good swimmers. However, they cannot stand for more than a few moments or walk more than a few steps because their legs, positioned right at the rear of the body, cannot bear the body weight.

Plumage Uniformly brownish-black, gray, or dark above, and paler or white below. The sexes are alike, and juveniles are similar to adults. White-bellied Storm Petrel (*Fregetta grallaria*) is reported to be polymorphic, with paler and darker morphs.

The pale carpal bar on the upperwing of this **Wilson's Storm Petrel** (*Oceanites oceanicus*), and the white band around the rump, are typical of the species. The bird is hovering over the sea with its feet dangling in the water, searching for prey. Atlantic Ocean.

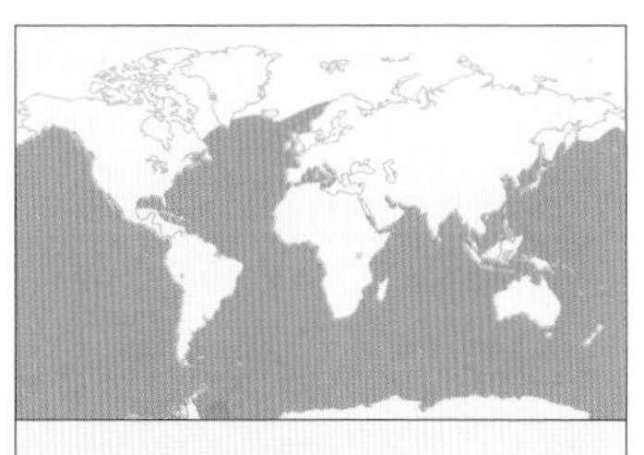

Number of genera 7
Number of species 21
Conservation Status 1 species likely now Extinct, 2 species Critically Endangered, 1 species Endangered
Distribution Worldwide: pelagic, coming to shore only to breed

Chicks hatch with sparse down, which is replaced by a second downy plumage. The molt for most species is not well known, but all presumably have a complete annual molt, which occurs during or after breeding. The plumage has a strong, musty odor derived from stomach oil.

Behavior The birds are gregarious at colonies, but at sea they are solitary or occur in small flocks or feeding aggregations, usually around concentrations of food resources.

They rest on the water singly or in groups ('rafts') between feeding bouts, which can occur day or night. Agile and active fliers, their flight may be fluttering or strong and direct, depending on species and weather conditions. Most 'walk' or patter on the surface as they forage, and either pick food from the surface or grab it in shallow dives. Both adults and chicks eject stomach oil from the mouth and nostrils as a form of defense. The birds' difficulty in walking makes them vulnerable on land, so, with the exception of Wedge-rumped Storm Petrel (*Oceanodroma tethys*) at the Galápagos Islands, they come ashore only in twilight (high northern latitudes) or darkness.

Voice Very vocal at colonies, where adults make a variety of whistles, coos, twitters, or groans; less vocal at sea but will peep or chatter during interactions while feeding.

Habitat Oceans worldwide. The birds come to land only to breed on remote oceanic or coastal islands, from sea level to 1,500 ft. (450 m.).

Movements The movements of many species remain poorly known. A few are sedentary, remaining in waters close to their breeding colony year-round. Most are at least partially migratory, and some are long-distance or even trans-equatorial migrants: for example, Wilson's Storm Petrel (*Oceanites oceanicus*) nests on sub-Antarctic islands and migrates as far as 77°N in the Atlantic Ocean.

Wedge-rumped Storm Petrel breeds on the Galápagos Islands. Unusually for storm petrels, it attends the nest by day, and feeds during the night. Galápagos Islands, Ecuador.

Diet Storm petrels eat planktonic crustaceans, small fish and fish eggs, and squid. They also take scraps of fish, fat, oil, or offal, which they scavenge from feeding predatory fish or cetaceans, or from refuse discarded by boats.

Leach's Storm Petrel (*Oceanodroma leucorhoa*) ranges widely across the northern Pacific and Atlantic oceans.

Breeding Colonial. Birds arrive at the colony well in advance of breeding, to establish territory, engage in displays, and reunite with or obtain a mate. They are monogamous, probably pairing for life. They nest in burrows dug by themselves, other seabirds, or animals, and lay one proportionately large egg. Incubation takes about seven weeks, and the chick is brooded for another one to seven weeks. Both parents feed and care for it, taking turns every day or every few days. The chick gains weight rapidly on a high-calorie mix of the parent's diet with stomach oil, eventually exceeding adult weight by up to 80 percent; this weight gain is necessary to support juvenile plumage growth. It continues to be fed until it is ready to leave the burrow at 7–10 weeks of age, although feeding bouts decline as it nears fledging. Young birds return to the colony as 'spectators' at two to four years old, but most probably do not breed until three to five years.

DIVING PETRELS *Pelecanoididae*

A **Common Diving Petrel** flies low and fast on whirring wings. These birds will sometimes fly through the crest of a wave. As with other diving petrels, this species swims and dives expertly, and it will often dive when approached by a ship. Kaikoura, New Zealand.

The diving petrels are a family of small seabirds comprising four species in a single genus, *Pelecanoides*. The most widespread, Common Diving Petrel (*P. urinatrix*), has six subspecies, some of which may represent cryptic species. All species are very similar in shape but are separated by slight plumage differences and, more importantly, by bill morphology.

Structure Diving petrels are 7–10 in. (18–25 cm.) in length, with a wingspan of 12–15 in. (30–38 cm.). Like petrels and shearwaters, they have a hooked bill with tube nostrils on the culmen ridge; however, the tube opens upward rather than forward. Diving petrels have a small, stocky body, a thick neck, and short wings. The feet are webbed, and the legs are laterally compressed and placed far back on the body. While this position facilitates swimming and diving, it makes the birds awkward on land; they usually move by 'tobogganing' on their belly, pushing themselves along with their wings. They have to patter along the water surface, into the wind, to get airborne, and ground landings are clumsy. The birds also have gular skin that can expand to form a pouch, which the birds use to carry food to their young.

Plumage The sexes are alike, and juveniles are similar to adults. The plumage is dark gray to blackish-brown above and white below. All the flight feathers are molted at once, so the birds are flightless for a few weeks. However, this does not affect their ability to dive and capture prey. Chicks hatch with a sparse coat of gray down, which is replaced with adultlike plumage.

Behavior Diving petrels come ashore only to breed. They are gregarious at their breeding colonies, but at sea are generally solitary or found in small groups.

The flight is fast, with steady wing beats, reminiscent of auks, and the birds' strong wings enable them to literally fly through waves. When pursuing prey, they dive from flight or from the surface, then 'fly' underwater to more than 200 ft. (64 m.).

Voice Reportedly silent at sea. On breeding grounds, make a variety of whistles, coos, and groans while airborne and in the nest burrow.

Habitat Marine, inshore, near-shore coastal, and pelagic waters of the southern hemisphere. Diving petrels breed along coasts, and on oceanic and inshore islands.

Movements The birds are generally sedentary, or disperse into waters adjacent to breeding colonies.

Diet Planktonic crustaceans such as krill and copepods; small fish; squid.

Breeding Diving petrels breed from two to three years of age. They are colonial and active at night, roosting at sea or in burrows. The nest, dug by both members of the pair, comprises a short burrow and a chamber lined with feathers, vegetation, or small stones. The entrance is camouflaged with vegetation. The female lays a single egg each season. Incubation is seven or eight weeks, with both sexes alternating on a nightly basis. The chick is then brooded for another 15 days. At about eight weeks old, it leaves the burrow without further parental assistance and flies out to sea.

Taxonomy Although some studies suggest including the diving petrels in family Procellariidae (petrels and shearwaters), the body shape is very different: more similar to some auks (family Alcidae).

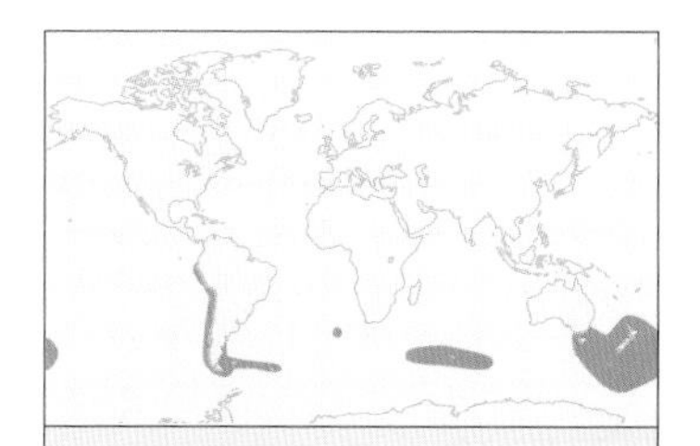

Number of genera 1
Number of species 4
Conservation Status 1 species Endangered
Distribution South America, Australia, New Zealand

GREBES *Podicipedidae*

Small or medium-sized diving birds, grebes spend almost all their time on the water. They are found on lakes and marshes across the world, but reach their greatest diversity in the New World.

Structure The sexes are similar. The smallest species is Least Grebe (*Tachybaptus dominicus*), which is just 9.5 in. (24 cm.) in length and weighs 4 oz. (115 g.). At the other end of the scale, Great Grebe (*Podiceps major*) is up to 31 in. (79 cm.) long and weighs up to 4.4 lb. (2 kg.).

The bill varies from short and chunky, as in Pied-billed Grebe (*Podilymbus podiceps*) to medium-long and heavy in Red-necked Grebe (*Podiceps grisegena*) and Great Crested Grebe (*P. cristatus*) to long and very narrow, as in Western Grebe (*Aechmophorus occidentalis*) and Clark's Grebe (*A. clarkii*). The nostrils are narrow slits, and there is a bare patch of skin on the lores. All species have a relatively long neck; in Great Grebe, the neck is slightly longer than the body. The birds have relatively small, narrow wings and are not strong fliers. They are virtually tailless. The legs are set far back on the body and have lobed toes, the front three of which are connected by small webs at the base.

A **Horned Grebe** in breeding plumage; both sexes show bright body coloration, a black head with a yellow facial streak, and a golden crest. Scotland, U.K.

Plumage The flight feathers are curved, and when the wings are folded, they fit close to the body, hidden under the contour feathers. There are 12 primaries and 17–22 secondaries. These feathers are molted simultaneously, resulting in a short period of flightlessness. The tail feathers are very small and are hidden under the tail coverts.

In nonbreeding plumage, grebes are somber in color, mostly shades of gray, brown, and buff, and generally darker above and paler below. Breeding plumage is brighter, in some cases markedly so, and with lateral or crown crests. For example, in breeding plumage Horned Grebe (*Podiceps auritus*) has golden-yellow lateral crest tufts on a contrasting black head, a rufous neck and flanks, a white belly and underwings, a scaly dark gray back, and dark upperwings with white patches. Eared Grebe (*P. nigricollis*) also has distinctive head markings in breeding plumage: wispy yellow plumes fan out behind the bright red eyes and contrast with the black head. Non-breeding Red-necked Grebes have a dull, pale brown neck, but in the breeding season the neck becomes bright rufous, contrasting with white cheeks and a black crown and lores.

Behavior Grebes are almost entirely aquatic, leaving water only to access their nest. They rarely fly except to migrate. Short-winged Grebe (*Rollandia microptera*) is entirely flightless, as is Junín Flightless Grebe (*Podiceps taczanowskii*) and the probably-extinct Atitlan Grebe (*Podilymbus gigas*). The birds are generally uncomfortable on land, although Short-winged Grebe will make wing-assisted runs, sometimes over considerable distances.

A **Great Crested Grebe** (*Podiceps cristatus*) in a 'cat' pose; the bird performs this action in 'discovery' displays, as its mate, having dived, resurfaces in front of it.

Grebes forage in water. They often dive in pursuit of prey, with their wings held close to the body, and propelling themselves with their legs. Some can move at more than 6 ft. (2 m.) per second underwater.

Most species are shy. If threatened, the birds can expel all the air from their body and sink rapidly; otherwise, they will either swim to the safety of shoreside vegetation or dive and swim away underwater.

In the breeding season, grebes are generally seen only in pairs or family groups, but at other times of the year some species may form large feeding concentrations at sites rich in food.

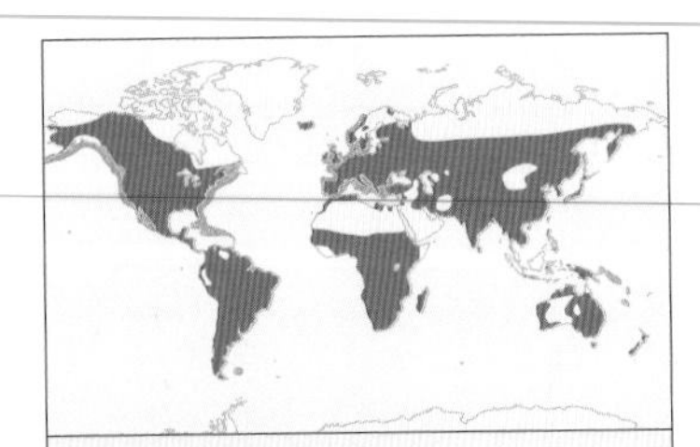

Number of genera 6
Number of species 22
Conservation Status 1 species probably Extinct, 2 species Critically Endangered, 2 species Vulnerable, 1 species Near Threatened
Distribution All continents apart from Antarctica

The courtship displays of **Western Grebe** pairs include 'rushing,' in which both birds rise out of the water and propel themselves along rapidly with their feet. U.S.

Voice Generally quiet outside the breeding season. During courtship, grebes make a variety of wailing, trilling, whistling, or barking calls, which differ according to species. For example, Least Grebe gives loud trilling calls, and Pied-billed Grebe has a wailing *eeow-eeow-eeow-keeow*.

Habitats In the breeding season, grebes occupy clean freshwater lakes, ponds, marshes, or rivers. Some species prefer very small ponds, while others favor larger waters. They avoid intensely cold water, and require sufficient emergent vegetation for nest sites. Some species breed at altitudes of more than 13,100 ft. (some 4,000 m.) on Andean lakes, notably Silvery Grebe (*Podiceps occipitalis*) and Junín Flightless Grebe. Outside the breeding season, some species occupy larger freshwater bodies without emergent vegetation, or sheltered coastal waters.

Movements Some species are sedentary. Others evacuate lakes that are liable to freeze in the fall, sometimes wintering in sheltered coastal waters. For example, many Horned Grebes migrate from their central Canadian breeding lakes to winter farther south, in the U.S. or along the Pacific coast of Canada. Similarly, many Eared Grebes move from breeding sites in the U.S. to winter inland or on the Pacific coast of Mexico.

Diet Mostly aquatic invertebrates, fish, and amphibians. Larger grebes take a bigger proportion of fish. Great Grebe eats mostly fish, but smaller species mostly predate insects and their larvae, mollusks, and crustaceans. Atitlan Grebe had a powerful bill and was adapted for catching and crushing crustacea.

Breeding Grebes form a monogamous pair bond lasting at least one breeding season; pairs sometimes remain together during the winter. Courtship may be elaborate, involving highly ritualized displays. Great Crested Grebes have three main display behaviors: head shaking, when the male and female face each other in the water and vigorously shake their head; 'discovery,' when one bird approaches the other under water, then suddenly emerges almost vertically in front of it; and the weed ceremony, when one or both of the pair collect aquatic vegetation underwater and emerge to face each other, breast to breast, with the plant matter hanging from the bill. The birds are territorial in the breeding season. Some are colonial nesters, but most are not. Usually, both partners build the nest, a shallow pile of aquatic leaves and stems, usually on a floating mass of vegetation (or rarely on land). The clutch of two to seven eggs is incubated by both parents for 20–28 days. Often, if the incubating bird has to leave the eggs it will cover them with vegetation. The young are cared for by both parents, and fledge after 44–79 days.

In many species, such as this **Pied-billed Grebe**, parents carry the chicks on their back for the first few weeks. Often, the parents take two or three chicks each. Texas, U.S.

FLAMINGOS *Phoenicopteridae*

The tall, pink flamingos are an instantly recognizable group of birds, which have been known from earliest times and often celebrated in popular stories. They belong to one of the oldest bird families, dating back at least 30 million years, when their range extended to North America and Australia.

Structure All species have a long, slender neck and tall, spindly legs, a fairly small body, and a large, specially adapted, drooping bill.

Plumage Coloration varies between pale and deep rose-pink, with crimson and black wings.

Behavior Flamingos are extremely sociable and usually occur in large flocks. At times, they form the biggest concentrations of any non-passerine birds: on occasions, more than a million gather at feeding sites. The birds forage by wading knee-deep at the edges of lakes or lagoons. They turn their head upside down and sweep the bill through the water, sucking in mouthfuls. As they squeeze out excess water with their tongue, comblike structures called lamellae trap tiny particles of food.

Voice Typically a deep, gruff grunting or growling. In flight, flamingos have a gooselike honk; feeding flocks make low murmurs.

Greater Flamingos (*Phoenicopterus ruber roseus*) in flight: their crimson and black plumage, distinctive shape, and close formations create a spectacular sight. They often form very large feeding flocks. Ngorongoro Crater, Tanzania.

American Flamingo (*P. r. ruber*), a subspecies of Greater Flamingo, is found in the Caribbean. Location unknown.

Habitat Although widespread, flamingos occur at very few places within their range: mostly at shallow lakes and lagoons. The water in these places can be alkaline or saline, and sometimes with a salinity level twice that of sea-water. The birds have a high tolerance of toxic substances including chlorides, sulfates, and fluoride. They may often be the sole inhabitants of otherwise inhospitable soda-lakes, and can drink water up to 150°F (65°C) around hot springs—all this while enduring midday ground temperatures of 155°F (68°C) or more. Three South American species occur on lakes at altitudes up to 14,764 ft. (4,500 m.) and remain there through the winter. Nighttime temperatures drop to -22°F (-30°C), but the birds are sustained largely by hot springs: their roost areas stay largely icefree while the rest of the lake is entirely frozen.

Movements Mostly sedentary.

Diet Mostly algae, invertebrates, and small shrimps. Each bird needs to eat about a tenth of its own body weight per day to survive.

Breeding Flamingos breed in large colonies at the edges of lakes or on islands. The nest (which may last for several years) is a circular mound of mud baked hard by the sun, into which the female lays one white egg.

The chicks look like fluffy ducklings on hatching, and are fed on a milky mash regurgitated by their parents. After a few days, they join a large crèche of youngsters within the colony, but continue to be fed by their parents for about 10 weeks longer, until they can fly and become fully independent.

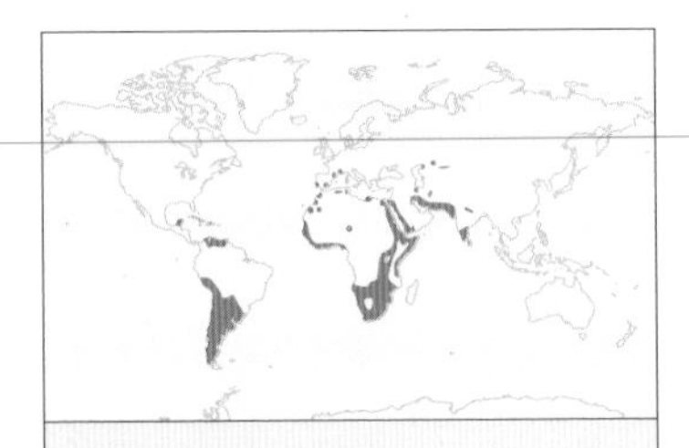

Number of genera 3
Number of species 5
Conservation Status 1 species Vulnerable, 3 species Near Threatened
Distribution Tropical and subtropical: mostly the Caribbean, South America, the Mediterranean, Africa, southern Asia

STORKS *Ciconiidae*

Tall, striking birds, storks are a familiar sight around wetlands. They resemble cranes and herons, but are generally more heavily built. The birds are usually obvious in open country when feeding, nesting, or migrating, but numbers of several species have declined drastically in recent years.

Structure Species range from 27 to 59 in. (76–150 cm.) tall, the smallest being Abdim's Stork (*Ciconia abdimii*) and the largest Greater Adjutant (*Leptoptilos dubius*). The sexes are alike, although males are slightly larger than females.

Most species are similar in structure. They have a long, heavy bill, decurved at the tip in four species and slightly uptilted in another three; a long neck; long legs with unwebbed toes; large bodies; large, broad, rounded wings; and a fairly short, rounded tail. Two species, Jabiru (*Jabiru mycteria*) and Marabou (*L. crumeniferus*), have an inflatable neck sac; in the latter's case, the sac often hangs below the bill. The African and Asian openbills (genus *Anastomus*), as their name implies, have a bill that is partly open along its length. This feature seems to be an adaptation for removing mollusks from their shell.

A **European White Stork** (*Ciconia ciconia*) in flight. Storks' long, broad wings enable gliding flight, which conserves energy while the birds are in the air. Spain.

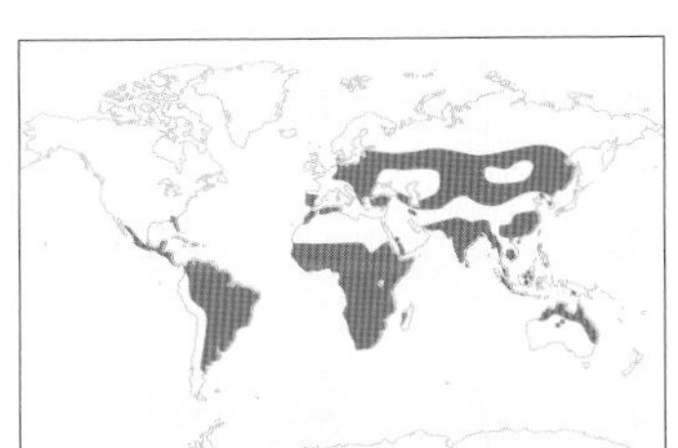

Number of genera 6
Number of species 19
Conservation Status 3 species Endangered, 2 species Vulnerable, 2 species Near Threatened
Distribution Worldwide apart from polar regions, northern and far eastern Asia, New Zealand, most of North America

Plumage All storks have bold coloring. The body plumage is usually black and white, although in two species the upperparts are pink. Bills range from bright yellow to red or black, and the legs and feet may also be red or black. At least half the species have bare areas on the face, head, or neck, with yellowish to bright pink skin. These areas are most extensive on Marabou, including the head, neck, and throat pouch; while they may be attractive on some species, on Marabou they only add to the overall ugliness.

Behavior Some species are solitary, especially when not breeding, but most storks are highly social. The birds often feed or roost together in flocks occasionally numbering thousands of individuals.

Foraging methods vary between the different genera. The four *Mycteria* storks, which have a slightly decurved tip to the bill, feed mostly in shallow water, where they walk with the bill slightly open, swaying the head from side to side, and occasionally also opening the wings as if to shade the surface of the water; as soon as the bill touches prey, it snaps shut. Openbill storks take large apple snails, frogs, and crabs from muddy water edges or soil turned by plows. The other species feed by walking slowly and watching the ground or shallows in front of them for movements, before pouncing and stabbing with the bill. Abdim's Stork follows locust swarms or outbreaks of insect larvae, and gathers in large numbers at grass fires to feed on fleeing insects and other animals.

All storks are graceful fliers and often soar up to considerable heights on rising currents of air. Almost all species fly with the neck and legs outstretched, but Marabou flies with the head retracted.

Voice Largely silent except when breeding. Mated pairs become very vocal at the nest, with loud hisses, whistles, or dry, hoarse croaks. They also throw back the head and clatter the bill rapidly. This behavior is most pronounced in European White Stork. Marabou makes a similar but slower clattering as a threat display. Openbill storks cannot make this noise on their own, but males rattle the bill against the female's during courtship displays.

Habitat Mostly wetlands, including swamps, marshes, flooded grasslands and savannas, rice paddies, open woodlands with occasional pools, banks of slow rivers, and mangroves. In Africa, Abdim's Stork and Marabou also widely forage in dry grasslands and agricultural plains. Both Marabou and Greater Adjutant often forage on rubbish dumps at the edge of towns and villages.

A **Saddlebill Stork** (*E. senegalensis*) catches prey. This species, one of the largest of the storks, derives its name from the saddle-shaped shield, colored orange, at the top of its bill. Etosha National Park, Namibia.

A **Jabiru** wades slowly through wetlands, watching the water in front of it, to hunt its prey. Brazil.

Movements Most tropical-breeding species are largely sedentary or make short-distance dispersal movements away from breeding areas in winter. Abdim's Storks breed north of the equator and move south, often in huge flocks, to winter in southern Africa. The truly migratory storks are the European White, Oriental White (*C. boyciana*), and Black Stork (*C. nigra*), which move south at the end of the breeding season to winter in Africa, India, and eastern China. The European White Stork, which often moves in flocks of more than 10,000, has very well defined migration times and routes, including narrow crossing-points between land-masses. Some birds, especially those moving from southern Africa to western Russia, travel more than 12,500 miles (20,000 km.).

Diet Mostly fish, insects, frogs, crabs, and small animals including rodents, snakes, and lizards. Larger species, including Jabiru, take mudskippers, young caimans, and freshwater turtles. Marabou and Greater Adjutant are mainly carrion-eaters. In company with vultures, they are often the first to arrive at a carcass, where they pull lumps of meat from it with the huge bill and compete with jackals and hyenas for scraps.

Breeding Jabiru, Black-necked Stork (*Ephippiorhynchus asiaticus*), Black Stork, Woolly-necked Stork (*C. episcopus*), and Storm's Stork (*C. stormi*) breed in isolated pairs. All other storks breed mainly in large or loose colonies. White Storks occupy a single nesting pole or rooftop, but are usually fairly close to other pairs. Nests are huge piles of sticks with grass tussocks and earth, up to 9 ft. (3 m.) deep, and may be re-used in subsequent years. Often, sparrows, weavers, starlings, or rollers take the chance to nest in the lower layers. Mated pairs greet each other every time they arrive at the nest, by calling loudly, bobbing the head and neck, and lifting the wings. Eggs vary from one to five, and are incubated by both parents for up to 30 days. The young fledge after about 50 days in smaller species and possibly at about 100 days in larger ones.

IBISES AND SPOONBILLS *Threskiornithidae*

Scarlet Ibis (*Eudocimus ruber*) occurs in northern South America and Trinidad. It gains its dazzling color from pigment ingested as it eats red crabs. Caroni Swamp, Trinidad.

This family of elegant wading birds comprises two distinct subfamilies. The ibises (subfamily Threskiornithinae: 12 genera and 26 species) are generally heron-like in shape but with a long, slender, decurved bill. The spoonbills (Plataleinae: 2 genera and 6 species) derive their name from their peculiar, long, spatula-like bill.

Black-faced Spoonbill (*Platalea minor*) is listed as Endangered, with only around 900 birds left in the world. Taiwan.

Structure The species vary greatly in length, from 17 to 40 in. (46–110 cm.). The sexes are generally similar, but in some cases males are slightly larger, with a longer bill; immatures are often smaller, with a shorter bill. All species have a small head, a long bill, and a long neck. The body is slender, with relatively long, broad wings and, in most species, a short tail. The legs are medium to long, with long-toed feet.

All species have bare skin on the head. The bare areas are most extensive in genus *Threskiornis*, comprising Sacred Ibis (*T. aethiopicus*) and relatives, in which the head, neck, and part of the breast are bare, with loose, pendulous skin. Genus *Pseudibis* has a bare head with warty skin; Wattled Ibis (*Bostrychia carunculata*) has a dangling throat wattle.

The chicks of Scarlet Ibis (*Eudocimus ruber*), Hadada Ibis (*Bostrychia hagedash*), and Sacred Ibis have vestigial claws on the wings.

Plumage The sexes are similar. Coloration in ibises varies dramatically from uniformly blackish or dark gray, some distinctly patterned with white, brown, gray, or buff, to predominately white or scarlet. Many species have highly iridescent feathers. The majority of spoonbills are mostly white, but, as its name suggests, Roseate Spoonbill (*Ajaia ajaja*) is predominately pink. Most members of the family have ornamental plumes, such as an elongated crest or chest feathers; genus *Threskiornis* has filamentous tertials, which hang over the bird's rump.

Irides may be white, pale blue, brown, or red. Bills and legs range from black or dark gray, through shades of brown or red, to orange or straw-colored. Areas of bare skin may be gray, blackish, dull green, or brilliant red or orange; in most species, these areas become brighter at the start of the breeding season.

Depending on species, the chicks hatch naked or covered with thin down. Juvenal plumage is different from that of adults, sometimes with more extensive feathering on the face or neck. Generally, the birds have one partial prebreeding molt and one complete postbreeding molt per year.

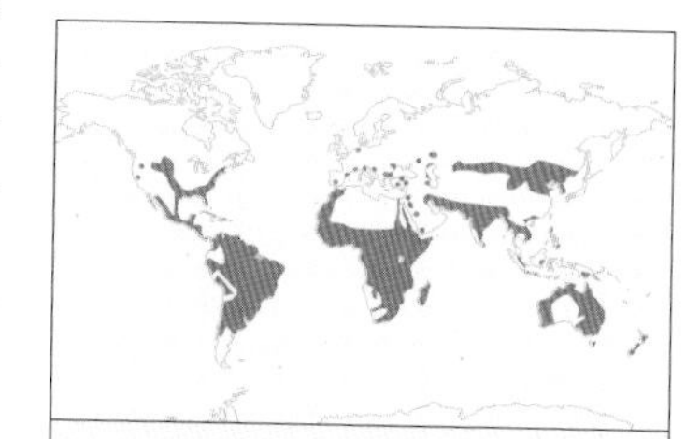

Number of genera 14
Number of species 32
Conservation Status 4 species Critically Endangered, 3 species Endangered, 1 species Vulnerable
Distribution All continents except Antarctica

Behavior Species range from solitary to highly gregarious. Many join feeding and roosting flocks with conspecifics as well as with other ibises or spoonbills, herons, and storks. They are primarily diurnal, but spoonbills may also be active at night. Most species perch or roost in trees and shrubs. In flight, the long neck is held outstretched and the legs extend behind the body. Spoonbills constantly flap, whereas ibises typically flap several times and then glide. The large wings enable both spoonbills and ibises to soar effortlessly on thermals.

The birds forage in water or soft soil. They may locate their prey visually, but more usually by feel, picking or probing with the bill. Most species, but especially the spoonbills, also use a swishing motion, swinging the bill back and forth. The birds may look for food while standing still or while walking or wading. Some employ more active running or flushing techniques, while others flip over objects. Prey is generally swallowed whole, but some species break larger prey into smaller pieces.

Voice A variety of grunts, groans, or whistles. Most species are generally silent away from breeding colonies or feeding aggregations. Many use the bill to produce nonvocal sounds such as rattles, clatters, or snaps.

Habitats Primarily wetlands, including marine, brackish, or freshwater marshes and swamps. However, the birds may also be found in arid savanna, rocky slopes, forest, and high, moist meadows or grasslands, at elevations from sea level to more than 15,700 ft. (4,800 m.).

Movements Tropical species tend to be sedentary, while subtropical species may be dispersive, primarily when searching for food resources or after breeding. Temperate-zone birds are migratory, and include austral migrants. Migrating birds often move in groups, many species flying in single file or in a 'V' formation.

Sacred Ibis is so named because the Egyptians used to worship it. The species has died out in Egypt, but still inhabits sub-Saharan Africa and Iraq. Tanzania.

Diet Mainly invertebrates, especially insects and crustaceans, as well as gastropods and worms. Also the adults, young, and eggs of fish, amphibians, reptiles, and small birds and mammals. Some species also take aquatic plants, berries, shoots, and rhizomes. Nestlings' diet is similar, but very young chicks are fed partially digested regurgitate.

Breeding Timing is generally synchronized with seasonally low water levels and peak food resources. Most species are colonial breeders, often nesting with other waders such as storks, herons, cormorants, or anhingas, and with the nest often situated very close to neighbors. Others, including genera *Pseudibis*, *Bostrychia*, *Mesembrinibis*, and *Lophotibis*, are solitary or loosely colonial. The birds are monogamous; many pair at the nest site, while others have a longer-lasting pair bond. Both sexes participate in nest-building. The nest is typically a woven platform of sticks and other vegetation, placed in a tree or shrub, on top of matted marsh vegetation, or on the ground; bald ibises (genus *Geronticus*) and Wattled Ibis nest on cliff ledges. The birds usually lay a single clutch, of one to seven eggs, but may lay a replacement set if they lose a clutch. Incubation takes 20–31 days. The chicks hatch with their eyes closed, and remain in or near the nest until they fledge in 28–56 days; the fledglings achieve independence after another 7–28 days.

Northern Bald Ibis (*Geronticus eremita*) is Critically Endangered; conservation programs are under way to protect or increase the wild populations. Morocco.

HERONS *Ardeidae*

Many herons are instantly recognizable by their shape: an S-shaped neck, long, bare legs, and daggerlike bill. This group of wading birds has four subfamilies, which are classified by similarities of morphology, powder down configuration, and behavior: Ardeinae (herons, night herons, egrets, and pond herons); Tigrisomatinae (tiger herons); Botaurinae (bitterns), and Cochleariinae (Boat-billed Heron, *Cochlearius cochlearius,* only).

The family has faced a variety of threats from human activity. Historically, adults, chicks, and eggs were collected for food, and fish-eating species were also killed as vermin (especially at fish hatcheries). During the late 19th and early 20th centuries, the use of herons' plumes in the millinery trade decimated or exterminated populations of many species. Pesticides (such as DDT) and other environmental contaminants (such as mercury and selenium) have also adversely affected breeding success.

Structure Medium to large birds, the Ardeidae range from 11 to 55 in. (27–140 cm.) in length, with wingspans from 16 to 91 in. (40–230 cm.). The sexes are alike. The subfamilies differ from each other in shape and proportions: members of Ardeinae are generally more slender, with a longer neck; more nocturnal or crepuscular species, such as the night-herons, have proportionately large eyes. Most members of the family have a dagger-shaped bill, whose length varies from short (about the length of the skull) to exceptionally long and slender (as in Agami Heron, *Agamia agami*). Boat-billed Heron, uniquely, has a wide-based, spade-shaped bill with a slightly hooked tip. The neck has a characteristic S-shape, with the sixth vertebra elongated. The wings are proportionately long and broad, and the tail very short. The long, slender legs have partially bare tibiae, and the feet have long toes, the middle with a pectinate (comblike) claw.

Plumage In most species the sexes are similar, but a few are strongly dichromatic, especially some small bitterns (genus *Ixobrychus*), in which the males are more colorful. The feathers are generally soft. Coloration ranges from uniform blackish-gray to entirely white. The plumage may also have patterns in shades of gray, blue, green, rufous, buff, and brown. A few species are cryptically colored, especially large bitterns (genus *Botaurus*) and tiger-herons. Several species, such as Reddish Egret (*Egretta rufescens*), have both dark and white morphs.

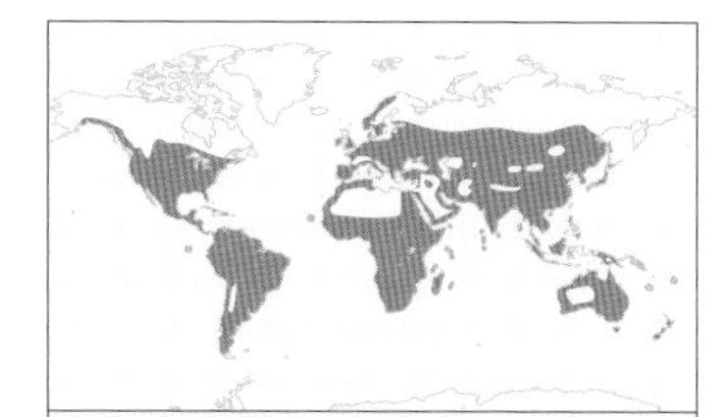

Number of genera 19
Number of species 65
Conservation Status 1 species Critically Endangered, 5 species Endangered, 2 species Vulnerable
Distribution Almost worldwide

Goliath Heron (*Ardea goliath*) is the world's largest heron, standing up to 53 in. (140 cm.) tall and with a wingspan of up to 84 in. (200 cm.). South Africa.

Most species grow elongated ornamental plumes on the head, chest, breast, or scapulars prior to the breeding season. These plumes may be lanceolate (featherlike) or filamentous (hairlike). The most extreme form are 'aigrettes,' which are wispy and ethereal in appearance. Agami Heron has unique, sickle-shaped plumes on the sides of the neck.

All species have paired patches of powder down, primarily on the breast and rump; the number of pairs varies by subfamily. These continuously growing feathers deteriorate at the tips, forming a powder that the bird spreads over its feathers when preening, to help maintain the feathers in good condition.

Molting is complex. Most species have a complete postbreeding and partial prebreeding molt, but some remain poorly studied. Exceptionally, Malagasy Pond Heron (*Ardeola idae*) molts into a distinct all-white plumage for the breeding season but is brownish for the rest of the year.

Juveniles are generally similar to adults, but in a few species can be dramatically different: for example, the juvenile Little Blue Heron (*E. caerulea*) is white. Adult plumage is typically acquired during the first year, but exceptionally takes up to five years to grow, as in Rufescent Tiger Heron (*Tigrisoma lineatum*).

The irises may be white, yellow, brown, or red. Bill colors include black, dark gray, yellow, or ivory. In some species, the bill is bicolored or with a paler base to the mandible, which is yellow, gray, blue, or even bright pink. The lores or face often have bare skin in an identical or similar color to the iris or the base of the bill. This skin typically becomes more brightly colored during courtship. Legs may be black, brown, gray, yellow, or red; a few species have contrastingly yellow feet.

Behavior Most species are conspicuous. Many are highly social or gregarious, while others are generally solitary. However, the behavior of many species can be extremely flexible to take advantage of fluctuating food availability, with individuals able both to be solitary, to defend a feeding territory, or to participate in conspecific flocks or mixed-species flocks of other wading birds (including ibis and storks). Some species are reclusive and less easily observed; in particular, bitterns are known to 'freeze,' relying on their cryptic coloration to avoid detection.

All species spear or seize prey with their bill; however, they employ a variety of foraging strategies. Most stalk by stealthy, slow walking or wading, or by waiting poised and motionless at the water's edge. Others hunt in shallow water, wrapping their wings around their head to form a shaded canopy to attract prey, or actively flapping their wings to flush the prey. Some species stir their feet to flush prey from beneath the substrate, or even entice prey closer by moving their colorful toes. Birds in genus *Butorides* have been documented to use live bait (such as insects and worms) or inanimate items to lure fish. Herons do not swim, but individuals can learn to forage over deeper water by hovering or even landing on the water's surface, or by shallow diving to spear prey under the surface. The birds can compensate for light refraction in water by moving their head back and forth to ascertain the distance to the prey. Most species are aquatic hunters, but a few are more terrestrial: for example, Cattle Egret (*Bubulcus ibis*) often forages alongside large grazing animals or harvesting or plowing farm equipment.

Reddish Egret (*Egretta rufescens*) has two morphs: dark, with a blue-gray body and reddish head and neck (shown right), and white (center). Juveniles (left) are brown and lack the striking pink and black bill.

Herons have a continuous, flapping flight; larger species have slower wing beats, and some soar for short periods. In flight, all except Whistling Heron (*Syrigma sibilatrix*) pull the neck in close to the body. All species walk or run well.

Voice A variety of squawks, grunts, groans, or coos. Some species 'sing' (such as small bitterns, *Ixobrychus*).

A **Boat-billed Heron**. The large, flat bill enables the bird to scoop up prey such as shrimps and fish. Location unknown.

Large bitterns (*Botaurus*) have a modified esophagus that enables them to produce booming sounds. Nonvocal sounds, including pops or clacks, may be made with the bill.

Habitat Herons occupy a wide variety of habitats, but most species are typically associated with wetlands, especially marine, brackish, and freshwater marshes and swamps, from sea level to 15,478 ft. (4,800 m.). The birds are now protected in most countries because they face severe threats from loss, degradation, and pollution of their habitats, and disturbance to nesting colonies.

Movements Species and populations range from sedentary to migratory. Birds gathering at roosts or colonies may travel long distances to find ephemeral food resources affected by water levels or prey concentrations. Additionally, most (including sedentary species) undergo post-breeding dispersal from nesting grounds. Migration occurs either individually or in flocks (including mixed-species), usually at night. Some species cross large expanses of open water (such as the Mediterranean Sea or the Gulf of Mexico), and there are many instances of moderate to long-distance vagrancy.

American Bittern (*Botaurus lentiginosus*). The most northerly-breeding birds migrate south to avoid harsh winters. U.S.

Diet Arthropods, mollusks, and other aquatic invertebrates; fish, and a wide variety of other small to medium-sized vertebrates. Birds' eggs, and rarely carrion, are also taken. Prey is swallowed whole; indigestible insect exoskeletons, fur, or feathers are regurgitated as pellets. Chicks are fed on the same diet, but smaller items are chosen for small chicks; meals are regurgitated, sometimes partially digested, for them.

Breeding Most species are monogamous. Polygyny has been recorded occasionally in a few species: for example, male Eurasian Bitterns (*Botaurus stellaris*) have a territory with several females that assume all parental duties. Many species are colonial, nesting with conspecifics, other herons, or other water birds. Others (such as *Ixobrychus* and *Butorides*) are solitary or loosely colonial. Breeding coincides with the highest availability of food and water (summer at higher latitudes, or rainy season in tropics). At colonies, males typically arrive first, initiate nest-building or nest-repair, and attract females to the nest site or a nearby display platform by various displays, postures, or vocalizations; some engage in mutual aerial displays. Boat-billed Herons pair prior to their arrival at the nest site.

Nests are simple, loosely constructed platforms of sticks, reeds, and other vegetation, located in a tree, bush, or reed bed, or on the ground. Some species reuse nests year after year, with the nests growing larger and bulkier. Depending on species and breeding location (temperate or tropical), herons lay one to three clutches per season; if nest or eggs are lost, the birds will lay a replacement clutch. Clutches comprise one to seven eggs, exceptionally more. Incubation takes 14–30 days and is performed by both parents. Hatching is asynchronous; the chicks are altricial, and hatch with very sparse down and eyes closed. The oldest have a size advantage, and siblicidal aggression has been documented for many species. Chicks are cared for by both parents on or near the nest site until they fledge, at about 25 days (in small bitterns) to 13 weeks (in larger herons).

A **Black Heron** (*Egretta ardesiaca*) fishing. The bird employs a hunting technique known as 'mantling,' wrapping its wings around its head to form a darkened canopy. The shaded water attracts fish, which the bird then seizes. Ethiopia.

TROPICBIRDS *Phaethontidae*

Like all tropicbirds, **Red-billed Tropicbird** is strong, graceful, and agile in the air; during courtship flights, the birds can even fly backward. Tobago.

A **White-tailed Tropicbird** nesting on a ledge. Nesting adults' tail streamers are often abraded or broken off, due to cramped quarters. Seychelles.

Plunge-diving marine birds of tropical and subtropical waters, tropicbirds recall terns (family Laridae) in body shape but have stronger flight. They are distinctive for their long central tail streamers and brightly colored bill.

Structure The birds are 31–43 in. (78–108 cm.) in length, including the central pair of rectrices, which can exceed 20 in. (50 cm.). Wingspans range from 35 to 43 in. (88–108 cm.). The head is large, with a heavy, sharp bill. The body is deep-chested; White-tailed Tropicbird (*Phaethon lepturus*) is more lightly built than the other two species. The small, webbed feet are set far back on the body; as a result, the birds cannot easily walk on land.

Plumage The sexes are alike. Adults are generally satiny white, with black markings around the eye and on the wings and mantle. Juveniles are heavily barred, with black on the upperparts. The central rectrices are white in Red-billed Tropicbird (*P. aethereus*), white to pinkish or pale yellow in White-tailed Tropicbird, or red in Red-tailed Tropicbird (*P. rubricauda*), and vary from ribbon-like (White-tailed) to stiff and wiry (Red-tailed). The adults' bill is yellowish in White-tailed Tropicbird, but red in the other two species. Juveniles have a black bill.

Behavior Flight is strong, with rapid, stiff wingbeats. The birds plunge-dive from heights up to 80 ft. (25 m.), to seize prey at or just below the water's surface. They also float on the water with their tail arched. Courtship includes synchronized tandem flights, one bird sometimes 'back-pedaling' around the other.

Voice Generally silent at sea. More vocal around breeding colonies, where courting birds give repetitive grating or shrieking calls.

Habitat The birds are strictly marine, foraging in warm waters far from land, but they nest on islands. Red-tailed Tropicbird occurs on oceanic islands, while Red-billed birds use offshore cliff ledges. White-tailed birds may nest on trees or cliffs well inland. No species is at risk, but introduced predators (such as rats and cats) have had an adverse impact on many island populations.

Movements The birds are pelagic but are not highly migratory, although some disperse to higher latitudes after breeding. Individuals occasionally stray, or are driven by storms, to latitudes above 40 degrees or, very rarely, well inland.

Diet Mostly squid and fish, especially flying fish as they leap out of water.

Breeding The birds nest in loose colonies. They lay a single egg in a scrape on the ground (often shaded by a shrub), a cliff ledge, or a hollow or fork in a tree. Incubation takes 42–47 days. The chicks are covered with dense down on hatching. They first fly at 10–13 weeks.

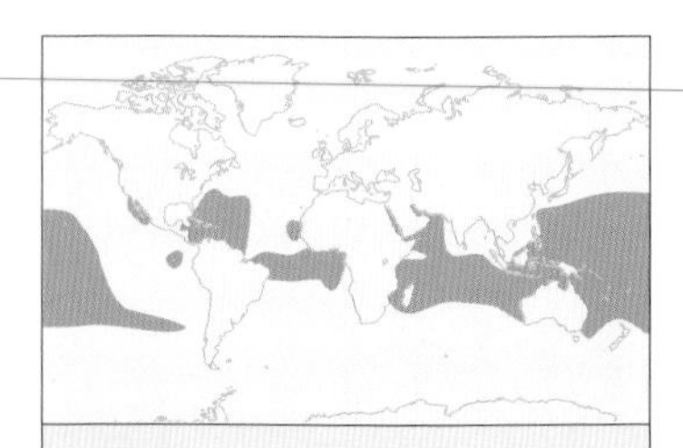

Number of genera 1
Number of species 3
Conservation Status Least Concern
Distribution Eastern Pacific, South Atlantic, Indian Oceans; Caribbean

FRIGATEBIRDS *Fregatidae*

A male **Great Frigatebird** (*Fregata minor*) with his red throat pouch fully inflated. Galápagos Islands.

A uniform group of striking seabirds, frigatebirds are easily recognized as they soar endlessly over warm oceans.

Structure The birds range from 28 to 45 in. (70–115 cm.) and are built for prolonged, effortless flight. They are lightweight, with the heaviest species only about 3.5 lb. (1.5 kg.). The long, pointed wings are sharply angled at the carpal joint, and the tail is long and deeply forked. The bill is long and hooked at the tip. The legs are small, with partially webbed feet. Males have a pouch of bare skin at the throat, which they inflate in displays.

Plumage Adult males are all black or with limited white on the belly and axillaries; the throat pouch is bright red. Females generally have more white on the breast. Juveniles have a white to rusty head and a variable brown bar on the upperwing.

Behavior Flight is buoyant and soaring, with rapid glides and remarkable maneuverability. Frigatebirds may chase other seabirds to steal their food; otherwise, they feed by dipping to the water's surface for prey or offal. They perch on trees, shrubs, wires, ship riggings, or cliffs, but do not land on the water.

Voice Birds at breeding colonies, particularly displaying males, give warbling, whinnying, and rattling calls, and also make rapid bill-clicking. Generally silent away from colonies.

Habitat The birds occupy tropical and subtropical coastlines and oceanic islands. They usually forage along coasts or in inshore waters, but also range widely at sea. Nesting colonies are usually sited on shrubby or wooded islands, but Ascension Frigatebird (*Fregata aquila*) nests on bare ground. Some populations are vulnerable due to a limited number of breeding areas; for example, the forest sites of Christmas Island Frigatebird (*F. andrewsi*) have been greatly depleted. Introduced predators and human disturbance also pose risks for some colonies.

Movements Not highly migratory, although nonbreeding birds disperse well away from breeding colonies and may be seen hundreds of miles out at sea. Juveniles tend to wander to temperate latitudes. Birds are sometimes blown well inland, or to unusually high latitudes, by storms.

Diet Fish (especially flying fish) and squid, taken at or close to the water's surface; also offal (for example, from ships) and sometimes the eggs and chicks of other seabirds.

Breeding The birds tend to breed every other year. They build a nest of sticks and lay a single egg, incubated for six to eight weeks. The young fledge in five to seven months, but depend on the parents for much of their first year.

Magnificent Frigatebirds (*F. magnificens*) steal from a Red-billed Tropicbird (*Phaethon aetherius*) by holding it upside down to make it regurgitate its catch. Galápagos Islands.

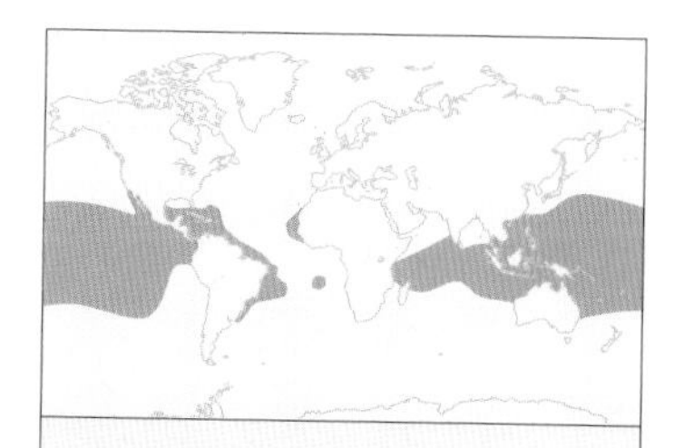

Number of genera 1
Number of species 5
Conservation Status 1 species Critically Endangered, 1 species Vulnerable
Distribution Tropical and subtropical areas of the Atlantic, Pacific, and Indian Oceans

HAMMERKOP *Scopidae*

A single-species family of African wading birds, containing only Hammerkop (*Scopus umbretta*). The common name comes from the Afrikaans *hamerkop*, meaning 'hammerhead,' and refers to the shape created by the bird's long bill and stiff crest. Hammerkops are notable both for their strange appearance and for their spectacularly large nests, which are often usurped by other bird species during construction, or used by other animals after being abandoned.

A **Hammerkop** captures a fish. The birds usually feed at dawn and dusk, and roost during the heat of the day. Lake Baringo, Kenya.

Structure Hammerkop is 19.5–22 in. (50–56 cm.) in length. The heavy head is set on a short neck. The bill is moderately long and heavy; the maxilla has a ridged culmen, concealed nostrils, and a slight hook at the tip. The body is slender, but the wings are broad and rounded. The legs are relatively short.

Plumage The sexes are alike. The plumage is generally a uniform brown or grayish-brown, with a purplish gloss on the upperparts and darker bands on the upper tail. The stiff crest extends back from the hindcrown and neck. The iris is dark brown, and the bill, legs, and feet blackish. Chicks hatch with gray down. Juveniles are like adults, but with a yellow iris, browner bill, and grayer legs. The birds are thought to have a single full molt, although molting has not been well studied.

Behavior Hammerkop is active by day and at dusk. Birds forage alone or in pairs, but may gather in roosts of up to 50 individuals. They prefer to forage by wading in shallow water. They may use their feet or flap their wings to flush out prey, which they then seize with the bill. They may also use their bill to probe the muddy substrate, like an ibis. In addition, they may glean prey from just under the water's surface as they fly. The birds may fly like storks, with the neck outstretched, or may tuck in the neck heron-style. They are often seen performing a unique 'false mounting' display, thought to have social significance.

Voice Generally silent, but when flocking the birds give a variety of loud, nasal cackling sounds. Typically, they give a shrill *kek* as a flight call, and also make croaking sounds during courtship flights.

Habitat A variety of wetlands, including freshwater marshes and the edges of lakes or rivers.

Movements Generally sedentary, but may engage in limited local dispersal to track food resources.

Diet Frogs, small fish, small mammals, and invertebrates (including crustaceans and worms).

Breeding Monogamous and rather territorial; nesting can occur year-round. Pairs often build several nests in a season, then select a new nest or one built the previous season. The nest is an elaborate, domed structure up to 6 ft. (1.8 m.) across, made of sticks and vegetation, fortified with mud, and placed in a large tree or on a cliff ledge. It has an entrance on the underside, with an entrance tunnel leading to a nest chamber. The clutch of three to seven eggs is incubated for 28–32 days. Chicks hatch asynchronously, fledge in about 47 days, and return to roost in the nest after fledging. Despite the fortified nest and fairly large clutches, broods seem to have a very low survival rate.

Taxonomy Hammerkop is traditionally of uncertain affinities, with similarities to storks, herons, and ibis. Recent genetic analyses suggest that it is most closely related to Shoebill (family Balaenicipitidae) and to the pelicans (Pelecanidae).

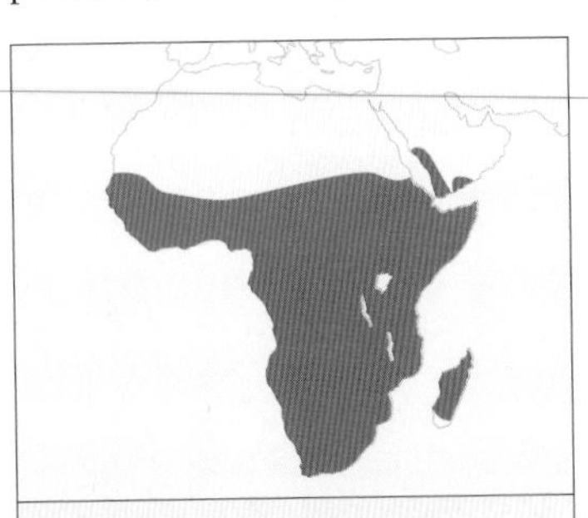

Number of genera 1
Number of species 1
Conservation Status Least Concern
Distribution Sub-Saharan Africa, Madagascar

SHOEBILL *Balaenicipitidae*

Shoebill has an enormous, flat, powerful bill, well adapted for catching fish such as lungfish, its most common prey. Uganda.

A single-species family, which is represented only by Shoebill (*Balaeniceps rex*). This large African wetland bird looks rather like a stork, except for its distinctive, huge, broad bill. The genus name means 'whale-headed,' and the bird is sometimes known as Whale-headed Stork.

Structure Shoebills are 43–55 in. (110–140 cm.) in length. They have a massive head set on a short, thick neck. The huge bill has a maxilla with a pronounced keel or culmen ridge, strongly hooked tip, and sharp edges. There is an area of bare gular skin under the bill. The body is slender, with long, broad wings, long legs, and long-toed feet.

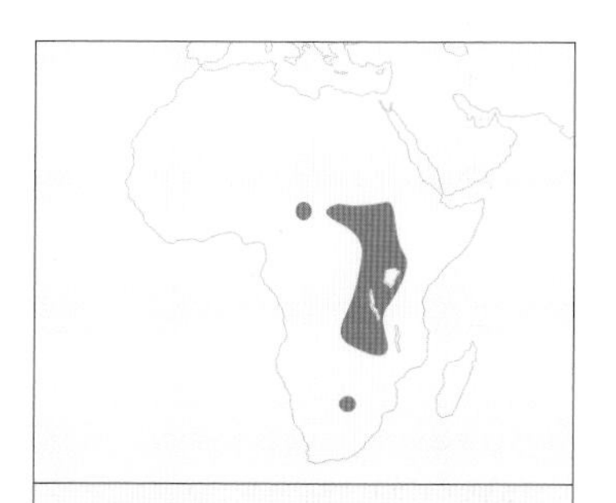

Number of genera 1
Number of species 1
Conservation Status Vulnerable
Distribution East and Central Africa

Plumage The sexes are alike. The plumage is mainly blue-gray; the back and upper-wing coverts are paler-edged, with a dull green gloss, the flight feathers darker gray, and the belly white. The bird has a stiff crest and elongated chest feathers. The iris is yellow, the bill horn-colored and mottled with brown, and the legs and feet brownish-gray. Chicks hatch covered with blue-gray down. Juveniles are similar to adults but browner, with a grayer iris and a pinker bill.

Behavior Shoebills are generally diurnal and typically solitary. They forage standing on floating or matted vegetation or at shorelines, or wading, sometimes in fairly deep water. To hunt, they use a 'watch and wait' or 'wade and pause' strategy. As soon as they see prey, they lunge at it with their entire body, gulping up a large amount of vegetation and water, which is then drained from the bill; they will occasionally make such a lunge from the air. The birds fly with deep, labored wing beats, the neck tucked in like a heron (family Ardeidae), and use thermals to soar. In hot conditions, they defecate on their legs, so that the evaporating liquid will cool them, and they also flutter the gular skin.

Voice Generally silent; adults make mewing calls at the nest. Nonvocal sounds include loud bill clapping, used as display or territorial defense. Nestlings make 'hiccup' sounds.

Habitat Papyrus swamps and freshwater marshes with dense aquatic and emergent vegetation; also wet grassland or meadows.

Movements Sedentary, although the birds may make limited local movements to track prey resources.

Diet Fish and other aquatic vertebrates (snakes, frogs, and turtles); also lizards, young birds and crocodiles, and small mammals.

Breeding Typically synchronized with low local water levels and highest prey concentrations, at the start of the dry season. Shoebills are monogamous and share parental duties. The nest is made from aquatic plants, and hidden on floating vegetation, a termite mound, or the ground. The clutch typically comprises two eggs; the parents keep the eggs cool by wetting them with water carried in their bill. Incubation takes 30 days, and hatching is asynchronous. The chicks fledge at 95–105 days. Normally, only one survives to fledging, and remains with the parents for a further week.

Shoebills usually regurgitate food for their chicks, or tear prey into small pieces with their sharp-edged bill. Zambia.

PELICANS *Pelecanidae*

A small group of fish-eating waterbirds, the pelicans are known for their unique 'pouch,' used for scooping up prey.

Structure These birds range from 40 to 70 in. (100–180 cm.) in length, with wingspans up to 10 ft. (3 m.). Their long bill has a hooked tip, and the lower mandible is joined to a large, highly extensible throat pouch. The body is stout, and the legs are short, with fully webbed toes.

Plumage The sexes are alike. Most species are mainly white, with black or dusky flight feathers. In contrast, Brown Pelican (*Pelecanus occidentalis*) is dusky-brown and gray with the head and neck marked yellow, white, and chestnut. The pouch is yellow to orange in white-plumaged species, and dark in the others. In most pelicans, the bill, pouch, and eye rings brighten in the breeding season; some species also have an ornamental crest or bill knob.

Behavior Pelicans typically fish while swimming, by scooping water into the throat pouch. Flocks often hunt cooperatively, herding fish into a circle and then scooping them up. Brown Pelican plunge-dives from the air to seize its prey.

A **Great White Pelican** captures a fish. The bird will then surface again, tip its bill to expel the water, and toss its head back to swallow the fish. Location unknown.

The birds fly with flaps and long glides, the head resting on the back and the bill on the foreneck, and rise high on thermals. Flock movements are highly coordinated, with birds flying in loose lines or 'V's.

Voice Generally silent, but in breeding colonies a variety of grunting, hissing, and squawking sounds may be heard, mainly from nestlings.

Habitat Pelicans occur across the world, from North America and the northern and western coasts of South America to southern Eurasia, Africa, and Australia. Most live on shallow inland lakes and estuaries; Brown Pelican occurs in coastal marine waters, but pelicans generally are not found far out to sea.

Movements Variable. Some northern populations of American White Pelican (*P. erythrorhynchos*), Great White Pelican (*P. onocrotalus*), and Dalmatian Pelican (*P. crispus*) migrate south to subtropical areas. Australian Pelican (*P. conspicillatus*) may disperse to newly flooded areas, departing in dry periods. Brown Pelicans wander northward, as far as southern Canada, after breeding.

Diet Almost exclusively fish.

Breeding The birds breed in colonies on islands or in marshes. Human disturbance at colonies has greatly reduced many populations, as in Dalmatian Pelican (now Endangered) and Spot-billed Pelican (*P. philippensis*), which have declined from millions to just a few thousand birds.

Nests vary from a small mound on the ground to stick structures placed in shrubs or trees. The clutch typically comprises two or three eggs, which the parents incubate on or under their feet, for about 35 days. The young first fly at 10–12 weeks of age.

A **Pink-backed Pelican** suns itself. One of the smaller pelicans, they breed in Africa and southern Arabia. Location unknown.

Taxonomy Pelicans are perhaps more closely related to the storks (family Ciconiidae) and the herons (Ardeidae) than to other 'pelecaniform' birds; their closest living relative appears to be the African Shoebill (*Balaeniceps rex*). One subspecies of Brown Pelican, *P. o. thagus*, is often recognized as a full species, Chilean or Peruvian Pelican.

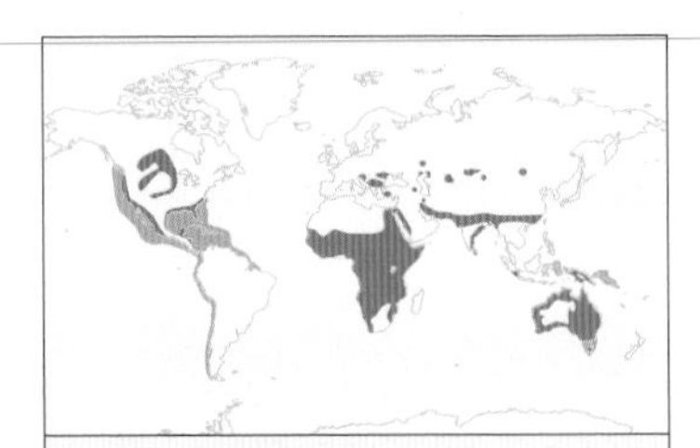

Number of genera 1
Number of species 7
Conservation Status 1 species Endangered
Distribution North and South America, Eurasia, Africa, Australia

GANNETS AND BOOBIES *Sulidae*

A **Northern Gannet** (*Morus bassanus*) plunge-diving. Gannets and boobies climb to about 165 ft. (50 m.), then fold their wings and dive; heavier birds dive deeper. U.K.

Supreme plunge-divers, gannets and boobies are found in most of the world's oceans except for the colder areas of the North Pacific.

Structure The birds are 28–38 in. (70–95 cm.) in length. They have bare facial skin, forward-facing eyes, a long, pointed, finely serrated bill, and a small gular pouch. The body is streamlined in shape, with long, pointed wings and a pointed tail. All four toes are connected by webs.

Plumage The sexes are similar. Coloration varies from mainly white to brown, with dark flight feathers. Most species have a white belly, except for dark-morph Red-footed Booby (*Sula sula*), which is pale brown. The bill, facial skin, and/or feet are often brightly colored, and the eyes are usually pale. Juveniles have brown plumage but attain adult plumage by five years old.

Blue-footed Boobies (*S. nebouxii*). Both sexes have blue feet, which they show off as part of displays. Galápagos Islands.

Behavior Flight involves flapping interspersed with short glides. The birds forage by plunge-diving. More lightly built species, such as Red-footed Booby, also swoop to take prey at the water's surface. Boobies roost on beaches, cliffs, trees, or ship riggings, while gannets may roost on the water.

Voice Highly gregarious, gannets are quite vocal, giving harsh, rasping calls, grunts, and moans. Most boobies are mainly vocal in the breeding colonies: males give wheezy whistles, and females quack or grunt.

Habitat Marine waters, from coastlines to open ocean. Boobies generally inhabit tropical and subtropical waters, but Peruvian Booby (*S. variegata*) occurs in the cool Humboldt Current off western South America. Gannets are found in temperate waters in the North Atlantic and off Australasia and southern Africa. All species nest on predator-free islands or sheer mainland cliffs. Humans and introduced predators pose threats to booby colonies, but most gannetries are well protected. Overfishing has impacted many species. Abbott's Booby (*Papasula abbotti*) is threatened by deforestation.

Movements Most species disperse along inshore continental waters or well out to sea outside the breeding season. Boobies generally remain at lower latitudes, but vagrants have been found as far north as the northwest U.S. and eastern Canada. Gannets breed at high latitudes, but some winter in subtropical waters.

Diet Gannets and most boobies feed mainly on schooling fish; Red-footed and Abbott's Boobies prefer squid and flying fish.

Breeding The birds breed in small to very large colonies. Nests are placed on the ground: gannets and most boobies build theirs in depressions rimmed with droppings, while Red-footed and Abbott's Boobies build stick nests in trees. The clutch of one to three eggs is incubated under the parents' feet for six weeks. The young (of which usually only one survives) fledge in 3–4 months, but are fed by their parents for longer.

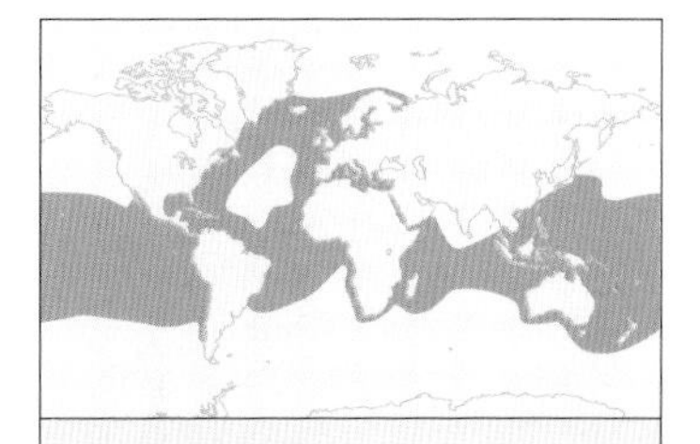

Number of genera 3
Number of species 10
Conservation Status 1 species Endangered
Distribution Atlantic, Pacific, and Indian Oceans

CORMORANTS *Phalacrocoracidae*

A **Great Cormorant** dries its wings. After fishing, cormorants spend long periods standing or perching with their wings spread to dry themselves before they can fly. U.K.

Cormorants are medium to large diving birds commonly seen along coastlines and in many interior wetlands. The main groups in this family include the true cormorants, typified by the widespread Great Cormorant (*Phalacrocorax carbo*); the small, long-tailed freshwater 'micro-cormorants;' and the marine species, known as shags.

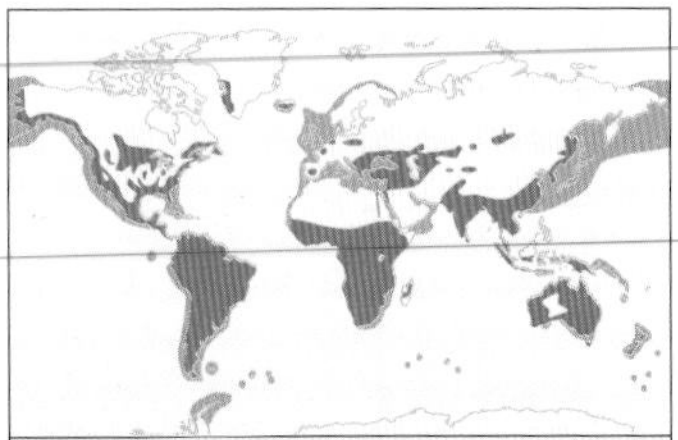

Number of genera 1
Number of species 36
Conservation Status 1 species Critically Endangered, 3 species Endangered, 4 species Near Threatened
Distribution Worldwide apart from tropical islands and northern Eurasia

Structure Species range from 22 to 40 in. (55–100 cm.) in length. Cormorants are long-necked, heavy-bodied birds. They have a thin, hook-tipped bill, a bare gular pouch, and, often, bare skin on the lores and around the eyes. Most, apart from Flightless Cormorant (*P. harrisi*), have medium-length wings with slightly rounded tips. The tail is stiff and moderately long. The feet are set well back on the body, and all four toes are connected by webs.

Plumage The sexes are alike. Adults are predominantly blackish and, in many species, glossed with green or blue. Some are spotted or scaled on the back and wing coverts. Most southern-ocean shags, and a few other species, are partly to mostly white below; some also show white markings on the wing coverts and flanks. A few species have dark and light morphs. Breeding birds may have a crest and short white plumes on the head and neck. Bare facial skin and eye rings are often brightly colored, and many species have blue or green eyes. Juveniles are mostly brown, and do not attain full adult plumage until two to four years.

Behavior Cormorants may feed solitarily, but may also forage in large flocks, which sometimes coordinate their movements to herd prey. The birds hunt underwater, propelled by their feet, and can reach depths of 65–100 ft. (20–30 m.). They also spend long periods resting on rocks, sea cliffs, pilings, or shorelines. If overheated, they flutter their gular skin rapidly to lose body heat.

Cormorants have a continuous, flapping flight. They usually fly low over the water, and glide or soar only briefly. The birds walk clumsily on land, but many species perch readily on tree branches or cables.

Voice Although generally silent, cormorants are noisy in breeding colonies, when a variety of guttural, croaking sounds may be heard.

A **Double-crested Cormorant** (*P. auritus*). When cormorants dive, air is expelled from under the feathers, and some feathers become wet, reducing buoyancy and helping them move faster. Florida, U.S.

Spotted Shag (*P. punctatus*) has prominent crests on the forehead and crown during the breeding season. New Zealand.

Habitat Cormorants are found nearly worldwide, apart from tropical oceanic islands and much of northern Eurasia. They mainly occur in colder coastal waters, with only a few species found along tropical coasts, and they are absent from the open ocean. About 10 species occupy freshwater lakes and marshes well into the interior of continents.

Breeding habitats vary from bare islands and cliff ledges to marshes and waterside trees and shrubs. Many species are vulnerable at nesting colonies, readily abandoning the sites when disturbed by humans or predators; most notably, the flightless Spectacled Cormorant (*P. perspicillatus*) of Bering Island was easily killed by humans, and was extinct by 1850. Some breed on only one or a few islands, increasing their vulnerability. Oil spills, pesticide contamination, fishing, and ocean current anomalies have also had severe impacts on populations, as has the unsustainable harvesting of guano at some species' colonies.

Movements Most southern shags are sedentary. Some temperate-zone species are short-distance migrants, moving from the interior to coastal areas at lower latitudes after the breeding season. Others disperse along coastlines. Some cases of long-distance vagrancy occur, such as Pelagic Cormorant (*P. pelagicus*) traveling to Hawaii, and Neotropical Cormorant (*P. brasilianus*) wandering north as far as Ontario.

Diet Almost exclusively fish, but aquatic invertebrates such as crustaceans and squid are important in diet of many cold-ocean shags.

Breeding Most species are at least loosely colonial, and many nest in large colonies. Nests are made of sticks, seaweed, and debris such as feathers or bones. A few species nest on tree branches or shrubs, but marine species generally nest on rocky ground or cliff faces. Clutches comprise two to six eggs, which the parents incubate on the webs of their feet for 25–35 days. Usually only one or two of the brood fledge. The young leave the nest and join crèches at 2–4 weeks of age, and in most species they first fly at 6–7 weeks old.

Flightless Cormorant has much smaller wings, as well as a smaller keel (breastbone), than species that can fly. Galápagos Islands.

Taxonomy The family's taxonomy is still in flux. Many authorities recognize from four to nine genera. There is also much uncertainty as to the number of species, at least regarding shags (many of which are island endemics) in the southern oceans.

Red-legged Cormorant (*P. gaimardi*) forms colonies high on sheer cliffs and builds its nest on a narrow ledge, using seaweed and other marine vegetation. Patagonia, Argentina.

ANHINGAS *Anhingidae*

This family comprises just the New World Anhinga (*Anhinga anhinga*) and the Old World Darter (*A. melanogaster*); the latter is sometimes split into two or three species. Both are diving birds that resemble cormorants (family Phalacrocoracidae), to which they are closely related. They are also known as 'water-turkeys' or 'snakebirds.'

Structure The anhingas are 32–38 in. (80–97 cm.) in length. They have a small head, with a slender, sharply pointed bill. The long, thin neck has a unique joint between the eighth and ninth vertebrae that allows quick strikes at prey. The wings are fairly long and broad, and the tail is long and fan-shaped. The legs are short, with large, fully webbed feet.

Plumage Male Anhingas are generally black, with white markings on the scapulars and wing coverts; females are similar, but with a pale brown head, neck, and breast. Darter males are similar to Anhingas, but variously marked with chestnut or brown and white on the throat and neck; females are extensively brown to whitish below. Juveniles are duller and browner than adults, attaining adult plumage in their third year. The bill is yellowish; male Anhingas have a bright blue-green eye ring.

Behavior The birds are generally solitary, but may form small, loose groups. They hunt underwater, propelling themselves with their feet, and impaling their prey with their sharp bill. At the surface, they often swim semi-submerged. Otherwise, they spend a great deal of time roosting on tree branches overhanging water, often with their wings spread to dry themselves. The birds can take off from land or water, and use thermals to travel farther.

A **Darter** in flight. The birds fly with their neck straight, alternately flapping and gliding. They have a distinctive, cross-shaped outline. Mauritius.

An **Anhinga** swimming. Both species have a distinctive style of swimming, with only the head and neck clear of the water. U.S.

Voice Generally silent outside the breeding season. At the nest, adults give guttural, chattering calls. Chicks can be highly vocal, giving a variety of clicking and squeaking calls.

Habitat Both species occur in tropical and subtropical regions. They are seen on still or slow-moving freshwater lakes, ponds, in marshes with abundant vegetation, and in wooded swamps, as well as in mangrove-lined coastal estuaries. Some populations have declined due to draining of wetland habitats, disturbance at nesting colonies, and the impacts of pesticide residues. Darters in India and Southeast Asia are of greatest conservation concern: one form, Oriental Darter (*A. m. melanogaster*), is considered Near Threatened.

Movements Most populations are sedentary, but in some regions the birds make local movements in response to changes in water levels. The northernmost-breeding Anhingas in the U.S. move south to the Gulf Coast or Mexico in winter. Occasional New World vagrants are found at higher latitudes, such as southern Ontario (Canada) and Tierra del Fuego (Argentina).

Diet Mainly fish, but also amphibians and aquatic invertebrates.

Breeding The birds usually nest singly or in small colonies, and build stick nests in trees, often over water. Clutches comprise three to five eggs, which the parents incubate (partly on their feet) for four weeks. The young may leave the nest and even swim in two weeks, and make their first flight at about six weeks.

Number of genera 1
Number of species 2
Conservation Status 1 species Near Threatened
Distribution Americas, Africa, Southeast Asia, Australasia

NEW WORLD VULTURES *Cathartidae*

The New World (or cathartid) vultures are found from southern Canada to Tierra del Fuego. They look similar to their counterparts in Europe, Africa, and Asia. Like Old World vultures, they are scavengers, with mostly featherless heads, broad wings, and excellent vision. The family includes the spectacular condors and King Vulture (*Sarcoramphus papa*).

Structure The birds range from 22 to 53 in. (56–134 cm.), with a wingspan of 63–126 in. (160–320 cm.); the condors are the largest. The sexes are similar. All have a powerful, hooked bill for tearing flesh, a large crop so that they can gorge themselves, strong legs and chest muscles to hold down carcasses, and long, broad wings for soaring flight. The birds can walk or run, but their feet are weak and the talons not well developed. The head is more or less naked; this feature may be an adaptation to help them stay clean while scavenging, and may also be important for temperature regulation and display. California Condor (*Gymnogyps californianus*) has an air sac at the throat, which it can inflate during courtship displays. Turkey Vulture (*Cathartes aura*) and Greater and Lesser Yellow-headed Vultures (*C. melambrotus* and *C. burrovianus*) have an exceptional sense of smell, enabling them to find carrion.

Plumage The skin on the head ranges from black to red, to yellow, or even multicolored in the case of King Vulture. Immature birds have more feathering on the head and especially the neck; their bare skin is dark gray or blackish. Two species, King Vulture and Andean Condor (*Vultur gryphus*), have elaborate skin folds, fleshy crests, or wattles.

In most species, the plumage is uniformly dark with slight to pronounced pale or white patterns on the upper- or underwings. The exception is King Vulture, with bold black and white patterned plumage. Northern species undergo a prolonged annual molt of the flight feathers; the molting pattern is not well known in tropical species. Chicks hatch with a thick, downy coat. In Yellow-headed Vulture, Turkey Vulture, and Black Vulture (*Coragyps atratus*), juvenal plumage is like that of adults, although duller. King Vulture and the condors take four to six years to acquire their adult plumage.

A **California Condor**. The bare skin on the head and neck may become redder during courtship displays, or if the bird gets excited or alarmed. California, U.S.

Number of genera 5
Number of species 7
Conservation Status 1 species Critically Endangered
Distribution North, Central, and South America

Behavior New World vultures spend considerable time on the wing, in search of food. Their gliding flight is assisted by wind or thermal updrafts; they may be grounded in windless conditions that require laborious flapping. They find food by scent or sight. Feeding behavior varies from solitary to highly gregarious, and species may feed in mixed flocks. Vultures roost either alone or communally; Black Vulture is the most gregarious. In hot climates, the birds defecate on their legs, so that moisture evaporating from the feces will cool them down. Sunning with outstretched wings is also important in temperature regulation.

Voice Although New World vultures lack a syrinx and are generally silent, all can make various hissing or rattling sounds. Most are vocal during courtship and at the nest.

Habitat The birds occupy a range of habitats, from lowland desert and tropical forest to alpine zones, at altitudes from sea level to the high Andes. Turkey Vulture and Black Vulture have adapted to human environments and are attracted to road kills, dead livestock, and garbage. The primary threats to vultures are poisoning and habitat loss. The species most at risk is California Condor. In 1987, the last 21 of these birds were captured for a breeding program; reintroductions to the wild began in 1992, and the current wild population numbers 125.

Movements Most species are sedentary, although movements of tropical species are poorly understood. Northern Turkey Vulture populations are highly migratory, moving south to northern South America, where their range overlaps with those of South American subspecies.

Diet A wide variety of dead animals. Condors prefer large mammals, and both species are known to feed on the carcasses of marine mammals. Only condors and King Vulture can breach tough animal hides; for this reason, a hierarchy may form at such carcasses, with smaller species having to wait their turn. Some species, most notably Black Vulture, occasionally kill weak or defenseless prey such as nestlings or baby sea turtles. Less often, vultures may eat bird eggs, dung, or fruit.

Breeding The vultures are solitary and monogamous, and first breed at three to eight years (depending on size). Breeding in northern populations is strongly seasonal. Smaller species breed annually, but condors raise only one brood every other year due to extended care of their young. No species builds a nest: instead, the single egg (in condors and King Vulture) or two-egg clutch is laid on the ground, on a ledge, or in a cave or cavity. Incubation takes 38–60 days; after the eggs hatch, the parents return to the nest only to regurgitate food to the young. The adults do little to defend the nest, but the foul smell of the young and nest site may deter predators, and the young of some species will also regurgitate food at intruders. Fledging occurs from 70 days (in Turkey Vulture) to six months (in California Condor). The young stay with their parents for a further month, or several in the case of condors.

Taxonomy Placement within the Falconiformes is debated: morphological, genetic, and behavioral evidence has been put forward to suggest a closer relationship to storks (family Ciconiidae). The Cathartidae may perhaps best be treated as a separate order allied to the Falconiformes.

A pair of **Turkey Vultures** soar on the wind, searching for prey. Like all New World vultures, they can glide for hours with little or no effort. Falkland Islands.

FALCONS *Falconidae*

This adult female **Red-footed Falcon** (*Falco vespertinus*) has just caught an item of prey. This falcon is a long-distance migrant. Sweden.

The most familiar members of this diverse family are the 'true falcons,' such as Peregrine Falcon (*Falco peregrinus*), the world's fastest aerial predator, which is found worldwide. Others include Black-thighed Falconet (*Microhierax fringillarius*) of southeast Asia, the tiniest bird of prey; the more terrestrial, omnivorous caracaras, comprising 11 New-World species; the forest-dwelling, *Accipiter*-like forest falcons (*Micrastur*), and Laughing Falcon (*Herpetotheres cachinnans*), both of which are found in the Neotropics. The family is united by a variety of structural features, a unique primary molt progression, and the similar chemical composition of their eggshells.

Structure Species vary in length from 6 to 24 in. (14–60 cm.), with wingspans of 12–53 in. (30–135 cm.). They feature pronounced sexual dimorphism, with females significantly larger than males.

Body shapes vary with hunting style, habitat, and prey size. Generally, true falcons have long, pointed wings and a fairly short tail, and are built for speed and aerial pursuits in open country. Forest-falcons have shorter, more rounded wings and a long tail, giving maneuverability and enabling the pursuit of birds in forested areas. The ground-foraging caracaras have long, rounded wings, a long tail, and a relatively slow and deliberate flight style. All species can soar well, and some kestrels frequently hover.

Eyes are large, with dark to reddish-brown irises. Bills vary in size but are generally stout, with a hooked maxilla. Falcons have tomial 'teeth' on the cutting edge of the maxilla and corresponding notches on the mandible, which aid in killing prey. These features are lacking in the caracaras and forest falcons. All falcons have strong legs and feet, and, with the exception of caracaras, long, sharp, and strongly curved talons. Forest falcons, which, unlike other members of the family, use auditory cues to locate their prey, have large, asymmetrically positioned ear openings. Laughing Falcon has rough, hexagonal tarsal scales, which are thought to provide protection against snakebites.

Plumage The sexes are similar, but males are more brightly patterned in some species. Polymorphism is rare: the best-known example is Gyrfalcon (*Falco rusticolus*), which varies from dark brown to almost white. Mostly, species that are cryptically colored black, brown, rufous, or gray are generally darker and uniform above with paler underparts, many patterned with barring, mottling, spots, or streaks. Some species, such as American Kestrel (*Falco sparverius*), are more colorful, with blue-gray and reddish tones. Many *Falco* species have a prominent mustachial streak on the face. The tail

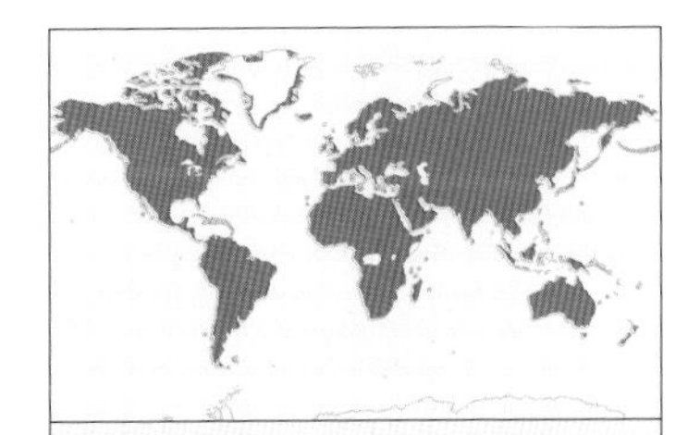

Number of genera 11
Number of species 64
Conservation Status 1 species Endangered, 4 species Vulnerable
Distribution Worldwide, apart from Antarctica

Laughing Falcon (*Herpetotheres cachinnans*) spends much of its time sitting motionless on a perch. This species specializes on snakes; when it sees prey, it swoops and pounces with great force, often biting off the head. Venezuela.

Aplomado Falcons (*Falco femoralis*), found from Mexico to South America, fly with deep, rapid wingbeats in pursuit of prey. They are highly maneuverable, able to turn, swoop, hover, and soar. The bird to the left is an adult; that to the right is a juvenile.

often has dark or paler interior or terminal bands. Forest falcons have specialized feathers forming an ear ruff, and some caracaras and Laughing Falcon have a crest. Chicks hatch with a sparse downy coat, which is replaced by a denser second down plumage and followed by juvenal plumage, which is fairly distinct in most species. There is one complete molt each year. The molt of the primaries and secondaries is ascendant and descendent; primary molt is unique, molted sequentially in pairs in each direction, beginning with primary number four.

Some caracaras have colorful bare facial skin. The cere and legs are yellow, but a few species have gray, green, or red coloration.

Behavior Usually solitary or found in pairs. Most species are territorial during the breeding season, and some also maintain a nonbreeding territory. A few species form loose aggregations when prey is abundant, and two, Red-throated Caracara (*Ibycter americanus*) and Lesser Kestrel (*F. naumanni*), are highly gregarious. A few species roost communally during migration or the nonbreeding season; the most spectacular are large migratory roosts of Lesser Kestrels, which can number 70,000 birds. All family members are diurnal hunters, but some, such as Bat Falcon (*F. rufigularis*), are crepuscular. Basic foraging strategies include searching (either from a perch or hovering) and then pouncing, and capturing prey in the air.

Voice Fairly vocal, especially during breeding season. Calls are typically a loud, strident series of chattering or cackling whistles, squawks, croaks, whines, or barks; the birds also make defensive hissing sounds. The young make peeping or yelping notes, mimicked by females soliciting food from their mate. Forest falcon and Laughing Falcon pairs duet.

Habitat A variety of habitats, from desert to arctic tundra, and from sea level to 9,500 ft. (3,000 m.).

Movements Sedentary to highly migratory. Some species are nomadic in response to diminished food resources. Peregrine Falcon is one of the longest-distance migrants, some individuals making a 15,500-mile (24,800-km.) round trip from their northern breeding grounds to winter in southern South America. Perhaps the most remarkable migration of all is that of Amur Falcon (*Falco amurensis*), which breeds in far eastern Russia and northeastern China, and migrates across southern China, India, and the Indian Ocean to winter in East Africa: a journey of 7,500 miles (12,000 km.).

Diet Falcons take a wide variety of prey depending on the species, including birds, mammals, reptiles, amphibians, fish, insects, and other arthropods. Several species steal nestlings from nests. Lesser Kestrels flock to follow insect swarms. Caracara diets are most diverse, and include carrion, ticks, worms, fruit, grain, other vegetable matter, and even dung. Red-throated Caracara specializes on wasp and bee larvae.

Breeding Generally monogamous; many probably mate for life. However, males are more attached to nest sites than females, and tropical species are poorly known. First breeding is at one year of age for smaller species, and two to three years for larger species. Breeding is usually seasonal, even in the tropics, and most species raise one brood per year. A few species nest colonially, including Eleanora's Falcon (*Falco eleonorae*) and Lesser Kestrel. Red-throated Caracaras nest cooperatively, and the group shares parental duties including brooding and nest defense. For most, the nest is merely a hollow in a natural cavity, on a ledge, or even on the ground; many use the old nests of other birds. Nest sites may be reused from year to year. Caracaras build their nests from sticks. In most species, the female incubates, feeds the young, and defends the nest while the male captures and delivers food. Clutch size varies by species, ranging from one to six; eggs are typically buff, spotted with reddish-brown. Incubation is 28–35 days (longest in the largest species), and eggs hatch asynchronously. Chicks fledge in as few as 27 days for smaller falconets to 53 days for Gyrfalcon. In Crested Caracara (*Caracara cheriway*), the fledgling period may last for up to three months. Depending on species and individual variation, the young stay with their parents, learning to hunt, for several weeks or months.

These **Crested Caracaras** are eating carrion. The structure of all caracaras, especially their long legs and long, broad wings, is very different from that of true falcons and is far better suited for foraging and scavenging on the ground. Texas, U.S.

HAWKS, KITES, AND EAGLES *Accipitridae*

Members of this extremely diverse family are superbly well adapted for life as raptors, with features such as a sharp, hooked bill and strong jaw muscles, strong legs and feet with sharp talons, and excellent vision. They include the huge Old World vultures, which spend hours soaring at great heights in search of carrion; the small, fast-flying accipiters, which hunt birds in forest habitats; the unique, long-legged Secretary Bird (*Sagittarius serpentarius*); and the most powerful avian predators of all, the monkey-eating eagles.

Structure Variation in body size is dramatic across the family: species range from 8 to 59 in. (20–150 cm.) in length, with wingspans from 21 to 122 in. (54–310 cm.). Females are larger than males. At one end of the scale is the petite male Pearl Kite (*Gampsonyx swainsonii*) of South America, which weighs just 3 oz. (80 g.), while at the other is the huge female Himalayan Griffon (*Gyps himalayensis*), at 26 lb. (12 kg.). The group is equally diverse in their overall body structure, according to hunting style and primary prey. This variation is manifested in bill structure; wing and tail shape; and leg, toe, and talon length and development.

The powerful bill has a strongly hooked maxilla. Size and shape vary from small and delicate in insectivorous kites, to large, strong, and capable of breaking dense mammal bones in large monkey-eating eagles, such as Philippine Eagle (*Pithecophaga jefferyi*). In some species, it is highly specialized to extract prey. For example, Snail Kites (*Rostrhamus sociabilis*) have a bill adapted to sever apple snails' muscles to extract them from their shell.

Wing shape varies from long and pointed to short and broad, very large and broad, and every variation

Black-collared Hawk (*Busarellus nigricollis*) has a range extending across much of South America. It is a bird of lowland tropical or subtropical forests and swamps, where fish form an important part of its diet. Location unknown.

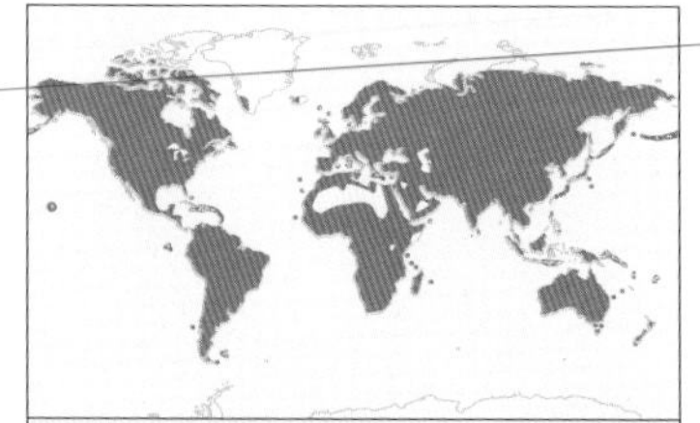

Number of genera 67
Number of species 233
Conservation Status 9 species Critically Endangered, 8 species Endangered, 24 species Vulnerable
Distribution Worldwide apart from Antarctica

between, depending on the species' style of flight. In general, large, broad wings allow effortless soaring; shorter, rounded wings and a long tail facilitate rapid maneuverability in tight spaces; and long, pointed wings offer speed. Tail shapes (primarily responsible for maneuverability) range from very short and square to broad, long, and wedge-shaped, or long and forked.

Prey is grasped with the feet. The legs and feet are strong, but the legs vary in length, with Secretary Bird having by far the longest legs. The strongly hooked talons range in size and development depending on the type of prey; they are most formidable in species that capture and kill large mammals.

A bony protective shield over the eyes of many species imparts a fierce look. The eyes are proportionately large, and the iris color varies from dark brown to white; in some species the iris color is indicative of age. All members of the family have a distensible crop to store food and maximize the value of a large meal; the fullest development of the crop is found in Old World vultures.

An adult **Cooper's Hawk** (*Accipiter cooperi*) in flight shows the distinctive 'blunt-winged' and long-tailed shape of its genus. Juveniles lack the rufous barring on the underparts.

Plumage In most species the sexes are alike, but a few show pronounced dimorphism. Several are polymorphic. Coloration is generally cryptic shades of black, gray, brown, or rufous on the upperparts, and often paler buff or white below, some species having streaked, spotted, or banded underparts. A few are all white, or nearly so. The tail is often banded. The cere may be yellow, green, gray, or bright red, and the legs, if unfeathered, are generally yellow, greenish to gray, or bright red. A few species have elongated head feathers that form a crest. Bearded Vulture (*Gypaetus barbatus*) has stiff, downward-pointing feathers that resemble a beard. The head and neck of Old World vultures is naked or only sparsely feathered. In general there is just a single molt each year, although in the largest species the completion of wing molt may span two or three years.

Chicks hatch with a coat of white down, which is replaced by a second coat prior to the growth of juvenal plumage. Juveniles of most species differ from adults and, depending on species, take from one to five years to attain adult plumage.

Behavior The majority of raptors are solitary, but live in pairs during the breeding season. Some are gregarious when food supplies are plentiful; a few are highly social and not only feed in groups but breed in colonies. Some have colonial or semicolonial roosts outside the breeding season, and some migrate in flocks. Members of the family have a poor sense of smell, so hunt mostly by sight. Old World vultures, unlike their New World counterparts (family Cathartidae), locate carcasses by sight or by observing the behavior of other birds and animals. A few species practice cooperative hunting, and some engage in kleptoparasitism: stealing the prey caught by other birds or animals.

Voice Most species are vocal during the breeding season, with a wide range of calls, including whistles, screams, barks, cackles, and grunts. Different calls are used for different situations and may include contact, courtship, alarm, or food calls.

Habitat Members of the family occupy a wide variety of habitats, including tropical rain forest, temperate forest, taiga, desert, grassland, tundra, and even the center of large cities. Altitudes range from sea level to 16,400 ft. (5,000 m.).

Movements The birds can be sedentary, dispersive, or highly migratory.

Black-shouldered Kites (*Elanus axillaris*) hunt from a perch or from midair, hovering up to 165 ft. (50 m.) above the ground. Mice and other mouse-sized animals make up the great majority of their diet. Australia.

Montagu's Harriers (*Circus pygargus*) fly low over open country in search of prey. They have graceful, buoyant flight, their wings held in a shallow dihedral profile. This bird's unmarked rufous-brown underparts identify it as a juvenile. Sweden.

The longest-distance migrants move from temperate breeding grounds in North America and northern Eurasia to tropical and subtropical nonbreeding areas in Central and South America and Africa. Many species move during the same periods, and large diurnal flights (sometimes of many thousands of individuals) can be observed at traditional sites. Migrants use thermals or the prevailing wind to glide so they use less energy. Movements are generally tied to the seasons and the availability of prey; a few species are essentially nomadic, moving according to the abundance of prey and completely evacuating areas once they have become devoid of food. The movements of some species living at high latitudes are closely tied to cyclic prey populations (such as grouse and small rodents); if population densities of these animals crash, it can result in large-scale movements or invasions to lower latitudes.

Egyptian Vultures (*Neophron percnopterus*) eat carrion, decaying fruits and vegetables, and eggs. They sometimes drop eggs onto the ground to crack them open, or drop stones onto the eggs to achieve the same result. Socotra, Yemen.

Diet Most of these birds are strictly carnivorous, but diets vary greatly from species to species. Food includes a wide range of invertebrates (especially insects and their larvae, arthropods, large snails, and earthworms) and vertebrates (fish, birds and their eggs and fledglings, mammals, reptiles, and amphibians). Some species are specialized predators, whereas others are more opportunistic. Some can kill prey larger than themselves. Prey is eaten whole or torn to pieces, and undigested material is regurgitated as pellets. Carrion is the primary diet of many Old World vultures, but other species will opportunistically take this food if necessary. Bearded Vulture specializes on the large bones of mammals, ingesting both the marrow and the bone itself. A few species also consume fruit: the diet of Palm Nut Vulture (*Gypohierax angolensis*) is primarily the fruit of *Elaeis* and *Raphia* palms. Nestlings are fed the same diet as adults, but depending on their age the food is often first torn to pieces, plucked, or decapitated; the young may also be fed partially digested and regurgitated food (for example, in Old World vultures in the genus *Gyps*).

Breeding Depending on species (and general size), the birds first breed at one to two years of age (for example, in genus *Accipiter*) up to as old as six to nine years (large vultures and eagles). Most are monogamous, and many pair for life. Polygyny is frequently observed in some harrier species (genus *Circus*), and polyandry is also recorded for a few, most commonly Galápagos Hawk (*Buteo galapagoensis*). Snail Kite is known to practice both, serially, to maximize the production of young during periods of abundant food. Nest helpers are reported for Harris's Hawk (*Parabuteo unicinctus*).

Most species are solitary breeders. Colonial breeders include some Old World vultures, such as Rüppell's Griffon (*Gyps rueppelli*) and Cape Griffon (*G. coprotheres*), and some kite species: for example, Mississippi Kite (*Ictinia mississippiensis*) and Swallow-tailed Kite (*Elanoides forficatus*) are loosely colonial. All are territorial, and defend the home range around the nest. Many have spectacular aerial courtship displays.

The nest is a platform built from sticks and lined with finer material. It may be located in a great variety of sites, including on the ground, in a tree, on a cliff ledge, or at the entrance to a cave. Nests may be reused in subsequent seasons, and those of large species that are enhanced year on year can become enormous.

The timing of breeding is coordinated to maximize food for young; thus most species breed seasonally and only once per year. However, some large eagle species with long periods of parental care breed only every other year. Harpy Eagle (*Harpia harpyja*), whose young remain with the parents for a year after fledging, may produce a brood every third year. In contrast, *Elanus* kites are known to raise two broods during periods when prey is exceptionally abundant. Harris's Hawks may also raise two broods when food is plentiful. In general, the largest species lay the fewest eggs (one or two), and the smallest lay the most (five or six).The incubation period varies from 28 days to as long as two months, in the largest vulture species. Incubation begins after the first egg is laid, so the young hatch asynchronously; as a result, the first-hatched have a size advantage, and infanticide has been recorded. The female usually incubates and broods the young, but in some species the male takes over during daylight hours, and Old World vultures share the duties equally. Fledging occurs at 24–30 days in the smaller species, and at up to 148 days in the larger vultures and tropical eagles. Depending on the species, after fledging the young remain with the parents for several weeks, months, or up to one year, being fed while they learn hunting skills.

Bald Eagles (*Haliaeetus leucocephalus*) are large raptors. Females may weigh up to 14 lb. (6.4 kg.) and have a wingspan of up to 90 in. (229 cm.). U.S.

Taxonomy Recent DNA studies have helped resolve relationships muddied by ecological and morphological convergence within the family, and several basic 'types' previously identified by morphology have been found not to be closely related. Based on DNA analysis, 14 subfamilies are now recognized: booted eagles (Aquiliinae), sea eagles (Haliaeetinae), harpy eagles (Harpiinae), snake eagles (Circaetinae), Old World vultures (Aegypiinae and Gypaetinae), kites (Milvinae, Perninae, and Elaninae), harriers (Circinae), buzzards (Buteoninae), and the unique Gabor Goshawk, *Melierax gabar* (Melieraxinae); African Harrier-hawk, *Polyboroides typus* (Polyboroidinae); and Secretary Bird (Sagittariinae).

BUSTARDS *Otidae*

A male **Great Bustard** (*Otis tarda*) displays to attract a mate. The courtship display involves the bird inflating his neck sac, cocking his tail over his back, and extending his wings down and backward. Note the secondaries, held out from the body. Spain.

This family of terrestrial birds has origins dating back more than 70 million years. Most species are similar in structure but vary greatly in size and plumage. Although they prefer walking, the family includes some of the heaviest flying birds. Bustards are generally shy or elusive but become more visible when displaying or breeding.

Structure Generally large, but can be divided into small (11), medium (six), and large (seven) species. Lengths range from 16 to 48 in. (40–120 cm.) and weights from 1–40 lb. (0.5–19 kg.). The sexes are generally alike, but in the large species males are larger than females. Bustards are stoutly built with a long neck (often slender in the smaller species), and a fairly short, dagger-shaped bill. The wings are large, broad, and rounded, with 10 primaries and 16–24 secondaries almost as long as the primaries. The tail is short and broad, with 18–20 feathers. The legs are long and sturdy; the feet have three toes, thick soles, and flat claws.

Plumage The crown, neck, and upperparts are cryptically colored to blend in with the surroundings. Colors vary from black or brown to sandy, gray, or buff. Several species have light brown upperparts finely barred darker, while others have irregular dark bars or spots. Several others have a two-toned neck with sandy hindneck and gray or black foreneck. Two have long black stripes down the neck, and the male Little Bustard (*Tetrax tetrax*) has a distinctive white 'V' and a broad white band on the otherwise black neck. In the larger species, the underparts are mostly white, and in the other two groups the underparts are almost equally divided into black and white. Blue Bustard (*Eupodotis caerulescens*) has a blue-gray neck and underparts. Two species, Arabian Bustard (*Ardeotis arabs*) and Kori Bustard (*A. kori*), have a short crest, and a further four (Houbara, *Chlamydotis undulata*; McQueen's, *C. macqueenii*; Red-crested, *Lophotis ruficrista*; and Buff-crested, *L. gindiana*) have loose crown feathers that they raise as a crest when displaying. The larger species have inflatable neck sacs and long, loose neck feathers, and several others also have vestigial crests. In flight, large bustards show extensive white in the wings.

Number of genera 11
Number of species 26
Conservation Status 1 species Critically Endangered, 2 species Endangered, 2 species Vulnerable, 6 species Near Threatened
Distribution Old World: Europe, Africa, Asia, Australia

A **White-quilled Bustard** (*Afrotis afraoides*) in the species' typical semi-arid habitat. Etosha Pan, Namibia.

Behavior Typically, bustards move with a slow, deliberate, sedate walk, holding the head high and rocking gently backward and forward. However, they can also dash nimbly after prey. The birds take flight infrequently, the smaller species rising vertically, and the larger ones needing a running start to become airborne. Most are social, living in pairs or loose aggregations of up to 50 for part of the year. Some of the smaller ones live almost entirely in family groups, but five species, mostly those living in semi-desert areas, are more solitary. In the larger bustards, flocks are usually all of the same sex, and the sexes rarely meet except at a common food source or at the start of the breeding season.

Voice Largely silent except when breeding. Displaying males have calls including croaks, barks, booms, and drumming notes; the smaller species have a wider range, with snores, yelps, popping, and quacking noises.

Habitat Temperate and tropical grasslands including low semi-desert scrub, scattered thornbush, dry wadis and steppes with short grasses, and salt-steppes; also the edges of cultivation, in particular traditionally managed farmlands.

Movements Largely sedentary, but the birds may sometimes make short-distance nomadic movements. Northern-breeding populations of Great Bustard, McQueen's Bustard, and Little Bustards in Central Europe and Asia move south between September and April.

Diet Mostly insects and vegetable matter. Insect prey includes ants, termites, beetles, locusts, grasshoppers, crickets, and mantises, plucked from the ground or vegetation, or flushed from long grass. Vegetable matter includes leaves, buds, and shoots of leguminous plants. The birds may also eat scorpions, snails, small mammals, snakes, and lizards. Bustards very rarely seem to drink, presumably obtaining most of their liquid requirements from their food.

Breeding The courtship displays by males of several species are remarkable. Some have traditional display or lek areas where the males display to and mate with several females without forming any pair-bond. Male Great Bustards inflate their neck sac, fluff out their neck and body feathers, raise the fanned tail, and twist their wings to show the white undersides; they are visible at a great distance. Male McQueen's Bustard and Houbara have a similar display, with the head sunk back into the back, the chest puffed out, and the long neck plumes covering the whole head and front. They simultaneously trot or run around the display area while swaying from side to side to show off their seemingly headless state, often crashing into bushes. Several of the smaller species leap high into the air, while making far-carrying nasal calls and vibrating their wings. In all species, except the African species of *Eupodotis* bustards, the female selects the nest, incubates the eggs, and rears the young. Clutches usually comprise one or two eggs, although some species may lay up to six. Incubation takes 20–25 days. Fledging periods vary, but the mother may care for the young for up to a year.

A pair of **Kori Bustards**. These elegant birds have a short crest, which projects from the rear of the crown. Etosha National Park, Namibia.

MESITES *Mesitornithidae*

An Old World family endemic to Madagascar, the mesites comprise two genera with three species. All resemble thrushes in shape and rails in behavior. However, they appear to be unrelated to any other living bird families and are most likely a remnant of a family that was formerly more widespread.

Structure Mesites are slim birds, 10–12 in. (25–30 cm.) in length. They have a small head, short wings, a long, rounded tail, and fairly long, strong legs and feet. In White-breasted Mesite (*Mesitornis variegatus*) and Brown Mesite (*M. unicolor*) the bill is short and straight, similar to that of a thrush, but in Subdesert Mesite (*Monias benschi*) the bill is longer and distinctly decurved.

Plumage Subdesert Mesite is sexually dichromatic, but in the other two species the sexes are alike. Brown Mesite is almost entirely dark brown on the head and upperparts with paler or pinkish-brown underparts. The other two species are gray-brown to rufous-brown on the upperparts, with white extending from the lower face to the breast or belly; these white areas are heavily or sparsely spotted darker. Female Subdesert Mesite has duller or buffish underparts and darker spots on the flanks. Subdesert and White-breasted Mesites also have a strongly striped face pattern. All three species have small to large white streaks behind the eye.

A **Subdesert Mesite** freezes still on a branch, in its typical defense posture. Ifaty Spiny Forest, Madagascar.

White-breasted Mesite lives in the north and east of Madagascar. Ankarafantsika Nature Reserve, Madagascar.

Behavior Shy or secretive terrestrial birds, mesites forage in small family groups. They move around with quick, pigeonlike steps, bobbing the head and slightly wagging or lowering the tail, flicking over leaves and decaying ground vegetation in search of invertebrates. They are generally poor fliers, due to their short wings and small collarbones. When alarmed, they run rather than take flight; however, Subdesert Mesite escapes danger by flying up to a low perch and adopting a defense posture, becoming motionless with the head down and tail raised.

Voice All three species are highly vocal and most often call in the early mornings. The songs consist of a series of whistled phrases or gruff chuckling notes given rapidly or excitedly; these may either be repeated or conclude with a short trill. White-breasted Mesite often indulges in song duets. Calls include hissing and clicking notes, and are usually given when the birds are disturbed or alarmed.

Habitat White-breasted and Brown Mesites inhabit forest, while Subdesert Mesite occurs in scrubland. All three species are considered vulnerable due to small, fragmented ranges and decreasing habitat; it is thought likely that their populations will decline further without stronger protection measures.

Movements Almost entirely sedentary, but Brown Mesite may move altitudinally when not breeding.

Diet Mainly small insects and their larvae; also seeds and small fruits.

Breeding Not well known. Nests are platforms of sticks, bark strips, and leaves, placed up to 6.5 ft. (2 m.) from the ground in shrubs (which they reach by climbing). Clutches comprise one to three eggs. The downy young leave the nest soon after hatching, and accompany the adult birds, remaining with them for up to a year.

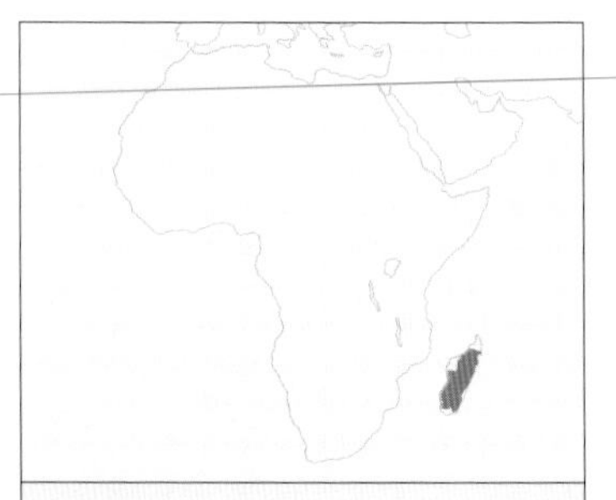

Number of genera 2
Number of species 3
Conservation Status 3 species Vulnerable
Distribution Madagascar

SERIEMAS *Cariamidae*

The seriemas are a tiny South American family comprising just two species in two genera: Red-legged Seriema (*Cariama cristata*) and Black-legged Seriema (*Chunga burmeisteri*). Their body shape and ecological niche parallel those of the African Secretary Bird (family Accipitridae), although the two families are not related. Seriemas are thought to be related to the extinct 'Terror Birds' (Phorusrhacidae): giant carnivorous birds that roamed South America until 2.5 million years ago.

Structure The seriemas are 30–40 in. (75–100 cm.) in length. The bill is like that of a chicken, but more robust and broad-based, with a definite hook. The birds have a long neck, a big, long body, and a long, graduated tail. Their most prominent feature is the long, powerful legs. The inner toe has a long, sickle-shaped claw that can be held retracted or brought down to hold or kill prey.

Plumage Red-legged Seriema is brownish-gray, while Black-legged Seriema is more evenly grayish with a noticeably pale supercilum. Both species are paler below than above. The head and neck, in particular, are finely streaked. The long tail is usually held closed, but is strongly patterned, with bold white tips to the feathers. Red-legged Seriema has a long, bristly crest and a bright red bill and legs, while the soft parts of Black-legged Seriema are blackish.

Black-legged Seriema is rather smaller than the Red-legged species, with a more restricted range. Location unknown.

Behavior Seriemas are most comfortable on the ground. They move in a slow, elegant manner, but when they need to they can run very fast, reaching speeds of at least 15 miles an hour (25 km/h.). They evade danger by running rather than flying, although they can fly at least short distances. When they do, they take off with quick bursts of flapping, followed by long glides.

Seriemas are seen alone or in pairs, sometimes accompanied by fully grown young; they do not flock, and are strongly territorial.

Voice Seriemas give loud yelping sounds that resemble a cackling laugh or high-pitched bark. Pairs often duet, both sexes sounding essentially alike. The songs carry a long distance and can be heard several miles away. When singing, the birds will often perch on a high termite mound or a small bush.

Habitat Red-legged Seriema prefers savanna or open scrub. Black-legged Seriema inhabits thicker scrub, open woodlands, and forest margins.

Movements Sedentary; there is no evidence for any major movements.

Diet Seriemas are omnivorous. They eat large insects and some seeds, but show a marked specialization on snakes and lizards. They will take poisonous snakes; they are not immune to the venom, but their long legs may help protect them from bites. The birds secure their prey with the bill and inner claw, or often hold the prey in the bill and beat it on the ground.

A **Red-legged Seriema** gives a threat display, with crest raised, mouth open, and wings spread. Brazil.

Breeding Seriemas build a bulky nest in a low tree. The clutch is usually two eggs, which are incubated by both parents for 25–30 days. Both parents feed the young. The chicks remain in the nest for about two weeks, and then are led away by their parents, completing their development in the open.

Number of genera 2
Number of species 2
Conservation Status Least Concern
Distribution Central South America

KAGU *Rhynochetidae*

Kagu spends all its time on the forest floor, among the dark undergrowth and leaf litter. New Caledonia.

An enigmatic, monotypic bird, Kagu (*Rhynochetos jubatus*) is endemic to the island of New Caledonia. It resembles several other species from which it is entirely isolated. The bird was originally thought to be closely related to herons and egrets on the basis of similarities in bill structure and plumage: for example, it has fine powder-down feathers, which are also a characteristic of herons in breeding plumage. However, it is now considered to be more closely related to Sunbittern of Central and South America and to other members of the order Gruiformes, which includes cranes and rails.

A **Kagu** extends its wings in a threat posture. New Caledonia.

Structure Kagu is about 22 in. (55 cm.) in length, and resembles a small, pale heron. It has large, dark red eyes, which are essential for living in the shaded light of the forest undergrowth. The head has a long, tapering crest, which extends from the back of the head to the lower back but is usually difficult to see unless it is raised in excitement or alarm. The rounded wings are about 29.5 in. (75 cm.) long when extended, but the bird is flightless: it uses its wings mainly for balance when fleeing from predators and gliding in long downhill descents.

Plumage Kagus are pale ash-gray except for their bright orange-red bill and legs, and their wings. The wings are broadly barred with alternating black and white bands, overlaid with fine brown and grey stripes. Like Sunbittern, Kagu uses its wings in displays and when defending its territory and young, holding them out horizontally from the body to show the pattern.

Behavior Entirely ground-dwelling, the Kagu lives on the forest floor. It forages for food in the leaf-litter, in the same manner as thrushes. This process involves watching, waiting (often on one leg), edging closer to the prey, then standing and aiming the bill directly at prey before striking. The bird also uses its strong, pointed bill to dig into soft earth in search of insects. Some birds also use vantage points such as rocks or fallen trees to locate their prey.

Voice Kagu's powerful song has been likened to a crowing cockerel or the bark of a young dog, and can be heard for up to 1.25 miles (2 km.). The song is given by paired birds mostly in early morning and predawn duets, but can also be given by lone birds establishing a territory. In areas where more than one pair of Kagu is present, the singing can last for up to an hour.

Habitat Lowland rainforests with little or no ground vegetation. Occasionally, the bird is also found in adjacent shrubby areas, but it usually roosts within the forest.

Movements Entirely sedentary, but birds may move around within their territories using paths made by man or feral pigs.

Diet Mostly small insects, worms, and lizards.

Breeding The breeding season is from June to December. Nests are made from leaves and situated on the ground, and clutches comprise just one egg. Both parents incubate the egg, for 33–37 days. The chick leaves the nest about three days after hatching, but continues to be fed by both parents.

Number of genera 1
Number of species 1
Conservation Status 1 species Endangered
Distribution New Caledonia, Western Pacific

SUNBITTERN *Eurypygidae*

A monotypic family containing only Sunbittern (*Eurypyga helias*). The bird is notable for the striking 'eyespot' pattern on the upper surface of its wings.

Sunbittern's striking threat display involves stretching its wings to reveal a yellow, chestnut, and black pattern designed to startle predators. Venezuela.

Structure Sunbittern is 17–19 in. (43–48 cm.) in length. The shape resembles that of a heron, except that Sunbittern has a horizontal posture. The bird has a small head, with large, red eyes, and a long neck. It has a slender body, whose size is exaggerated by large, broad wings and a long tail. The bill is long and daggerlike; the top is blackish, while the maxilla tip and mandible are orange to reddish-orange. The legs are short in proportion to the body, and are yellow to orange, with unwebbed feet and long toes.

Plumage The sexes are similar, and juveniles superficially resemble adults. The head is black with two white stripes and a white throat. When the bird is at rest or foraging, the visible plumage is cryptically colored in gray and brown, with black barring on the upperparts and prominent white spots on the wing coverts. The three recognized subspecies vary in body coloration: *E. h. helias* is heavily barred above and more heavily streaked on the breast, whereas *E. h. major* and *E. h. meridionalis* are more uniformly dark above and lack breast streaks. In all subspecies, the wings reveal a large, bold 'eyespot' pattern when opened: on the upper surface, there is a black and chestnut central patch surrounded by a pale, yellowish area. The tail is finely barred gray and white, with broader black and chestnut basal and subterminal bands. Sunbitterns undergo an annual molt after breeding.

Behavior This wary and reclusive bird is mainly solitary, but is sometimes seen in pairs. It hunts rather like a heron, stalking its prey at the water's edge, then stabbing or grabbing the prey with its bill. When threatened, it has a display that involves spreading the wings to show the spots, fanning the tail, and tilting forward, to give the illusion of larger size and intimidating 'eyes.' The breeding display involves a flight above the canopy, followed by a vocal descent with wings outstretched and bowed to show off the wing pattern.

Voice Soft, plaintive, drawn-out whistle; also, peeping calls and nonvocal bill rattles. During courtship displays, the bird utters several sharp *kah* notes, followed by a trill.

Habitat Sunbittern occurs in low population densities near streams, rivers, and oxbow lakes. Its habitat is tropical or lower subtropical forest, up to 6,000 ft. (1,830 m.).

Movements Generally sedentary.

Diet Fish; also crustaceans, insects, and other aquatic invertebrates.

Breeding Both sexes are involved in nest-building and raising chicks. Sunbitterns build their nest at the beginning of the rainy season. The large, bulky structure is made of sticks, leaves, and mud, and is situated in a tree, 12–20 ft. (3.7–6 m.) above the ground. The clutch comprises one or two eggs. Both parents incubate the eggs and then feed the young, which remain in the nest for several weeks. The chicks are initially clad in black-striped, brownish down. Fledging occurs in about a month, but the chicks may remain with their parents for an additional two months.

Number of genera 1
Number of species 1
Conservation status Least Concern
Distribution Central and South America

RAILS AND COOTS *Rallidae*

Okinawa Rail (*Gallirallus okinawae*), endemic to Okinawa Island, is almost flightless. Okinawa, Japan.

Small to medium-sized birds, the Rallidae mainly inhabit wetlands. Most members of the family are known as rails, but other names for particular species or subgroups include coots, crakes, moorhens, flufftails, gallinules, and swamphens. Some species are widely known, but others are extremely secretive, inhabiting dense vegetation. The birds face a variety of hazards. Many have been hunted as sport-shooting birds. Many continental species have declined in numbers due to ongoing habitat loss. Additionally, flightless species endemic to oceanic islands are particularly at risk from introduced predators; 19 species have become extinct since about 1600.

Number of genera 33
Number of species 141
Conservation Status 13 species Critically Endangered, 12 species Endangered, 18 species Vulnerable
Distribution Worldwide (except polar regions), including many oceanic islands

Structure The birds vary in length from 4 to 24 in. (10–61 cm.). The terrestrial, flightless Southern Takahe (*Porphyrio hochstetteri*) of New Zealand is the largest extant species, with males weighing up to 7 lb. (3.25 kg.). In most species, the sexes are alike, or the males are slightly larger. However, in a few, such as Watercock (*Gallicrex cinerea*) and Weka (*Gallirallus australus*), the males are substantially larger than the females.

Bill shapes vary, and are generally indicative of diet: they range from small, short, and stubby, in more herbivorous or omnivorous species, to long and slender in those birds that mainly glean for insects or probe for worms and crabs. All species have fairly short, rounded wings and well-developed, strong legs. The rails, flufftails (*Sarothrura*), and gallinules (*Gallinula* and *Porphyrio*) have a laterally compressed body, which helps them move easily through dense vegetation. Some species, including the gallinules, have very long toes, which help distribute their weight when they walk on floating plants. The coots (*Fulica*) have lobed toes, which help them swim strongly.

Plumage Generally cryptic, the plumage broadly matches ground-level vegetation. Typical coloration comprises variations of black, brown, gray, and rufous, interrupted by stripes, bars, or spots. A few species are brightly colored, especially the *Porphyrio* gallinules, which are largely blue or green. There is little sexual or seasonal variation; a few species have somewhat brighter breeding plumage, including more colorful soft parts (bill, frontal shield, facial skin, and legs). After breeding, the birds undergo a full molt; in some species, the rapid loss of flight feathers results in temporary flightlessness. Juveniles generally resemble adults, but their plumage and soft parts are duller; juvenal plumage is replaced fairly quickly by the first basic plumage.

Behavior Most species are terrestrial or semi-aquatic. All of them can swim, and most can dive or sink underwater to escape predators, using their wings to propel them through the water. Moorhens (*Gallinula*) and coots are the most aquatic, foraging while swimming; when disturbed, they patter along the water's surface to get airborne. Other species regularly climb upward into vegetation, while a few forage in trees. Rails and flufftails rarely forage in the open, preferring the cover of vegetation to hide from predators, and prefer to run rather than fly to avoid danger. If flushed, rails typically fly a short distance and then drop back into cover. However, once airborne, most are surprisingly strong fliers.

Voice As a group, the Rallidae are very vocal (especially during the breeding season and at night), which aids communication in dense vegetation. Calls and songs vary from soft cooing to harsh, monosyllabic series of mechanical sounds; some species engage in duets.

Habitat Most species inhabit wetlands, including fresh, brackish, or salt marshes, mangrove swamps, bogs, wet meadows, or flooded croplands. The birds are adapted for life in low, dense vegetation on moist ground.

Movements Vary from sedentary to highly migratory. Many species, especially those living in unpredictable habitats, may disperse over short to long distances in search of new habitats. Most temperate species are short- to long-distance migrants to warmer regions. Species occurring in tropical and subtropical regions are mainly sedentary but may disperse to new habitats, or make modest seasonal movements.

Diet Some species, including coots and moorhens, are largely herbivorous. However, most are omnivorous, consuming a wide variety of plant material and animals, including invertebrates, amphibians, small mammals, and eggs. A few species have been reported taking carrion.

Breeding For many species, the breeding biology is poorly known. Most are presumably monogamous; some are promiscuous; at least one species is polygynous, and one polyandrous. Coots and moorhens are known to perform nocturnal display flights. Nests are open or globular, built of plant material on the ground or on vegetation just above the ground; some birds use the old nests of other bird species. Clutch sizes range from one to 19 eggs (typically 15–19). Incubation takes 13–31 days per egg, with synchronous or asynchronous hatching depending on the species. The chicks hatch with thick black down. Some have bold patterns, as in Nkulengu Rail (*Himantornis haematopus*) and Madagascan Forest Rail (*Canirallus kioloides*); colorful plumes (*Gallinula*); naked skin on the head (coots and some gallinules); or a brightly colored bill. The down is replaced by juvenal plumage before the young become full-sized. Chicks can swim and leave the vicinity of the nest within a few days and, depending on the species, are fed for a few days to several weeks.

The long toes of **Black Crake** (*Amaurornis flavirostra*) enable it to walk across water plants such as this lily leaf. South Africa.

Both parents attend the nest and chicks, and, in some species, young from a previous brood act as nest-helpers. Fledglings typically remain under parental care for three to four weeks, although Giant Coot (*Fulica gigantea*) cares for its young for up to four months. Most species raise one or two broods per season.

A pair of **Eurasian Coots** (*Fulica atra*) build a raftlike nest of twigs in shallow water. Coots are sometimes confused with moorhens, but can in fact be distinguished by their white beak and the white shield on their head. U.K.

FINFOOTS *Heliornithidae*

Secretive and poorly known, finfoots are grebelike tropical birds. There are three species, each found in a different region of the world: African Finfoot (*Podica senegalensis*), the Neotropical Sungrebe (*Heliornis fulica*), and the Southeast Asian Masked Finfoot (*Heliopais personatus*).

Structure The species range from 12 to 23 in. (30–58 cm.) in length, with Sungrebe being the smallest. In African Finfoot, the males are larger than the females. Finfoots have a daggerlike bill and a long, thin neck. Their feet have lobed toes.

African Finfoot is known to have a single well-developed claw on the bend of the wing, which is thought to aid the birds in climbing. Male Sungrebes have a shallow 'pocket' of skin under each wing, in which they carry their young in flight—the only known case of this feature in any bird. Breeding male Masked Finfoots develop a fleshy knob on their forehead.

Plumage Sexual dichromatism is slight, typically involving differences in head patterning. All species show bold patterns of black or gray and white on the head and neck. They are generally brown on the upperparts and whitish on the underparts. African Finfoot is variably spotted and barred; the West African subspecies are much darker, or entirely blackish, on the upperparts. Bill colors vary: in African Finfoot the bill is red or yellow, and in Masked Finfoot it is yellowish. In the breeding season the female Sungrebes' bill turns red. African and Masked Finfoots have stiffened tail feathers, while Sungrebe does not.

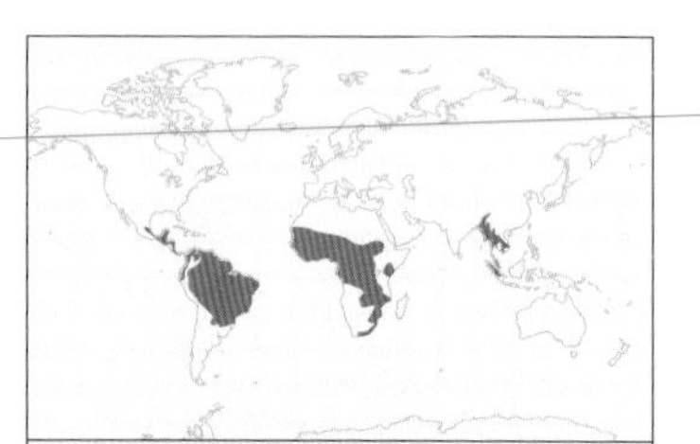

Number of genera 3
Number of species 3
Conservation Status 1 species Vulnerable
Distribution Central America, South America, Africa, Southeast Asia

African Finfoot comprises several highly distinct races. Some have whitish underparts (top), while the race *camerunensis* (below) is entirely dark brown.

Behavior Finfoots are very secretive. They are most often seen on water, but also occur on land and even climb into vegetation. They do not dive after prey but glean food items from the surface of the water and from vegetation. They often swim partially submerged, with just the head and neck above water.

Voice Calls are seldom heard and vary between the species, including booming or bubbling notes. Sungrebes' calls are described as similar to those of some grebes.

Habitat Beside rivers and along the shorelines of slow-moving bodies of water, in forested or shrubby areas with abundant overhanging vegetation. Masked Finfoot is listed as Vulnerable due to habitat loss.

Movements Generally sedentary, with low population density. They have been known to wander as vagrants; records from Trinidad and west Java show individuals can even cross stretches of open ocean.

Diet A variety of insects, crustaceans, fish, and other items.

Breeding Very little is known. The birds are typically seen in single pairs or family groups, which appear to defend large territories. Courtship has been observed only a few times. Nests are typically placed above water in dense shoreline vegetation, at seasons when the water level is high. African Finfoot and Sungrebe lay two or three eggs, while the clutches of Masked Finfoot apparently comprise five to seven eggs. Incubation periods are very poorly known; periods of only 10–12 days have been reported, but need confirmation. The young of African Finfoot are precocial, while those of Sungrebe are altricial. Fledging periods are apparently unknown.

Taxonomy The species are so different that most taxonomists place each in its own genus. African Finfoot shows enough variation to warrant division into four subspecies.

The family is currently placed closest to the rails (family Rallidae). The lobed toes suggest affinity with coots (genus *Fulica*) and moorhens (*Gallinula*), although this feature is also seen in the unrelated grebes (family Podicipedidae) and phalaropes (Scolopacidae).

TRUMPETERS *Psophiidae*

Three medium-sized, mainly terrestrial, South American species in a single genus. As their name suggests, trumpeters are noted for their resonant calls.

Structure The birds are 18–21 in. (45–52 cm.) in length; males are slightly bigger than females. They resemble chickens in size and shape. All species have a small head, with large, dark eyes and a stout, curved bill. They have short, rounded wings, which they hold in a way that gives them a hunch-backed appearance, and a very short tail. The legs are long and stout, and the hind toe is raised (as in many terrestrial birds).

Plumage Mainly black with green, gray, buff, or white on the secondaries and in patches on the inner wing coverts. The head and neck feathers are short and very closely spaced, giving a velvety appearance. The inner wing patches vary by age and play a role in social displays. The secondaries are long and plumed, overhanging the rump and concealing the tail.

Behavior Social birds, trumpeters are found in groups of 3–13 individuals that defend large territories together. Subordinate group members give 'wing-spread' displays to dominant members, crouching and making high-pitched, twittering sounds like chicks' calls. Dominant birds respond with a 'wing-flick' display. Territorial fights between groups may occur. The birds forage on the ground for fallen fruit; they often take fruit dropped by foraging monkeys. They are poor fliers, but roost and nest in trees.

All three species, like this **Dark-winged Trumpeter** (*Psophia viridis*), have long legs and are able to run fast.

Voice Trumpeters have a unique, coiled windpipe that enables them to give loud, resonant calls. They usually make a series of low hoots or booms that may accelerate into bubbling or gurgling sounds. Alarm calls include loud clacks and sometimes grating or clucking noises.

Habitat Confined to lowland forests in northern South America. Each species has a separate range, divided from those of the others by major rivers in the Amazon and Orinoco basins; these rivers are thought to have been a factor in the species' differentiation. No species is threatened, but they could face risks due to large-scale destruction of their dense forest habitat, as well as from the fact that they are often hunted for food by humans.

Movements Sedentary within their range. Family groups roam through fairly large territories year-round, following available food sources.

Diet Mainly large fruit. The birds may also take invertebrates and grubs, and occasionally vertebrates such as small snakes and lizards.

Breeding Trumpeters breed in cooperative groups, with a dominant female that copulates with several males and lays all the eggs. This breeding system, called cooperative polyandry, is rather rare. The nest is most often placed in a large cavity or hole in the trunk of a tree.

A **Gray-winged Trumpeter** (*P. crepitans*). These birds are tame around humans and are often kept as pets. Ecuador.

Clutches comprise three eggs, incubated by all group members for about four weeks. Hatching is synchronous. The young leave the nest within a day of hatching, but are fed by the group for several weeks.

Taxonomy The trumpeters are considered to be most closely related to the cranes (family Gruidae) and Limpkin (Aramidae). However, the exact relationships remain unclear: recent evidence suggests relationships with Kagu (Rhynochetidae) and Sunbittern (Eurypygidae).

Number of genera 1
Number of species 3
Conservation Status Least Concern
Distribution South America

CRANES *Gruidae*

Very tall, elegant birds, cranes are similar to herons and storks but have a more streamlined body. They are widely found in shallow wetlands and open grasslands around the world.

Structure The species are very similar in structure, and differ only in size and plumages. All cranes are large birds, standing at 35–72 in. (90–180 cm.) tall and weighing 5.9–26.4 lb. (2.7–12 kg.), with males being slightly larger than females. They have a fairly small head and a long, slender neck. The bill is medium to long, straight, and pointed. Four species have a bare face patch or bare head and upper neck. Three have facial wattles; two of these species also have distinctive 'crowns,' and two others have elongated head feathers, used in display. The wings are large, broad, and rounded, with long tips to the inner secondaries, which overhang the wingtips. The tail is short and rounded. The legs are long and each foot has four toes, which are unwebbed with a raised hind toe.

Adult **Whooping Crane** (*Grus americana*) and chick. The young of this species fledge after about 90 days.

Eurasian Crane (*G. grus*) breeds in wet forest clearings in taiga, from northwestern Europe to eastern Siberia. Sweden.

Plumage Siberian Crane (*Grus leucogeranus*) is entirely white apart from the bare red face, and Blue Crane (*Anthropoides paradiseus*) is almost entirely uniform bluish-gray. In all other species, the head or face is usually the most colorful part, with red on the face or crown, black crown, nape or malar stripes, or long white stripes on the sides of the neck. Shades of gray and white predominate as body colors, with the neck either the same color or black. The two crowned cranes have prominent yellow crowns, white or red and white cheek patches, and contrasting white and yellow wing patches.

Behavior Highly social, most species of crane spend at least part of the year in flocks. Some species migrate and winter in flocks of several thousand; smaller species usually form small groups or family parties. In contrast, some of the larger species nest in isolated pairs.

Cranes forage in shallow wetlands, digging with their long bill into muddy edges and bottoms of rivers, as well as in fields and open grasslands. The birds also spend long periods each day bathing and preening. They walk with a slow, stiff-legged gait. Their flight is slow and graceful, with the neck and legs outstretched and wings flat.

Voice Cranes' mellow, fluting calls are widely considered to be one of the most evocative of all bird calls. The birds make a variety of elaborate and often far-carrying calls and vocal displays, from soft, guttural honking by the crowned cranes; low, gruff or rasping honks in Demoiselle Crane (*A. virgo*) and Blue Crane; low, rattling notes in Sandhill Crane (*G. canadensis*); and musical, fluting, or trumpetlike calls in most of the other species. Displaying birds give a combination of loud piping or trumpetlike notes.

Habitat Mainly grasslands with shallow freshwater pools, ranging from tundra to edges of deserts, and shallow rivers or large, reed-fringed lakes in forests. Cranes may also feed in open grasslands and edges of cultivated land but return to wetlands to roost and breed. One species, Brolga

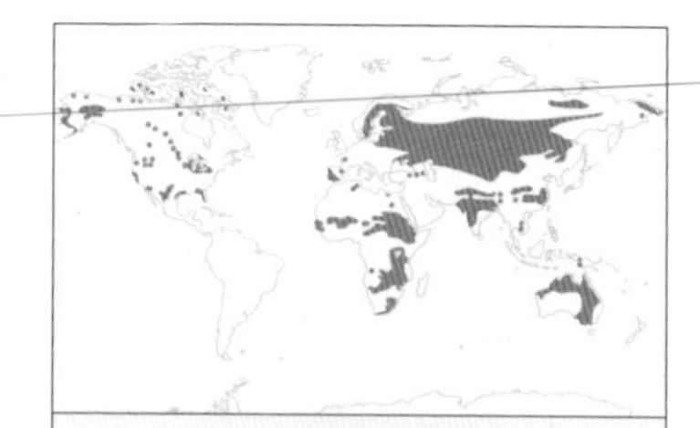

Number of genera 4
Number of species 15
Conservation Status 2 species Endangered, 5 species Vulnerable
Distribution Worldwide apart from polar regions and Central and South America

A group of **Sandhill Cranes** on a lake. The birds are the most common crane species in North America. Huge flocks gather to migrate to their wintering grounds in the southern U.S. and Mexico. New Mexico, U.S.

Crane (*G. rubicunda*), is adapted to feed in saline pools and saltmarshes.

Movements Six of the species are entirely sedentary or make only short-distance movements to nonbreeding areas. The rest are migratory, traveling south to winter in warmer or icefree areas. Prior to migration, flocks gather at traditional staging areas; flocks mostly comprise family units, but total thousands of birds altogether. They fly at up to 6,560 ft. (2,000 m.), in tight 'V' formations, and often call to each other. Cranes can fly at 37–50 miles an hour (60–80 km/h.), and cover 186–310 miles (300–500 km.) before landing at traditional sites; Whooping Crane has been known to travel up to 497 miles (800 km.) in a single day. The most impressive migrations are made by Eurasian Crane, which moves from Siberia to northern India, crossing the Himalayas at more than 32,808 ft. (10,000 m.) in doing so. The longest journey is undertaken by Sandhill Cranes, which breed in northeast Siberia and move east to North America, then south to winter in northern Mexico.

Diet Cranes eat a wide variety of foods, including the roots, bulbs, tubers, and shoots of plants, as well as beetles, grasshoppers, aquatic insects, worms, snails, mollusks, crustaceans, fish, frogs, lizards, small snakes, and birds. They occasionally take spilled grain from fields.

Breeding One of the best known aspects of cranes is their pair-bond, which usually lasts until one partner dies. Part of the bond is the dancing display that the pair performs; the birds bow, arch their necks, make leaps and short bouncing runs with wings partially spread, and make short flights before circling each other, all accompanied with loud, bugling calls. In some species these dances are very short, but in others, such as Blue Cranes, they may last for up to four hours. Both parents build the nest and raise the young. The flat nest, usually made of sedges and other wetland plants, is usually sited on the ground, although the Gray Crowned Crane (*Balearica regulorum*) nests in trees. The clutch, usually of one or two eggs, is incubated for 28–32 days. The young, once hatched, soon leave the nest and follow the parents to feeding areas. They remain in the family group until the following year.

A pair of **Gray Crowned Cranes** performs a display. These 'dances' include bowing, wing-flapping, jumping, and stick-tossing. It is thought that the birds dance to strengthen their pair bond as well as for courtship. Kenya.

LIMPKIN *Aramidae*

A **Limpkin** captures a water snail. The birds forage by walking through water and finding their prey either visually or by feeling with the bill. Florida, U.S.

A tiny family containing only Limpkin (*Aramus guarauna*). The bird's common name derives from its 'limping' gait.

Structure The bird is 22–28 in. (56–71 cm.) in length, and resembles a heron or an ibis, with a long bill, slender neck and long legs. The tip of the lower mandible is twisted and has a sharp edge, which the bird uses to extract snails from their shells. The wings are fairly small and rounded; the outermost primary is modified to a sickle shape with a clubbed end, which makes a winnowing sound during Limpkin's flight display. The feet have a long hind toe at the same level as the front toes, and long, sharp claws.

Plumage The sexes are alike. The head is fully feathered. The plumage is dark brown, striped and spotted with white; the intensity of markings vary with geographical location. Juveniles are similar but their markings are more delicately spotted.

Behavior Generally solitary, but Limpkin is sometimes found in small groups. It can be active at any time, but is mainly diurnal, and roosts in trees at night. The bird forages in shallow water, often on floating vegetation. It swims well, floating high on the water, and either flies or runs to escape danger.

Voice Males utter a loud, strident, three-noted, wailing scream, usually at night but sometimes in the early morning. Females are generally silent, but during the breeding season they produce a single or double-note *gon*, also made during short duets with males. In addition, adults make clucking notes, and the young give chipping notes.

Habitat Wooded swamps, lake edges, freshwater marshes, and mangrove swamps. The species' distribution correlates strongly with that of its main food, the apple snail (*Pomacea caliginosa*). In more arid areas such as the West Indies, Limpkin remains near water, where this preferred food is found.

Movements Generally sedentary; any local movements are usually tied to water level and food abundance. Post-breeding dispersal has been observed for some populations.

Diet Primarily feeds on large apple snails, augmented by a variety of other food items, including other mollusks, insects, and seeds.

Breeding The timing of breeding is associated with food abundance and water level. Birds may be monogamous, or a female may pair with successive males (serial polyandry). Pairs are territorial.

The large, bulky nests are built from plant material. They can be located in rafts of marsh vegetation, shrubs, and even in trees up to 15 ft. (4.6 m.) above the ground. One or two clutches may be produced in a season, comprising four to eight eggs (typically six); both sexes incubate the eggs. The young hatch covered in brown down. They take their first flight at about six weeks, and disperse from their family groups at 15–17 weeks.

Taxonomy Four subspecies are currently recognized. Recent genetic studies indicate that Limpkin is more closely related to cranes (Gruidae) than to rails (Rallidae).

Number of genera 1
Number of species 1
Conservation Status Least Concern
Distribution Southern U.S., Central America, northern South America, Caribbean

BUTTONQUAILS *Turnicidae*

An Old World family of small, plump, ground-living birds of grassland, scrub, and forests. Most species are similar in shape and are cryptically colored. Although similar in behavior and structure to the true quails (family Phasianidae), they are unrelated—a good example of convergent evolution. Several species show reversed sexual roles, with females being larger and more dominant than males.

Structure Buttonquails range from 4 to 9 in. (10–23 cm.) in length. They have a small head and short neck. The bill is also fairly short, like that of a small rail or quail, slightly arched, and often strong for the size of the bird. The body is plump, with short, rounded wings and a short, round-tipped tail. The strong legs and feet are adapted for running and scratching in dusty soil. Unlike true quails, buttonquails lack a hind-toe; for this reason, they are sometimes called 'hemipode' or 'half-foot.'

Plumage Sexual dichromatism is common but usually fairly slight. Plumage is mostly a variety of buffy browns, often in a finely barred or cryptically intricate pattern with darker lines or chevrons, and closely matching that of the surrounding vegetation, especially in dry-country areas. The underparts may be paler or whitish with bars, streaks, or lines of chevrons down the flanks. Several species have prominent buffish to bright orange patches on the breast, while others have the entire underparts heavily barred. The wings are usually uniform in color, but several species have prominent pale or yellowish wing-coverts. The enigmatic Quail-Plover (*Ortyxelos meiffrenii*) is mostly cream to rufous-brown, but other species show more marked plumage variation.

A male **Black-breasted Buttonquail** (*Turnix melanogaster*) forages on the forest floor, using its strong feet to dig through the leaf litter. Queensland, Australia.

Behavior Shy and secretive, and consequently very difficult to see. These terrestrial birds prefer camouflage or running to escape detection, and fly only as a last resort. They forage on the ground, among leaf litter and vegetation; some species dig in light soil to search for seeds.

Voice Mostly silent, although several species have a low-pitched, booming note, which is given by the female during courtship.

Habitat Open grassland, scrub, forest undergrowth, and cultivated land. One species, Buff-breasted Buttonquail (*Turnix olivii*), is considered endangered due to continuing destruction of its habitat.

Movements Most species are entirely sedentary; some, however, make local movements in search of feeding areas. Quail-Plover and Common Buttonquail (*T. sylvaticus*) are intra-African migrants, moving north with the rains. One species, Yellow-legged Buttonquail (*T. tanki*), which breeds from India to northern China, is truly migratory, with northern-breeding birds wintering in southeast Asia.

Diet Seeds and small insects.

Breeding The females (which are polygamous in some species) court the males, and establish and fiercely defend the breeding territory. The nest is built by both parents; structures vary from a pad of leaves and dry grass on the ground, in the open or under a rock, to a domed shape with a tunnel-like entrance. The male incubates the clutch of three to seven eggs for 12–14 days, and cares for the young. Chicks leave the nest shortly after hatching, and are almost independent by 14 days.

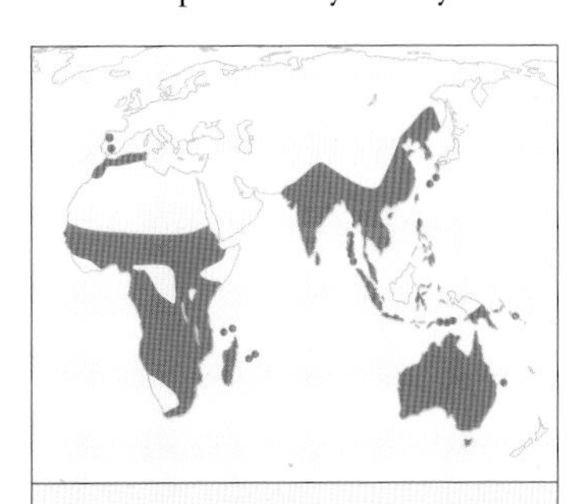

Number of genera 2
Number of species 16
Conservation Status 1 species Endangered
Distribution Mediterranean, Africa, Asia, New Guinea, Australia

THICK-KNEES *Burhinidae*

So named because of their prominent leg joints, thick-knees are also known by various other colloquial names, such as 'stone curlew' and 'stone-runner.' Two African species are known as 'dikkops,' an Afrikaans word meaning 'thick-head.'

Structure Thick-knees are 14–22 in. (35–57 cm.) in length. They have a broad head with large, round, yellow eyes, which are well adapted to help the birds see in the dusk or dark. In most species, the bill is straight and about the same length as the head, but in some it is larger, longer, and slightly upturned.The birds resemble waders in shape, with a slim body and long legs and wings.

Plumage All species have cryptically colored plumage. Most thick-knees are largely brown, streaked or spotted with darker brown, although some are grayer, with whiter underparts. In several species, the head is boldly striped black and white, with a pale base to the bill that continues as a curving line below the eye. The lightly scaled legs and toes vary from pale gray to bright yellow.

Behavior Tropical species are most active at dusk and after dark, spending most of the day in bushes or tall grass. More northern species are active during the day. The birds forage by adopting a slow walk and alert watching pose, before making a short dash to pounce on their prey. They generally allow close approach and walk rapidly or run from danger, rather then flying. When they do take to the air, they fly at low levels, with their feet trailing.

Voice Generally silent by day, thick-knees become most vocal after dark. They give penetrating, far-carrying, wailing cries and whistles, which in Eurasian Stone Curlew (*Burhinus oedicnemus*) rise and accelerate into a prolonged series of *curlee* notes before dying away. African species have similar or more nasal calls.

Thick-knees, like this **Eurasian Stone Curlew**, have cryptically colored and patterned plumage, which provides ideal camouflage in their surroundings. Canary Islands.

Habitat Mostly flat, dry, open grasslands with sparse, shrubby vegetation, but the birds also frequently occur in semidesert areas. Water Dikkop (*B. vermiculatus*) of southern Africa and Great Stone Curlew (*Esacus recurvirostris*) of southern Asia live almost entirely by lakes and streams. In Indonesia and northern Australia, Beach Stone Curlew (*E. magnirostris*) occurs on beaches and occasionally in mangroves.

Movements Most species are entirely sedentary. However, Eurasian Stone Curlew is a migrant to northern parts of the range in Europe and central Asia, moving south to winter around the Mediterranean, Arabia, and North Africa.

Diet Insects, lizards, and other small animals; occasionally birds' eggs.

Breeding In most species, pairs establish territories on areas of open ground. The nest is a shallow scrape in the ground, in which the birds generally lay two eggs. The eggs are incubated by both parents for 24–27 days, one parent sitting on them while the other stands guard. Chicks leave the nest about a day after hatching; the parents lead them away to live in a new area, and feed them for the first few days. The chicks fledge at 36–42 days.

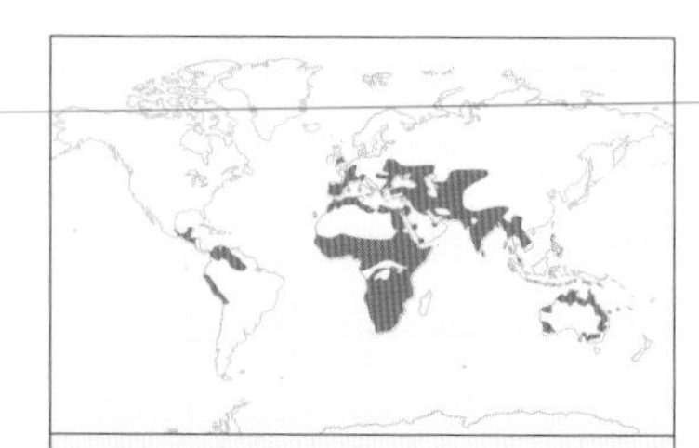

Number of genera 2
Number of species 9
Conservation Status 2 species Near Threatened
Distribution Worldwide, mainly in tropical areas; most numerous in Africa and Asia

SHEATHBILLS *Chionidae*

A **Snowy Sheathbill**. Sheathbills are the only Antarctic birds to lack webbed feet. Patagonia, Argentina.

The sheathbills comprise two species: Snowy Sheathbill (*Chionis albus*) and Black-faced Sheathbill (*C. minor*). Both are white, pigeonlike, scavenging shorebirds, found mainly in association with penguins. Their name comes from the saddle-shaped sheath at the base of their bill.

Structure Sheathbills are 14–16 in. (38–41 cm.) in length, with males being larger than females. They have a small head, and a short, stout bill that has a horny sheath covering the end of the upper part. They also have a short neck, a large, plump body, and robust legs and feet.

Plumage The birds have thick, pure white plumage with a dense layer of gray underdown. They have a bare face, with wattles and caruncles. The face is pinkish in Snowy Sheathbill and black in Black-faced Sheathbill. Juveniles may be similar to adults, or may be gray or brownish at first and have smaller wattles.

Behavior Sheathbills are usually loosely colonial: they congregate at breeding colonies of penguins, cormorants, and albatrosses, feeding on anything remotely edible. They usually grab food as soon as it becomes available, with pairs often working together to harass and rob their host. Sheathbills occasionally forage alone among rotting piles of seaweed.

Voice Mostly silent except when breeding. Pairs contact each other by harsh, strident, or grating calls. They also make threat calls to other encroaching birds, comprising a series of rising *kek kek kek* notes.

Habitat Almost entirely tied to the locations of penguin, cormorant, and albatross colonies, on coastal headlands, bays, and islands.

Movements Black-faced Sheathbill is entirely sedentary, as are some Snowy Sheathbills. However, most of the latter move north to the tip of South America between April and early October, to take advantage of the penguin breeding season there; they cross the Drake Passage, probably in a single flight, but may also stop on icebergs or passing ships.

Diet Sheathbills mainly eat carrion and fish scraps, which they steal from other birds regurgitating food to their young, but will also steal eggs and chicks from unwary parents if nothing else is available, and eat the blubber and hides of rotting or putrefying seal carcasses if necessary. At seal colonies, they gather around pupping females to feed on placentas, feces, blood, and dead or dying seal pups.

Breeding Sheathbills' breeding season is entirely dependent on that of their host species. Those breeding around penguin colonies often arrive and set up territories before the penguins, so that by the time the penguins have produced their eggs and young there will be food for the sheathbills' chicks. The untidy nest of feathers, bones, algae, and grass is placed in a crevice, cave, or under rocks. The clutch of two or three eggs is laid in December; the chicks hatch from mid-January and are fledged by the end of March.

Black-faced Sheathbills are found only on certain subantarctic islands in the southern Indian Ocean. Crozet Islands.

Taxonomy Magellanic Plover (*Pluvianellus socialis*) is now included as a subfamily (Pluvianellinae) within this family on the basis of genetic analyses, morphology, behavior, and chick development.

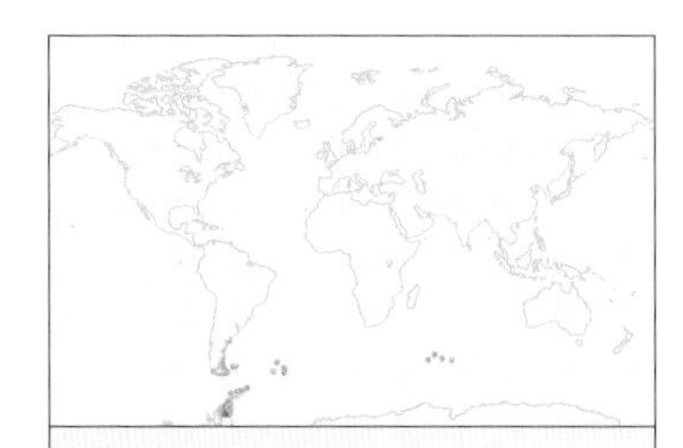

Number of genera 2
Number of species 3
Conservation Status Least Concern
Distribution Antarctic Peninsula, sub-Antarctic Atlantic and Indian Ocean islands

OYSTERCATCHERS *Haematopodidae*

A **Eurasian Oystercatcher** in breeding plumage. The white bar on the upperwings extends almost to the tips. Scotland, U.K.

Striking, noisy shorebirds, oystercatchers occur around the world. Despite the name, they rarely, if ever, catch or eat oysters.

Structure The birds are 16.5–18.5 in. (42–47 cm.) in length. They are sturdy or plump, with a thick neck, long, daggerlike bill, and relatively short legs. The wings are fairly long and pointed, and the tail is square.

Plumage Five species are entirely black. In the others the head, breast, most of the upperparts, and the tip of the tail are black, with the remainder white. The wings are similarly pied black and white, varying in extent and pattern. In Variable Oystercatcher (*Haematopus unicolor*) of New Zealand, about two-thirds of the population is entirely black, with the remainder, apart from a small number of intermediates, being pied. All species have a long orange-red bill, deepest in color during the breeding season, and deep pink legs and feet. The eye and eye ring are mostly red or yellow in three of the species.

Behavior Usually territorial, the birds live in pairs during the breeding season, often close to neighboring pairs. In winter they feed in loose flocks and gather, sometimes in hundreds, at high-tide roosts.

Voice Mostly loud or shrill, piercing, piping calls used variably in differing contexts. The usual contact note is a single high-pitched note, repeated at intervals. During the breeding season, the call is given as a long series of rapidly repeated calls rising to a contined piping trill.

Habitat Rocky and sandy shores as well as muddy estuaries, saltmarshes, and shingle beaches. In eastern Europe and parts of Russia, Eurasian Oystercatcher (*H. ostralegus*) is widespread in areas of freshwater lakes, rivers, and gravel pits and in agricultural areas. Similarly, in southern Chile and Argentina, Magellanic Oystercatcher (*H. leucopodus*) occurs on upland grasslands and the edges of freshwater lakes and pools.

Movements Inland-breeding birds move to coastal areas in fall. Eurasian Oystercatcher is the only true migrant, moving south to winter mostly within southern areas of its breeding range. Birds breeding in Iceland fly 500 miles (800 km.) to the U.K. in a single flight.

Diet Oystercatchers forage in all types of habitat but favor rocky areas, where they collect mollusks, especially mussels and limpets. They may carry the food off a little distance and then break up the shells by repeatedly stabbing them with the tip of the bill.

Breeding Oystercatchers have an elaborate territorial display in which they give piping calls, either while adopting a 'bowing' posture at the edge of their territory, running up and down in pairs, or while flying in a slow, deliberate fashion (often referred to as 'butterfly flight').

The clutch of two or three eggs is laid in a scrape on the ground and incubated by both parents, for 24–35 days. The chicks can walk within a day of hatching, and fledge in 28–35 days, but stay with their parents through their first winter to learn feeding skills.

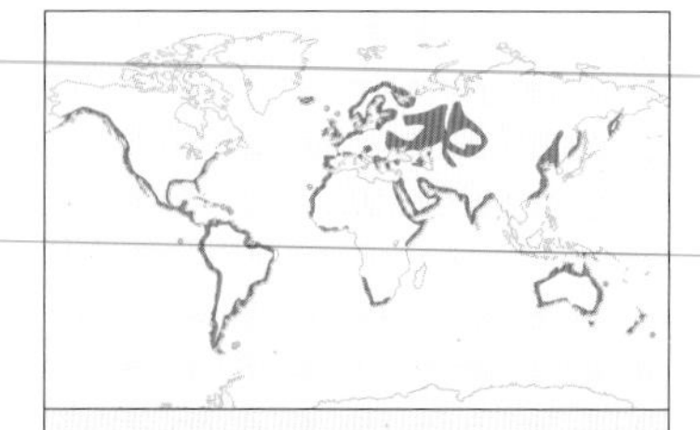

Number of genera 1
Number of species 11
Conservation Status 1 species Endangered, 1 species Near Threatened
Distribution North and South America; Europe; tropical Africa south of the Sahara; South and Southeast Asia; Australasia

CRAB PLOVER *Dromadidae*

A **Crab Plover** can swallow its crab prey more easily by first removing the crabs' legs. United Arab Emirates.

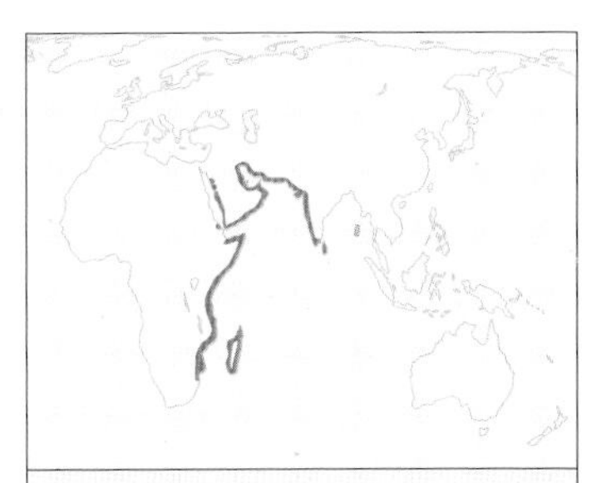

Number of genera 1
Number of species 1
Conservation Status Least Concern
Distribution Coasts of East Africa, Madagascar, Arabia, South Asia

A fairly large, distinctive, black-and-white shorebird, Crab Plover (*Dromas ardeola*) is the only member of its family.

Structure The birds are 13–16 in. (33–41 cm.) in length. They have a short, stout, daggerlike bill with an upward-angled lower mandible, which is used for breaking open tough crab shells. The large, dark eye reflects this species' ability to feed in poor light conditions, even at night. The legs are long and the feet are partially webbed. In flight, the legs and feet extend well beyond the tip of the tail, and the wings are held stiffly in a similar manner to the thick-knees (family Burhinidae).

Plumage Mainly white except for the black mantle, flight feathers, and primary coverts. The tail is pale gray. The legs are slate-gray, and the bill is black.

Behavior Often very social, sometimes joining high-tide roosts of 1,000 or more. Birds fly up to 12–13 miles (20 km.) to these traditional communal roosts. At other times, they can be seen feeding on mudflats or sandy beaches, usually in flocks of up to about 20. The birds stalk crabs in shallow water or on the open sandflats. They grab their prey after a short dash, then either swallow it whole or stab it with hammerlike blows from the bill and shake it to remove claws and legs.

Voice Generally a noisy and very vocal bird. Flocks keep up a regular chattering commotion, occasionally interrupted by more strident *ha-how* or *crow-ow-ow* calls, or more raucous notes during the breeding season. When disturbed, birds give a sharp *kiep* alarm cry.

Habitat Entirely coastal, usually mudflats, sandflats, estuaries, sandy beaches, and lagoons; also on coral reefs. Breeds in burrows in sandbanks and low-lying dunes.

Movements At the end of the breeding season (August), most birds breeding in the Persian Gulf move south along the coast of East Africa, reaching Madagascar and the Seychelles. Some may make longer journeys across the Indian Ocean to winter along the west coast of India and around Sri Lanka, and regularly reach the Andaman Islands. However, some move only short distances. Those that breed in the United Arab Emirates move about 155 miles (250 km.) to winter on large coastal lagoons around Dubai.

Diet Mostly crabs, particularly small burrowing varieties. Also, mollusks, shrimp, and mudskippers.

Breeding Breeds in large colonies, which may number hundreds off pairs. Uniquely for a shorebird, it digs a downward-sloping nest burrow up to 6.5 ft. (2 m.) long in sand. A single egg is laid in the burrow. The length of incubation is not known. Both parents bring food to the growing chick, and even after fledging the young birds are fed by the parents.

When feeding opportunities on intertidal sand or mud are limited by a rising tide, **Crab Plovers** congregate at traditional roost sites, such as Khor al-Beidah in the United Arab Emirates.

IBISBILL *Ibidorhynchidae*

Ibisbill is named for its long, downcurved bill, which recalls that of an ibis. The bill is slightly longer in females.

A unique monotypic shorebird, Ibisbill (*Ibidorhyncha struthersii*) is found only in river valleys in the mountains of Asia. Despite its striking coloration, it can be hard to see, often blending in against its rocky background.

Structure A robust, bird, Ibisbill is 15–16 in. (39–41 cm.) in length. The sexes are very similar, but females are, on average, slightly larger. The bill is long and strongly decurved. The bird also has rounded wings and longish legs.

Plumage Adults have a white belly, underwings, and undertail coverts, with blue-gray upperparts and white flashes at the base of the upperwing primaries. The black face, crown, and breast band are bordered by contrasting narrow white lines. The bill and legs are red, and are brightest in the breeding season. Juveniles are duller and paler, with scalloped brown fringes on the upperparts, and no black on the face or breast; their bill and legs are pinkish-gray.

Ibisbill in flight. The white patches at the base of the primaries are visible only when the wings are spread. Nepal.

Behavior Ibisbill occurs in pairs or small groups. It feeds by using its long bill to probe and peck under boulders and pebbles along stream beds, occasionally using a raking action to dislodge food. In flight, the bird holds its head upright and outstretched, like an ibis. However, the flight pattern recalls oystercatchers, with deep, slow wingbeats.

Voice Usually silent outside the breeding season. The typical call is a penetrating *klew-klew*, like a drawn-out, higher-pitched version of those by larger sandpipers in genus *Tringa*, such as Common Greenshank (*T. nebularia*). Ibisbill also has a faster, ringing *tee-tee-tee-tee* like the calls of Whimbrel (*Numenius phaeopus*).

Habitat Found only in the mountain systems of central and southern Asia, including the Himalayas and the high plateaus of Tibet and central Asia. Confined to valleys with rocky rivers and stony floodplains from 5,577 ft. (1,700 m.) to 14,435 ft. (4,400 m.). Ibisbill is currently not considered at risk, although population densities vary across its range.

Movements The bird is largely sedentary within its range, although it makes localized movements to lower elevations in winter (October to March), with some birds descending as low as about 330 ft. (100 m.).

Diet Aquatic invertebrates and small fish; terrestrial insects are also taken.

Breeding Occurs from late April to early June, depending on altitude. Ibisbill is highly territorial in the breeding season, when rival birds perform circling displays. The usual clutch is four eggs, laid in a pebble-lined scrape, usually on an islet or shingle bank. The parents share incubation, but the incubation period is unknown. The young can fly at 45–50 days. Adults sometimes feign injury in a ploverlike display to draw predators away from their chicks.

Taxonomy The species is a taxonomic curiosity. It may be related to the oystercatchers (family Haematopodidae) and to the stilts and avocets (Recurvirostridae), so is usually placed between them, but experts are still divided on its true classification.

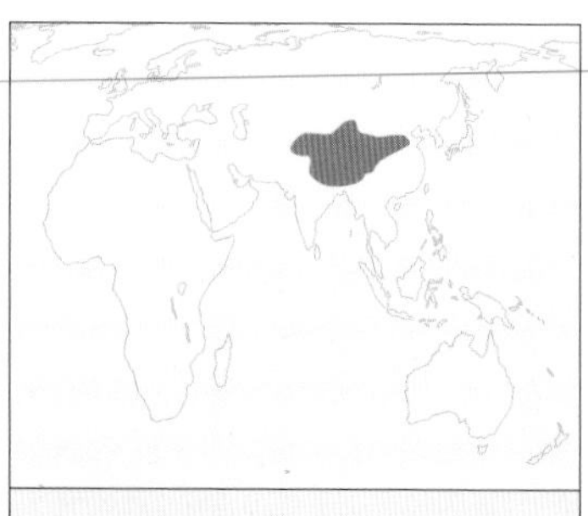

Number of genera 1
Number of species 1
Conservation Status Least Concern
Distribution Central and southern Asia

STILTS AND AVOCETS *Recurvirostridae*

Both avocets and stilts typically fly with their legs trailing straight out behind them, like this **Black-winged Stilt**. Location unknown.

Slim, elegant waders, stilts and avocets have very long legs and a fine, straight or upturned bill. They are similar to other wading birds, such as snipe and plovers, but are generally more streamlined, with less cryptic plumage. All are birds of shallow wetlands, estuaries, and coastal lagoons.

Structure The birds are 12–20 in. (30–50 cm.) in length, and most species are fairly similar in structure. The sexes are alike. The bill is long and fine (slightly longer in males), straight in stilts and slightly upturned at the tip in avocets. The avocets' bill, and that of Banded Stilt (*Cladorhynchus leucocephalus*), is flattened toward the tip, and has fine, comblike structures called lamellae inside it, which the birds use together with their broad tongue in filter-feeding. Stilts and avocets have long, slender, pointed wings and a short, rounded or pointed tail. The legs are long and slender, and the feet have partly webbed toes, with a reduced or vestigial hind toe.

A **Black-necked Stilt** (*H. himantopus mexicanus*). Stilts plunge their head and neck underwater to search for prey. Florida, U.S.

Plumage Most species are variably black and white. Black Stilt (*Himantopus novaezelandiae*) is entirely black; Banded Stilt has a broad chestnut breastband; Red-necked Avocet (*Recurvirostra novaehollandiae*) has a chestnut-red head and upper neck; and American Avocet (*R. americana*) has a soft orange-brown head, neck, and breast. All stilts have bright reddish to orange-pink legs, while all avocets have blue-gray legs and feet.

Behavior Stilts and avocets often breed in large colonies, and usually live in smaller groups at other times of the year. All species forage more by touch than by sight, characteristically probing or sweeping the water with the bill, and can feed at night. Stilts wade, often up to their knee-joints, in shallow water, and dip the head and neck underwater while probing the mud or bottom substrate and grasping prey. Avocets scoop or sweep through the surface of water or soft mud to filter out smaller aquatic insects. Both stilts and avocets swim, but avocets more frequently feed while swimming over deeper water. Feeding patterns in coastal areas are often determined by tidal pattern: the birds spend high tides roosting, usually on one leg with the head and bill tucked under folded wings. When alarmed or disturbed by a possible predator, they bob the head rapidly before taking off in rapid, dashing flight. Nesting birds also have a display in which one parent feigns a broken wing and leads potential danger away from the eggs or young.

Voice Fairly noisy when in defense of nest or young. The contact calls of stilts are either a sharp yelp or a piping *kep* or *kek*. Avocets' calls are more disyllabic: a rising *chowk*, *kleet*, or nasal *kluit*, all of which become

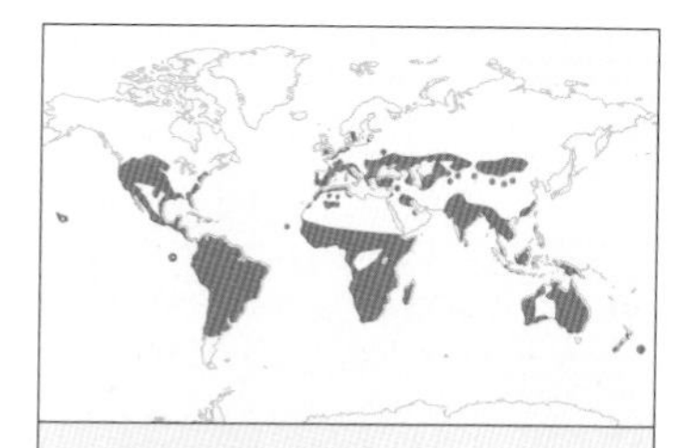

Number of genera 3
Number of species 7
Conservation Status 1 species Endangered
Distribution Worldwide apart from polar regions and northern temperate zones

An **American Avocet** in breeding plumage. During the winter, the head and neck lose their striking rust-brown color and turn gray. U.S.

more high-pitched or shrill when used in alarm or defense.

Habitat Shallow freshwater and saline wetlands, usually adjacent to low sandy beaches, dunes, or sandbanks for nesting. Some stilts also feed alongside fast-flowing rivers, salt-pans, marshes, and irrigated fields. Most species inhabit lowlands, but Andean Avocet (*R. andina*) occurs year-round on saline lakes and pools, mostly at elevations above 16,500 ft. (5,000 m.).

Movements Most species are sedentary, or make local movements in response to feeding and breeding requirements. In Australia, large numbers of Red-necked Avocets and Banded Stilts appear and breed in temporary or newly formed lakes following seasonal rains. Andean Avocet is largely resident at high altitudes, but small numbers wander to the coasts of northern Chile and Peru. Three of the northerly species are more truly migratory. Black-winged Stilt (*H. himantopus*) and Pied Avocet (*R. avosetta*) breed widely across Europe and Asia and move south to winter in southern Spain, around the coast of Africa, and, in the east, in southern China to the Philippines. American Avocet moves south and east of the breeding range to winter in Florida, southern California, and south into Mexico and Guatemala.

Diet Stilts feed mostly on small mollusks, crustaceans, fish, worms, and insect larvae, and occasionally also take seeds of aquatic plants. Avocets feed mostly on small crustaceans. Both also catch flying insects.

Breeding Mostly colonial or loosely colonial when nesting, except for the endangered Black Stilt, whose population is now less than 100 birds, and which nests in solitary pairs. Some colonies are utilized regularly, but others seem to be used opportunistically and only when all conditions for breeding and feeding young are suitable. Some species, such as Red-necked Avocet, often move several hundred miles to form new colonies, possibly only for the duration of the season. Banded Stilt only breeds when heavy rains have provided sufficient water on otherwise dried-out inland salt-lakes for huge numbers of brine-shrimps to emerge; at these times, colonies numbering hundreds of thousands of birds spring up nearby.

All species are monogamous. However, the bond rarely endures for more than a single season, except in the case of Black Stilts, which pair for life. The nest is usually a shallow scrape in sand, sparsely lined with local vegetation. The clutch of three or four eggs is incubated by both parents for 19–26 days. The long-legged young can feed themselves within hours of leaving the egg, and become fully independent after a further two to three weeks.

Two **Pied Avocets** fighting. All members of the family appear to be fairly fierce in defending their territories and nests, noisily driving away intruders. U.K.

PLOVERS *Charadriidae*

Small to medium-sized shorebirds, plovers are found on all continents apart from Antarctica. There are three recognized subfamilies: Charadriinae (mainly the 'ringed' plovers); Pluvialinae (the 'golden plovers'); and Vanellinae (chiefly the lapwings). Most species have subtle colors or markings that help them blend in with the sand, rocks, or grassland of their environment. In contrast, lapwings are generally gaudy in color.

Structure The species range from 5 to 15 in. (12–38 cm.) in length, with a wingspan of 13–34 in. (33–87 cm.). The sexes are similar. Plovers are generally plump, with a rounded head and a short neck. Their large eyes have a high proportion of retinal rods; these cells provide excellent visual acuity in low light. In most species, the bill is straight, short to medium-length, and typically swollen toward the tip. A few species have a longer, more slender bill: for example, Diademed Plover (*Phegornis mitchellii*), has a slightly decurved bill more reminiscent of a sandpiper's bill. Wry-bill (*Anarhynchus frontalis*) has an unusual bill that is bent to the right at its midpoint. Wings may be long and pointed (especially in long-distance migrants) or rounded (in lapwings and Diademed Plover). Male lapwings have a spur or bony knob on the carpal joint of each wing, which is used in fighting rivals. Plovers have medium to long legs, and feet with a small or vestigial raised hind toe (plovers generally do not perch on vegetation). A few species have partially webbed toes.

Plumage Plovers are sexually dichromatic: usually, the males are more colorful or boldly patterned than the females, but the reverse is true for Eurasian Dotterel (*Charadrius morinellus*).

Black-bellied Plovers (*Pluvialis squatarola*) in breeding plumage. Both the male (at the front) and the female (seen behind) have striking black-and-white coloration, although the female is a little paler. Nonbreeding and juvenile birds are a soft gray-brown.

Most species have cryptically colored upperparts in various shades of gray, brown, or olive; the coloration may be either uniform or patterned, with feathers edged or scalloped in white, yellow, or buff. The head, and especially the underparts, are usually paler white or buff; however, several species of lapwing are more extensively black below, and Diademed Plover has finely barred black and white underparts. In many species, the head, neck, and chest are contrastingly colored in black, gray, brown, or rufous, with stripes, masks, collars, or chest bands. Several of the lapwings have iridescent plumage, and three species have elongated crest feathers. African Wattled Lapwing (*Vanellus senegallus*) has a white crown and narrow streaks of black on the sides of the head and neck.

The onset and timing of molts varies. Most species have a full postbreeding molt. Some plovers have distinct breeding and nonbreeding plumages; the breeding plumage is typically acquired by a partial molt often just prior to courtship. Chicks have cryptically colored, dense, fluffy down; the young of Inland Dotterel (*Peltohyas australis*) resemble coursers (family Glareolidae). Most species also have a distinctive juvenal plumage.

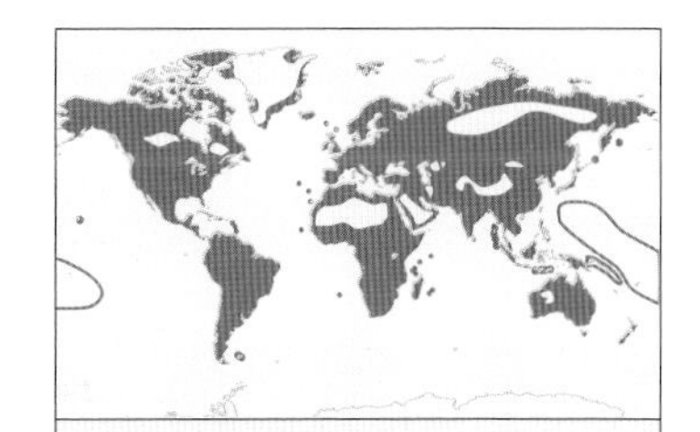

Number of genera 10
Number of species 66
Conservation Status 3 species Critically Endangered, 2 species Endangered, 3 species Vulnerable
Distribution Worldwide

Iris color may be brown, red, yellow, or white. The bill is usually black or dark gray, but in some species it is bicolored or with a pale base in pink, red, orange, or yellow. A few species have bare, fleshy facial skin: some lapwings have red, yellow, or orange wattles, lappets, lores, or comb, while some plovers have colorful yellow, orange, or red eye rings, which are particularly fleshy in species such as Three-banded Plover (*Charadrius tricollaris*). Leg colors include various shades of gray, olive, red, pink, yellow, or black.

Behavior Plovers may be diurnal or nocturnal. They forage singly or in small flocks, taking prey from on or just below the ground surface using a distinctive 'run, stop, and pick' action; Wrybill probes beneath stones. Prey is generally located by sight, but the birds may also use auditory detection. Several species use 'foot-trembling,' rapidly tapping one foot on the ground, apparently to elicit movement of prey beneath. Some lapwings forage near wild or domestic animals. Individuals of several species defend feeding territories during migration and outside the breeding season. During migration, the birds may occur in large flocks.

Voice Most species have a variety of calls, including loud, strident notes used in alarm, aggression, distraction, courtship, or contact. The calls usually comprise one- or several-note peeps, whistles, rattles, or trills. Longer songs are often given during display flights. Chicks have distinct calls to signal their location, or to show distress or contentment. A few species make nonvocal sounds with the wings during flight displays.

Habitat Open habitats ranging from Arctic tundra, steppes, inland or coastal playas, shorelines, and grasslands to sparsely vegetated deserts, and at elevations from sea level to 16,400 ft. (5,000 m.). In migratory species, habitat preference can vary greatly between breeding and nonbreeding seasons. Loss and degradation of habitat, and disturbance during the breeding season, are the greatest threats for plovers; island populations may also be threatened by introduced predators.

Movements Some species or populations are migratory. Long-distance migrants make transoceanic flights: for example, Pacific Golden Plover (*Pluvialis fulva*) breeds in northern Alaska and Russia and winters along southern coasts of Australia, making a journey that includes a nonstop 2,800-mile (4,500-km.) flight from Alaska over the north Pacific Ocean to the Hawaiian Islands. Other species have relatively short-distance

Northern Lapwing (*Vanellus vanellus*) is one of the most colorful of the plovers, notable for its iridescent upperparts and its long, spiky crest. U.K.

Eurasian Dotterel (*Charadrius morinellus*) breeds on tundra and mountainsides in high northern latitudes, but winters in northern Africa and western Asia. Norway.

migrations. The remaining species undergo seasonal elevational movements, are locally dispersive depending on the availability of resources, or are sedentary.

Diet Mainly invertebrates, including insects and larvae; spiders; crustaceans; worms; snails; and small vertebrates and their young or larvae (including fish, reptiles, and amphibians). Plant seeds or leaves may also be taken. Indigestible material is regurgitated as pellets.

Breeding Plovers generally breed during periods of highest food abundance (such as the Arctic summer, the onset of the tropical rainy season, or following dry-season fires). Species vary from territorial to nonterritorial and solitary to loosely colonial. Most are seasonally monogamous (although a few may have prolonged pair bonds), and share incubation and parental duties. Polygyny is rare; one example is Northern Lapwing. Multi-clutch breeding systems or sequential polyandry have been reported for a number of species. Female Eurasian Dotterels may produce clutches with two or three males and leave them to carry out all subsequent parental duties. Some females have been documented to leave their partners after the eggs have hatched, and then bond with another mate.

The nest is a mound of plant material or a simple scrape in sand. Plovers usually have one brood per season, although some tropical species are double-brooded. Clutches comprise two to five eggs, typically four. Additional eggs are not uncommon; this could be due to multiple females laying in the same nest. Incubation takes 21–30 days. The chicks are precocial; they move away from the nest after one or two days, and are brooded or guarded by the parents for a month before they become independent. Many species use distraction displays to lure predators away from the nest or young.

Taxonomy Magellanic Plover (*Pluvianellus socialis*) has traditionally been placed in Charadriidae, but recent genetic analyses, morphology, behavior, and development of chicks indicate that it has closer affinities to the sheathbills (family Chionidae).

Feeding and Breeding in the Harshest of Conditions

Plovers that inhabit barren or harsh habitats have developed a variety of adaptations to help them survive. One such adaptation is camouflage, which is vital in habitats with little or no vegetation. The upperpart coloration of many species blends in with the ground; when roosting in a depression or sitting on a nest, individuals are almost invisible and can hide in plain sight. Nest concealment is critical to reproductive success. The 'nest' may be no more than a depression in the ground, and the eggs are cryptically colored to escape detection when the adult is away. After the chicks have hatched, the adults remove the egg shells from the vicinity. The chicks are also cryptically colored: for example, the dorsal coloration of Eurasian Golden Plover (*Pluvialis apricaria*) chicks, speckled with black and yellow, matches tundra vegetation. When camouflage proves insufficient, plovers scare off predators, or lure them away from the nest or chicks, by spreading their wings or tail to expose bold patterns, or by deftly executing a 'broken wing' display.

Crowned Plover (*V. coronatus*) is constantly alert for danger in its arid, bare environment. South Africa.

Species inhabiting windswept, very hot, cold, or saline environments require additional physical or behavioral adaptations. For example, removal of excess salt from the body is critical for survival in marine and arid habitats; species inhabiting such regions have highly developed salt glands above the eyes, which excrete excess salt into the nasal cavity. Species inhabiting hot deserts, such as Black-headed Lapwing (*Vanellus tectus*), are more nocturnal, resting during the heat of the day. To keep their eggs from overheating if shading is inadequate, some species, such as White-headed Lapwing (*V. albiceps*), wet the eggs with water, which they collect by soaking their belly feathers. Other species, such as White-fronted Plover (*Charadrius marginatus*), cover the eggs with sand, which also serves as camouflage. ■

PAINTED-SNIPES *Rostratulidae*

The painted-snipes are wading birds that look rather like true snipes, but are more brightly and intricately patterned. There are only two species: Greater Painted-snipe (*Rostratula benghalensis)* and South American Painted-snipe (*Nycticryphes semicollaris*).

A female **Greater Painted-snipe** feeding. Painted-snipes forage by probing in muddy water with their bill, using a short, semicircular sweeping motion. Location unknown.

Structure The birds are 8–10 in. (20–28 cm.) in length, and stoutly built; males are slightly smaller than females. They have large eyes and a fairly long, slightly drooping bill, which is slightly more decurved in the South American species. They also have short, rounded wings, long legs, and slender toes.

Plumage Greater Painted-snipe is sexually dichromatic, but the South American species is not. The female Greater Painted-snipe has cryptically patterned upperparts of deep metallic green, broad white to golden scapulars that run into a white stripe on the shoulders, and white underparts. The face and neck are barred bronze. There is a broad white eye ring that extends along the side of the head, and a golden stripe along the crown. The male is paler or less well marked, with browner wings spotted or edged in golden-brown. South American Painted-snipe is much darker brown or reddish-brown on the breast, with narrow white lines along the crown and brow line, and the scapular line is golden-yellow to whitish. In both species, the legs and feet are yellow.

Behavior The birds may be solitary, found in pairs, or loosely social: groups may gather at feeding areas during the dry season, or at nesting areas. Painted-snipes are mostly active in the late evening and early dawn, and also at night when there is a full moon. By day, they roost in tall sedges or clumps of aquatic vegetation in swampy areas. They forage in shallow water and soft mud.

Voice Mostly silent unless breeding, although they may give a short *kek* if disturbed. Female Greater Painted-snipes, when displaying to males, have a large range of mellow hoots; males may respond with growls. South American Painted-snipes have a short, plaintive whistle.

Habitat Lowland, marshes, swamps, rivers and pools with muddy edges and dense cover.

Movements Most populations are migratory. Those breeding in northern China and Japan migrate to Southeast Asia. In Africa and Australia, the birds are nomadic, and movements are largely seasonal. Greater Painted-snipe in Africa is mostly a wet-season visitor to the Sahel zone between northern Nigeria and Chad. In South America, the birds make similar movements in southern Brazil, and are nonbreeding visitors to northern Argentina.

Diet Mostly aquatic insects, worms, snails, and crustaceans; also crickets, grass seeds, and rice.

Breeding Painted-snipes breed either monogamously or polygamously. Female Greater Painted-snipes court males, mating with at least two in succession, and leaving each male after laying two to four eggs. The nest is a shallow scrape of a few leaves, stems, and sticks. The male carries out all parental care. The young leave the nest after a few days and follow their father until they can fend for themselves.

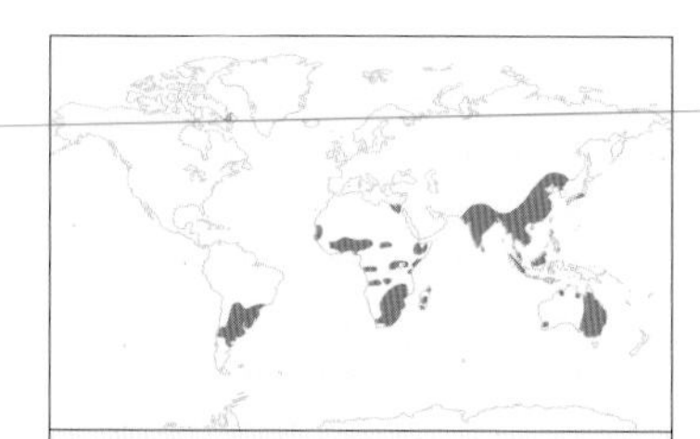

Number of genera 2
Number of species 2
Conservation Status Least Concern
Distribution Worldwide apart from North and Central America, western Palearctic, polar regions

JACANAS *Jacanidae*

Jacanas, also called lilytrotters, are colorful, long-legged waterbirds that resemble rails and are found almost exclusively in tropical regions. Their long, spidery toes enable them to walk easily over lily pads or other floating plants, giving them the appearance of walking on water. All species have similar structure but vary in plumage.

Structure Lengths vary from 6 to 23 in. (15–58 cm.), with the largest species being Pheasant-tailed Jacana (*Hydrophasianus chirurgus*) and the smallest being Lesser Jacana (*Microparra capensis*). Females are larger than males: in some species, they may weigh two-thirds more. All except Pheasant-tailed Jacana have a frontal shield or facial wattle. The birds have elongated toes and claws. Three species have carpal spurs.

Plumage Bronzy-brown and glossy black predominate in many jacanas, often with contrasting bright yellow flight feathers. Five species also have white or yellow on the front or back of the neck. Pheasant-tailed Jacana is the only long-tailed species. It has a white face and neck, and mostly white wings with black primaries; outside the breeding season, the bird has a much duller, greenish-brown plumage and a much shorter tail. Frontal shields or wattles may be yellow, red, or blue, becoming brighter during the breeding season.

Two **Comb-crested Jacanas** (*Irediparra gallinacea*). Conspicuous birds, jacanas spend much of their time in the open. Northern Territory, Australia.

Behavior Jacanas spend long periods foraging in aquatic vegetation. They often feed in association with larger animals that stir up muddy water; African Jacanas (*Actophilornis africanus*) will reportedly stand on hippopotamuses to watch for prey.

Voice The birds are often noisy or vocal in courtship or defense of eggs and young, with a variably pitched series of thin, piping squeaks, hoarse cackles, and rattles. Calls between adults and young are much softer.

Habitat Shallow freshwater wetlands; also reedbeds, swamps, and areas of deeper water with suitable surface cover. Occasionally, the birds are also seen in fields and agricultural areas near wetlands.

Movements Largely sedentary, but Pheasant-tailed Jacanas from the Himalayas and eastern China move south for the winter, either within their breeding range or into Oman, Malaysia, and Java.

Diet Jacanas feed on aquatic insects, such as water beetles; also snails, crustaceans, and small fish.

An **African Jacana's** long toes spread the bird's weight so it can walk over water plants. South Africa.

Breeding In all species apart from Lesser Jacana, the male builds the nest, incubates the eggs, and cares for the young. The females are often polygamous.

The nest is a flat platform of aquatic vegetation, situated at or just above the waterline, or occasionally located on a small, muddy island. The male incubates the clutch of four eggs for 22–28 days. Males of several species carry their young around under their wings, with the chicks' legs and feet dangling down.

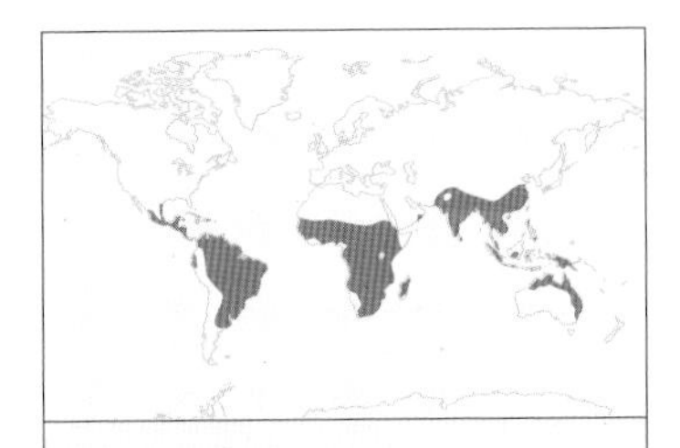

Number of genera 6
Number of species 8
Conservation Status Least Concern
Distribution Central and South America, Africa, Madagascar, India to Eastern China, Southeast Asia, Australia

PLAINS WANDERER *Pedionomidae*

A shy and enigmatic species, Plains Wanderer (*Pedionomus torquatus*) is confined to parts of eastern Australia where its habitat still survives. In this species, the roles are reversed: the female engages in courtship behavior, and the male raises the young.

Structure Plains Wanderer resembles a quail, but is slimmer, with a longer neck and legs. The bird is 5.9–7.5 in. (15–19 cm.) in length; females are larger than males.

Plumage The plumage has a cryptic brown coloration, patterned with blackish crescents and pale feather edges on the upperparts. The female has a distinctive rusty breast band and black collar marked with white. The male is far more nondescript, blending perfectly with its background while nesting. Both sexes have a distinctive pale yellow eye, and yellow legs and feet.

Behavior Secretive and difficult to observe. Plains Wanderer was formerly believed to be most active by night, but radio-tracking studies have revealed activity both during the day and at dusk. It avoids detection by remaining motionless and relying on its camouflage to disguise it from predators. To escape danger, the bird runs in a crouched posture, and will only take flight as a last resort; it is so reluctant to fly that it will sometimes remain frozen and allow approaching dogs or humans to pick it up. Females may stand on tiptoe to keep watch for predators, usually while perched on a tussock. The species is usually solitary outside the breeding season.

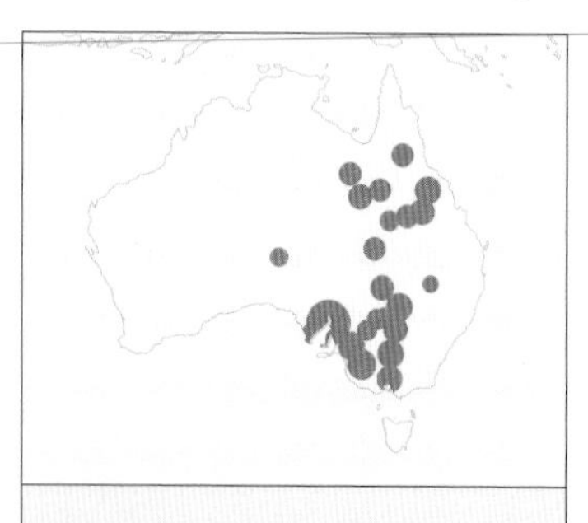

Number of genera 1
Number of species 1
Conservation Status Endangered
Distribution Eastern Australia

Voice Silent except in the breeding season, when the birds (possibly only females) make a low moaning *moo* or *oo* sound. The male has two calls for the chicks: a low, two-part clucking response of alarm, and a longer, repeated, piping contact call.

Habitat Large, open, treeless expanses of native grassland and saltbush. Today, this habitat survives only in interior eastern Australia as far north as Queensland and the Northern Territory. As a result, the species' distribution is limited. Its main remaining stronghold is the Riverina area of southern New South Wales; the most recent records are from the Delinquin area. Plains Wanderer is vulnerable to population falls during prolonged droughts. Its occurrence is fragmented and poorly known across wider ranges; the species is now on the verge of extinction in Victoria.

Movements Mainly sedentary. The bird is only nomadic when forced to disperse by habitat change, prolonged drought, or exceptionally heavy rains (which cause an explosion in the growth of non-native plants, reducing the quality of food).

Diet The species is omnivorous. Small insects and other invertebrates make up around 50 percent of the diet; other staple foods include a variety of seeds.

Breeding Breeds broadly between spring and early winter: August to November in the south, but usually March to June in the north. The female calls and displays to the male. The nest is little more than a hollow lined with grasses, usually hidden in taller vegetation, and occasionally shaded by a woven canopy. Clutches comprise two to four eggs, pointed in shape and camouflaged with blotches and speckles. The male carries out most of the parental duties, incubating the eggs for 23 days and tending the young for about two months. Second broods are raised if there is adequate rainfall.

The female **Plains Wanderer** is more colorful than the male, with a black-and-white spotted collar and rufous breast patch. Australia.

Taxonomy A monotypic family. Although Plains Wanderer resembles a buttonquail (and was formerly classified as a relative of the Turnicidae), its nearest relatives are now thought to be the seedsnipes (Thinocoridae) of South America.

SEEDSNIPES *Thinocoridae*

Although they resemble game birds, the seedsnipes are actually shorebirds that have adapted to a terrestrial existence. Despite the name, they bear no relation to true snipes (family Scolopacidae). The larger two species are like ptarmigans (genus *Lagopus*) in body shape, habitat, and behavior. The smaller species are ecologically more similar to larks. The family is closely allied to Plains Wanderer (family Pedionomidae) as well as to the painted-snipes (Rostratulidae) and the jacanas (Jacanidae).

Rufous-bellied Seedsnipe is the largest of the seedsnipes. The head and upperparts are scalloped in black, brown, and cream, while the underparts are pale rufous. Ecuador.

Structure Seedsnipes range from 7 to 12 in. (18–30 cm.) in length. They are rotund birds with a small bill like that of a chicken, short legs, and large feet. In contrast, they have the long, pointed wings of a shorebird. The tail is short. The birds also have a thin flap of skin covering the nostrils, to keep out dust.

Plumage In the larger species (genus *Attagis*) the sexes are similar, but in the smaller ones (*Thinocorus*) there is marked sexual dichromatism. All seedsnipes, however, are well camouflaged, having intricately marked plumage with marbled patterns. Rufous-bellied Seedsnipe (*A. gayi*) is rusty below, while the southern White-bellied Seedsnipe (*A. malouinus*) is white below. Both species have a dark bill and grayish-yellow legs. *Thinocorus* males have a gray face and breast; the females are patterned like White-bellied Seedsnipe, although they show white wing stripes and black wing linings in flight. Both *Thinocorus* species have yellowish bills and yellow legs.

Behavior Seedsnipes are strongly terrestrial. The *Attagis* species often walk in a crouched manner, but when startled they noisily take to the air. *Thinocorus* males spend much of the spring and summer singing from a high mound or rock, or while performing aerial displays.

Voice *Attagis* species give quavering whistles, particularly when startled into flight. Male *Thinocorus* species have a cooing, repetitive song like a fast-paced dove call.

Habitat Mainly southern South America, with one species extending to the equator along the peaks of the Andes. Most are upland birds, except for Least Seedsnipe (*T. rumicivorus*), which lives in flat, arid, or sandy terrain. Gray-breasted Seedsnipe (*T. orbignyianus*) is associated with highland bogs (*bofedale*).

Movements Migratory movements are not well studied in this group. Some Gray-breasted and Rufous-bellied Seedsnipes are year-round residents, while other populations either descend to lower elevations or move north for the winter. Some Least Seedsnipes living in southern coastal areas migrate north for the winter, but where they go is not known. The two *Attagis* species are strictly alpine when breeding, but in winter, White-bellied Seedsnipe descends to open, windswept Patagonian grasslands to escape the extreme weather in its breeding areas.

Diet Seedsnipes forage largely on plant matter, including seeds and berries. During the breeding season, they may also take invertebrates.

Breeding Little is known. The nests are placed on the ground and well camouflaged, as in other shorebirds. The clutch of four spotted eggs is incubated by the female. The young are precocial; both parents care for them until they fledge.

Number of genera 2
Number of species 4
Conservation Status Least Concern
Distribution Mainly southern South America

SANDPIPERS AND SNIPES *Scolopacidae*

A **Western Sandpiper** (*Calidris mauri*) in breeding plumage. The distinctive rufous and brown head and upperparts, and the bold patterning on the wings, are very different from the winter plumage, which is a muted gray. Alaska, U.S.

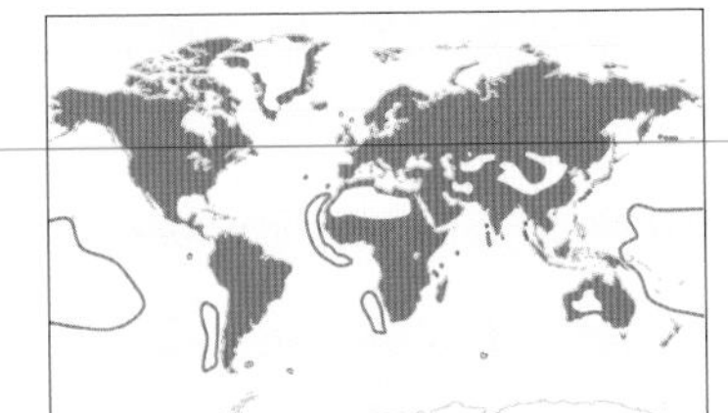

Number of genera 23
Number of species 92
Conservation Status
2 species Endangered,
2 species Vulnerable
Distribution Worldwide, including very high latitudes in the Arctic

The largest family of wading birds, this group comprises several subfamilies: sandpipers (Calidrinae); turnstones (Arenariinae); shanks, godwits, and curlews (Tringinae); phalaropes (Phalaropodinae); snipes and dowitchers (Gallinagininae); and woodcocks (Scolopacinae). Some species are well known for undergoing long migrations, or for gathering in huge, spectacular flocks.

Structure The birds range in length from 5 to 26 in. (13–66 cm.). Most species are of similar body shape, but otherwise they vary greatly in structure. Many are sexually dimorphic in size: in some species, males are larger than females, while in others, females are larger. The head is streamlined, and in some species the neck is long. The eyes, well placed to give a wide field of view, are particularly large and slightly bulging in woodcocks, which have all-round vision. The bill may be short, long, or very long. The shape may be straight, or slightly or heavily decurved. In most species it is pointed, but Spoon-billed Sandpiper (*Eurynorhynchus pygmeus*) has a unique, spatulate bill tip. The wings are short in sedentary species, rounded in snipe and woodcocks, and longer and pointed in migratory species. The tail may be square or rounded at the tip. The legs also vary from fairly short to long, and in some species the toes are partly webbed.

Plumage Coloration varies between genera and species. The sandpipers, shanks, godwits, and curlews are predominantly mottled gray and brown to chestnut on the head and upperparts, and often cryptically marked with fine bars or spots. Some have broad white wingbars or panels, or white on the back and rump. The underparts are often paler, and in some species also spotted or barred. Most species have more colorful summer or breeding plumage. Godwits become much brighter, turning from gray to rich orange on the body and wings. Several sandpipers change from their largely gray and whitish winter colors to a brighter red on the head, breast, and upperparts, especially Red Knot (*Calidris canutus*), Spoon-billed Sandpiper, Curlew Sandpiper (*C. ferruginea*), Sanderling (*C. alba*), Red Phalarope (*Phalaropus fulicarius*), Little Stint (*C. minuta*), and Red-necked Stint (*C. ruficollis*). Ruddy Turnstone (*Arenaria interpres*) has a summer plumage of bright orange heavily mottled with black spots, and a white head and face with a blackish line running from the face over the breast. Male Ruff (*Philomachus pugnax*) become variably but very spectacularly marked for three months, with a mottled pattern of fine grays, browns, and chestnut on the upperparts. They also develop an elaborate ruff of long feathers from the back of the head and around the neck to the breast, richly colored from orange to black or white, coupled with a bare face of the same color and a short tuft or crest. This plumage is seen to best effect when the males display at a lek (to the rather dowdy-colored females). Snipe and woodcocks are more cryptically colored to match their surroundings. Their plumage is rich, dark brown, scalloped or broadly barred in paler brown or orange, and lined bright yellow or golden. Three species, including American Woodcock (*Scolopax minor*), have bright, unbarred orange underparts. The head and face are often striped (in snipe) or barred (in woodcocks) with broad light and dark lines.

Behavior The family overall includes species that are largely solitary, birds found in pairs or small groups, and those that spend most of their lives in large flocks. Those that occur in large flocks are often renowned for their fast, direct flight and aerial maneuvers when on passage or flying to feeding areas, or, in the case of some species, the twists and turns they make to evade predators.

Sandpipers feed in several ways. They may 'pick' at the surface of water or intertidal mud for small crustaceans, or probe into muddy substrate, occasionally with the head entirely submerged. Several of the smaller sandpipers and shanks regularly swim, but only the phalaropes do for prolonged periods, as they spin around stirring up aquatic insects. Snipe forage largely by probing in shallow water, often with a rapid digging motion of the head and bill. It seems they rarely see their prey, but detect it only by the sensitive tip of the bill. Woodcock forage in similar manner in soft earth and leaf-litter. Turnstones, as their name suggests, turn over stones in search of invertebrates, but actually spend more time searching among seaweed on rocks or beaches, or occasionally among mollusk-beds.

A **Spotted Sandpiper** in breeding plumage (foreground) and an unspotted juvenile bird (flying).

Two **Red Knots** in breeding plumage. Knots breed in high northern latitudes, but migrate far into the southern hemisphere to reach their wintering areas. Sweden.

A **Eurasian Curlew** (*Numenius arquata*) in flight. Curlews can often be heard calling plaintively while on the wing. Sweden.

Voice Most calls are a series of shrill whistles, trills, or twitters. Some species have a dry or hoarse, rasping note. This is particularly true of snipe, which give a dry cough or cry as they rise, and Eurasian Curlew, whose calls often include a distinctive *cur-loo* or *cur-lee*. The songs of several species are more musical, and are often given as part of a circular display flight. Upland Sandpiper (*Bartramia longicauda*) makes a gurgling water sound followed by an eerie *whip-whel-ee-ooo*, and Red Knot has ringing *poor-me poor-me* sounds, considered to be one of the finest sandpipers' songs.

Habitat Different genera occupy different habitats. Some of the sandpipers spend much of their time in freshwater and coastal marshes. Snipe are birds of open bogs and moorland areas, and woodcocks are ground-dwellers of woodlands and subtropical forests.

Movements Most species breed in the north or far north, some nesting within 560 miles (900 km.) of the North Pole, and winter in tropical to southern latitudes including the open ocean. They are most commonly seen on spring and autumn passage. Certain sandpipers are particularly well known for being highly migratory, and include some of the longest-distance migrants of all bird species.

Diet Sandpipers feeding along the shoreline in muddy, sandy, or rocky areas take a variety of small crustaceans, invertebrates, eggs, and larvae, together with mollusks such as tellin and spire snails. The larger and longer-billed species take marine worms including lugworms and ragworms. In the breeding areas, flies, crane-flies, and midges form a large part of the diet. Snipe also feed on aquatic insects and earthworms, while the diet of woodcocks mostly comprises earthworms and other invertebrates, such as beetles, bugs, slugs, and leeches, as well as some fruit, seeds, and leaves of grasses. Turnstones have a wide diet and seem to eat almost anything.

Breeding All species perform elaborate song-flights and wing-flapping displays prior to forming pairs and establishing territory. The nest is a shallow cup or scrape of grasses, plant fibers, and down. It is usually well concealed amongst vegetation (including partly floating plants), in tussocks, or occasionally under tree-roots. However, some species, such as Green Sandpiper (*Tringa ochropus*), Wood Sandpiper (*T. glareola*), and Solitary Sandpiper (*T. solitaria*), use the abandoned nests of other birds in trees. The clutch comprises two to four eggs. They are usually incubated by both parents. The exceptions are species such as Ruff, Curlew Sandpiper, and Pectoral Sandpiper (*C. melanotos*), where males play no part at all; the dowitchers, where the female incubates and the male looks after the young; and Sanderling (*C. alba*), in which the female lays two clutches, the second incubated solely by the male. Incubation times range from 16–20 days for the smaller sandpipers to 35–45 days in the curlews. The chicks can walk almost as soon as they hatch; with their parents, they then leave the nest and make for feeding grounds.

Three species, including two curlews, are listed as Critical, and are most likely extinct; populations of the third, the highly engimatic Spoon-billed Sandpiper, have declined drastically in recent years and may now only consist of several hundred individuals.

A **Common Snipe** (*Gallinago gallinago*) perches on a rock. Snipe are most often seen in moorlands and wetlands. County Durham, U.K.

COURSERS AND PRATINCOLES *Glareolidae*

Temminck's Courser (*Cursorius temminckii*) often forages in areas where fires have recently occurred. South Africa.

An Old World family ranging from southern Europe through Africa and Asia and into Australia. There are two subfamilies: the coursers (subfamily Cursoriinae) and the pratincoles (Glareolinae). Species within each of the two groups are very similar and vary mostly in size and plumage.

Structure Lengths range from 7–12 in. (18–30 cm.) and weights from 1.25–5.75 oz. (35–170 g.). Both groups have a short, slightly arched bill and long, slightly pointed wings. The coursers are upright, ploverlike birds, more terrestrial than the pratincoles, and well adapted for running, with long legs and short toes, although Egyptian Plover (*Pluvianus aegyptius*) has slightly shorter legs. Coursers also have a square-tipped tail. The pratincoles (apart from Australian Pratincole, *Stiltia isabella*) are more horizontal in posture. They are more aerial, with longer wings, shorter legs, and a forked tail (except for Australian Pratincole, which has a square-tipped tail).

The coursers have no hind toe. The pratincoles do have a hind toe, which is fairly small and slightly raised. All species, apart from Egyptian Plover and Australian Pratincole, have comblike serrations of the front claw. The three species of courser in genus *Rhinoptilus* are almost entirely nocturnal and have large eyes.

Egyptian Plovers frequent the shores of lowland rivers in tropical Africa. They nest on bars of sand and gravel. The Gambia.

Plumage In both subfamilies, the sexes are alike. The coursers, which are mainly ground-dwelling, have cryptic plumage that varies in color from entirely sandy to gray or gray-brown, with broad pale edges or scalloped patterns on the upperparts in several species. They also have broad or prominent black eyestripes and long white supercilia, which often meet in an inverted 'V' on the rear of the head. The two species of banded courser have broad double breastbands, and Heuglin's Courser (*Rhinoptilus cinctus*) has a longer third band extending down from the sides of the neck. The two larger species, Bronze-winged Courser (*R. chalcopterus*) and Jerdon's Courser (*R. bitorquatus*), have subtle shades of grays and brown but both show prominent breastbands. Egyptian Plover shares many of these features except for a more extensive area of black running from the nape to the back and across the breast above the white breastband.

Pratincoles are brown, gray, or pale gray with black flight feathers, white from belly to rump, and a black-and-white tail. Three of the species have a pattern of blackish lores and border, extending below the eyes and around the creamy throat patch; three other species have a white or black-and-white line extending backward from the eye to the nape or curving behind the ear-coverts. Both Gray Pratincole (*Glareola cinerea*) and Little Pratincole (*G. lactea*) are pale gray, the latter entirely so and the fomer mainly gray with a pale chestnut nape and buffish breast.

Behavior Coursers may occur either alone, in pairs, or in small groups. They forage in open country, mainly on insects; the birds spend long periods surveying the area around them for large insects such as locusts and beetles before chasing and seizing their prey. They also escape predators by running. Pratincoles

Number of genera 5
Number of species 18
Conservation Status 1 species Critically Endangered
Distribution Southern Europe, Africa, Asia, Australia

take most of their food in flight, in a similar way to terns, but are just as able to catch insects on the ground. They often feed in large numbers in wetlands, alongside marsh terns.

Voice Pratincoles have high-pitched whistles and trilling notes and frequently give contact calls, especially in flight. Coursers are generally less vocal, but may give soft or mellow nocturnal calls.

Habitats The family is widely distributed and fairly numerous, although several species are only seasonally common within their range. Pratincoles occur mainly in wetlands and at the edges of lagoons, lakes, and wide rivers in open country. Coursers are drier area specialists, ranging from desert margins to open grasslands and scrub woodland. One species, Jerdon's Courser, is restricted to a small area of the Eastern Ghats in India, and is listed as Critically Endangered. It was considered extinct until 1986, when a small population was rediscovered in an area where there is also increasing cattle grazing, disturbance, and quarrying; it is completely nocturnal and its thornscrub habitat makes it extremely difficult to detect or survey. The entire population may number only a few hundred birds.

Movements Coursers are mainly sedentary, but some Cream-colored Coursers (*Cursorius cursor*) move south across the Sahara for the winter. Most pratincoles are highly migratory, moving short or long distances from breeding to wintering areas. Madagascan Pratincole (*G. ocularis*) moves west to winter in East Africa. The longest distances are covered by Black-winged Pratincoles (*G. nordmanni*), which breed in Central Asia and winter in South Africa, and Oriental Pratincoles (*G. maldivarum*), which breed in Australia and move north for the winter, reaching parts of Indonesia.

A **Collared Pratincole** (*Glareola pratincola*) in breeding plumage, with the characteristic creamy-yellow throat patch encircled by a black ring. Greece.

Diet Both coursers and pratincoles are insectivores.

Breeding All species are monogamous. Coursers are largely solitary, while pratincoles are more social and nest colonially. Nests are placed on the ground, in a scrape that may be bare or sparsely lined with grass; courser nests are usually in shade. Clutches comprise up to two eggs in coursers but can be up to four or even five for pratincoles. Incubation is performed by both parents, and takes about 20 days. The young leave the nest, accompanying the parents, after a further 25–30 days.

Taxonomy The two subfamilies are largely thought to be linked through Australian Pratincole, in the monotypic genus *Stiltia*, due to its longer-legged, more courserlike shape and its habits. The Glareolidae's ploverlike or ternlike shape and behavioral characters place them clearly within the overall grouping of the large family of waders (order Charadriiformes), but their exact relationships have not been satisfactorily resolved: several species are frequently placed within the true plovers or, in the case of Egyptian Plover, with the thick-knees.

A **Little Pratincole** swoops low over grass in pursuit of insect prey. Pratincoles' long, pointed wings give them great maneuverability in flight. Rajasthan, India.

GULLS, TERNS, AND SKIMMERS *Laridae*

This large seabird family contains some of the world's most familiar birds, including the 'seagulls' that are ubiquitous in many regions. Many have developed a very close association with humans and have prospered spectacularly as a consequence. There are about 50 species of gulls (subfamily Larinae), 44 terns (Sterninae), and three closely related species of skimmers (Rynchopinae). Gulls are largely similar in appearance, but there is a rough divide between the larger species and the smaller ones, many of which have a black head in summer. Terns are slimmer and daintier than gulls, with narrower wings, a pointed bill, and a forked tail. They, too, can be divided into two basic types: the more gull-like sea terns, which have a black cap in summer, and the three or four species of marsh terns, which are black, or virtually so, in summer. Another group of terns is the noddies, most of which are dark brown with a white cap. Skimmers are similar to terns in shape, but are larger and more robust, with a uniquely shaped lower mandible. Few species are endangered; Chinese Crested Tern (*Sterna bernsteini*) is the only one on the critical list.

Herring Gull (*L. argentatus*) is one of the most familiar of all gulls. Breeding adults (flight figure) have a white head, while those in winter plumage (standing figure) have a 'hood' of pale brown streaking.

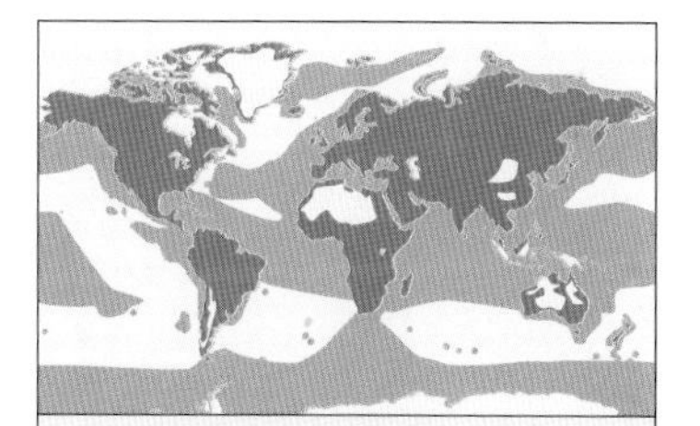

Number of genera 15
Number of species 97
Conservation Status
1 species Critically Endangered,
3 species Endangered,
5 species Vulnerable,
10 species Near Threatened
Distribution Worldwide

Structure Gulls are long-winged but evenly proportioned birds, ranging in size from the diminutive Little Gull (*Larus minutus*), at 9.5–11 in. (24–28 cm.), to Great Black-backed Gull (*L. marinus*), at 24–29 in. (61–74 cm.). Males tend to be larger than females. The bill is slender, although the larger species usually have a much heavier and more powerful bill, with a marked gonydeal angle. All species have webbed feet, two have a forked tail, and one has a markedly graduated tail.

The largest tern species, Caspian Tern (*S. caspia*), at about 20 in. (50 cm), is the size of a gull, but most terns are small to medium-sized, the smallest being Little Tern (*S. albifrons*), at about 9 in. (23 cm.). All species have long, pointed, slender wings and a forked tail; in some, the tail has elongated outer feathers.

The skimmers are among the most peculiar of all birds. They have an elongated lower mandible that facilitates their remarkable feeding method. The head is supported by a stout neck and elongated, modified anterior cervical vertebrae, to hold the hypertrophied neck muscles.

Plumage Most gulls and terns are simply colored, with gray upperparts and a white head and underparts. Some gulls, such as Great Black-backed Gull and Slaty-backed Gull (*L. schistisagus*), are virtually black above. Two South American species—Lava Gull (*L. fuliginosus*) and Gray Gull (*L. modestus*)—are almost completely gray in adult plumage. Some species, particularly Slender-billed Gull (*L. genei*) and Ross's Gull (*Rhodostethia rosea*), have pink-tinged underparts in summer. Many smaller species, and also Great Black-headed Gull (*L. ichthyaetus*), have a black head when in summer plumage, while many of the medium and large gulls have grayish streaking on the head in winter.

Most species have black tips to the primaries, the pigment melanin giving additional strength to these feathers. Within the black, there are often prominent white spots or 'mirrors.' Only a few species, mainly from the Arctic, have white primary tips. Sabine's Gull (*Xema sabini*) and Swallow-tailed Gull (*Creagrus furcatus*) have bold black, white, and gray triangles across the upperwings. These species also have a forked tail.

Most larger gulls have a yellow bill with a red spot on the lower mandible (which the young peck at when begging to be fed). Smaller species tend to have a black or red bill when in breeding plumage, although a few have a yellow bill. Eye rings and irises are variously colored, the former apparently serving as a species-recognition character in some of the closely-related larger gulls.

Gulls take a relatively long time to reach maturity: two years in the smaller species, and up to four or five in the larger ones. All start with a juvenal plumage that is predominantly brown, gradually becoming whiter and grayer in a series of twice-yearly molts. Most first-year gulls have a black band at the tip of the tail, and black outer primaries, although some small species, such as Bonaparte's Gull (*L. philadephia*) and Black-headed Gull (*L. ridibundus*), have white 'wedges' on the primaries at all ages. A few species, such as Black-legged Kittiwake (*Rissa tridactyla*), have a large black 'W' across their upperwings in immature plumages. The plumages of young gulls are both varied and complex, and their appearance is subject to wear and bleaching, so much so that identification, particularly of the larger species, is often extremely difficult.

Sea terns (genus *Sterna*) resemble the smaller 'black-headed' gulls in plumage, except that they have a cap rather than a hood, and they do not show clear-cut black primary tips (although the outer primaries often darken with wear). As in smaller gulls, the bill is mainly red in breeding condition. The marsh terns (*Chlidonias*) are much darker. Black Tern (*C. niger*) and White-winged Black Tern (*C. leucopterus*) have a black body with paler wings, while Whiskered Tern (*C. hybrida*) has a dark gray body (with white cheeks) and a black cap. Four tropical species are much darker above, Sooty Tern (*S. fuscata*) being jet black, while three of the four species of noddy have a very dark brown body with a white cap, like a negative of a typical *Sterna* tern. Gray Noddy (*Procelsterna albivitta*) is pale gray, while the tropical White Tern (*Gygis alba*) is white. Inca Tern (*Larosterna inca*), from the west coast of South America, is unique, being almost completely blue-gray but with a black cap and curly, white, mustache-like facial plumes.

Pacific Gull (*L. pacificus*) has a strikingly large, thick bill with a scarlet tip and an especially prominent gonydeal angle. Australia.

Like young gulls, most juvenile terns have mainly brown plumage, which is gradually lost after they fledge. They take two years to reach maturity. Both Sooty Tern and Inca Tern are unusual in that the juveniles are completely dark.

Skimmers are completely black above and white below, with a white forehead, a prominent white trailing edge to the wing, and a predominantly orange or red bill.

Behavior The Laridae are one of the few families of birds to be equally at home in the air, on the ground, and on water. They are highly opportunistic and versatile in their feeding behavior. Although many species pursue their traditional habits, foraging in lakes, marshes, and the sea or along the seashore, others have successfully exploited ar-

Courtship feeding, and begging for food, are common behaviors in breeding pairs of gulls, as seen in these two **Laughing Gulls** (*L. atricilla*). Florida.

Bridled Tern (*Sterna anaethetus*), found in tropical oceans, is darker than most other tern species. It can be recognized by the black head markings from which its name is derived, gray-brown upperparts, and deeply forked tail. Location unknown.

tificial environments. Several species take advantage of human untidiness, scavenging in cities and on garbage dumps. Many more have adapted to agricultural land, where they may feed almost exclusively on earthworms. In suitable conditions, both gulls and terns hawk flying insects high in the sky, while several species become food pirates. Many large gulls take shellfish, dropping them onto rocks to break open the shells.

Wherever they forage, in winter most gulls form communal, mixed-species nocturnal roosts, in the tens or hundreds of thousands, on the sea, estuaries, or large lakes or rivers. Several species, such as the kittiwakes and Sabine's Gull, are almost exclusively pelagic outside the breeding season, while two northern species—Ivory Gull (*Pagophila eburnea*) and Ross's Gull—eke out a living around Arctic pack ice.

An **Arctic Tern** (*Sterna paradisaea*) plunge-diving to capture prey. The birds typically hover 30–40 ft. (9–12 m.) above the water, before diving head-first. Scotland, U.K.

Most terns hunt fish by making spectacular plunges from the air into the water. Marsh terns, however, usually pick insects and other invertebrates from the water's surface; some *Sterna* terns also feed like this in inland areas. Gull-billed Tern (*S. nilotica*) often forages over dry land and mud flats, where it swoops on insects and vertebrates such as mudskippers (genera *Periophthalmus* and *Scartelaos*). The skimmers, as their name implies, skim low over water in flight, with the tip of the lower mandible in the water, and snap the bill shut when it comes into contact with prey.

Voice Most species give a variety of calls, depending on circumstances. In general, the larger the bird, the deeper the calls. Great Black-backed Gull gives loud, deep, but rather muffled bellowing calls. Herring Gull gives higher-pitched trumpeting and other 'mewing' calls, gruff, chattering alarm calls, and so on. The smaller species give softer mewing, rasping, and grating calls, while some species have fairly musical, higher-pitched notes. Begging juveniles often make a persistent whimpering sound. Terns' calls tend to be harsh and rasping; the smaller the species, the higher-pitched the calls. The large Caspian Tern has a loud, harsh call that is reminiscent of Gray Heron (*Ardea cinerea*).

Habitat Many species, particularly the gulls, can live in almost any habitat, including inland city centers, but their most natural environments are coastlines, offshore islands, and the sea. Most gulls are found at temperate latitudes in the northern hemisphere, although Ivory Gull breeds in the high Arctic. The terns, in contrast, are more concentrated in tropical and subtropical regions. Many of the gulls and terns disperse into offshore waters outside the breeding season, and some, such as the kittiwakes and Sooty Tern, become pelagic. The marsh terns and the skimmers may be found in marshland and by freshwater lakes and rivers.

Movements Many gulls are reasonably sedentary, but some make relatively short seasonal movements to the nearest milder areas. Only a few species are long-distance migrants. Franklin's Gull (*L. pipixcan*) and Sabine's Gull are true transequatorial migrants. The former breeds in central Canada and the U.S., and migrates overland to winter on the west coast of South America. The latter breeds in the northern U.S. and Canada, wintering at sea off the west coasts of South America and South Africa. Terns, on the other hand, are great trans-globe travelers: Arctic Tern (*S. paradisaea*) has one of the longest migrations of any bird, from the Arctic to the Antarctic. Skimmers are more sedentary, although Black Skimmer (*Rynchops niger*) disperses widely around the coasts of South America.

Diet Fish and shellfish; also insects, earthworms, and other invertebrates. Near human habitation, gulls may feed on garbage.

Breeding Gulls have a variety of displays that involve ritualized calling, posturing, strutting, begging, and courtship feeding. Terns' displays are basically similar, but in many species, such as Common Tern (*S. hirundo*) and Arctic Tern, aerial displays are an important element, with high display flights that often culminate in a ceremonial food pass.

Most gulls, terns, and skimmers are communal nesters, often forming large colonies on islands and in marshes. Gulls also breed on cliffs, and many of the larger species have adapted to nesting on buildings; this behavior is spreading. A few species, such as Mew Gull (*L. canus*) and Bonaparte's Gull, may even nest in trees. The birds form monogamous pair bonds, which are lifelong in some species, such as Herring Gull, but usually seasonal in the smaller species. They may be aggressive in defending their territories, even attacking humans. Some species, such as Arctic Tern, may draw blood and defecate on human intruders.

The normal clutch comprises two to three eggs, although some skimmers may lay four. In the smaller gulls, incubation takes three to four weeks and fledging about five, whereas in Great Black-backed Gull, incubation takes four weeks and fledging occurs after seven or eight weeks. First-years of most tern species usually remain in their winter quarters for their first summer.

Taxonomy Great advances have been made in recent years in understanding the taxonomy of large gulls, and major changes have been proposed in species definitions. For example, Herring Gull was traditionally considered a 'ring species,' with a range that circled the sub-Arctic latitudes. The species appeared to become darker from east to west, with the two ends of the ring overlapping as Herring Gull and Lesser Black-backed Gull (*L. fuscus*). However, molecular studies have revealed a more complex picture. For example, Great Black-backed Gull, Herring Gull, Yellow-legged Gull (*L. cachinnans atlantis*), and Armenian Gull (*L. armenicus*) evolved in the Atlantic, but Caspian Gull (*L. c. cachinnans*), Lesser Black-backed Gull, and American Herring Gull (*L. argentatus smithsonianus*) evolved from ancestors that lived around the Caspian and Aral Seas in Central Asia.

The relatively rare **Indian Skimmer** (*R. albicollis*) lives around the rivers and lakes of the Indian subcontinent and the Mekong Delta in Vietnam. India.

JAEGERS AND SKUAS *Stercorariidae*

Strong, swift, and fierce, jaegers and skuas are the pirates of the high seas and the seemingly ever-present predator at the edge of tern and penguin breeding colonies. They are similar to gulls, their nearest relatives, with long, pointed and angled wings.

In **Long-tailed Jaeger**, the central tail feathers are about 9 in. (22 cm.) long, making up nearly half the bird's total length. Norway.

Structure The species vary in size: jaegers are smaller than skuas. The smallest species is Long-tailed Jaeger (*Stercorarius longicaudus*), at about 21 in. (53 cm.) including the tail. The largest species are Chilean Skua (*S. chilensis*) and Brown Skua (*S. antarctica*), at about 25 in. (64 cm.). All jaegers and skuas have a strong bill with a hooked tip, adapted for tearing flesh, and webbed feet with sharp claws.

Plumage The coloring in the largest species is generally dark brown or gray, with darker streaks. The smaller birds—Long-tailed Jaeger, Parasitic Jaeger (*S. parasiticus*), and Pomarine Jaeger (*S. pomarinus*)—are polymorphic, with dark and pale morphs. Pomarine Jaeger has extensive white underparts, black cap, and cream or buff neck patches. All species have broad white flashes at the bases of the primaries. The smaller species have medium-length tails. Long-tailed Jaeger has very long central feathers, while Pomarine Jaeger in breeding plumage has unique spoon-shaped tips to the tail.

Pomarine Jaegers, showing dark morph (left) and light morph (right).

Behavior Usually solitary outside the breeding season, but occasionally found in numbers: the birds may temporarily form loose groups, or small flocks may gather around prey species. All species aggressively harry other seabird species, which vary in size from small terns to gannets. The birds chase their victims vigorously and aerobatically in the air until the prey disgorges any food they are carrying (or have eaten), at which point the jaeger or skua swoops on it. The larger species also usually attend breeding colonies of alcids, auks, or penguins, and swoop in to seize any unprotected eggs or young.

Voice Skuas make a variety of yelps and screams, while jaegers produce mewing calls.

Habitat Most breed in coastal moorland or tundra, or on grassy islands. Brown Skua nests on open ground near penguin colonies. Outside the breeding season, they spend most of their time on the ocean.

Movements Most species disperse from their breeding areas during winter. In one or two species, most birds only disperse into adjacent coastal areas, and some stay around the colonies all year. Three species are transequatorial migrants. One of the longest movements recorded for a ringed bird was made by a juvenile South Polar Skua (*S. maccormicki*), which was ringed on the Antarctic Peninsula and recovered (shot) north of the Arctic Circle in Greenland six months later.

Diet Fish, and the eggs and young of other birds. May also take adult birds and small mammals.

Breeding They usually pair for life. The nest is a scrape in the ground, with a clutch of two eggs (or occasionally just one). Incubation takes 23–28 days in smaller species, and about 30 in larger ones. The female stays with the nest and the young, while the male brings in food. Jaeger chicks fledge at 24–32 days, while those of skuas fledge at 45–55 days.

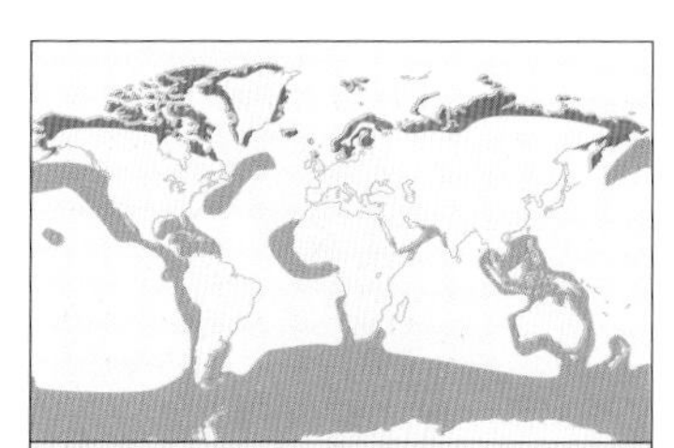

Number of genera 1
Number of species 7
Conservation Status Least Concern
Distribution Worldwide in all oceans; breeds mainly in extreme northern and southern latitudes

AUKS *Alcidae*

The auk family, also known as alcids, are social seabirds that resemble the penguins (family Spheniscidae) of the southern hemisphere, and are similarly well adapted for life in the water. The two families are not related, but share these similarities as a result of convergent evolution. The family once included Great Auk (*Pinguinus impennis*), the largest species, which was hunted to extinction in the mid-19th century.

Dovekie (*Alle alle*), also known as Little Auk, nests on the Arctic coasts of Europe, North America, and Greenland. In winter, the birds disperse across the north Atlantic. Norway.

Structure The largest living auk is Thick-billed Murre (*Uria lomvia*), at 18 in. (45 cm.) in length. The smallest is Least Auklet (*Aethia pusilla*), at 6 in. (15 cm.), and weighing 3 oz. (85 g.). Auks have a bulky, torpedo-shaped body, short wings, a stubby tail, and webbed feet. They are awkward on land, but highly agile in flight and underwater.

Bill shapes vary widely between genera and according to dietary specialization. The fish-eaters in genera *Uria* and *Cepphus* have a relatively long, thin bill. Puffins (genus *Fratercula*) have a large, high-ridged, triangular bill. The smaller auklets and murrelets have a small, stubby bill. Parakeet Auklet (*Aethia psittacula*) has a distinctive, upturned lower mandible, while Rhinoceros Auklet (*Cerorhinca monocerata*) is named for the small, pale 'horn' that it sports at the base of the upper mandible in the breeding season. Regional variations occur within some species; for example, the Arctic-dwelling population of Black Guillemot (*Cepphus grylle*) is noticeably shorter-billed and paler than its Atlantic counterpart.

Plumage Many species have dark upperparts and white underparts. The murrelets have white underparts and soft gray upperparts, and are often marked with white patches on the wings. Ancient Murrelet (*Synthliboramphus antiquus*) has hoary white streaks on its nape and eyebrows, from which it derives its name.

In the breeding season, Crested Auklet (*Aethia cristatella*) shows a curly crest, while the similar Whiskered Auklet (*A. pygmaea*) has a white 'V' line on the face, and another long white streak tailing the eye. Puffins have a striking multicolored bill. In Atlantic Puffin (*F. arctica*), the bill has an orange tip, a

Parakeet Auklet has a bright red bill; the stubby shape is an adaptation that helps the bird feed on jellyfish and crustaceans.

The monotypic **Razorbill** (*Alca torda*) is one of the largest auks. Its heavy, strikingly marked bill has very sharp edges.

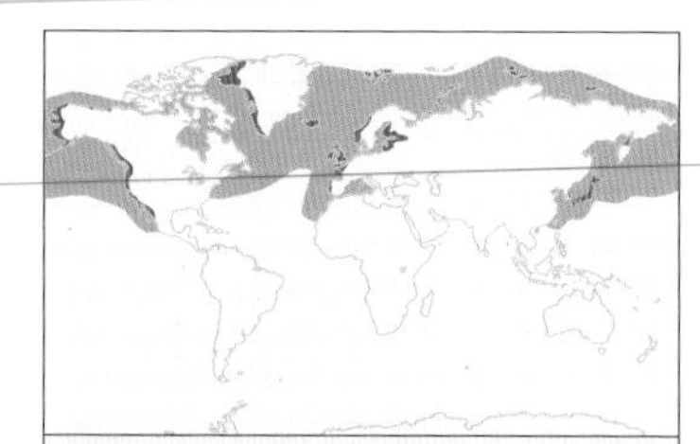

Number of genera 11
Number of species 24
Conservation Status
1 species Critically Endangered, 1 species Endangered, 3 species Vulnerable, 1 species Near Threatened
Distribution Holarctic, including most of northern latitude oceans

Tufted Puffins. In the breeding season, both sexes have a long double crest and a white face. Alaska.

thin yellow stripe, and a blue base; there are also rosettes of yellow skin at the sides of the gape. Tufted Puffin (*F. cirrhata*) has an orange and yellow bill with a slaty-blue spot at the base of the lower mandible. It also has long, straw-yellow tufts that hang loosely at the nape, contrasting with its white face and smoky gray-black plumage.

In winter, alcids appear duller and lose their breeding ornamentations. Perhaps the most dramatic change occurs in the guillemots of genus *Cepphus*, which molt their smart summer plumage of velvet-black offset with large white upperwing patches, changing to white and mottled gray; this gives them a very pale appearance at long range.

Behavior All species are monogamous and often mate for life. They rely on ritualized displays to build and maintain pair bonds. These displays are most complex in the auklets, but most species have walking and mutual billing displays that they use as a greeting when partners are reunited at the nest.

Alcids often swim and fly in lines; all have fast, whirring wingbeats, and some species patter across the water before takeoff. Feeding behavior varies between species, but all are fast-moving, efficient hunters that use their wings to propel them underwater in pursuit of prey. Larger species can dive to greater depths; *Uria* murres have been recorded as deep as about 330 ft. (100 m.).

Voice Although generally silent while at sea and outside the breeding season, all alcids are vocal at their nest site, some especially at night. Calls play a vital part in keeping paired birds and their young together in crowded colonies, particularly when fledglings first leave the nest for open water. Larger alcids make hoarse, groaning calls. Atlantic Puffin has a low, moaning call that has been likened to the sound of a distant chainsaw. *Cepphus* guillemots produce loud whistles and piping calls, while murrelets make chirruping, mewing, and peeping notes. Least Auklet produces a harsh, high-pitched trill.

Habitat All species are restricted to cooler northern waters, and to areas of the western U.S. seaboard where cold-water upwelling occurs. This is because their hunting method, pursuit diving, is less efficient in warmer waters, where their cold-blooded prey can move faster. Most auks are found in the Pacific, with just six species occurring in the Atlantic.

Movements Alcids' movements are complex and vary according to population, region, and species. Most species are at least partially migratory: outside the breeding season, they disperse away from nesting territories in response to greater availability of food elsewhere.

Some species remain close to home year-round. Examples include the North Pacific populations of *Aethia* auklets and Thick-billed Murre, and the Craveri's Murrelets (*Synthliboramphus craveri*) of the western U.S. coast. By contrast, Dovekies (*Alle alle*) are notable long-haul travelers; the population that breeds in Greenland winters off Newfoundland, covering a distance of some 2,175 miles (3,500 km.) each fall. Bad weather at sea during migration sometimes causes Dovekies to 'wreck,' and exhausted birds may turn up at coastal locations or occasionally many miles inland.

The most extreme journeys are those occasionally undertaken by wandering murrelets. Long-billed Murrelet (*Brachyramphus perdix*) breeds in eastern Russia and Japan, yet vagrants have been recorded across the U.S. as far south as Florida. Remarkably there are also records from Europe, including a

A **Black Guillemot** in summer plumage. In addition to the white patches on the upperwings, the bird has white underwings with black-tipped primaries. Norway.

bird found drowned in a fishing net on a Swiss lake in 1997.

Diet Alcids take fish, squid, crustaceans, and a wide range of planktonic organisms such as copepods. Larger species such as Common Murre (*Uria aalge*) prey more extensively on fish. Plankton-feeding specialists include smaller species such as Dovekie, *Aethia* auklets, and Cassin's Auklet (*Ptychoramphus aleuticus*). Areas of cold-water upwelling at sea offer rich feeding grounds, and auks converge on these hot spots from great distances (along with other seabirds). In most species, the composition of the diet alters seasonally in response to the availability of different prey types.

Breeding Auks nest on rocky coastlines and oceanic islands. Most species, except the *Brachyramphus* murrelets, are colonial breeders: *Uria* auks form dense colonies on cliff ledges, while puffins and auklets nest in burrows. Some colonies number many thousands or more than a million birds, and comprise several species. To avoid predation, most murrelets and small auklets are only active at night when on land.

The breeding season varies with species and location, and is timed to coincide with burgeoning food supplies and the retreat of winter ice and snow farther north. Most auks lay one or two eggs. Incubation periods vary widely across the family, from 27 to 46 days, being longest in puffins and shortest in Dovekie, *Cepphus* guillemots, and *Brachyramphus* murrelets. Ancient Murrelet is notable for the parents' prolonged incubation shifts, with each remaining at the nest for up to 72 hours at a time. Its young are among the most precocial of alcids, leaving the nest within one or two days of hatching. Atlantic Puffins abandon their chicks when they are ready to leave the nest, and hunger drives them out into the sea.

The unique breeding strategy of Marbled Murrelet (*Brachyramphus marmoratus*) remained a mystery until 1979, when the first nest was discovered in California. Uniquely for an alcid, the bird nests in the canopy of old-growth coniferous forests up to 65 miles (105 km.) inland, returning only at night; how the adults are able to locate their nests in the dark remains unknown. Sadly, logging of its breeding forests have placed Marbled Murrelet on the Endangered list in the U.S.

Taxonomy The auks are currently placed in the order Charadriiformes, but their precise ancestral affinities remain unclear. Within the family, the *Synthliboramphus* murrelets are now treated as a distinct tribe by some experts, as DNA analysis has revealed a closer relationship with the Alcini group of murres and auks. Other genetic studies conducted in the 1990s split Long-billed Murrelet from Marbled Murrelet, and revealed the latter to be more closely related to Kittlitz's Murrelet (*Brachyramphus brevirostris*).

Research has also shown that the extinct Great Auk was most closely related to Razorbill (*Alca torda*), and to Dovekie, despite the latter species having evolved along a radically different path.

Colonial Breeders: Life on the Ledge

Murres and Razorbill nest in large colonies on steep sea cliffs, often sharing space with other seabirds such as gulls (family Laridae), Northern Fulmar (*Fulmarus glacialis*), and cormorants (family Phalacrocoracidae). There are subtle differences in nest-site preference: Thick-billed Murre uses narrower ledges than its close congener, Common Murre, while Razorbill prefers wider, more sheltered ledges and will nest much lower down the cliff if the colony is sufficiently remote and predator-free. Colony life is complex, and with space at a premium, nest sites are always defended vigorously by both sexes; the cliffs frequently ring with the noise of disputes between neighbors.

Common Murre (*Uria aalge*). Their tightly packed, precarious colonies give protection against predators. Scotland, U.K.

Cliff-nesters produce just one egg per year, which is rarely left unattended; the egg is pear-shaped (pyriform), which helps prevent it rolling off the ledge. The chick is fed by both parents for 15–23 days, after which time it accompanies the male out to sea, where he feeds it for another 4–8 weeks until it is fully grown.

By contrast, *Cepphus* guillemots tend to nest in smaller, much looser colonies (or occasionally even singly), using boulder-strewn islets, burrows, and crevices on low cliffs. Their hatchlings are semi-precocial, and scramble down to the sea before they can fly; they are not accompanied by either parent, and once in the water they remain independent. ■

SANDGROUSE *Pteroclididae*

Lichtenstein's Sandgrouse (*Pterocles lichtensteinii*), like all sandgrouse species, has cryptically patterned plumage that enables it to blend in perfectly with the ground on which it lives. Socotra Island, Yemen.

Found in the deserts and semi-arid parts of the Old World, sandgrouse are medium-sized, ground-loving birds. They are well adapted for living in such harsh environments: not only do they blend in perfectly with their surroundings, but they are also known for making daily and often long flights to drink. Most species are of similar size and shape, differing only in the length of their tail and in their cryptically-colored plumage; males are slightly larger than females and generally more brightly colored. Despite their name, the birds are not related to grouse; their closest relatives are the pigeons and doves.

A young male **Burchell's Sandgrouse** (*P. burchelli*) squats still, with wings ready for instant flight. Kalahari Desert, South Africa.

Structure Most species are 9–14 in. (24–35 cm.) in length, but male Tibetan Sandgrouse (*Syrrhaptes tibetanus*), Pallas's Sandgrouse (*S. paradoxus*), and Pin-tailed Sandgrouse (*Pterocles alchata*) are longer due to their elongated central tail feathers. Sandgrouse have a small head and neck, with a short, pigeonlike bill. The body is fairly large or squat. The wings are broad at the base and pointed, with 11 primaries and 17 or 18 secondaries. The legs are strong and the toes short and broad, spreading the bird's weight evenly and helping it walk across sand. The two *Syrrhaptes* species have feathered legs and toes and lack a hind toe.

Plumage The plumage forms an effective defense against the extremes of temperatures: the feathers are tightly packed, and are underlain with a thick, downy undercoat. Even the nostrils have feathers surrounding them, to keep out sand. The birds are colorful but cryptically camouflaged, and blend in well in their habitat; females are slightly duller or less colorful than males. On the upperparts, predominant colors are buff, yellow, orange, and chestnut, with a mixture of fine to broad bars, spots, and mottling. Several species have pale sandy yellow, chestnut, or pale blue faces. Most have long, prominent, crescentlike breastbands, which, in some species, are contrastingly black and sandy or white. Several species have dark chestnut to black bellies, but in two species the belly is pure white.

Behavior Generally social, sandgrouse are often seen in flocks, although several species more usually occur in pairs when breeding. Flocks of some African species usually number hundreds, but exceptionally may comprise thousands of birds. The birds spend long periods foraging out of sight on the ground among low vegetation; they usually move around at a slow shuffle or

Number of genera 2
Number of species 16
Conservation Status Least Concern
Distribution Old World: Europe, Africa, Asia

Safety in numbers: a large flock of **Namaqua Sandgrouse** (*P. namaqua*) gathers at a watering hole. While some birds drink, others can keep a look-out for predators. Kgalagadi Transfrontier Park, South Africa.

walk, head down, pecking on the ground. Fairly shy and unobtrusive, they rely greatly on their camouflaged plumage to escape detection. When disturbed, they fly up only at the last minute, taking off by springing vertically upward. Their flight is strong, swift, and waderlike, with regular wingbeats.

With the exception of Tibetan Sandgrouse, which lives in areas of plentiful water supply, most species need to spend a lot of time seeking water. They make daily flights in the early morning, dusk, or even after dark, to drink at favored waterholes. In especially dry areas, or at times of severe drought, these flights can be up to 50 miles (80 km.).

Voice Sandgrouse are usually most vocal when traveling in pairs or small flocks to drink. Their contact calls are loud and far-carrying, and comprise one to three dry, churring or guttural syllables, varying from *khattar-khattar* to a repeated *quitoo.* The alarm calls are mostly clucking notes. Pallas's Sandgrouse emit a loud whistle in flight; this sound is actually produced by the long, narrow outer primaries.

Habitat Mostly dry, arid or semi-arid savannas, desert edges, and oases, often in some of the areas with the lowest rainfall in the world. The birds also usually require some vegetation for food and occasional shelter. Several species occur in stony or rocky areas with scattered thornscrub. The endemic Madagascan Sandgrouse (*P. personatus*) lives in open savanna and lightly wooded plains. Around the Mediterranean, at least one species occurs along the edges of newly cultivated areas.

Movements Mostly sedentary, or make local nomadic movements. Some populations of Pin-tailed Sandgrouse and Black-bellied Sandgrouse (*P. orientalis*) that breed in Central Asia move south to the Middle East and into Pakistan to avoid severe winters. Pallas's Sandgrouse are also partially migratory from the northern areas for the same

A **Burchell's Sandgrouse** takes off from a water hole. Despite flying long distances to find water, sandgrouse take only a few seconds to drink. Kalahari Desert, South Africa.

reasons. This species has also seen occasional but spectacular long-distance irruptions of large numbers of birds east into northern China and west to Europe, although none have occurred in recent years.

Diet Principally seeds, occasionally buds and leaves; more rarely, the birds take insects. Most seeds are taken from a small number of preferred but usually widespread plants; for example, Namaqua Sandgrouse (*P. namaqua*) appears to feed mostly on seven species of leguminous plants that have a high protein content. Elsewhere, Black-bellied and Pin-tailed Sandgrouse feed on cultivated pulses and cereals, including oats, wheat, and barley.

Breeding The breeding season for sandgrouse usually coincides with the most abundant time of year for seeds. Most species breed monogamously in isolated pairs, or semicolonially; the birds pair for life but have no elaborate courtship displays. The nest is a shallow scrape in the ground, often in the footprint of an animal; some species may embellish their nest with small strips of grass or small stones. The clutch of two or three eggs is cryptically colored to match their surroundings, and is incubated for 20–25 days by both parents: males usually during the night, and females, being more heavily camouflaged, by day. The chicks are able to feed themselves within a day of hatching, and follow their parents when foraging. The young can fly, although poorly, from around four weeks, but they are not strong enough fliers to accompany their parents to waterholes until they are about two months old.

Survival in Arid Environments

A **Chestnut-bellied Sandgrouse** (*P. exustus*) collects water in his belly feathers. This process can take up to 20 minutes. Oman.

Living in some of the driest places on earth, most sandgrouse species need to be supremely efficient at collecting and conserving water. One of the birds' most unique features is the males' ability to collect water in their specially adapted belly feathers, like a sponge, and return with it to the chicks. Male sandgrouse perform this task every day until the chicks are able to fly to the waterholes by themselves.

Before walking slowly into the water, the adult male removes the waterproof covering of preen oil from his feathers by rubbing his belly vigorously in dry soil or sand. On entering the water, he holds his wings and tail clear and bobs his body up and down to work the water into his belly feathers. Once the feathers have become saturated, he flies back to the chicks. As he stands over them, they drink the water from a vertical groove running centrally through his plumage. When the chicks have finished drinking, the adult rubs his belly on the ground to dry his feathers again. Unlike regurgitation, which would deplete the adult male's own store of moisture, this soaking process enables him to supply his young at little or no cost to himself. ■

PIGEONS AND DOVES *Columbidae*

A **Woodpigeon** (*Columba palumbus*) takes a bath. This species is common in Europe and western Asia, and is accustomed to being around humans. U.K.

The huge, cosmopolitan Columbidae family comprises five subfamilies: Columbinae (typical pigeons), Gourinae (crowned pigeons), Treroninae (fruit doves), and the monotypic Otidiphabinae, comprising Pheasant Pigeon (*Otidiphaps nobilis*), and Didunculinae, with Tooth-billed Pigeon (*Didunculus strigirostris*).

The Columbidae are related to the flightless 'giant pigeons' of the Mascarene Islands—Dodo (*Raphus cucullatus*), Réunion Solitaire (*R. solitaria*), and Rodriguez Solitaire (*Pezophaps solitaria*)—although these birds belonged to a unique, separate family, Raphidae. All three species were driven to extinction from c.1680 to c.1800 by hunting and introduced mammals that preyed on their eggs and young.

Structure Lengths range from the sparrow-sized Plain-breasted Ground Dove (*Columbina minuta*), at 5.5 in. (14 cm.), to the bulky Southern Crowned Pigeon (*Goura scheepmakeri*), which reaches 33.5 in. (79 cm.). Sexual dimorphism is rare, although males are marginally larger than females in some species.

Pigeons and doves are characterized by a small head that bobs back and forth rapidly while the bird is walking; a short bill; short legs; a plump body; and comparatively large, broad wings. The eye is often surrounded by a bare patch of colored skin. The bill has a distinct shape in many species, with a bulky cere and bulbous tip. Highly developed flight muscles enable considerable prowess in the air. Many species have a long tail. Digestive tract adaptations reflect a predominantly frugivorous or granivorous diet.

Unlike most bird families, the columbids are able to drink without tilting their head back, instead sucking up water in a rapid, continuous flow. It is believed this ability enables them to exploit water from a wider variety of marginal sources. However, Tooth-billed Pigeon cannot drink in this way. This difference, together with the bird's thick, notched bill, are among several characteristics that separate it from all other columbids. Pheasant Pigeon of New Guinea has adapted to a forest-floor niche usually occupied by pheasants, partridges, and quails, and has evolved a similar morphology, with longer legs and tail, and a bulky, pheasantlike body.

Plumage Sexual dichromatism exists, but usually extends no further than differences in coloration on the head, neck, and breast.

Columbids display a great deal of variation in plumage, from the bold markings and rainbow colors of some tropical species, such as the bleeding-hearts (genus *Gallicolumba*), to the sober gray, buff, and brown tones of many genera in the Columbinae subfamily. Fruit doves and green pigeons are predominantly leaf-green with yellow, bronze, blue-gray, and purple present in varying degrees. Some fruit doves, such as the male Many-colored Fruit Dove (*Ptilonopus perousii*), have more dramatically contrasting coloration, with striking crimson on the forehead, breast, mantle, and undertail coverts. Some of the large imperial pigeons (genus

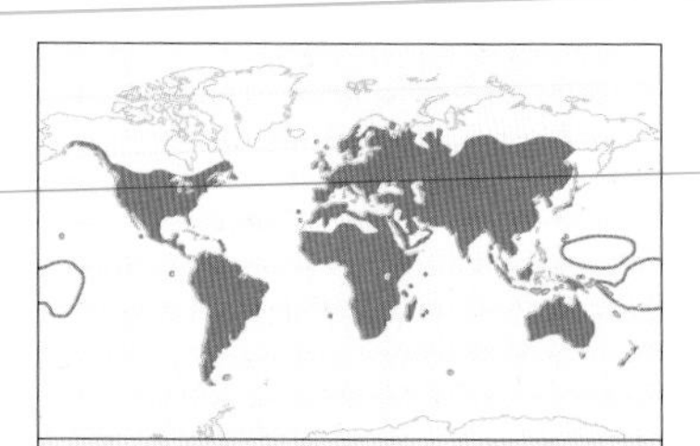

Number of genera 42
Number of species 308
Conservation Status
10 species Critically Endangered,
15 species Endangered,
34 species Vulnerable,
38 species Near Threatened
Distribution Worldwide apart from Antarctica

Crested Pigeons, with their characteristic black, spiky, erect crest, are commonly found in Australian grasslands, brush, and even gardens. Australia.

Ducula) have rich metallic-green upperparts that contrast with pale to smoky gray or wine-colored underparts. In contrast, Pied Imperial Pigeon (*D. bicolor*) is among the very few columbids to have solely black and white plumage. Pink Pigeon (*Nesoenas mayeri*) of Mauritius has a white to pale pink and buff head, neck, breast, and underparts, and a pinkish-brown mantle, with a chestnut rump and tail.

Somber-colored genera include the cuckoo-doves (genus *Macropygia*), which are predominately rufous and brown, with fine barring in a few species, and the ground doves of genera *Columbina* and *Metriopelia*. Inca Dove (*C. inca*) and Scaled Dove (*C. squammata*) show some darker barring or spotting on their wings, and reveal contrasting coloration on their wings and tail when in flight.

Uniquely, the crowned pigeons (genus *Goura*) have a striking fan of tall head-dress plumes on the crown; a similar characteristic was displayed by Choiseul Pigeon (*Microgoura meeki*) of the Solomon Islands, which is now believed extinct. Two Australian species, Crested Pigeon (*Ocyphaps lophotes*) and Spinifex Pigeon (*Geophaps plumifera*), have a slender, pointed crest, while the male Topknot Pigeon (*Lopholaimus antarcticus*) has a curious gray and ginger 'bouffant hairdo' of elongated feathers that curl upward on the crown. Nicobar Pigeon (*Caloenas nicobarica*) has long, glossy, purplish hackles that extend from nape to breast, giving it an almost rooster-like appearance.

In species with bare skin around the eyes, the skin color varies but is usually red, blue, white, or yellow.

Behavior Social behavior varies from highly gregarious, as in Flock Bronzewing (*Phaps histrionica*), which is almost never seen alone, to the shy and solitary habits of Ruddy Quail-dove (*Geotrygon montana*), which lives in dense forests. Most species are usually seen in pairs or small flocks, which can swell to thousands of birds on migration and outside the breeding season.

Feeding behavior varies between genera. Some, such as the ground doves, quail-doves, crowned pigeons, and Pheasant Pigeon, forage almost exclusively on the ground, while fruit-loving species tend to be more arboreal, foraging in the lower, middle, and upper stories of the canopy. Some of the frugivorous species can be fairly agile, particularly the *Ptilinopus* fruit doves, which often hang upside down to pluck fruit from less accessible branches. However, similar attempts by more ungainly species such as Woodpigeon (*Columba palumbus*) often end in ignominious failure.

In flight, most species are agile, powerful, and surprisingly swift, often capable of outmaneuvering predators such as falcons and accipiters. Some species have a characteristically noisy take-off when flushed, with whistling or clapping sounds produced by the flight feathers.

Voice The soporific, rhythmic, and often repetitive cooing of columbids are among the most easily identifiable of bird calls in many parts of the world. The majority of species are at their most vocal in the breeding season, when defending territory and attracting or bonding with a mate; many have additional calls used only while at the nest. Some of the most social species also have a variety of shorter calls, used to keep in contact or signal aggression or alarm while flocking. Mourning Dove (*Zenaida macroura*) has a sorrowful-sounding *oowoo-woo-woo-woo*, which gives the bird its common name. Similarly, Laughing Dove (*Streptopelia senegalensis*) takes its name from its hollow-sounding *do-do-DU-DU-do* song.

Some species produce sounds rather atypical of the family as a whole, such as the quacking and chattering calls of Pink-necked Green Pigeon (*Treron vernans*), the froglike calls of Croaking Ground Dove (*Columbina cruziana*), and the 'bongo-drum' notes of Bare-eyed Pigeon (*Columba corensis*).

Habitat Pigeons and doves occupy a wide variety of terrestrial habitats, including forests, scrub, deserts, and mountains up to 16,400 ft. (5,000 m.). A number of species have adapted to living closely with humans, some thriving even at the heart of the world's largest cities.

Movements Many doves and pigeons are sedentary, although local movements between feeding, drinking, and roosting areas can be quite extensive. Other species are nomadic, moving around in response to abundant food supplies and favorable breeding conditions. Eared Dove (*Zenaida auriculata*) forms colonies of many thousands of birds in semi-arid northeastern Brazil every two or three years, where supplies of their favored croton seed are abundant; when this foodstuff dwindles, the doves disperse to seek new territory. Similarly, Flock Bronzewing depends on supplies of grass seed and moves around Australia's arid interior in response to erratic rainfall patterns that prompt grass growth.

A **Speckled Pigeon** (*Columba guinea*) calls from a rooftop. The call, produced by expanding the throat area, is a loud *doo-doo-doo*. South Africa.

Truly migratory species often travel great distances between their breeding and nonbreeding territories. For example, each fall, Mourning Doves from the northern U.S. may fly up to 2,485 miles (4,000 km.) to their nonbreeding quarters in Central America. European Turtle Dove (*Streptopelia turtur*) is unique among columbids for migrating along a broad front: around three million birds have been estimated moving south along a single 60-mile (100-km.) stretch in Iraq during fall. Some high-elevation species conduct altitudinal migration, such as Snow Pigeon (*Columba leuconota*) in the Himalayas, which moves from its breeding territory, above 9,840 ft. (3,000 m.), down to a more hospitable 4,920–2,460 ft. (1,500–750 m.) in winter.

Yellow-legged Green Pigeon (*Treron phoenicopterus*) is a South Asian species that feeds on a variety of fruits, such as figs, balancing skillfully in tree branches as it does so. India.

One of the most spectacular of all columbids is **Victoria Crowned Pigeon** (*Goura victoria*), with its blue and maroon plumage, dark face mask, and ornate, lacy crest. New Guinea.

Migrating flocks of columbids can be an impressive sight. Famously, the now-extinct Passenger Pigeon (*Ectopistes migratorius*) was reported migrating in flocks numbering millions of birds during the 18th and 19th centuries. (Incredibly, by 1914 the species had dwindled to a single bird, which died in captivity at Cincinnatti Zoo. Combined habitat loss and overhunting are attributed to its spectacular decline—the same hazards that face many threatened columbids today.)

Diet Some species are almost exclusively frugivorous, including the fruit doves (genus *Ptilinopus*) and green pigeons (genus *Treron*). Others show a marked preference for seeds, particularly genera in the subfamily Columbinae. Many species will also supplement their diet with small invertebrates such as snails, worms, and insects. Rarely, small vertebrates such as lizards are also taken, notably by Atoll Fruit Dove (*Ptilinopus coralensis*), a resident of the Tuamotu Islands, where little other food is available. The forest-dwelling quail-doves (genus *Geotrygon*) of Central and South America often eat more invertebrates than either fruit or seed; the fledglings of some species are also fed on higher quantities of live food. Some species, particularly Woodpigeons, congregate in large, voracious feeding flocks, and are often considered to be agricultural pests.

Breeding Most species appear to be monogamous, at least for the breeding season; some may pair for life. Many columbids are highly visible in the early part of the breeding season: males perform display flights with steep ascents, gliding dives, wing-clapping, and drawn-out calls.

Nest sites vary between species and are diverse: cliffs and buildings, tree hollows and branches, low bushes, or even on the ground. Nests are often no more than a rudimentary platform of twigs and other vegetation. Clutch size is usually small, with one or two white or buff-colored eggs as the norm; granivorous species tend to lay more than their frugivorous and larger ground-nesting relatives. Both parents usually share incubation and chick-rearing duties. Incubation lasts for 11–30 days depending on species and size. Columbid fledglings, or 'squabs,' grow more rapidly than almost any other young birds, due to a diet of nutrient-rich crop-milk, which is produced by both parents and is often the sole food given to the chicks in the first few days after hatching. Fledging occurs at about 12 days in the seed-eating species, but can take as long as 22 days in some frugivorous doves.

Mourning Doves usually have two chicks per brood, but may produce up to five or six broods in a single year. Arizona, U.S.

PARROTS AND COCKATOOS *Psittacidae*

A large and diverse family that includes well-known birds such as parrots, cockatoos, macaws, and parakeets. Members of this family vary greatly in size, shape, and coloration, with many being admired for their attractive or even dazzling plumage. The cockatoos of Australia and South Pacific islands are sometimes considered to be in a separate family, but they are treated here as a subfamily of the Psittacidae; the term 'psittacids' is used below for all members of the family, including cockatoos. Many species feed primarily on fruit, form species-specific flocks, and are famous for pairing for life. Psittacids are also commonly kept in captivity because of their social behavior and ability to 'talk'—a famous example being African Gray Parrot (*Psittacus erithacus*). Many species are raised successfully in captivity, but the capture of birds from the wild has led to a large number of species being threatened or endangered; some have even become extinct.

Rainbow Lorikeet (*Trichoglossus haematodus*) is one of the most colorful psittacids. Its range extends as far south as Tasmania. Australia.

Structure Psittacids vary widely in size, from tiny parrotlets in the genus *Forpus*, no more than 4.5 in. (12 cm.) long and 1 oz. (30 g.) in weight, to Hyacinth Macaw (*Anodorhynchus hyacinthinus*), at 35–39 in. (90–100 cm.) long, and the large, ground-dwelling Kakapo (*Strigops habroptila*), weighing up to 6.5 lb. (3 kg). However, all species share certain characteristics. They are rather short-necked and large-headed. Their heavy bill has a curved, pointed upper mandible; together with the thick, muscular

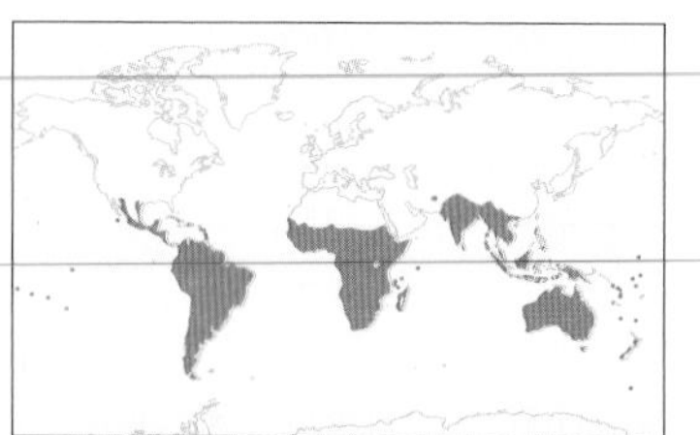

Number of genera 85
Number of species 364
Conservation Status 17 species Critically Endangered, 33 species Endangered, 44 species Vulnerable
Distribution Mostly tropical and subtropical regions, including Central and South America, Africa, India, Southeast Asia, Australasia

tongue, it is well adapted for feeding on fruits, nuts, and seeds. Some species have a tongue with brushlike structures for feeding on nectar. Psittacids also have relatively stocky legs with four thick, fleshy toes (two facing forward and two backward), adapted for holding food items while feeding, and for climbing or walking along branches.

Plumage While some parrots are mainly green in plumage, and blend in well in forested environments, others are very brightly colored. The most spectacular species include the bright Red-and-green Macaw (*Ara chloropterus*) and the all-blue Hyacinth Macaw, both found in South America; the scarlet Eclectus Parrot (*Eclectus roratus*) from New Guinea; the golden-yellow Golden Parakeet (*Guarouba guarouba*), which lives in Amazonian Brazil; the multicolored Scarlet-chested Parrot (*Neophema splendida*) of Australia; the lories (genus *Lorius*) from New Guinea; and Yellow-faced Parrot (*Poicephalus flavifrons*), which lives in Ethiopia. A number of species show bright yellow or red in the wings, which is visible only in flight.

Some parrots, as in those of genus *Amazona*, have short, blunt wings and a squared-off tail, while others, such as macaws in genus *Ara*, have very long, pointed wings, and a very long, pointed tail. Birds in one genus, *Prioniturus* of the Philippines, have a unique, racquet-shaped tail.

The cockatoos of Australia are unique in having an elaborate crest, which they can raise or lower. Their coloration varies from black to white, to the amazing Major Mitchell's Cockatoo (*Cacatua leadbeateri*), with its pink body and elaborate scarlet and white crest.

A pair of **Pacific Parrotlets** (*Forpus coelestis*) enjoys a mutual grooming session. Grooming helps strengthen pair bonds. Ecuador.

Behavior Psittacids are very social birds. They often occur in species-specific flocks, probably a behavioral adaptation for finding food items such as fruit and seeds that occur only patchily in the environment. Many species have large communal

A **Military Macaw** (*Ara militaris*) in full flight. This species flies with rapid and direct flight below the canopy. Macaws' long wings make them very powerful fliers, and the long tail gives them agility as they maneuver through the forest. Location unknown.

Feeding at Mineral Licks

In Amazonia, many species of psittacids congregate at areas of clay that are often exposed by rapid erosion along the banks of rivers and inside the forest. There, the birds swallow soil containing minerals that are hard to find elsewhere in the forest. The minerals neutralize the acids from the fruit that makes up the bulk of their diet. Such mineral 'licks' may attract hundreds of noisy, squabbling parrots and macaws. They will take to the air in an explosion of wings if a bird of prey appears above the forest canopy. Some scientists suggest that these congregations may provide additional benefits for the parrots: an opportunity to pass on and receive information about good feeding areas.

Red-and-green, Scarlet, and **Blue-and-yellow Macaws** congregate at a clay lick. Peru.

night roosts, and can be seen flying to and from these roosts at dusk and dawn. Many forage in flocks and seek out fruiting trees: the birds walk along branches, then pull off and hold fruit with one foot as they eat. In Amazonia, various species eat clay from the soil (see box, above). The hanging-parrots in genus *Loriculus* have the unique habit of hanging upside-down from tree branches.

Voice Psittacids are very vocal birds with a highly complex repertoire of calls and songs, mostly nonmusical in nature, which may be given both from perches and while in flight. They are often extremely noisy and raucous when flying, particularly the macaws. In contrast, while feeding they are often quiet and difficult to locate. Some species are excellent mimics, such as the *Amazona* parrots of South America, which have been kept in captivity because of their excellent ability to talk. African Gray Parrot is perhaps the most famous of the talking parrots, and is also well known for its intelligence.

Habitat Psittacids fill most available niches, from rain forests to deserts, and from sea level to high mountains. However, nearly one-third of all parrot species face the threat of population decline. All the main causes of recent declines in parrot populations, from habitat fragmentation and deforestation to the illegal pet trade, are connected with human activity.

Several species have estimated populations of fewer than 50 individuals and are on the verge of extinction, including Kakapo, found on offshore islands of New Zealand, and Puerto Rican Parrot (*Amazona vittata*), endemic to Puerto Rico. Many of the species most threatened around the world have tiny populations within a very limited range: for example, Yellow-eared Parrot (*Ognorhynchus icterotis*) is now found only in a small area of humid mountain forest in the Colombian Andes, where it is very dependent on wax palms for food and nesting. Spix's Macaw (*Cyanopsitta spixii*) formerly lived in caraiba forest in a tiny part of northeastern Brazil, but destruction of this habitat has led to a decline in numbers and no individuals have been seen in the wild since 2000; this species may now be extinct in the wild.

A courting pair of **Keas** (*Nestor notabilis*). The male regurgitates food for the female before the pair mates. South Island, New Zealand.

A **Major Mitchell's Cockatoo** (*Cacatua leadbeateri*), or Pink Cockatoo. This striking bird has suffered due to loss of its habitat and to trapping. Australia.

Movements Most species are resident, and although they do not 'migrate,' many make seasonal or regular daily movements. Various species travel long distances between their feeding and roosting areas; one example is Scarlet Macaw (*Ara macao*) in Costa Rica, which forages in forest reserves along the Pacific coast, yet roosts in coastal mangroves.

Diet Psittacids vary considerably in diet, eating fruit, seeds, flowers, nectar, and occasionally insects. Most species are generalized fruit-eaters, taking advantage of locally fruiting and flowering trees. Others specialize on locally abundant resources, such as Barred Parakeet (*Bolborhynchus lineola*), which specializes on seeding bamboo in the genus *Chusquea*. One species, Kea of New Zealand, has been seen eating carrion. Red-bellied Macaw (*Orthopsittaca manilata*) specializes on the fruit of *Mauritia* palms. Some species take advantage of seeding cultivated agriculture, and can be quite destructive to individual crops.

Breeding The majority of psittacids breed in isolated pairs or small colonies. Other species, such as Budgerigar (*Melopsittacus undulatus*), form enormous colonies comprising thousands of birds. Most species are cavity nesters, often hollowing out holes in the trunks of old trees and palms, or using old termite nests. Some, such as Monk Parakeet (*Myiopsitta monachus*), build a nest from sticks. Psittacids are monogamous, and many are thought to pair for life. Females lay between two and eight eggs, which are usually white. Incubation usually lasts 20–30 days, and in most species both parents take responsibility for this task. The young take four to eight weeks to fledge, depending on the size of the bird, and adults continue to feed the immature birds after fledging.

HOATZIN *Opisthocomidae*

A **Hoatzin** clambers through foliage, using its wings to balance itself. Although arboreal, the bird moves awkwardly in trees, and is a reluctant flier. Guyana.

The only member of this family, Hoatzin (*Opisthocomus hoazin*) has a prehistoric appearance, looking like a cross between a pheasant and a cuckoo. Adults have a musty smell; the bird's various local names translate to 'stink bird.'

Structure The bird is 24–27 in. (61–69 cm.) in length and similar to a pheasant in its size and shape. The head is small in proportion to the body, with a long, wispy crest, bare blue facial skin, and red eyes with mammal-like eyelashes. The bill is short, stout, laterally compressed, and decurved. As in parrots, the bill is not attached to the skull. The neck is long and slender. Hoatzin has broad, rounded wings, but is a weak flier due to the undersized keel and associated flight muscles. It has strong legs and rather large feet, with three toes in front and one behind. Due to its vegetarian diet, Hoatzin has evolved an enormous crop and a foregut superficially like that found in ruminant mammals, with large muscles designed to grind down food.

Plumage The birds have a long crest (slightly shorter in females), which is buff at the front and brown-edged at the rear. The upperparts are olive-brown, with white streaking on the neck and back, and white shoulders and wing bars. The outer nine primaries are bright rufous. The chin and breast are buff, and the belly and underwing coverts rufous. The broad, rounded tail is dark, with a buff tip. Juveniles resemble adults.

Behavior A gregarious bird, Hoatzin may be found in groups sometimes numbering 100 or more. It is most active at dawn and dusk, but also feeds at night. To escape predators, it clambers through the branches and then flies short distances. It spends a great deal of time resting on its sternum.

Voice The adults make hissing, monotonous grunts; chicks peep.

Habitat Vegetation at the edge of fresh, brackish, or salt water bays, lakes, oxbows, streams, and rivers.

Movements Sedentary.

Diet Mainly new-growth leaves and buds from a wide variety of plants; also seasonal fruit and flowers.

Breeding The bird nests in colonies during the rainy season. It is primarily monogamous, but pairs may be assisted by young from a previous season. Nests are made of sticks placed in a tree over water. Clutches usually comprise two eggs, which are incubated for about 30 days. The chicks hatch sparsely covered with down, and it takes 24 hours for their eyes to open fully. A unique claw-wing, with fully functional claws, allows the chicks to climb around in vegetation. To avoid predators, they scramble away from the nest or even drop into water. They can then swim underwater back to overhanging vegetation and climb back up to the nest. Chicks are fed on regurgitated, half-digested vegetation; they first fly at about two months, and often depend on their parents for food for a further month.

Taxonomy The family has been linked with turkeys, chickens, and relatives (order Galliformes) and with cuckoos (order Cuculiformes). Recent genetic studies support a closer relationship with the latter.

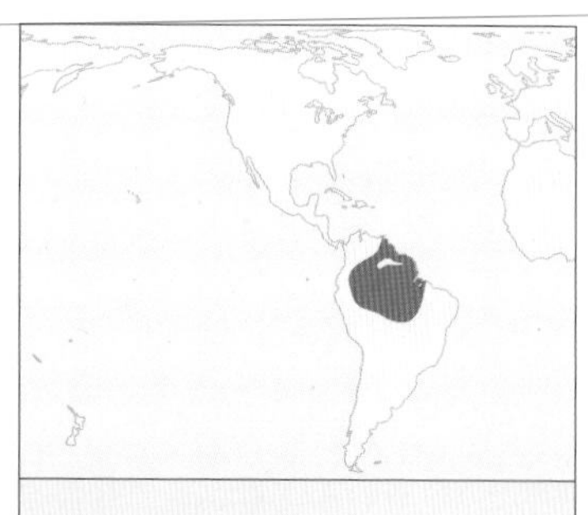

Number of genera 1
Number of species 1
Conservation Status Least Concern
Distribution South America

TURACOS *Musophagidae*

An African family of colorful, vocal birds, the turacos comprise the 'true' turacos, sometimes called louries; plantain-eaters, named for the food that they were presumed to favor (although they rarely eat plantain); and go-away birds, named for the distinctive calls that they make.

Structure Most species are 15–18 in. (38–46 cm.) in length, except for Great Blue Turaco (*Corythaeola cristata*), at 27.5–29.5 in. (70–75 cm.). They have a short, stout bill with a decurved maxilla. The wings are relatively short and rounded, and the tail is long and slightly rounded at the tip. The feet have four toes; the hind toe can be brought forward, almost touching the third toe, to help the bird perch, or can be moved backward to aid climbing.

Plumage Most of the true turacos are predominantly deep or glossy green with varying shades of blues, purple, or yellow to buff. Great Blue Turaco is mostly pale blue with yellow breast and flanks, while Ruwenzori Turaco (*Ruwenzorornis johnstoni*), Ross's Turaco (*Musophaga rossae*), and Violet Turaco (*M. violacea*) are deep purple or violet. All species except Great Blue Turaco have scarlet flight feathers, visible only when the bird is in the air. The greens and scarlet in the plumage are unusual in that they are created by copper-based pigments in the bird's body.

All species are characterized by prominent crest feathers and bright head patterns. The head markings vary from black and white to deep orange, with white lines, commas, or spots around the eye and brightly colored eye rings. Ross's and Violet Turacos have a large yellow shield on the bill, extending into a broad eye ring on Violet Turaco.

The go-away birds and plantain-eaters, by contrast, are dull gray or dowdy brown and white; Gray Go-away Bird (*Corythaixoides concolor*) is uniformly smoke-gray.

Many species, such as this **Fischer's Turaco** (*Tauraco fischeri*), have mainly green and red coloring. Kenya.

Behavior Turacos live mainly in small groups or family parties in the upper canopy of forests. They are remarkably agile birds, climbing easily through foliage and moving from branch to branch with great leaps and bounds, coming down to the ground only rarely to bathe or drink. Their flight is generally weak and limited to short distances. They move from one group of trees to another in single file, waiting for the first to land before the next one takes off. They fly with broad wing-flaps and a short glide and usually land well within the foliage, where they quickly climb up to the canopy. The go-away birds and plantain-eaters are much less agile in comparison, and spend more time in the outer foliage of trees and bushes. However, they have a stronger and more buoyant flight, and cover longer distances over open country.

Violet Turaco inhabits treetops in the humid tropical forests of West Africa. Gambia.

Voice Most turacos have a gruff, barking note and a higher-pitched hooting. Pairs sing in duet, with one partner having a higher-pitched call

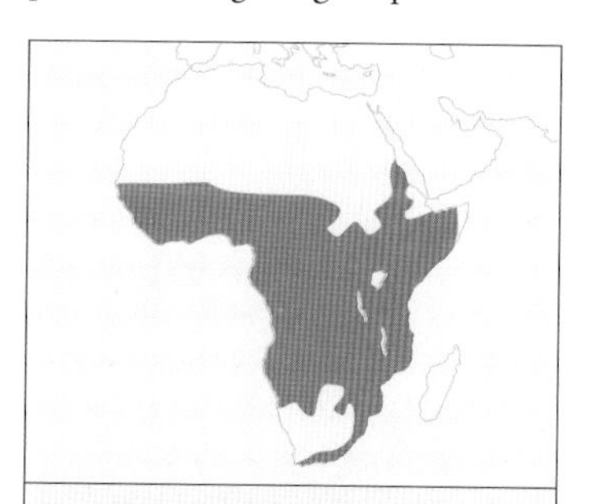

Number of genera 6
Number of species 23
Conservation Status 1 species Endangered, 1 species Vulnerable
Distribution Africa south and east of the Sahara

A **Gray Go-Away Bird** sunbathing on bare ground. Unlike true turacos, go-away birds are more widely found in relatively dry, open country. South Africa.

than the other. The call of Great Blue Turaco is a loud *kok-kok-kok*, while Ross's and Violet Turacos have a series of deep, rolling *cou-cou-cou* notes; when given by a duetting pair, these notes are repeated to become a rapid, vibrant sound. The go-away birds are so named for their *g-way* or *go-waay* call, which is repeated on a rising scale.

Habitat Turacos occur in all types of lowland, equatorial, and montane forests, including riverine and gallery forests and large or substantial woodlands. In parts of the range, they also occur in well-wooded suburban gardens. Most species of true turacos are dense rain forest dwellers but go-away birds and plantain-eaters occur in more open country. The go-away birds are found mostly in open savannas with scattered *Acacia*, thornbush, and scrub, and the plantain-eaters occur in a wide variety of open grasslands with scrub.

Movements Largely sedentary, but the birds make nomadic movements away from breeding areas in search of fruiting trees.

Diet Most turacos are mainly fruit-eaters. They prefer parasol and waterberry fruits; staple foods also include other fruiting plants with large crops, such as olives, figs, and guavas. In South Africa, Knysna Turaco (*Tauraco corythaix*) eats the fruits of Bushman's Poison, which are highly toxic to other animals. In addition to fruit, go-away birds eat *Acacia* buds, leaves, and pods; *Aloe* flowers; and *Erythrina* blossoms. Several species also eat insects, including adult and larval moths, beetles, and termites; slugs; and snails. Whether the birds forage for them or take them as they eat fruit is not well known, but Yellow-billed Turaco (*T. macrorhynchus*) joins mixed-species flocks following ant swarms.

Breeding The birds are monogamous, but some go-away birds have helpers, probably young from a previous brood, to help defend the territory and feed young. In displays, turacos raise and lower their crest, bow, and spread the tail and wings, showing off the scarlet patches. The nest is a flimsy platform of sticks: turacos build theirs high up in the foliage of trees, while go-away birds locate theirs in the upper branches of a tree or bush. True turacos lay two eggs, but go-away birds and plantain-eaters often lay three. The eggs are incubated by both parents, for 16–18 days in the smaller species or 30–31 days in the larger ones. The young are fed in the nest for several weeks; those of Great Blue Turaco continue to be fed for up to three months after leaving the nest.

Two **Purple-crested Turacos** (*T. porphyreolophus*) clamber along a branch. The scarlet flash of the flight feathers can just be seen on the wing of the nearer bird. Kenya.

CUCKOOS *Cuculidae*

A highly diverse family, the cuckoos are known chiefly for the parasitic breeding habits of many members, in which they lay eggs in another species' nest and leave the young to be reared by the host. As well as the cuckoos, the family Cuculidae includes the coucals, couas, koels, anis, malkohas, lizard-cuckoos, ground-cuckoos, and roadrunners. The family name is derived from the most commonly heard call of Common Cuckoo (*Cuculus canorus*).

Structure Cuckoos are medium to large birds, some with a long tail; species range in length from 6.5–28 in. (16–70 cm.). The bill of all species is basically the same, varying only in size: fairly short, strong or stout, and slightly decurved. The lizard-cuckoos have a slightly longer bill with a hooked tip, while the anis have a laterally compressed bill, with a high, arching maxilla in one species. Tree-dwelling cuckoos have a slender body; pointed or slightly rounded wings, with 10 primaries and 9–13 secondaries; a tail that is often longer than the body, with 10 feathers (but only eight in the anis); short legs; and feet with a reversible (zygodactyl) outer toe, with the toes usually arranged so two face forward and two back. Ground-dwelling cuckoos tend to be large, short-winged, and very long-tailed, and have long legs with flexible toes.

Squirrel Cuckoo (*Piaya cayana*) gets its name from the squirrel-like way in which it moves through trees, running along branches and leaping from one branch to another. Brazil.

Common Cuckoo. Adults are usually blue-gray, but females occasionally exhibit a rufous, or 'hepatic' morph. Hungary.

Plumage The cuckoos show a wide diversity of plumage, with colors including bright metallic and emerald green, violet-blue, and bronze, and patterns varying from bright colors to plain or dowdy and cryptic combinations. Most of the terrestrial cuckoos have a dark head, variably gray, black, and brown coloration on the upperparts, and chestnut to yellow or white underparts. Several species have prominent black and white barring and bright yellow eye rings. Some African and European cuckoos, which are very similar, have a distinctive rufous morph in which all the gray is replaced by deep rufous-brown and black barring. Several of the tropical species living in Africa, southern Asia, and Australia have bright emerald-green or bronze upperparts and a mixture of chestnut, brown, and yellow in the wings and underparts. In south-

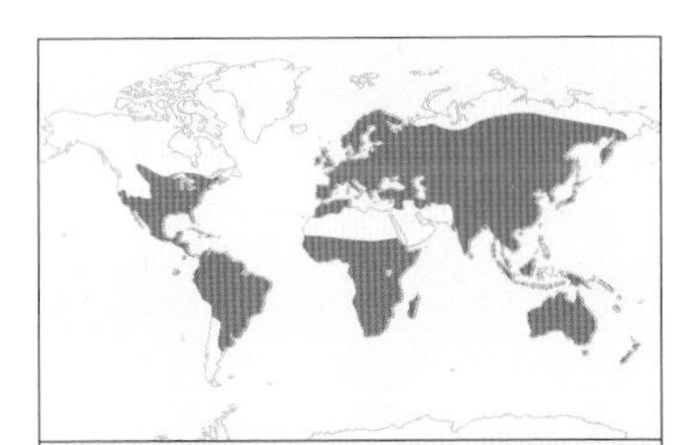

Number of genera 35
Number of species 138
Conservation Status 2 species Critically Endangered, 1 species Endangered, 6 species Vulnerable, 9 species Near Threatened
Distribution Worldwide

east Asia, the uniquely colored Violet Cuckoo (*Chrysococcyx xanthorhynchus*) is deep, glossy violet-blue on the head and upperparts.

The couas (genus *Coua*), endemic to Madagascar, range from buff-brown to olive-gray, greenish-brown, and blue, and have a patch of bare skin reaching from the bill to around the eyes. The coucals (genus *Centropus*) are mostly glossy black with chestnut or rufous wings, although two species are pale yellow on the head, neck, and underparts.

Lizard-cuckoos (genus *Saurothera*) are mostly gray and buff or chestnut, with bold black and white tips to the tail feathers. The anis (genus *Crotophaga*) of Central and South America are entirely black, often with a glossy purple, blue, green, or brown sheen on the upperparts. The two species of roadrunners (genus *Geococcyx*) are heavily striped on the head and upperparts, as is Striped Cuckoo (*Tapera naevia*). The malkohas of southern Asia include buff, brown, and gray species and those with deeper green, chestnut, and blue or purple. Most malkohas also have a bright yellow or red bill, and several have broad, pale blue eye rings or larger patches of bright red bare skin around the eyes.

White-browed Coucal (*Centropus superciliosus*) spends most of its time skulking in low shrubbery or foraging on the ground. Kenya.

The three Asiatic species of ground-cuckoos have similarly dark gray or green upperparts with pale green to bright blue patches around the eyes and a pale green to bright coral-red bill. The South American ground-cuckoos have bright metallic green, blue, or purple upperparts, a finely barred head and neck, and warm buff to white underparts.

Behavior Most species are solitary or live in pairs, but some are more frequently found in small to large flocks, while others are loosely colonial. The birds spend long periods foraging in foliage at all levels, from leaf litter on the ground to the canopy level of trees; at other times, they sit quietly or motionless in trees and bushes. Cuckoos are also known for their sunbathing habits, often perching on a prominent branch in the early morning sun, or warming themselves after a shower before they can feed; sunbathing is widely recorded in the glossy-plumaged tropical cuckoos, as well as in the anis, couas, and roadrunners.

All species fly, although some of the ground-cuckoos do so rarely or infrequently. The method and speed of flight varies from the flapping and gliding flight of the round-winged malkohas and coucals, to the fast and direct flight of the long-distance migrants, with their longer and more pointed wings.

Voice The most widely known call is the distinctive *cuck-oo* of Common Cuckoo. Several other species are also known from their calls: Diederik Cuckoo (*Chrysococcyx caprius*) has a rising and accelerating *dee dee dee diederik*; koels have a rapid, rising *koel, ko-el, ko-el*; and couas give a loud or resonant *kookookoo* or *koo ha*. Other cuckoos' calls and sounds include hisses, whistles, and screams to more guttural grunts and snores; roadrunners have a low-pitched hooting note like that of an owl.

Habitat Mostly forests and woodlands, including tropical rain forests, secondary woodlands, plantations, and mangroves. Couas live in dry country and semi-arid scrub, and roadrunners in hot, dry deserts and desert edges. Northerly breeding species are found in a wide range of open habitats including heathland, the margins of agricultural land, city parks, and large gardens—anywhere that has food and host species for breeding.

Different Forms of Brood-parasitism

Cuckoos are renowned for their parasitic habit of laying their eggs in the nests of other birds and having nothing more to do with the raising of their young. However, while this is true for many species, including the best-known Old World species, Common Cuckoo (*Cuculus canorus*), it is certainly not the case for all members of the family. A little under half of the family—about 50 Old World and three New World species—are brood-parasitic, but the rest live in pairs, build nests, and raise their own young.

A **Common Cuckoo** chick begs for food from its diminutive foster-parent, a Eurasian Reed Warbler (*Acrocephalus scirpaceus*). U.K.

Brood-parasitism is not restricted to cuckoos: a number of other families, including the New World cowbirds, honeyguides, and Black-headed Duck (*Heteronetta atricapilla*) of South America, use foster parents for raising their young. Those cuckoos that are parasitic, however, are perhaps better adapted than any other bird to this system of reproduction.

Female cuckoos spend long periods watching potential host species nest-building. For example, a female Common Cuckoo will select a particular host species, most frequently a warbler, pipit, or Dunnock (*Prunella modularis*), and visit the nest when the host is away to check on progress. She lays her egg once the host bird has laid one of her own. If there are several eggs in the nest, she may even remove and eat one of the host's eggs. Striped Cuckoo (*Tapera naevia*) selects host species such as ovenbirds or spinetails, with dome-shaped, covered nests. In Australia and Indonesia, Channel-billed Cuckoo (*Scythrops novaehollandiae*), the largest of all the brood-parasitic species, selects butcherbirds and crows as host species, and occasionally more than one cuckoo fledges from a nest.

The cuckoo's egg closely matches those of the hosts in size and pattern (which, again, may indicate an adaptation to a particular host species). It is usually accepted by the hosts and incubated along with their own eggs, but in cases where the hosts recognize a different egg, it is rejected. The nestling cuckoo, in turn, has an instinctive urge, from only a day or so old, to remove all other eggs or young from the nest. This action is vital for the cuckoo's survival, because it will need all the food that its foster parents can supply. With a seemingly insatiable appetite, it requires constant feeding and grows rapidly, often dwarfing the unfortunate host species. Perhaps surprisingly, the foster parents seem unable or unwilling to recognize the young cuckoo as anything other than their own. As the nestling grows, it becomes too large for the nest and calls loudly and persistently from any nearby perch; by now the foster parents may have to perch on its back to feed it.

The most successful cuckoos are those that select the hosts that will not only accept the eggs and rear the youngster but will also provide it with the appropriate food. Most host species are insectivorous, as are cuckoos, but several species that breed in the tropics parasitize host species that are predominantly fruit- and seed-eaters, although the foster parents will provide some insects. ■

The green and white **Klaas's Cuckoo** (*Chrysococcyx klaas*) is one of the smallest but most striking species in the family. Ethiopia.

Movements Most of the tropical species, including the coucals, couas, malkohas, ground-cuckoos, and roadrunners, are entirely sedentary. However, in Africa and parts of southern Asia, several species move with the seasonal rainfall; these birds are sometimes called rainbirds because they arrive and start calling at the onset of the rainy season.

Certain species are long-distance migrants. Common Cuckoo breeds in northern Eurasia and winters in southern Asia and South Africa; the longest-distance migrants travel at least 1,900 miles (3,000 km.). Yellow-billed Cuckoo (*Coccyzus americanus*) breeds in North America, as far north as southern Quebec, and winters as far south as northern Argentina. Lesser Cuckoos (*Cuculus poliocephalus*) breed from Japan through China to Afghanistan, and winter in India, but a considerable number continue across the Indian Ocean to winter in East Africa. Channel-billed Cuckoos (*Scythrops novaehollandiae*), which breed in north and east Australia, move north to winter in New Guinea.

Diet Cuckoos feed almost entirely on insects, such as large cicadas, locusts, beetles, and stick insects. They specialize on caterpillars, which they often eat in great quantities: several species regularly eat the hairy caterpillars of certain moths that are considered poisonous or are avoided by most other birds. Cuckoos also take centipedes and spiders, worms, snails, crabs, and tree-frogs. Some species, including the coucals, regularly feed on other birds' eggs and nestlings. The lizard-cuckoos, roadrunners, and some of the couas eat lizards; the first two also take scorpions, small snakes (including rattlesnakes), small rabbits, birds, and mice. Some cuckoos take fruit including figs, tamarinds, berries, and prickly pear and cactus fruits in the desert-living species.

Breeding Many cuckoos practice brood parasitism (see box on previous page). Those that do not include malkohas, couas, coucals, ground-cuckoos, and most of the New World cuckoos. Anis and Guira Cuckoo (*Guira guira*) are the most social, living in noisy groups that defend a large territory and communal nest in which two or more females lay eggs. Female African Black Coucal (*Centropus grillii*) is polygamous and mates with up to three males, each of which holds a separate territory, builds the nest, incubates the eggs, and cares for the young. Most nonparasitic cuckoos build shallow platform nests, although coucals build large, dome-shaped nests with a side entrance, while roadrunners' nests are large, untidy bundles of twigs, bones, snake skin, and scraps of paper. Clutches comprise two to five eggs, incubated for 11–16 days. The young fledge in 16–24 days.

A young **Channel-billed Cuckoo** begs its foster-parent for food. Unlike other parasitic cuckoos, the young of this species do not eject other eggs or nestlings, but instead monopolize the food supply, starving the other nestlings. Australia.

BARN OWLS *Tytonidae*

A family of medium-sized owls distributed worldwide. Barn owls are characterized by their large face, soft plumage, silent flight, and eerie cries. These features, together with their habit of nesting in old buildings, including churches and barns—especially in the case of Common Barn Owl (*Tyto alba*)—has given rise to tales of such places being haunted.

Structure Barn owls are 13–15.5 in. (33–39 cm.) in length, with a large head. Most species have an oval, rounded, or heart-shaped face, although Oriental Bay Owl (*Phodilus badius*) has a squarish face. The stiff feathers around the sides of the face act to amplify sound and reflect it to the ears. The body tapers to a fairly short, square-tipped tail. The wings are broad and rounded, giving silent, buoyant, stiff-winged flight. The legs are long, with strong feet and long, sharp talons.

Plumage Light and very soft, the plumage is often cryptically marked with subtle tones or bars, providing camouflage. The face may be white to cream, buff, or grayish, and is usually defined by a narrow, dark border. The head and upperparts vary from golden to rufous-brown, gray, or blackish; Greater Sooty Owl (*T. tenebricosa*) is almost entirely sooty-black. All species are spotted with white on the upperparts and black on the underparts.

Behavior The birds are mostly nocturnal and can hunt in total darkness, although several species are active by day, especially when feeding young. Barn owls may be solitary, or may occur in pairs that hold a specific territory throughout the year; territory size is largely determined by location, habitat, and the availability of prey.

A female **Common Barn Owl** with her chicks. Barn owls nest in holes or cavities; they use very little material to line their nests. U.K.

Voice Mostly various screeches, screams, hisses, or sibilant whistles. Breeding birds make a wider range of cackling, croaking, and chuckling noises, as well as grating rasps, snores, twitters, and wheezes.

Habitat Forests, including tropical and temperate forest, old-growth, and plantations; swamps; woodlands; and open edges of cultivated land. In Europe, Common Barn Owl occurs in farmland with scattered trees and unoccupied buildings.

Movements Mostly sedentary, although the young disperse after fledging, usually less than 12 miles (20 km.) from the nest. Tropical-breeding Common Barn Owls in Africa move farther: ringed birds have been found more than 625 miles (1,000 km.) away. In Australia, inland-breeding Eastern Grass Owls (*T. capensis longimembris*) become nomadic outside the breeding season or irruptive during rodent plagues.

Diet Mostly small mammals, supplemented, when seasonally available, by small birds, frogs, and reptiles. Also take large insects in flight.

Breeding The nests are usually located in holes in trees; Common Barn Owl may also nest in buildings, and occasionally in nest-boxes or on the ground. Clutches number between four and seven eggs, and are incubated by the female for 27–34 days. The young fledge at between six and seven weeks. In years when the food supply is poor, the youngest chicks in the nest are often eaten by their older siblings.

Number of genera 2
Number of species 15
Conservation Status 1 species Endangered, 4 species Vulnerable
Distribution Australia, South America, large parts of North America, Africa, and Europe; also, much of southern Asia

OWLS *Strigidae*

Owls, like this **Eurasian Eagle-Owl** (*Bubo bubo*), are instantly identifiable by their huge, brightly colored eyes and, in most species, the facial disk. Location unknown.

Owls are a large group of predatory birds, most of which are nocturnal and solitary. There is much variation in size between species, but all have the same basic structure. Among the most familiar of all birds, owls have a rounded facial disk—a clearly defined circle of feathers around each eye—framed by a narrow ruff of feathers. The disk acts as a funnel to maximize sound-gathering, and so enhances the birds' ability to detect prey by sound. Owls' large eyes give acute vision in poor light.

Structure Sizes vary markedly. The smallest species of owl is the diminutive Elf Owl (*Micrathene whitneyi*) of the southwestern U.S. and Mexico, at just 5.1–5.5 in. (13–14 cm.) long. The largest are the eagle-owls (genus *Bubo*), such as Eurasian Eagle-Owl, the female of which may reach up to 29.5 in. (75 cm.) long, with a wingspan of about 6.6 ft. (2 m.), and weigh about 10 lb. (4.5 kg.). Females are larger than males in some species, and polymorphism occurs in species that have a widespread range.

Owls are bulky birds with a large, rounded head, large, forward-facing eyes, a hooked, raptorlike beak, and powerful legs and feet equipped with razor-sharp talons. They are skilled in flight, and the majority have broad, rounded wings. Most owls have a compact, comparatively short-tailed appearance, but a few species, such as Northern Hawk Owl (*Surnia ulula*), have a longer tail that permits increased agility in flight through forest.

Plumage Most owls are brown, rufous, or buff, and they are often heavily marked with cryptic barring, spots, or streaks that provide camouflage. Snowy Owl (*Nyctea scandiaca*) is a notable exception: males are almost pure white, while

Number of genera 27
Number of species 180
Conservation Status
6 species Critically Endangered,
8 species Endangered,
12 species Vulnerable,
24 species Near Threatened
Distribution Worldwide except Antarctica

Great Horned Owl (*Bubo virginianus*) is a solitary species that occurs in a wide range of habitats, from subarctic tundra to tropical rain forest. U.S.

females and juveniles are white with narrow blackish barring over the head and body.

Ear tufts are a feature of several species, including Great Horned Owl (*B. virginianus*) and Northern Long-eared Owl (*Asio otus*). These feathery extensions have no hearing function, but may actually be used to signal the bird's mood. The odd-looking, monotypic Crested Owl (*Lophostrix cristata*) has long white 'eyebrow' feathers that extend above the head, giving the bird a horned appearance. Some pygmy owls and owlets have false 'eye spots'—dark feather patches on the nape, which create the impression that the back of the owl's head is actually the face. This feature may ward off potential aggressors approaching from the rear while the bird is roosting or feeding.

As hunters relying on stealth and surprise, most owls have a highly effective plumage adaptation: serrations on the leading edge of the flight feathers, which muffle the sound of the wings and enable the birds to fly in virtual silence.

Behavior Most owls are active either at twilight or after dark, although a few species are diurnal, such as Short-eared Owl (*Asio flammeus*) and Burrowing Owl (*Athene cunicularia*). Nocturnal species spend daylight hours roosting in cover. However, they may be forced to move if detected by other birds; these other species join forces to mob the predator with constant scolding alarm calls, displays, and occasionally physical attacks. Such noisy harangues are often the only clues that an owl may be perched nearby, concealed in foliage.

Owls are highly skilled predators that use their talons to seize prey. The prey is then either swallowed whole, if it is small enough, or torn apart with the bill and talons. The indigestible parts, such as fur, bones, and shells, are regurgitated in the form of pellets.

Owls are generally solitary or occur in pairs. In contrast, a few species, including Marsh Owl (*Asio capensis*) and Northern Long-eared Owl, form communal roosts outside the breeding season.

Voice Owls rely heavily on vocal communication, particularly species that are nocturnal and live in dense forest where visual communication is difficult. Calls are used primarily to advertise territorial boundaries, attract and pair-bond with a mate, or advertise a nest site. Larger species produce booming hoots and barks that may carry for several miles, while smaller species give higher-pitched hoots, whistles, peeps, and rhythmic trills that often sound similar to insects or frogs.

Paired birds often duet, particularly at the start of the breeding season when owls are at their most vocal. In some species the calls are radically different, but the response calls of the female Eurasian Scops Owl (*Otus scops*) are so similar to those of her mate, and synchronized so closely, that they appear to come from a single bird.

The often eerie sounds made by owls have contributed a great deal to the birds' place in folklore. The calls of numerous species have been considered bad luck omens. Today, the calls of Tawny Owl (*Strix aluco*) are

Ferruginous Pygmy Owl (*Glaucidium brasilianum*), one of the smallest of all owl species, occurs in open woodland and often hunts by day. Mexico.

A year-old **Snowy Owl**. Immatures have blackish scalloping, spots, or bars, but grow paler with age. Females, however, retain more dark markings than do males. Canada.

used as a classic method to create suspense on film and television soundtracks: males produce a 'typical' resonant hooting, and females give a high-pitched *kee-wick!* call.

Habitat Owls occupy most terrestrial habitats, including Arctic tundra north of the treeline, elevations up to 15,420 ft. (4,700 m.), and islands. A number of limited-range rain forest species have only recently been described, and remote regions may still hold species as yet unknown to science. Forest Owlet (*Heteroglaux blewittii*) was rediscovered in India only in 1997, more than 100 years since the last sighting.

Movements Most species are sedentary and tend to remain territorial year-round. However, owls with more northern populations are often nomadic, partially migratory, or (in the case of insectivorous species) fully migratory. One such species, Eurasian Scops Owl, breeds in Europe and Asia but migrates south to winter in Africa, India, and the Middle East, returning in the spring when insects are once more readily available. Similarly, Flammulated Owl (*Otus flammeolus*) breeds in the west of Canada and the U.S., but winters in Central America.

Great Gray Owls often sit on a perch such as a tree branch or fence post to watch for prey, and then silently pounce. Alternatively, they may hunt by making short flights, low to the ground. Sweden.

Nomadic and irruptive species of the boreal zone, such as Great Gray Owl (*Strix nebulosa*), are far less predictable. Their movements are linked to cycles of abundance of key prey species such as lemmings and voles, but are complex and highly variable in size and extent. If prey are numerous, owls may appear south of their usual range and even stay to breed, yet will have gone by the following season. In eastern North America, Great Gray Owls have occasionally been reported as far south as Massachusetts and Connecticut.

Burrowing Owls are capable of digging their own tunnels for nesting in, but usually prefer to take over burrows abandoned by animals such as paririe dogs.

Diet Owls prey on a wide variety of vertebrates, such as rodents, birds, reptiles, and amphibians, as well as invertebrates such as earthworms and insects. Smaller species such as scops owls (genus *Otus*) tend to specialize on insects such as crickets and beetles, while pygmy-owls (genus *Glaucidium*) are fiercely efficient predators of small birds. The larger species, including the eagle-owls, are capable of handling much bigger prey, and have been recorded taking rattlesnakes, hares, foxes, cats, and even small deer. The fish owls (genus *Scotopelia*) are piscivorous, but also eat other aquatic vertebrates; Pel's Fishing Owl (*S. peli*) is capable of seizing fish of up to 4.5 lb. (2 kg.). The presence of particular species can often be ascertained from the pellets of undigested food (see above) that they leave behind.

Breeding Owls are usually monogamous, and pair-bonds are often long-lasting. They do not build their own nests; some species prefer to adopt an old crow's, hawk's, or even stork's nest, while others use tree hollows, cavities, or caves. Some owls also take advantage of barns, derelict buildings, and nest boxes. A few species are ground-nesters. Burrowing Owl, as its name suggests, uses underground burrows made by mammals such as prairie-dogs.

Clutch sizes vary from two to around 12 eggs; smaller species tend to have smaller clutches. Egg-laying occurs at intervals of one to three days, so hatching is staggered, hence an often marked variation in nestling size. This strategy is believed to improve the survival chances of older, stronger chicks if food becomes scarce. Cannibalism among chicks at such times has been observed in a number of species.

The incubation period varies from 22 days in smaller species to around 35 days in larger species. The female performs all incubation and brooding duties, and is fed by the male until the chicks are large enough for her to leave the nest for short intervals. In some species, the young emerge from the nest while still downy and unable to fly—a strategy that reduces the risk of predation. Several of the larger species can be highly aggressive when protecting their young, and may attack intruders that approach too closely. The female Ural Owl (*Strix uralensis*) is well known for this behavior; the Swedish name for the species is *Slaguggla*, meaning 'punch owl.' The offspring of larger owls may remain dependent on their parents for up to five months after leaving the nest.

The defensive threat display of a young **Long-eared Owl** (*Asio otus*). Older nestlings typically respond this way when disturbed at the nest. Germany.

FROGMOUTHS *Podargidae*

Named for their extremely wide mouth, the frogmouths are an ancient family of secretive, nocturnal birds. Like owls, nightjars, and potoos, they have cryptic plumage that gives ideal camouflage in their forest habitat.

Structure Frogmouths vary in length from 9 to 21 in. (24–55 cm.); Australasian species (genus *Podargus*) are larger, while Asian species (*Batrachostomus*) are smaller. All frogmouths have a large head and eyes; a wide bill with a short, hooked tip; broad, rounded wings; a long tail; and relatively small legs and feet.

Plumage Mostly a mixture of browns to reddish-browns and grays, with fine, dark stripes and bars. This patterning creates a cryptic mottled effect, which helps them blend in with the tree trunks on which they roost in daytime. The paler areas of the plumage often perfectly echo the color of bark or lichens. All species have short but prominent bristles on the forehead and ear-coverts. The function of these bristles is unclear, but they may help in detecting prey.

Tawny Frogmouth is the species most familiar to human viewers. Australia.

All species, including this **Marbled Frogmouth** (*P. ocellatus*), have a very wide gape. Papua New Guinea.

Behavior The birds are entirely nocturnal, and, apart from Tawny Frogmouth (*P. strigoides*), their behavior remains little known. Most roost alone. Their preferred roosting site is a high branch, where they sit upright or at an angle, to look like a broken stump and blend in with the surrounding branches. If they become aware of a predator, they freeze into an upright position with the head pointing skyward and the eyes narrowed to fine slits, watching the approaching danger, and take flight only as a last resort. They also appear to bathe during showers or downpours of rain, fluffing out their plumage, spreading their wings and tail, and swaying from side to side. Afterward, they dry themselves with their wings and tail spread in a similar fashion to cormorants.

Voice Frogmouths are often most easily located by their calls or song. Those species whose vocalizations are known have a soft hooting or booming voice, with a series of variably pitched phrases often repeated. In several species, duets have been heard between paired birds.

Habitat Evergreen tropical forests. Overlapping species are separated by habitat niches either within or at the edge of forests. Tawny Frogmouth is widespread throughout Australia: it occurs in dry woodlands, eucalyptus groves, and partly open country, and is also seen in suburban areas.

Movements Entirely sedentary.

Diet Frogmouths are adapted to feed on fairly large, tough, hard-shelled insects, such as beetles and cicadas, as well as small animals.

Breeding Paired birds occupy their breeding territory year-round. The nests are cup-shaped and situated on tree branches. Asian species build small nests lined with down, and lay one egg; Australasian birds build bulkier nests from twigs, and lay one to three eggs. The female usually incubates the eggs, and both parents feed the young. Chicks remain in the nest until they are able to fly.

Taxonomy In 2007, a new, monotypic genus was discovered in the Solomon Islands. Named Solomon Islands Frogmouth (*Rigidipenna inexpectata*), this bird occurs on Bougainville, Isabel, and Guadalcanal Islands.

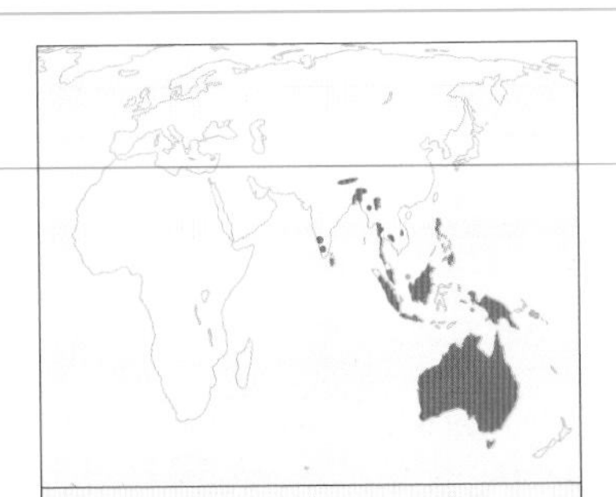

Number of genera 3
Number of species 13
Conservation Status 5 species Near Threatened
Distribution South and southeast Asia; Australasia

OILBIRD *Steatornithidae*

An **Oilbird** with chicks. Hatching is asynchronous, so the chicks are different sizes. While on the nest, they grow exceptionally fat. Aripo Caves, Trinidad.

The sole member of this family is Oilbird (*Steatornis caripensis*). The name 'oilbird' refers to the traditional practice of some indigenous peoples in taking nestlings from nest sites, and rendering their fat to produce oil for cooking and lamps.

Structure The bird is 16–19 in. (40–49 cm.) in length; males are slightly larger than females. It has large brown eyes, giving good night vision. The bill is hooked, like that of a raptor, but broad-based, with a wide gape, and edged with sensitive rictal bristles. The wings are large in proportion to the slender body; their size, and the deeply slotted primaries, are thought to enable birds to carry heavy loads of fruit from foraging sites back to nestlings. The long, graduated tail provides maneuverability when approaching and hovering during foraging bouts. Oilbird has short legs, set forward on the body, and long anisodactyl toes (three pointing forward), causing the bird to perch awkwardly.

Plumage Similar for both sexes: chestnut with black-edged white spots, which are largest on the wings and tail. It has been suggested that these spots help the birds see each other and avoid collisions when in caves, or during nocturnal foraging bouts. A single, protracted annual molt occurs after breeding. Chicks hatch with sparse gray down, replaced by a second dense coat of brown down at three weeks; adult-like feathers emerge at one month.

Behavior Oilbird is nocturnal and highly social. It roosts by day in caves, and exits at dusk to feed in the forest, traveling as far as 93 miles (150 km.). It has excellent night vision and a highly developed sense of smell, which helps it locate food. The bird's unique echolocation system, in which it emits clicking sounds at 7,000 cycles per second, allows it to navigate in the complete darkness of the caves.

Voice Loud, harsh calls and shrieks if disturbed in caves. The echolocation clicks can also be heard in caves. Outside, the calls are less diverse, mainly comprising clucking notes.

Habitat Tropical and subtropical primary forest, at elevations from sea level to 11,150 ft. (3400 m.). Oilbird requires caves for roosting and nesting: large caves can support colonies with thousands of birds.

Movements Most colonies disperse following breeding; movements and behavior at this time are largely unknown. Some populations are more sedentary, roosting in nesting caves outside the breeding season.

Diet Ripe palm and aromatic fruits up to 2.3 in. (6 cm.), usually taken from the forest canopy. Oilbird hovers as it plucks the fruit. The wide gape allows the bird to swallow fruit whole; the pits are later regurgitated.

Breeding The species is monogamous, and pairs share parenting duties. Colonies synchronize breeding, with the onset near the end of the dry season. The nest is a shallow bowl high on a ledge, often far from the cave entrance. It is made of fruit pits, pulp, and excrement, glued together with saliva. Nests are reused each year and increase in height through time. The bird lays a single clutch of one to three eggs. Incubation and fledging take four to five months. The extent of post-fledging parental care is unknown.

Taxonomy Oilbird's closest relatives are probably the potoos (family Nyctibiidae); it is more distantly related to the nightjars (Caprimulgidae) and frogmouths (Podargidae).

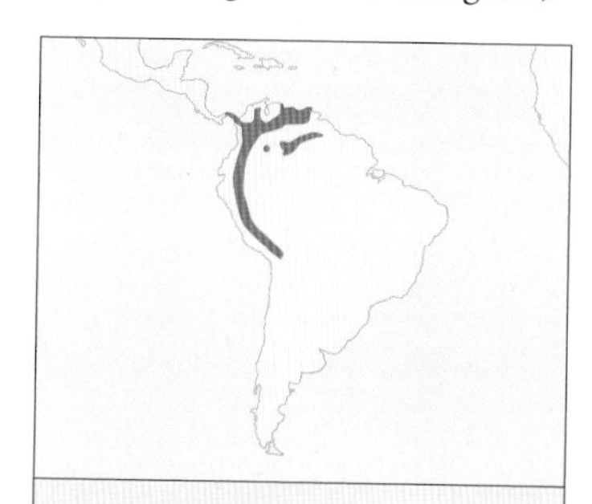

Number of genera 1
Number of species 1
Conservation Status Least Concern
Distribution Panama, northern South America

POTOOS *Nyctibiidae*

A small family of large-eyed, nocturnal, forest-dwelling birds of the Neotropics. Potoos are adept at concealing themselves from predators, as well as being agile insect-hunters.

Structure Potoos vary in length from 8 to 21.5 in. (21–55 cm.). They have a small, hooked bill, which belies a cavernous gape and mouth. The loral feathers are long and decurved, with bristlelike tips, but the birds have no rictal bristles. The eyes are huge, with unique small folds on the upper eyelid, which probably allow the birds to detect movement even with their eyes closed. Potoos have long wings and a long tail, making them highly maneuverable in flight. Their legs are very short, but the feet have strong toes and decurved claws, enabling the birds to perch easily.

Plumage The sexes are alike. The plumage is cryptically colored, with intricate patterns of gray, brown, black, russet, and white, providing excellent camouflage. There is a great deal of variation between individuals. Distinctive juvenal plumage has been described for some species.

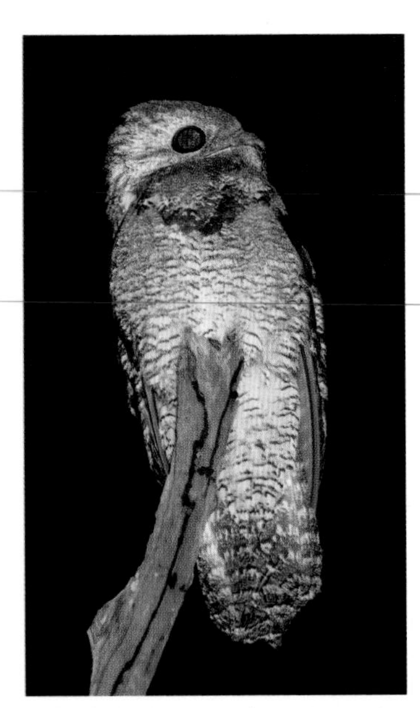

A **Great Potoo** (*Nyctibius grandis*) watches for prey. Potoos have soft, lax feathers, enabling silent flight. Brazil.

Behavior Potoos are nocturnal, solitary birds, whose habits are poorly known in general. Most species are forest canopy dwellers, using perches to roost by day and from which to make short foraging sallies by night. During the day, potoos usually stay totally still, assuming a stretched, upright posture with the bill pointed upward, so that they appear to be an extension of their perch. However, when undisturbed or actively foraging, the plumage is more fluffed and the head held normally.

Voice Potoos sing at night, from exposed perches. The voice varies between species, ranging from plaintive, clear, multi-noted whistles to grunting croaks. The chicks' begging calls have been described as buzzing notes.

Habitat A variety of forested tropical lowland areas; one species occurs in Andean cloud forest at altitudes up to 9,200 ft. (2,800 m.).

Movements Sedentary.

Diet Large insects, ranging from flying termites to large moths, which are usually caught in flight or gleaned from vegetation.

Breeding Potoos are monogamous and share parental duties. They do not build a nest; instead, they lay a single egg, which is precariously balanced in a crevice on top of a bare horizontal branch. Incubation takes about one month. One parent attends the egg or chick by day, and both take turns brooding or feeding at night. The chick is covered in down at first, and takes about two months to fledge; during this time, it remains on the branch. The nest site may be reused. Breeding periods are not well known, although Common Potoo (*Nyctibius griseus*) has been recorded breeding in all months of the year.

A **Common Potoo** in characteristic roosting posture. Brazil.

Number of genera 1
Number of species 7
Conservation Status Least Concern
Distribution Tropical areas of Central and South America; Caribbean

NIGHTHAWKS AND NIGHTJARS *Caprimulgidae*

A largely nocturnal family, nightjars and nighthawks look a little like owls, with a large head and eyes, and cryptic plumage. The family name, Caprimulgidae, means 'goatsuckers.' This may relate to their wide mouths, and a superstition that the birds suckled goats. The family is divided into two subfamilies: Chordeilinae (nighthawks: four genera and 10 species) and Caprimulginae (nightjars: 11 genera and 79 species).

A **Common Nighthawk** (*Chordeiles minor*) drinks on the wing, using the characteristic technique of scooping up water with the lower mandible. Texas, U.S.

Structure Length varies from 6 to 15.5 in. (15–40 cm.). The bill is tiny but the gape is very wide, revealing a cavernous mouth; the lower jaw is specially adapted to open both vertically and horizontally. The mouth not only engulfs prey efficiently but also allows birds living in hot climates, or sitting in direct sunlight, to dissipate heat by opening and fluttering the throat area. Nightjars (except genus *Eurostopodus*) have long rictal bristles, but nighthawks do not. Both subfamilies have very large eyes, adapted for low light conditions. The eyes have a tapetum, a reflective membrane that increases the amount of light entering the eyeball; its presence causes reflective 'eye-shine' when the eyes are illuminated by artificial light. The birds are primarily aerial, but are well adapted to perch and even walk short distances. Most species have long wings and a long tail, with a large wing ratio to body weight, giving them exceptional maneuverability so that they can capture insects in midair. The wings are pointed in nighthawks, and more rounded in nightjars. In some species, the males have extremely long inner primaries or outer rectrices, used in courtship displays. The legs are very short. The toes and decurved claws are also short; the middle toe has a longer pectinate (comb-like) claw, used to preen the feathers.

Plumage The body feathers are soft and cryptically colored, usually with variegated patterns of gray, brown, black, rust, and white, to camouflage the birds. Most species exhibit sexual dimorphism, involving minor differences in throat coloration; tail coloration, shape, and length; and wing pattern and structure. Most also have one annual (usually postbreeding) molt. Chicks have pale, downy plumage, later replaced by a distinctive juvenal plumage.

Behavior Most species are nocturnal or active at dusk, and are solitary, secretive, and retiring. They concentrate their foraging bouts during twilight hours. By day, they roost on exposed ground or rocks, in leaf litter, or on branches. When roosting, they adopt a horizontal posture, in contrast to owls. A few species are more diurnal; these also tend to be more gregarious and colonial, especially some nighthawks. The foraging height varies by species, from ground level to 575 ft. (175 m.). Nightjars sally from the ground or

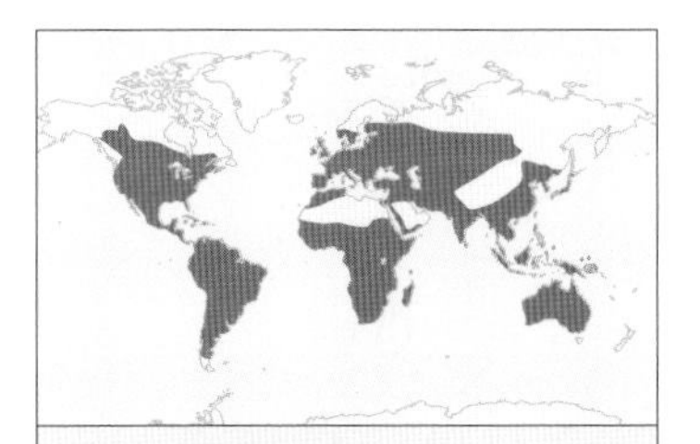

Number of genera 16
Number of species 89
Conservation Status
2 species Critically Endangered, 2 species Endangered, 3 species Vulnerable
Distribution Nightjars found worldwide except for polar regions; nighthawks restricted to New World

elevated perches and capture aerial prey with their mouth, swallowing it whole. A few species capture prey on the ground, gleaning it from vegetation. Some may hunt on foot. Most nighthawks drink on the wing, whereas most nightjars obtain liquid from their prey. Other aspects of behavior, in many species, remain poorly known. One, Common Poorwill (*Phalaenoptilus nuttallii*), tolerates cold temperatures by lowering its metabolism to achieve torpor.

Voice In many species, males utter a variety of simple to complex whistled notes. The songs of other species have been described as purrs, twitters, trills, or knocking or bubbling sounds. The birds primarily sing at dusk and dawn, from the ground, an elevated perch, or even during flight. Contact calls are given by both sexes; usually single or double-noted, they have been described as whistling, croaking, or chattering. Birds at the nest produce arrays of other vocalizations, including hissing to distract predators. Chicks make soft peeping notes. The males of a few species also make clapping or booming sounds with their wings during courtship; the females of some species may also make wing claps during takeoff. Otherwise, nightjars and nighthawks have fairly silent flight.

Habitat Members of the family occur at altitudes from sea level to 13,800 ft. (4,200 m), in a wide variety of habitats including barren, arid deserts, humid tropical forests, and mountain meadows.

Movements The species vary from sedentary to highly migratory; their movements are largely driven by the availability of prey. Short-distance and long-distance migrants move

A **Fiery-necked Nightjar** (*C. pectoralis*) rests on the ground. The bird's plumage helps to camouflage it and make it less visible to predators. Zambia.

Displays and Plumage: Extraordinary Tail Feathers and Wing Patches

Despite being so shy for much of the time, male nightjars are highly conspicuous in the breeding season, when they perform striking territorial and courtship displays in which they show off their plumage. Many species have conspicuous white patches on the throat, wings, or tail, which they flash. A few have evolved specialized plumage features, which they use to augment their displays. Five South American species in the genera *Hydropsalis*, *Uropsalis*, and *Macropsalis* have elongated outermost tail feathers. From the ground, the male Scissor-tailed Nightjar (*H. torquata*) displays by rapidly opening and closing his wings to make popping sounds, then launches into the air, flashing his 14-in. (36-cm.) tail. Lyre-tailed Nightjar (*U. lyra*) has an even grander tail, up to 29.5 in. (75 cm.) long, and a more impressive display, in which several males gather at a lek and take turns calling, whipping their tails about, and chasing females. Two African species (genus *Macrodipteryx*) have an elaborate elongated inner primary (primary number 2) when in their breeding plumage. Male Pennant-winged Nightjars (*M. vexillarius*) perform communal displays during which they flash their wings and vibrate their enormously long primaries from the ground or in flight. Most amazing, though, is Standard-winged Nightjar (*M. longipennis*), whose long inner primary is modified as a bare shaft with a 'racket' tip. At a display arena, several males will gather and engage in a display flight, circling the females with their 'standards' held high and vibrating, which makes the males appear as though they are being mobbed by small birds. ■

During display flights, the male **Standard-winged Nightjar** holds his unique, racket-tipped plumes vertically. Ethiopia.

This **Large-tailed Nightjar** (*C. macrurus*) has the wide mouth (extending to just under the eye), tiny bill, and stiff rictal bristles typical of the genus. Nightjars differ from nighthawks in that the latter have no rictal bristles. Thailand.

from temperate breeding areas to tropical wintering areas: for example, Common Nighthawk (*Chordeiles minor*) migrates from Canada to South America. Most species migrate at night, but nighthawks will move by day, especially under overcast conditions, and sometimes in large aggregations.

Diet Primarily insectivorous, taking a wide variety of aerial insects ranging from tiny flies and termites to huge moths. One species is known to consume *Euphorbia* flowers and small mice; another has been documented as taking small birds.

Breeding Most species are monogamous, but two are polygamous. Most are territorial; a few are semicolonial or colonial. The breeding season is usually correlated with local seasonal peaks of insect abundance. The birds do not build nests, instead placing eggs in a scrape or on dead leaves on the ground. Depending on species, the clutch size is one to three eggs, and incubation takes 16–22 days, with asynchronous hatching. The chicks hatch with downy plumage, and are brooded primarily by the female. They take their first flight at around 20 days, and are independent by 35 days. Some species are known to lay a second clutch when the first brood is two or three weeks old, whereupon the male takes over care of the first brood. If disturbed at the nest, some species perform a feigning-injury display. For many species, the breeding biology remains little known.

Taxonomy Divergent 'eared-nightjars' (genus *Eurostopodus*) possibly deserve separate familial status. The family's closest relatives are probably the swifts (order Apodiformes) and owlet-nightjars (family Aegothelidae), which similarly eat prey whole.

OWLET-NIGHTJARS *Aegothelidae*

As their name suggests, this secretive nocturnal family resembles owls, possibly their closest relatives, in structure and behavior, but also has some affinity with the nightjars.

Structure Sizes range from 8 to 12 in. (20–30 cm.). The birds have a large head, large, round eyes, and a flat, rounded bill with a wide gape. There are a number of stiff bristles on the forehead and around the base of the bill; their function is unclear, but they may assist in detecting prey or locating young in dark nest-holes. The wings are long and rounded. The legs are fairly long, but the feet have small, weak toes.

Plumage Mostly somber shades of gray to brown or warmer, rufous brown, generally with paler underparts. Several species have brown and rufous morphs that appear to match the predominating landscape; others are finely barred paler above or below. The largest two species have bold white spots or lines, while the smaller species have barred head patterns. Feline Owlet-Nightjar (*Euaegotheles insignis*) is intricately patterned and has long, catlike whiskers.

Behavior Entirely nocturnal, owlet-nightjars usually spend the day hidden in holes in trees, among dense foliage and vines, or occasionally in crevices in cliffs. Although usually solitary, most birds are thought to pair for life. Foraging seems to be concentrated into the periods immediately after dark and before dawn. The birds mostly take insects in flight from within the forest canopy. Australian Owlet-Nightjar (*Aegotheles cristatus*) hunts mostly from a low perch, pursuing moths in flight and gleaning insects from foliage; it also takes ants and earthworms either on the ground or without landing.

A **Mountain Owlet-Nightjar** (*Aegotheles albertisi*); the upright posture is typical of all species in this family. Papua New Guinea.

An **Australian Owlet-Nightjar** performs a threat display at the nest. Australia.

Voice For most species, little is known of their calls. Some appear to give gruff *owrr* or squeaky *kee* notes, or a series of eerie screams, whistles, and cackles. The better-known Australian species has a high-pitched rattling, 'chirring' note, repeated several times, as well as a loud hissing note given as a threat or display.

Habitat Mainly lowland primary tropical forests. Australian Owlet-Nightjar also occurs in more open woodlands and shrubby areas, and less frequently in rain forests.

Movements Entirely sedentary; pairs occupy their territory year-round.

Diet Mostly flying insects such as beetles, grasshoppers, and moths.

Breeding Little is known for any species apart from Australian Owlet-Nightjar. These birds nest in holes in trees or tree stumps, both parents lining the nest with a layer of dried leaves. Clutches comprise two to five eggs, which are incubated mainly by the female and hatch after about 26 days. Both parents feed the young, which fledge and leave the nest after 21–29 days.

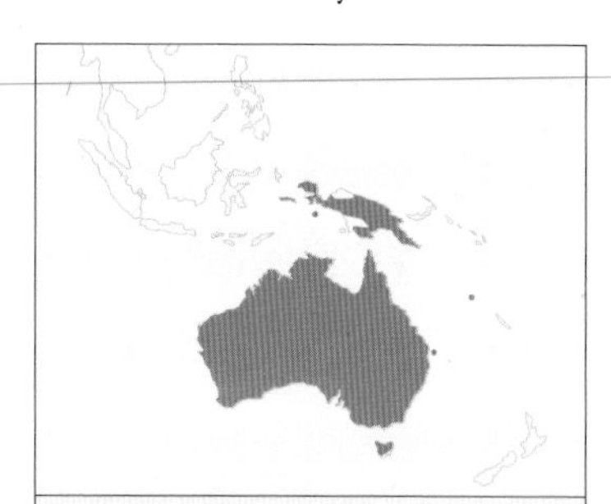

Number of genera 2
Number of species 9
Conservation Status 1 species Endangered
Distribution Australasia

SWIFTS *Apodidae*

Swifts, like this **White-throated Swift** (*Aeronautes saxatalis*), have sharp claws for clinging to vertical surfaces and a stiff tail to brace themselves. Montana, U.S.

With a mastery of the skies that is almost unparalleled among birds, swifts spend virtually all their waking hours on the wing. Some are even thought to sleep while in flight.

Structure Swifts range from 4 to 10 in. (9–25 cm.) in length, and weigh from just 0.2 oz. (6 g.) up to 6 oz. (180 g.). Their bill is small, but with a wide gape. The wings are long, with a distinctive shape: the 'arm' (humerus and forearm) bones, closest to the body, are very short, while the 'hand' (carpal and finger) bones are elongated; thus the primaries are long, while the secondaries are short and bunched. The tail is short, but may have elongated outer rectrices; many species have spinelike projections at the tip of the tail. The legs are very short.The small feet have three forward-facing toes and one to the rear; some species can bring all four toes forward, two to each side.

Pallid Swift (*Apus pallidus*). Swifts recall swallows in shape, but are smaller and fly higher and faster. Canary Islands.

Plumage The sexes are alike. Plumages are generally subdued in color. Most species are entirely blackish (sometimes glossed purple) or dark brown, often slightly or contrastingly paler on the throat and rump. A few, such as Alpine Swift (*Tachymarptis melba*), show bold black and white patterning, and some are extensively white on the underparts. Juveniles generally resemble adults, but their plumage may be duller or less contrasting, and the body feathers often have pale fringes. Migratory species usually undergo a full molt while on their winter grounds: flight feather molt may actually begin on the breeding grounds, be suspended during migration, then be completed on the winter grounds.

Behavior Feeding almost exclusively in flight, swifts usually stay in the air all day, and can travel long distances daily. They also drink water and collect nest material while on the wing. Some species forage at considerable height; others, especially some of the swiftlets and African spinetails, forage closer to the ground and often weave through trees in pursuit of prey. Most species fly extremely fast, with stiff, rapid wingbeats interspersed with glides; the largest may use long glides, while the smallest have more fluttering flight. Many species roost in large groups; their swirling, noisy pre-roosting flights can be impressive. Swifts cannot walk, and are rarely seen perched. They land mainly to roost at night or in bad weather, or to build nests and care for eggs and young.

Voice Lone swifts are often silent, but foraging groups, and especially birds gathering around nesting colonies or evening roost sites, can be highly vocal. Their calls vary from ear-splitting, high-pitched screeches to chippering or clicking calls. Some of the cave-dwelling swiftlets (genus *Collocalia*) use rapid dry, clicking trills to echolocate; these sounds are audible to humans.

Habitats Swifts are found in tropical and temperate regions; they are most diverse in tropical regions, but are generally absent from high latitudes. They need open areas for feeding. Foraging heights can be affected by weather conditions: highland species, such as Black Swift (*Cypseloides niger*), may be seen in lowlands in bad weather. Nesting and roosting habitats vary greatly; they include forests, waterfalls, caves, cliffs, and human settlements.

Most species are not at risk. However, some populations have suffered disturbance of nesting colonies and loss of nesting sites. Although many swifts have adapted well to human environments, large roosts in manmade structures may

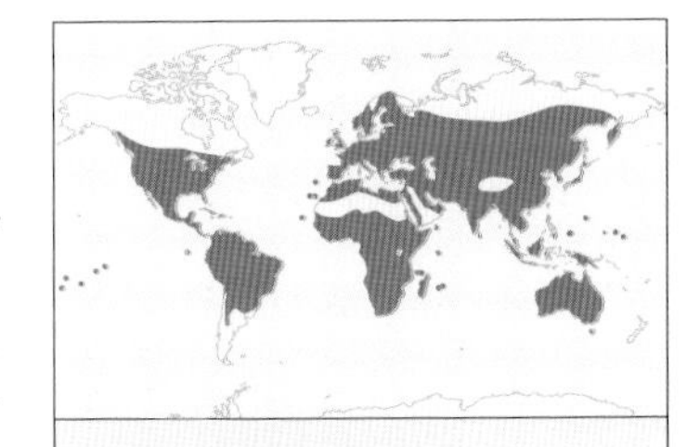

Number of genera 19
Number of species 94
Conservation Status 1 species Endangered, 5 species Vulnerable, 4 species Near Threatened
Distribution Worldwide in tropical and temperate regions, including many oceanic islands

A **Common Swift** (*Apus apus*) drinking. Swifts drink in flight, without stopping, by skimming low over water and dipping their lower mandible just below the surface. U.K.

be vulnerable to catastrophic loss. The aptly named Edible-nest Swiftlet (*Aerodramus fuciphagus*) has suffered from overharvesting of its nests to make 'bird's nest soup.' Overuse of pesticides can result in severe reductions of prey. Certain species are vulnerable due to very small ranges or restriction to tiny island groups.

Movements Swifts vary from generally resident, as in most tropical and subtropical species and island endemics, to long-distance migrants; some species have both resident and migratory populations. Some common northern-breeding species migrate across the equator: Common Swift winters in east Africa, and Chimney Swift (*Chaetura pelagica*) moves from Canada to the Amazon basin. Considerable long-distance vagrancy can occur (such as Old World species reaching the New World, and vice versa). Even outside migration periods, swifts may also undertake large-scale movements to avoid bad weather.

Diet Almost exclusively small arthropods, including insects (especially flies, ants, and beetles) and small spiders carried through the air on silk.

Breeding Swifts may breed singly or in small to very large colonies. Those that have been studied are monogamous. They build nests from twigs, feathers, and moss, usually cemented with saliva; Edible-nest Swiftlets build theirs entirely of dried saliva. The nests are placed in a cavity or crevice on a vertical surface such as a cliff, interior of a cave or chimney, a hollow tree, or even a palm frond. The birds court and mate on the wing. The clutch of one to seven eggs is incubated for 16–30 days. The young are altricial, taking their first flight at four to ten weeks. Some chicks can withstand several days' starvation, if bad weather prevents their parents finding food; they go torpid to save energy.

Taxonomy Most authorities unite the swifts and hummingbirds (family Trochilidae) in order Apodiformes; this order, in turn, may be closest to the nightjars and allies (Caprimulgiformes). The closely related treeswifts of Southeast Asia are usually placed in their own family, Hemiprocnidae.

A colony of **Great Dusky Swifts** (*Cypseloides senex*). These birds are known for roosting or nesting behind waterfalls, as protection from predators. Brazil.

TREESWIFTS *Hemiprocnidae*

A small family of birds in the single genus *Hemiprocne*, treeswifts live in tropical and subtropical forest. They are fast and extremely maneuverable in flight.

Structure The birds are 6 to 12 in. (15–30 cm.) in length: the largest species is Moustached Treeswift (*H. mystacea*) and the smallest, Whiskered Treeswift (*H. comata*). The most obvious features are very long, thin, scythe-shaped wings and a long, forked tail. The head is small, with large eyes, a short bill, and a broad gape surrounded by bristles; the body is slender, and the legs short.

Plumage Colors are dominated by shades of gray, from near-black to off-white. The upperparts are darker than the underparts. Adult males have chestnut, orange, or brick-red ear-coverts, but females' ear-coverts are dark. Adults of both sexes have a crest: in most species it can be held erect or relaxed, but Whiskered Treeswift has a nonerectile crest. Moustached and Whiskered Treeswifts have wispy white facial 'streamers' and bold white tertial patches. The tail is usually closed and spinelike but is sometimes spread, revealing the forked shape. Juveniles lack the crest, streamers, and bright ear-coverts, and are mostly shades of brown and buff.

Moustached Treeswift has a white 'moustache,' a deeply forked tail, and bold, pale tertial patches.

Gray-rumped Treeswift will come close to urban environments more readily than other members of the family. Singapore.

Behavior All species glide and soar in graceful arcs, and can fly very fast. When feeding, Whiskered Treeswift typically perches on a branch to watch for prey, then makes a brief sally to catch an insect in flight or on foliage. The other species more commonly take insects during long, gliding flights over the tree canopy or open country. Treeswifts are generally not social birds but sometimes forage in parties of six to 10, especially at the end of the breeding season. When perching, Whiskered Treeswift closes its wings in two or three distinct movements.

Voice Treeswifts are typically very vocal. Calls include repeated shrill chattering, a downward-inflected *kiiee*, nasal squeaks, and a piercing *ki*. Crested Treeswift (*H. coronata*) has a distinctive, harsh *kee-kyew* call.

Habitat Mostly primary and secondary forest, forest edge, and clearings. Moustached Treeswift often lives in open country with scattered trees. Crested Treeswift sometimes inhabits cultivated land and the vicinity of small towns. Gray-rumped Treeswift (*H. longipennis*) sometimes feeds in mangroves and gardens. Crested Treeswift has been recorded at altitudes up to 4,200 ft. (1,280 m.) in Nepal, while Moustached Treeswift has been recorded at 14,520 ft. (4,400 m.) in New Guinea.

Movements Crested and Moustached Treeswifts are sometimes nomadic, making seasonal movements to find food; Gray-rumped and Whiskered species are probably sedentary.

Diet Flies, beetles, ants, wasps, bees.

Breeding The birds are monogamous. Breeding usually coincides with the onset of the wet season, when food is more plentiful. The tiny, saucer-shaped nest, made of bark scales and feathers, is cemented to a high branch with saliva. The single egg is fixed inside with saliva and incubated by both parents. In Whiskered Treeswift, incubation lasts about 21 days and the chick fledges in about 28 days. It is fed by its parents until about 21 days after fledging, after which time it leaves the parents' territory.

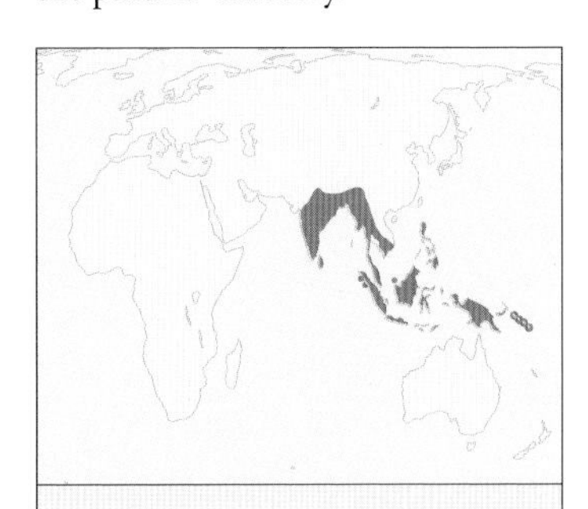

Number of genera 1
Number of species 4
Conservation Status Least Concern
Distribution South and southeast Asia; New Guinea

HUMMINGBIRDS *Trochilidae*

A male **Ruby-throated Hummingbird** (*Archilocus colubris*) demonstrates the hovering ability for which hummingbirds are renowned. U.S.

The tiny hummingbirds are widely known for their size, jewel-like colors, and ability to hover in midair, wings whirring, as they feed from flowers. This large, exclusively New World family has two subfamilies: the typical hummingbirds (Trochilinae), comprising the majority of genera, and the six genera of 'hermits' (Phaethornithinae), which are generally regarded as being more primitive. Several new species have been discovered in the past decade, including Gorgeted Puffleg (*Eriocnemis isabellae*), found in Colombia in 2007.

Structure Hummingbirds generally are among the smallest birds. Bee Hummingbird (*Mellisuga helenae*), weighing about 0.07 oz. (2.3 g.), is similar in size to a large bumblebee, and the largest species, Giant Hummingbird (*Patagonia gigas*), is about the size of a sparrow (family Emberizidae). Sexual dichromatism is common in typical hummingbirds, but practically absent in hermits.

The birds are characterized by their usually long, very narrow bill. Those of typical hummingbirds vary in shape from straight to decurved, and range in length from very short, in Purple-backed Thornbill (*Ramphomicron microrhynchum*), for example, to longer than the head and body combined in Sword-billed Hummingbird (*Ensifera ensifera*). The bills of two species, Fiery-tailed Awlbill (*Anthracothorax recurvirostris*) and Mountain Avocetbill (*Opisthoprora euryptera*), are slightly recurved. That of hermits is long and usually decurved. Two further species, Saw-billed Hermit (*Ramphodon naevius*) and Tooth-billed Hummingbird (*Androdon aequatorialis*), have toothlike serrations near the tip. The tongue is long, tubular, and forked at the end, with microscopic structures that draw liquids into the mouth by capillary action.

The wings have a distinctive bone structure, with a very short upper arm (humerus), longer hand bones, and a highly flexible shoulder joint that allows the wing to be rotated 180 degrees during hovering flight. The legs and feet are very small and weak, although some species that hang from flowers while feeding have stronger feet.

Plumage Hummingbirds are particularly renowned for their iridescent plumage. This iridescence is caused by microscopic structures within the feathers reflecting light of specific colors, rather than by pigment, with the result that the colors can be variable and are often transient and directional. Hermits are generally duller brown, gray, or reddish in plumage, sometimes with iridescent bronze or green usually restricted to the upperparts.

Typical hummingbirds are mostly highly iridescent, with green, blue, red, bronze, and purple dominating. Males of many species show highly iridescent throat (gorget) and crown patches, and some species have specialized adornments on the head, including crests and elongated throat and cheek feathers, all of which are

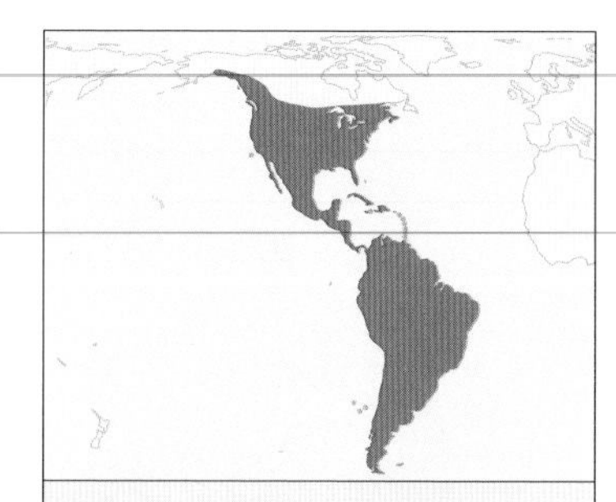

Number of genera 104
Number of species 332
Conservation Status 6 species Critically Endangered, 6 species Endangered, 12 species Vulnerable, 21 species Near Threatened
Distribution North and South America; concentrated near Equator

used in courtship displays. Males of some species have specialized flight feathers that produce mechanical sounds during courtship displays. Some species show small to large white tail patches, which they use to signal other individuals in territorial displays, mainly at food sources. Females are typically duller than males, often simply iridescent green above and whitish below. Many females show white tips on the tail feathers.

The tails of typical hummingbirds show many variations. Most are of moderate length, and squared, slightly notched, or forked at the tip. The variations include greatly elongated tails shown by members of genera *Topaza*, *Trochilus*, *Hylonympha*, *Ocreatus*, *Lesbia*, *Sappho*, *Aglaiocercus*, and *Thaumastura*. The male Marvelous Spatuletail (*Loddigesia mirabilis*) has one of the most unusual tails of any bird species, with only four feathers; the two central feathers are narrow and pointed, while the remaining two are very long and curved well forward, with a bare shaft and a large, spatulate tip.

Marvelous Spatuletail (*Loddigesia mirabilis*) is one of the rarest but most spectacular hummingbirds. In the male, the outermost two tail feathers are long streamers, with fan-shaped tips, which he can move independently in displays. Brazil (captive).

This **Fork-tailed Woodnymph** (*Thalurania furcata*) shows the iridescent plumage typical in hummingbirds. Location unknown.

Hermits have a significantly longer tail, with a whitish tip, and the longest feathers at the center.

Behavior Hummingbirds feed in flight by hovering in front of a food source (usually a flower). The rapid wingbeats of most species produce an audible hum, which gives the birds their name; the rate can vary from 70–80 per second in small species to 10–20 per second in the largest species. The birds' extremely flexible shoulder joint allows them to hover completely motionless, or even to fly backward and sometimes briefly upside down. Hummingbirds have a very high metabolic rate and specialized muscle tissue, which also facilitate hovering. Many species are able to enter a state of torpor, significantly reducing their metabolic rates, during periods of low temperatures, mainly at night and at high elevations.

Perhaps due to their small size, hummingbirds are highly aggressive, mainly in defense of nectar sources. Even migratory species can remain aggressive year-round. Males of typical hummingbirds defend fairly large territories, based on the availability of food sources, that often broadly overlap with those of other males, and often perform impressive display flights to attract females. These highly stereotyped displays usually involve showing off their colorful head and other adornments to best advantage. Hermits are generally not territorial. Males form leks where they court females with vocalizations and low-key visual displays.

Voice Hummingbird vocalizations are generally not melodious, consisting largely of high-pitched squeaks, squeals, chips, chirps, and buzzy or chattering notes. Although males of many species perch in the open and advertise their presence using simple vocalizations, few species are considered to actually sing. One species that does sing is Wedge-tailed Sabrewing (*Campylopterus curvipennis*), which has amazingly complex songs with much chattering, twittering, and even gurgling. In North America, the premier songsters are Blue-throated Hummingbird (*Lampornis clemenciae*) and Anna's Hummingbird (*Calypte anna*). Hermits give simple, somewhat monotonous songs at their leks, as do the few lekking species of typical hummingbirds. Often, the calls made when feeding, chasing, and fighting are more varied and complex than these advertising songs. Females are known to sing in only a few species. Among species nesting in more open areas, fledglings in the nest do not give begging calls. Some hummingbirds use specialized wing and tail feathers to produce nonvocal, mechanical sounds in flight.

Habitat The greatest diversity of species is found in Andean regions near the Equator, with diversity decreasing to the north and south. Single species occur north to the subarctic region, as in Rufous Hummingbird (*Selasphorus rufus*), and south to the subantarctic region, as in Green-backed Firecrown (*Sephanoides sephanoides*). The Andean cloud forest contains many species, while temperate and subtropical forests also show significant species diversity. High-elevation habitats, including *paramo* (mountain grasslands), have a number of species well adapted for life in these harsh environments. Lowland rainforest contains a relatively low diversity.

An unusually large number of hummingbird species are restricted in range, and as a result many are endangered, threatened, or vulnerable. Several range-restricted species not currently under threat could potentially face declines due to habitat destruction in the near future. Two species, Coppery Thorntail (*Discosura letitiae*) and Turquoise-throated Puffleg (*Eriocnemis godini*), are possibly extinct. However, in the western U.S., Anna's Hummingbird has expanded its range significantly in the past 30–40 years.

Movements The majority of hummingbirds are mainly sedentary. A few of the southernmost species move northward during the southern hemisphere winter. Some Andean species are altitudinal migrants, though the extent of these movements is poorly known.

Some northern-hemisphere species are complete migrants, with Rufous Hummingbird migrating the longest distance, making a round trip of about 3,730 miles (6,000 km.) from southern Alaska to southern Mexico. Ruby-throated Hummingbird (*Archilochus colubris*) migrates nearly as far; many individuals cross the Gulf of Mexico, traveling more than 500 miles (800 km.) nonstop, while others take a longer circum-Gulf route through Texas and Mexico. It was debated for many years whether trans-Gulf flights were possible for such a small bird, but laboratory experiments

A courting pair of **Black-throated Mangos** (*Anthracothorax nigricollis*) in flight. As with most hummingbird species, the male (at left) is much more brightly colored than the female. Trinidad.

A female **Anna's Hummingbird** (*Calypte anna*) at the nest. The species usually nests on top of a twig or branch; this bird has built its nest on a fir cone. Arizona, U.S.

proved that the birds have sufficient energy. Some recent field observations also suggest that by using a flight style like that of woodpeckers, beating the wings rapidly and then gliding with wings closed, the birds can save additional energy during migratory flights. Several other western North American species are also complete migrants, or nearly so, while some species from Mexico and the southwestern U.S. are partial migrants, withdrawing from northern parts of their range in winter.

Diet Hummingbirds are considered to be the ultimate nectar-feeders; in fact, many people put out special sugar-water feeders to attract them. Insects also feature prominently in the diets of many species. In species that have been studied extensively, nestlings are fed almost exclusively on insects, and migratory species feed heavily on insects to build up fat reserves for extended flights. Insect prey consists primarily of gnats, fruit flies, micro-moths, and others, captured in the air. Some species have been seen hover-gleaning spiderlings from their webs, and taking tiny insects from leaves and bark. Others may enter a swarm of insects in flight and snap up as many as possible before they disperse.

Breeding Northern-hemisphere species mostly breed during the northern summer. In tropical species, peak breeding activity is associated with peak flowering, which can be during either wet or dry seasons, and with times when insects are most abundant. Male hummingbirds are polygynous, mating with several females each season.

In almost all hummingbirds, nest-building and other parental duties are left entirely to the female. The nest is typically an open cup woven from fine plant fibers, animal hairs, or spider silk, and camouflaged with lichens, mosses, and other vegetation. Hermits typically attach their nest to the underside of a large leaf, where it hangs down, well hidden and protected. Typical hummingbirds place their nest in a well hidden but often fairly open location on a small branch within a tree or shrub. Sylphs (genus *Aglaiocercus*) build domed nests, metaltails (*Metallura*) build partially domed nests, and some high-Andean species build their nests on rock faces.

Clutches typically comprise two tiny white eggs, and very rarely one or three eggs. Incubation takes 14–23 days, with an average of 14–16 days. The young are altricial, hatching blind and nearly featherless. Their bill is very short, but grows quickly. Fledging occurs after 23–40 days. Almost nothing is known of the post-fledging period in most hummingbirds, even the well-studied North American ones.

Taxonomy The family is highly diverse, with 42 genera containing a single species and 24 genera having only two. In subfamily Phaethornithinae (hermits), the largest genus is *Phaethornis*, with 25 of the 35 species. The largest genus in subfamily Trochilinae (typical hummingbirds) was formerly *Amazilia*, which contained 31 species, but this has now been split into four genera.

Hummingbirds form a clearly defined group within order Apodiformes, together with their closest relatives, the swifts (Apodidae) and treeswifts (Hemiprocnidae).

MOUSEBIRDS *Coliidae*

A small family of exclusively African birds, the mousebirds get their name from their long tail and their ability to climb and run through dense bushes and along branches. The species are very similar in size and shape, differing only in plumage and facial features, and the sexes are alike.

Structure The birds range from 12–14 in. (30–35 cm.) in length; the total length includes the graduated tail, which is nearly twice the length of the body. They have a small head with a short, stubby, but strong and slightly decurved bill. The wings are short and rounded, and the legs are also fairly short. The birds have four highly mobile toes, with long, sharp claws; they can move their first and fourth toes forward, sideways, or backward to improve their grip on branches.

Plumage The body plumage is mostly brown or varying from buffish to gray-brown. In three species, the lower back or uppertail and rump are distinctively marked: white on White-backed Mousebird (*Colius colius*), reddish-chestnut in Red-backed Mousebird (*C. castanotus*), and blue-gray in Blue-naped Mousebird (*Urocolius macrourus*). The head has a short but conspicuous crest, often held partly raised, and patches of white on the cheeks or on the crown and crest. Blue-naped Mousebird has a bright blue patch on the back of the head. The two *Urocolius* species have bright red, bare facial skin. In all species, the legs and feet are deep reddish.

A **Red-faced Mousebird** (*Urocolius indicus*) forages for fruit at the top of a bush. Location unknown.

Behavior Highly social and gregarious, mousebirds are most often seen as family groups. Flocks forage together in bushes and thickets, where they eat fruit and buds; in some fruit-growing areas they are considered to be pests. When moving on, they often climb to the top of a bush before taking to the air; they fly with an explosive take-off and then whirring flight, with a long glide before crash-landing. Pairs roost and sunbathe facing each other, hanging down from a thin branch with their feet level with their head.

Voice The birds are highly vocal. Flock members keep in contact through clicking and twittering notes. Two species have more melodious, whistling or peeping notes.

Habitat Mostly open thornbush, light woodland, and savanna with *Acacia*, at elevations from sea level to about 8,000 ft. (2,400 m.). All species are fairly common within their range. Most are widespread; two have a small or restricted range, but none is considered to be at risk.

Movements Largely sedentary, although the birds travel short distances to find food.

Diet Mousebirds take a wide range of plant food, including buds, leaves, flowers, and fruit.

Breeding Monogamous and communal. Nests vary from loose, bulky structures of twigs to shallow cups or platforms made from plant fibers, stems, seed and flower heads, and

A pair of **White-headed Mousebirds** in typical roosting pose, hanging from their feet, with their bellies touching. Kenya.

feathers; they are usually well hidden in thornbushes. Clutches vary from one to eight eggs; often, more than one female lays in the same nest. Incubation takes 11–12 days, and is performed by both parents. Additional males may help feed the young. Chicks fledge in 10–14 days.

Taxonomy The family's relationships are obscure. Mousebirds were previously thought to be allied to both the trogons (family Trogonidae) and the cuckoos (Cuculidae), but are now considered unique and have been assigned to their own family.

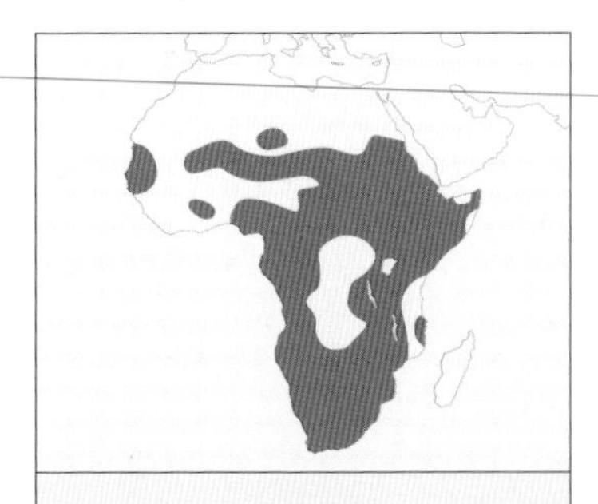

Number of genera 2
Number of species 6
Conservation Status Least Concern
Distribution Sub-Saharan Africa

TROGONS *Trogonidae*

The male **Resplendent Quetzal**, with his iridescent plumage and long, luxuriant tail, is the most spectacular of the trogons. Guatemala.

The trogons are a family of striking tropical birds. They include the spectacular quetzals of Central and South America, such as Resplendent Quetzal (*Pharomachrus mocinno*). Most species are neotropical, but three occur in Africa and 11 in Southeast Asia. Although they can often be difficult to see in the forests, both trogons and quetzals have long been admired for their beautiful plumage. The ancient peoples of Central and South America would harvest feathers from live birds, but in the last few centuries some species, notably Resplendent Quetzal, were killed for their plumage. Killing by humans, and habitat destruction, have led to population declines in this species.

Structure Trogons are medium-sized birds of 10–15.75 in. (25–40 cm.) in length, excluding the tail. They have a small, rounded head, large eyes, and a short neck. The bill is short and broad-based, with a curved maxilla; the color varies from black to yellow to red. The legs are short, and two of the four toes are turned backward.

Plumage All species are sexually dimorphic, with males having much brighter plumage than females.

Species in the Americas are generally iridescent green on the head, back, and breast, and have either a bright red or yellow belly. Tails have varying amounts of white on the tips of the feathers, forming either spots or barring when viewed from below. Different species have eye rings that vary in color from red to yellow to whitish. Asian species are also colorful, but generally have brown, rather than green, body plumage. Red-headed Trogon (*Harpactes erythrocephalus*) has an entirely red head, and Blue-tailed Trogon (*H. reinwardtii*) is spectacularly patterned, having green upperparts, a yellow throat and belly, and a blue tail. Trogons' tails are rather long, broad, and graduated.

Quetzals have highly modified wing and tail coverts: Resplendent Quetzal's uppertail coverts can be as long as 2–3 ft. (60–90 cm.).

Behavior Trogons and quetzals are rather inconspicuous forest birds that are usually difficult to find, as

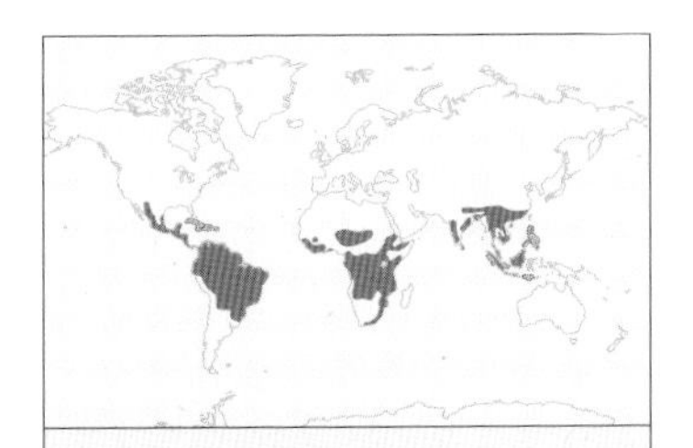

Number of genera 6
Number of species 40
Conservation Status 1 species Endangered
Distribution Mainly pan-tropical: Central and South America, Africa, Southeast Asia

Trogons, like this **Narina's Trogon** (*Apaloderma narina*), typically turn their back when another animal approaches, to conceal their bright underparts and thus make themselves less visible. Kenya.

they have a tendency to sit quietly for long periods. They are more conspicuous when vocalizing. Generally, trogons feed by sallying out from a perch to catch flying insects, or by hover-gleaning prey from leaves. Many trogons also hover-glean fruit from trees.

Voice Both males and females vocalize. The primary songs are generally a series of 'coos' or whistles, which vary between species in pattern and tone, and are sometimes quite harsh; for example, the call of Elegant Trogon (*Trogon elegans*) is often compared to the sound of a barking dog. Trogons' songs are usually loud, with the sound carrying through the forest for great distances. Calls are usually softer, resembling a *churrrr*. The birds often cock and lower their tail slowly while calling.

Habitat Trogons are found in a variety of habitats, from tropical rain forest at sea level to montane forests at high elevations in the Andes. Some are also found in arid forests. One species, Elegant Trogon, breeds as far north as canyons in Arizona. The highest diversity of trogons can be found in lowland tropical forests, where as many as five species may coexist at the same location.

Movements Most species are resident within their range, although two species are at least partially migratory. Elegant Trogons range from southeastern Arizona to Costa Rica, and while generally resident, most individuals in the northern portion of the range migrate south into Mexico. Eared Quetzal (*Euptilotis neoxenus*) is mostly resident in the highlands of northern Mexico, yet inexplicably disperses across inhospitable habitat and turns up occasionally in Arizona. Some species move short distances (often in altitude) in search of temporary increases in fruit availability.

Diet Trogons and quetzals feed on both insects and fruits, with the diet varying seasonally, at least in part. Quetzals are well known to feed on a certain type of avocado, ingesting the entire fruit and then regurgitating the seed. Trogons can be adept at catching flying insects, such as butterflies and moths, but they also fly out and glean prey, such as caterpillars, off leaves and branches.

A male **Blue-crowned Trogon** (*T. curucui*) singing. Trogons' songs are usually loud and far-carrying. Peru.

Diard's Trogon (*Harpactes diardii*), found in the lowland forests of Southeast Asia, is listed as Near Threatened due to logging and land clearance. Location unknown.

Breeding Trogons are monogamous, with males and females sharing in nest construction, incubation of eggs, and feeding of young. Nests are hollowed out of cavities in trees; sometimes, the birds may use old woodpecker holes or termite nests. Clutches consist of two to four eggs, which are incubated for 16–19 days. Male Resplendent Quetzals can sometimes be seen incubating, with their elongated tails sticking out of the nest hole. The young fledge at 15–31 days, but the adults continue to feed them for several weeks or months afterward.

ROLLERS *Coraciidae*

Rollers are widely distributed throughout the Old World, at least half of them in Africa. They are brightly colored birds, best seen in flight, especially during their dramatic, rolling courtship displays. Anatomical studies show that they are related to bee-eaters, motmots, and perhaps more distantly to kingfishers.

Structure The species range from 9 to 20 in. (22.5–50 cm.) in length. Rollers are thickset birds with a large head and a shortish, broad, powerful bill; in two species, the bill has a hook at the tip. The wings are fairly long. The broad, strong feet are syndactylous, having the middle two toes joined together.

Plumage The sexes are alike. Plumage colors are often bright or dazzling. They include pale to bright blue, greenish, and azure blue; also lilac, pink, brown, rufous, cinnamon, grayish, white and black. In all species, the plumage has contrasting shades of blue, including electric and paler blues, and brown in the wings and tail. One species (Dollarbird, *Eurystomus orientalis*) has pale blue or whitish patches on the wings. The bill is black in most species, but bright yellow in two and dark red or scarlet in Dollarbird. The tail is square-tipped or slightly notched. Some species, notably Racquet-tailed Roller (*Coracias spatulatus*), have elongated outer tail feathers or streamers.

A **Lilac-breasted Roller** (*C. caudatus*) watches for prey. Rollers typically perch at the top of trees or bushes. South Africa.

Behavior Rollers are mainly 'watch and wait' foragers. They spend long periods sitting on a bush or tree-top, scanning the vicinity for prey. Other feeding methods include catching large insects in flight, or taking insects, small amphibians, and reptiles fleeing from forest or grass fires. The birds usually carry large insects to a perch and beat them against it to remove the wings before eating them.

Voice Calls include a series of dry, harsh, abrupt *caw* sounds, like those given by crows but generally higher-pitched. Alarm calls include *kack-ack* and a more croaking *cruk*. The birds also give high-pitched screams when pursuing a predator.

Habitat Rollers are mainly found in wooded habitats, including open woodlands, plantations, shrubby hillsides, grasslands, and scrubby edges of cultivated land. At least two species occur in more densely wooded areas and forest edges.

Movements Mostly sedentary, but three species are migratory. Broad-billed Roller (*E. glaucurus*) is an intra-African migrant, moving north and south (and northwest from Madagascar) to winter within the equatorial forest belt. Dollarbirds travel from northern China and eastern Australia to winter in Southeast Asia and Indonesia. The longest journeys are undertaken by European Roller (*C. garrulus*), which breeds between Spain and central Russia and winters throughout Africa south of the Sahara.

Dollarbird is named for the dollar-sized, whitish patch on each wing, partially visible on this bird. Location unknown.

Diet Large insects, particularly crickets, cicadas and termites; scorpions; frogs, lizards, and small snakes.

Breeding The most spectacular aspect of rollers' breeding behavior is their fast, tumbling, wheeling flight during courtship displays. The nest is located in a cavity in a tree trunk, or in rock. Clutches comprise two or three eggs in tropical birds, or three to six in more northerly species, and are incubated for 18–20 days. The young fledge in 25–30 days; both parents feed them until about 20 days after fledging.

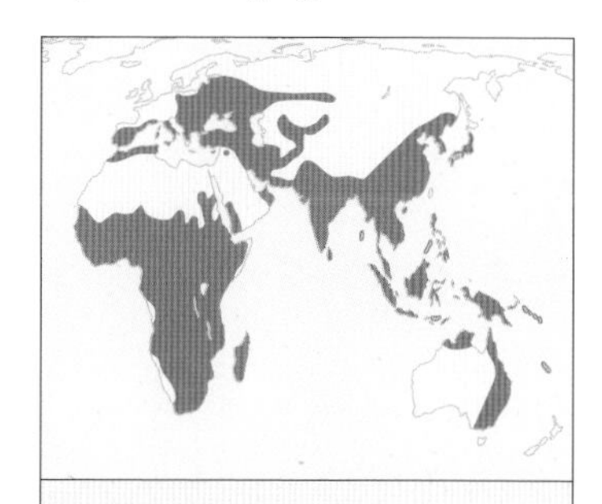

Number of genera 2
Number of species 12
Conservation Status 1 species Vulnerable
Distribution Southern Europe, Africa, Asia, Australasia

GROUND ROLLERS *Brachypteraciidae*

Rufous-headed Ground Roller (*Atelornis crossleyi*) foraging on the forest floor. Marojejy National Park, Madagascar.

A small family of shy, terrestrial birds, ground rollers are endemic to Madagascar.

Structure These stoutly built birds range from 9 to 18 in. (24–47 cm.) in length. They have a large head and large eyes, and a short, strong bill. All species have a long tongue with a brushlike tip. The birds have a long tail, and long, sturdy legs. The claws of the first and fourth toes are outward-pointing, while those of the middle two point inward.

Plumage Brightly but also cryptically colored, the plumage comprises a mixture of vivid green, browns, and yellow, with patches of light or deep blue. Two species have a predominantly scaly pattern, broken or interrupted by broader dark bars or scales or bands across the underparts; three have a white chin and throat bordered by a broad black, brown, or blue gorget; and one has a broad black half-collar finely striped with white. In some species, the sexes are alike; in others, the females are slightly smaller or duller.

Behavior Ground rollers are generally shy, retiring, and skulking. They spend long periods during the day roosting in thick vegetation, becoming more active toward evening and again in the early morning. One species, Long-tailed Ground Roller (*Uratelornis chimaera*), may even be nocturnal. The birds forage alone or in pairs, climbing through thick undergrowth and scratching in leaf litter on the forest floor. They also make short runs after prey. Short-legged Ground Roller (*Brachypteracias leptosomus*) actively forages in the middle and lower levels of trees and tall shrubs.

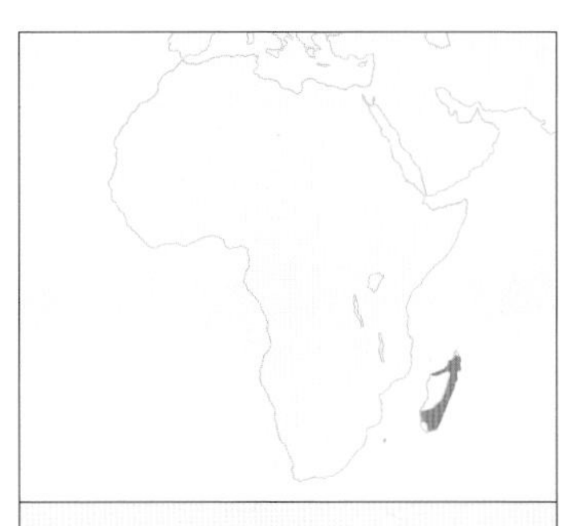

Number of genera 4
Number of species 5
Conservation Status 3 species Vulnerable
Distribution Madagascar

Voice Mostly silent except before and during breeding. Contact calls are a series of guttural whooping or softer *boobooboo* notes, or a chuckling series of *too-tuc* notes.

Habitat Most species live in pristine lowland and lower montane rain forests, to about 6,560 ft. (2,000 m.), with damp undergrowth and deep leaf litter. Long-tailed Ground Roller lives in thorn scrub.

Movements Largely sedentary, but may wander short distances from breeding territory when not nesting.

Diet Mostly insects and their larvae, including beetles, butterflies, moths, cicadas, grasshoppers, wasps, and praying mantises. The birds may also take snails, frogs, small lizards such as geckos, and small snakes.

Breeding Pairs breed from September to February but possibly defend territories all year. Most species nest in a small chamber at the end of a burrow up to 3.25 ft. (1 m.) long, which is usually dug into a steep earth bank. In contrast, Short-legged Ground Roller nests in holes in trees. Clutches comprise two to four eggs, which are incubated by the female for 20–26 days.

Unlike most ground rollers, **Long-tailed Ground Roller** lives in semi-arid conditions. Ifaty Spiny Forest, Madagascar.

CUCKOO-ROLLER *Leptosomidae*

A male **Cuckoo-Roller**. The birds have a large head with a crest and a distinctive tuft of feathers under the bill. Madagascar.

The monotypic Cuckoo-Roller (*Leptosomus discolor*) is neither a cuckoo nor a roller but shares anatomical and behavioral features with both, and with the closely allied order Coraciiformes, which includes kingfishers, bee-eaters, and hoopoes. It is also known by its French name, *Courol.*

Structure The bird is 15–20 in. (38–50 cm.) in length. It has a large head with a short, loose crest, steeply rounded forehead with upward-curving feathers, and eyes that appear to be set back in the middle of the head. The bill is short, stout, and broad-based, with a fine hooked tip. The wings are long and rounded, and the tail is moderately long and square-tipped. The legs are short, and the feet have zygodactylous toes: the third toe is reversible, and the bird usually perches with two toes facing forward and two backward.

Plumage The sexes differ. Males have a gray head, neck, and underparts; the upperparts and tail are deep, metallic green with a pronounced purplish gloss. A thin, blackish line runs from the base of the bill over the eye to the nape, and connects with a second band from the base of the bill over the crown; another band runs across the crown, joining the eyes. Females have a brown head, heavily spotted darker; blackish upperparts, heavily spotted with chestnut, as are the tips to the scapulars and wing coverts; and pale buff underparts, boldly and erratically spotted dark brown.

Behavior Cuckoo-Rollers are usually found alone or in pairs. They forage by sitting motionless and waiting, often for long periods, on a high but usually concealed perch within the foliage, and then making a short, dashing flight to seize prey. They occasionally make short aerial flights high into the air above tree canopy level, in pursuit of passing insects.

Voice The call is a loud, rising, whistled *qui-yuu* or *whee-oo*, repeated or followed by a short *wha-ha-ha.*

Habitat Restricted to Madagascar and the Comoro Islands. The species occurs in a range of forest, dry bush, and scrub, and occasionally trees at the edges of cultivation. On Madagascar it is widespread but local or uncommon. The subspecies *gracilis*, found on Grand Comoro, has a population estimated at around 100 pairs, in restricted areas of forest.

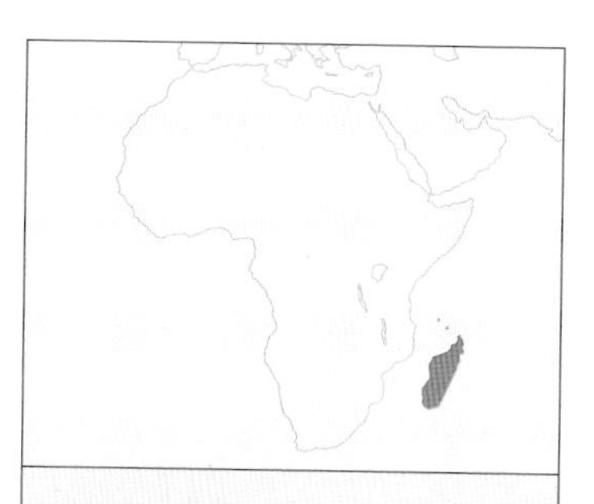

Number of genera 1
Number of species 1
Conservation Status Least Concern
Distribution Madagascar, Comoro Islands

Movements Largely sedentary.

Diet Includes beetles; large, hairy caterpillars, locusts; cicadas; chameleons; and geckos.

Breeding Cuckoo-Rollers' nests are unlined cavities in trees, up to 20 ft. (6 m.) from the ground. The clutch of four or five eggs is incubated by the female for about 20 days. Nestlings fledge after 30 days.

A **Cuckoo-Roller** with a gecko. Mainly arboreal birds, they usually forage in the canopy or above the outermost foliage of trees. Madagascar.

KINGFISHERS *Alcedinidae*

A large family of active and often colorful birds, the kingfishers are distributed throughout the world, absent only from very high northern latitudes. The greatest diversity appears to exist from Southeast Asia to New Guinea. Kingfishers have also spread to many of the islands in the Pacific Ocean. The family comprises three subfamilies: the Halcyoninae (60 species), Alcedinae (22 species), and Cerylinae (9 species). Several species, especially those in genus *Todiramphus*, have a number of subspecies: for example, Collared Kingfisher (*T. chloris*) has 49 subspecies, many of which are endemic to isolated Pacific islands.

Most continents have one well-known species, often seen near rivers. In much of North America, Belted Kingfisher (*Megaceryle alcyon*) is a common bird, while in Europe, Common Kingfisher (*Alcedo atthis*) is familiar as a fleeting dash of electric blue and deep orange. The family is also known for its ringing calls and, in Australia, the raucous laughing notes of the kookaburras (genus *Dacelo*). Most kingfisher species are similar in structure, but differ primarily in size, plumage, and habitat preferences.

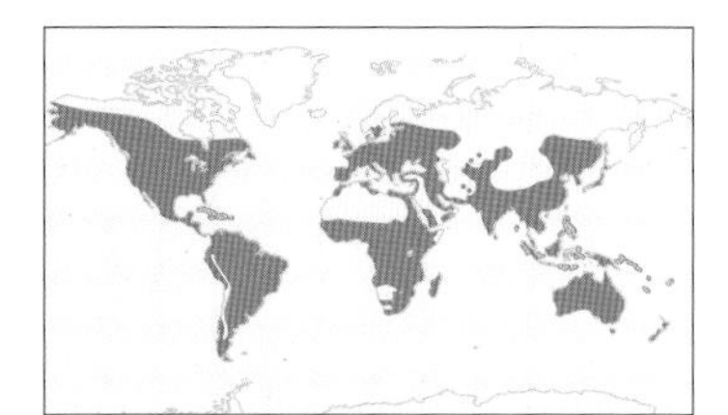

Number of genera 17
Number of species 91
Conservation Status 1 species Endangered, 11 species Vulnerable
Distribution Worldwide apart from polar regions

Structure Species vary in length from 4 to 18 in. (10–45 cm.), and in weight from 0.3–0.4 oz. (9–12 g.), in African Dwarf Kingfisher (*Ceyx lecontei*), to about 16.5 oz. (490 g.) in some female Laughing Kookaburras (*Dacelo novaeguineae*), with most species weighing about 1–3.5 oz. (30–100 g.). In most kingfishers, the sexes are alike in size, but in some species, one sex is larger; in two species of kookaburra, the females are significantly larger.

Kingfishers are characterized by a moderately long bill, which may be daggerlike, wedge-shaped, or shovel-shaped, or massive in the case of the kookaburras. Most species have a fairly broad head, set on a short neck. The body is plump, with short, rounded wings; tails are mostly short, except in the paradise kingfishers (genus *Tanysiptera*), whose tails have long streamers. The legs are short, and most species have

A **Common Kingfisher** captures a fish. As a kingfisher dives into water, it pulls its wings back, shooting downward like an arrow. On seizing its prey, it uses its wings first to 'brake,' then to power itself back up through the water and into the air. U.K.

syndactylous feet, with three of the toes facing forward and the fourth pointing backward.

Plumage A wide variety of brilliant or metallic colors is found across the kingfisher family; in most species, the sexes are alike in color or differ only a little. Deep azure to electric blue predominates in many species, mostly on the head and upperparts. In some, these areas are darker or blackish, and in others the upperparts are green, and often combined with white patches on the head or neck or orange-rufous underparts.

There are, however, numerous exceptions to this general pattern. Some species have a brown head or upperparts and a blue back or rump and tail. Others are heavily barred across the head and upperparts. One of the main exceptions is Ruddy Kingfisher (*Halcyon coromanda*), which is almost entirely rufous or darker chestnut on the upperparts, except for a blue or white rump patch. Birds in genera *Ceryle* and *Megaceryle* are mostly gray or black and white; some, such as Ringed Kingfisher (*M. torquata*), have bright chestnut underparts. The kookaburras are mostly whitish with blackish or bluish wings and tail.

The head has fairly loose or fluffy feathers on the crown and nape, and in several species these are developed into a full crest. The eight species of paradise kingfishers have central tail feathers that are elongated to form blue or white streamers, in some cases with a racquetlike tip.

Bill colors vary greatly across the family. They include deep red, in many species; black, or red on one mandible and black on the other; and yellowish or whitish.

Behavior The birds are usually solitary or found in pairs, although some species (such as the kookaburras) may form family parties that remain together for at least a year.

A male **Belted Kingfisher**. This species is unusual in that females are more colorful than males, with a rufous band across the chest and flanks. Texas, U.S.

All species forage by sitting and waiting before diving or plunging onto prey. They have extremely good eyesight, and can detect the movements of their prey at some distance, or through murky or light-reflecting water. Most wetland kingfishers hunt by watching and waiting from a prominent perch, before making a spectacular dive to snatch their prey underwater. Some species also hover over water to locate their prey before diving. Forest- and savanna-living species prey on various insects in the same manner, scanning from high locations before making a shallow or steep dive.

Voice Most kingfishers are very vocal, with loud, distinctive contact or advertising calls. These calls may comprise slow, rising or falling, ringing notes; high-pitched whistles; *zeet* notes; or harsh, coarse rattling sounds. Laughing Kookaburra has a unique series of laughing, chuckling, or gurgling notes.

Habitat Most kingfishers live beside rivers or streams, or are found at some stage along a watercourse. However, several species, mainly those in the tropics, have become forest birds, inhabiting open areas among lower-level trees and ground vegetation. Some have adapted to drier open country, including acacia savannas and margins of desert scrub. One species, Marquesan Kingfisher (*Todiramphus godeffroyi*), which is endemic to two small islands in the Marquesas group in the central Pacific, is considered Endangered because of introduced predators and habitat destruction within its small range. Eleven other species are regarded as Vulnerable, mainly due to habitat degradation and loss.

Movements Few species are truly migratory, but a number make some kind of seasonal movement. Several

Rufous-backed Kingfisher (*Ceyx rufidorsa*) is one of the smallest species. Japan.

A pair of **Pied Kingfishers** (*Ceryle rudis*) performs a courtship display; the female is at left and the male at right. In addition to the display, the male will offer food to the female. The breeding pair may have helpers assisting them with rearing the young. Botswana.

of the species that breed in Australia move north to winter in New Guinea and Indonesia. Ruddy Kingfishers breeding in Korea and Japan move south to winter in Sulawesi and the Philippines. Black-capped Kingfishers (*H. pileata*), which breed through eastern China to Korea, move south to winter in southeast Asia and Indonesia. The longest movements are made by Common Kingfishers that breed in Russia and move up to 1,865 miles (3,000 km.) to winter in southern Europe and Asia, and by Belted Kingfishers that breed in Alaska and across northern Canada and travel south to winter in Mexico, central America, and northern Venezuela.

Diet Kingfishers mainly eat fish and aquatic invertebrates. The birds may also take insects, earthworms, reptiles, small mammals, and birds. Kookaburras may eat snakes (including venomous species) up to 3.3 ft. (1 m.) in length, which they bash vigorously against the ground or a branch before swallowing head-first. They also visit garbage tips, picnic areas, and bird-feeders.

Breeding The birds are monogamous, territorial, and occasionally colonial. Several species, including most of those in genus *Halcyon*, have noisy courtship displays in which the pair perch close to each other and engage in duetting while raising and spreading their wings, or have spiraling flight displays in which both rise high into the air.

Nests are placed at the end of a tunnel up to 6.5 ft. (2 m.) long. The birds may dig the nest themselves, usually in an earth bank, or use a termite nest or a hole in a tree. The nest rarely contains anything except for excreta and discarded food items. Clutch size varies between species: the small dwarf kingfishers and the large kookaburras have two eggs, while most of the *Alcedo*, *Halcyon*, *Ceryle*, and *Megaceryle* kingfishers have six or seven and occasionally more, which may be the result of two females laying in the same nest. Both sexes incubate the eggs; incubation periods vary according to size and species, from 13–28 days. Fledging periods also vary, from 21–44 days. Kookaburras are cooperative nesters, with helpers, probably young from previous years, assisting with incubating the eggs and feeding the chicks.

Laughing Kookaburra can be seen—and heard—throughout eastern Australia. Victoria, Australia.

TODIES *Todidae*

Tiny and colorful, the todies comprise five very similar species in one genus. The family is confined to the Greater Antilles in the Caribbean, but the existence of fossil todies from Europe and North America suggests it was more widespread in the early Oligocene, 30–35 million years ago. Although they look like kingfishers, todies are more closely related to the Motmots (family Momotidae).

Structure The birds are 4–4.5 in. (10-11.5 cm.) in length. They have a large head and a long, narrow, flat bill. The bill has serrations on its edges, which are microscopic in adults but more prominent in nestlings. The body is rounded, with short wings and tail and small legs.

Plumage All species are very similar in appearance, being bright green on the upperparts and red on the throat. The breast, belly, and flanks vary between the species from white to yellow or pink. Iris colors vary from brown to yellow or white; these variations may be correlated with sex or age in some species, but are seemingly random in others.

Tody species can be told apart by the tints of color on their underparts. **Cuban Tody** (*T. multicolor*) is perhaps the brightest, with pinkish sides and yellow undertail coverts.

Behavior The birds are mainly found singly or in pairs. They forage using an 'underleaf-sally' technique: making short, darting upward flights and sweeping insects off the undersides of leaves, ferns, branches, or tree trunks with a sideways motion of their flattened bill. Todies are very acrobatic when hunting, similar to some of the smaller tyrant-flycatchers (Tyrannidae).

The birds perch with the bill angled upward, like hummingbirds (Trochilidae). They perform 'wing-snapping' displays during courtship and territorial chases. Both sexes also have a 'flank display' that precedes copulation and involves the birds fluffing out their flanks to become a nearly spherical ball of feathers.

Voice Todies most frequently make short, nasal buzzy *beep* notes, but sometimes give chattering calls, and even clear whistles in one species.

Habitat The birds are found in forested areas that have dense undergrowth, with higher population densities at lower elevations. Narrow-billed Tody (*Todus angustirostris*) occurs in montane forests. No species is thought to be at risk, but habitat destruction, overuse of pesticides, and introduced mammalian predators could have a detrimental effect on todies in the future.

Movements Sedentary within their ranges, todies are unlikely to move to different islands given their fairly weak powers of flight. Narrow-billed Tody apparently undertakes seasonal altitudinal movements.

Diet The birds take a great variety of insects and insect eggs, but also eat other small invertebrates including spiders, worms, and millipedes. Occasionally, the diet includes small vertebrates such as small lizards, and sometimes fruit.

A **Puerto Rican Tody** (*Todus mexicanus*) in typical tody pose: sitting quietly, with the bill tilted upward. Puerto Rico.

Breeding Todies are monogamous and single-brooded. Pairs defend relatively small breeding territories. They excavate nesting burrows in earthen banks, and lay one to four eggs, which are incubated by both parents for 21–22 days. The altricial young fledge in 19–20 days.

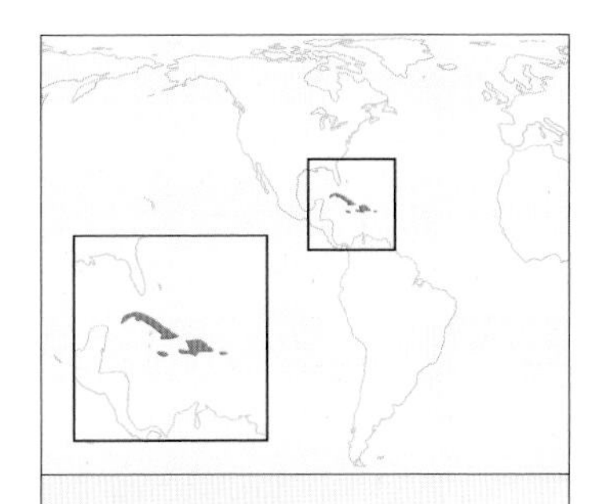

Number of genera 1
Number of species 5
Conservation Status Least Concern
Distribution Greater Antilles, Caribbean

MOTMOTS *Momotidae*

A **Blue-crowned Motmot** (*Momotus momota*). The weak barbs on motmots' central tail feathers are easily dislodged in everyday activities such as preening.

An exclusively Neotropical family, motmots gained their original Mexican name and current English name from their double-noted hooting calls. They are also distinctive for their spatulate-tipped tail feathers. Their closest relatives include the todies (family Todidae), kingfishers (Alcedinidae), and bee-eaters (Meropidae).

Structure Motmots range from 6.25 to 18.75 in. (16–48 cm.) in length. They have a stout, broad bill with serrated edges, and well developed rictal bristles. The wings are short and rounded, but the tail is long. The legs are short and the feet small and syndactylous, with the outer two front toes fused for much of their length: a feature seen in many species that dig nesting burrows.

Plumage Most species are colorful, with iridescent blue or green on the upperparts and head; several have rufous on the head or underparts, or a black face mask and one or two black spots on the breast. In most species, the central tail feathers grow significantly longer than the others do. The barbs near the tip are only weakly attached and readily drop off once the feather is fully grown, leaving a bare shaft with a spatulate tip.

Behavior Motmots are usually seen alone or in pairs. The birds typically perch upright and motionless for long periods on exposed perches. They often wag their tail like a pendulum, possibly to attract a mate or warn off rivals or predators. They sally out to capture prey from leaves, branches, on the ground, and occasionally in the air.

Voice Motmots seldom vocalize, mostly calling at dawn and dusk. Some larger species give deep double hoots, which may rise to a rapid bubbling sound. Other species give nasal twangs or hoarse honks.

Habitat Motmots are found in a fairly wide range of woodlands; primary and secondary, dense and open, moist and dry. Most occupy lowlands and foothills, but several species occur in highlands. Only one species, Keel-billed Motmot (*Electron carinatum*), is listed as Vulnerable, mainly due to its restricted range, low population density, and very secretive habits.

Movements Entirely sedentary within their range. Some Turquoise-browed Motmots (*Eumomota superciliosa*) have been recorded making seasonal movements in part of their range, in the Yucatan Peninsula of Mexico.

Diet Mostly large to medium-sized insects: particularly beetles, but also flies, dragonflies, and butterflies. Motmots may also take other arthropods such as small crabs and scorpions, and occasionally vertebrates including small reptiles and amphibians, and even small fish. Fruit is an important component in the diet of the larger species.

Breeding Courtship is not well studied; it appears to consist mainly of courtship-feeding and singing duets. The birds are monogamous. Most nest as single pairs, although some Turquoise-browed Motmots nest in small to large colonies. Motmots excavate nesting burrows in earthen banks and lay three to five eggs, incubated by both parents. Incubation periods have been determined for only two species; 21–22 days in Blue-throated Motmot (*Aspatha gularis*), and 18–20 days in Turquoise-browed Motmot. Both parents feed the altricial young. Fledging has been observed in only a few species, in which it occurred at 24–32 days.

Turquoise-browed Motmot is particularly striking, with bright turquoise on its brow, throat, primaries, and tail. Nicaragua.

Number of genera 6
Number of species 10
Conservation Status 1 species Vulnerable
Distribution Mexico and Central America, northern South America, and the Caribbean

BEE-EATERS *Meropidae*

An Old World family of agile, brightly colored birds, the bee-eaters are known chiefly for their graceful aerial pursuit of large insects. All are birds of warm, sunny lowland grasslands or forest edges. The family is closely related to kingfishers and motmots.

Structure The birds are 7 to 14 in. (17–35 cm.) in length. Most species are similar in shape. They have a broad head and a short neck. The bill is long, sharply pointed, and slightly decurved. The body is slender, the wings either short and rounded or longer and more triangular, and the tail usually long. The legs are short and the feet fairly weak.

Plumage Most species have vividly colored plumage. Bright green usually predominates on the upperparts, with blues, yellows, and reds on the head and underparts. European Bee-eater (*Merops apiaster*) combines a range of brilliant colors: chestnut mantle and back; bright yellow chin and throat; straw-yellow scapulars; deep greenish-blue underparts; deep bronze wing-coverts and secondaries; blue-green primaries; and a green tail. Carmine Bee-eater (*M. nubicus*) has a carmine-pink body, wings, and tail and bottle-green head, wingtips, and undertail. All species, apart from the bearded bee-eaters (genera *Nyctyornis* and *Meropogon*) and Black Bee-eater (*Merops gularis*), have a black facial mask that contrasts with the body.

The tail has 12 feathers. It usually has a square or slightly rounded tip, and elongated central feathers that form finely pointed streamers. However, in Swallow-tailed Bee-eater (*M. hirundineus*), the outer feathers are longer than the central ones. Males in all species have brighter plumage and a longer tail than females.

Like most bee-eaters, **Green Bee-eater** (*M. orientalis*) has a black face-mask and long central tail feathers. India.

Behavior Bee-eaters are social birds, occurring in pairs, small groups, or large foraging and breeding colonies. The bearded bee-eaters occur mostly in isolated pairs, as do some of the forest-living bee-eaters, but most of the open-country species are highly social and frequently nest in colonies. They forage from high vantage points including tree-tops, roadside wires, and telegraph poles, where they intently watch the area around and above them and dash out on swooping, gliding flights to grab passing insects. In savannas where there are few high perches, several species, especially Carmine Bee-eater, sit on the backs of Kori Bustard (*Ardeotis kori*) and occasionally antelopes or larger animals, which also undoubtedly assist with flushing insects. Bearded bee-eaters forage within the canopy of tall trees, moving sluggishly except when in pursuit of insect prey.

Voice All species have very similar contact notes: a soft, rolling, purring *prruip*, *pruup*, or *prrilp*, or a slightly higher-pitched *tree* or *kreee*. The two *Nyctyornis* bee-eaters have harsher or more gruff call notes. In some smaller species, pairs often call to each other in flight or when landing. Most of the open-country bee-eaters are highly vocal.

Habitat Most species occur where there is a plentiful supply of insects, and several are rarely found away from water. The bearded bee-eaters live mostly in forests, where they are found in the middle and canopy level of trees, but they occasionally venture into more open areas and large gardens with tall trees. Three of the Central African species are closely tied to lowland rain forests. All other species live in more open areas, including savanna; grassland with sparse woodland; thickets; and the edges of deserts.

Movements The bearded bee-eaters and tropical forest species are largely sedentary. Carmine Bee-eater and White-throated Bee-eater (*M. albicollis*), which breed in Africa, migrate up to about 1,240 miles (2,000 km.) to winter in the coastal grasslands and equatorial forest belt. The most highly migratory species

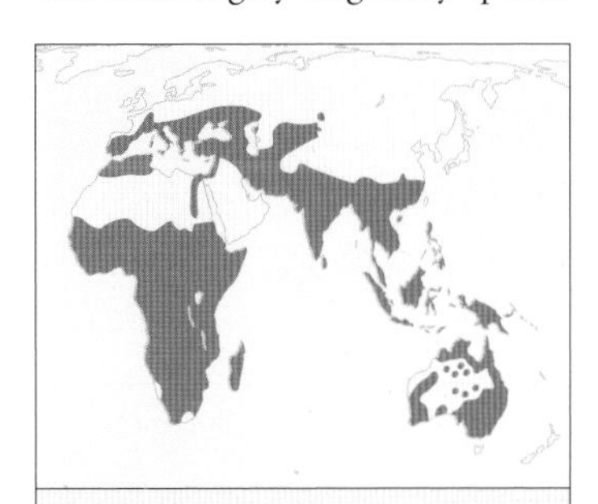

Number of genera 3
Number of species 25
Conservation Status Least Concern
Distribution Europe, Africa, Asia, Australia

Bee-eaters catch a wide variety of flying insects. This **Rainbow Bee-eater** has just captured a butterfly in midair. Location unknown.

are European Bee-eater, Blue-cheeked Bee-eater (*M. persicus*), and Rainbow Bee-eater (*M. ornatus*). The European species travels the longest distances. It breeds across Europe, central Russia, and North Africa, and most winter in southern Africa, with smaller numbers in West Africa. Migrating flocks travel up to 325 miles (520 km.) by day and congregate to roost at oases, on islands and headlands. Rainbow Bee-eater is an austral migrant, with birds from southern Australia moving north across the central desert areas to winter in northern Australia and New Guinea.

Diet As well as a variety of bees and wasps, bee-eaters take ants, termites, weevils, beetles, cicadas, dragonflies, moths, butterflies, crickets, and grasshoppers. They beat large insects against their perch to remove the wings, and rub the sting of larger bees against a branch to remove it before swallowing.

Breeding Most species are monogamous and pair for the duration of the breeding season. However, colonial breeders have a more complex structure, in which most pairs have helpers, either related or unrelated, to assist in the care and feeding of the young. Six species breed in loose to very dense colonies, those of Carmine Bee-eater and Rosy Bee-eater (*M. malimbicus*) often numbering tens of thousands together. All bee-eaters are hole nesters, excavating or re-using a nest chamber in a bank or cliff of soft earth. Riverbanks, road cuttings, and quarries are favored places, but some birds nest in flat, sandy ground. The nest has an entrance tunnel 4.25–9.75 ft. (1.3–3 m.) long, with a rounded chamber at the end, and takes up to three weeks to complete. The clutch of two to seven eggs is incubated for 18–23 days. The young leave the nest after about 30 days.

Carmine Bee-eater is highly social. The birds typically nest in large colonies, in cliffs of earth or soft rock, usually near river banks. South Luangwa National Park, Zambia.

HOOPOE *Upupidae*

The unique Common Hoopoe (*Upupa epops*) has long been admired for its beautiful plumage and distinctive crest. Although classified here as one species, hoopoes vary across their range in coloration and in the width of the bands on the back, wings, and tail. In addition, the Madagascar form has a different song. Eight subspecies are recognized.

Structure Hoopoes are 10–12.5 in. (25–32 cm.) in length. The sexes are alike. The bill is long, decurved, and laterally compressed. The wings are long, broad, and rounded; migratory subspecies have longer wings. The legs are short. The forward-facing middle and outer toes are fused at the basal joint; this feature helps the birds cling to vertical surfaces.

Plumage Coloration varies between subspecies. The head, upper back, and underparts are buff to bright cinnamon-chestnut; the lower back, wings, and tail are black and barred white or buff; and the crest is spotted black and white or buff. The bill is black and pink. Chicks hatch with downy plumage, then briefly molt into juvenal plumage similar to that of an adult bird.

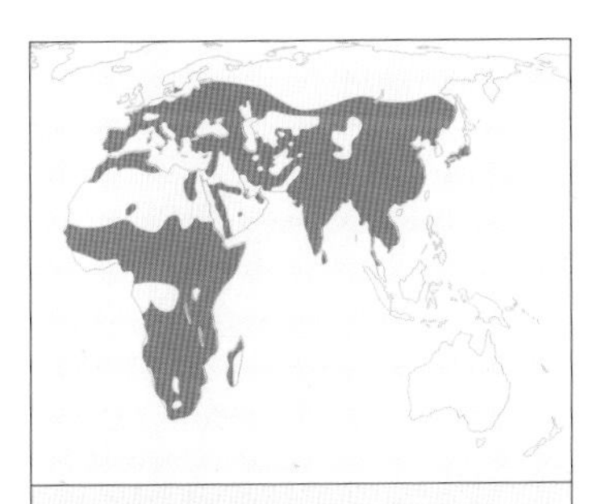

Number of genera 1
Number of species 1
Conservation Status Least Concern
Distribution Europe, Africa, Madagascar, south and east Asia

A **Common Hoopoe** returns to its nest. Despite their erratic 'butterfly' style of flight, hoopoes can be strong and agile in the air. Spain.

Behavior Hoopoes are usually found singly or in pairs, but may form small groups outside the breeding season. They are conspicuous birds, often using elevated perches, but they usually forage by walking or hopping on the ground, using their open bill to probe loose soil, vegetation, leaf litter, or dung. They may also use their feet to dig out tunneling prey such as mole crickets (genus *Gryllotalpa*). The crest may be raised and fanned when the birds are excited, or laid flat when they are calm. Typically, hoopoes 'freeze' when danger threatens. Their flight is undulating and sometimes erratic, but they are highly proficient at evading aerial predators such as falcons.

Voice The males' song, which gives the birds their name, comprises two to five resonant *hoop hoop hoop* notes. The Madagascar subspecies has a purring song. Hoopoes may also make caws or rasps.

Habitat Open country: beach dunes, savanna, woodland, and farmland, from sea level to about 9,000 ft. (2,800 m.). The Madagascar subspecies occurs at forest margins.

Movements Highly migratory to sedentary. Northern subspecies move into the ranges of more sedentary southern subspecies. In the Himalayas, migrants have been recorded at more than 20,000 ft. (6,400 m.).

Diet Arthropods, snails, and slugs; also vertebrates such as small lizards. Prey is swallowed whole; larger or harder items are softened by beating. Hoopoes may also take the fruit, seeds, leaves, or rhizomes of plants.

Breeding The birds are monogamous. The nest is a loose platform of vegetation, feathers, or other soft material, placed in a cavity in a rock or tree, a termite mound, a sand bank, or even a nest box; nest sites are often reused. Clutches comprise five to eight eggs and are larger in northern populations. The female incubates them for 15–18 days. Hatching is asynchronous. The chicks reportedly help defend the nest by spraying feces and emitting foul-smelling oil from their preen gland. They fledge at 25–30 days, but are attended by the parents for one or more weeks afterward.

Taxonomy The number of species varies, according to different authorities, from one to four. Hoopoes are closely related to the hornbills (family Bucerotidae) and wood hoopoes (Phoeniculidae), but the relationships remain unresolved.

WOOD HOOPOES *Phoeniculidae*

The wood hoopoes and scimitarbills form a distinctive family closely related to the hoopoes (family Upupidae) but separated by a number of differences including plumage and feeding behavior. All now occur solely within tropical Africa.

Structure Wood hoopoes are slim birds 9 to 18 in. (23–46 cm.) in length, with about half their length taken up by the long, graduated tail. The bill is long and decurved, especially so in the aptly named scimitarbills. The legs are strong, and the toes have long, sharp claws that are well adapted for climbing up trees and along branches.

Plumage Generally blackish or deep blue, with a strong metallic gloss that creates a green, bluish, or violet sheen. Forest Wood Hoopoe (*Phoeniculus castaneiceps*) has a bright chestnut head and breast. White-headed Wood Hoopoe (*P. bollei*) has a white face and throat. Five species have white bands in the wings and white tips to the tail. In several species the bill is bright red, and in the remainder it is black.

A **Green Wood Hoopoe** (*Phoeniculus purpureus*) grooms its companion. Mutual preening helps strengthen social bonds as well as remove parasites. Senegal.

Behavior The birds are almost entirely arboreal. They spend most of each day in pairs or family groups, moving from tree to tree. They examine the bark for food, often stripping away loose bark and probing behind it. Occasionally, the wood hoopoes (and more rarely the scimitarbills) descend to the ground, where they rather awkwardly hop or shuffle around on their short legs.

Voice Calls include a harsh dry, chattering *ka-ka-ka*, often given with the body bobbing up and down. Flocks performing territorial displays have a longer, cackling *kak-kak-kkkk*.

Habitat Open *Acacia* woodlands, thornbush, savanna with scattered trees, riverine trees, and wooded gardens. Forest and White-headed Wood Hoopoes occur in primary forest, secondary growth, and forest edges, the former species favoring the canopy of tall trees.

Movements Mostly sedentary: most birds breed in or close to the territory in which they were hatched. Fluctuations in the numbers of Black Scimitarbill (*Rhinopomastus aterrimus*) in southern Nigeria in the dry season suggest small-scale or local migratory movements.

Diet Mostly adult and larval spiders, millipedes, moth caterpillars, and small lizards. The birds may also take birds' eggs, and several species eat seeds and berries. In parts of their range, Green Wood Hoopoes visit birdtables in large rural gardens.

Breeding Wood hoopoes are monogamous. They build unlined nests in the hollows of trees, and lay two to five eggs. The female incubates the eggs for 17–18 days and feeds the chicks with the assistance of nest-helpers; these are often young from a previous brood. In some cases, an adult may forego breeding to help another raise her chicks, and the next year, the helper may breed and enjoy the same support from others.

This **Common Scimitarbill** (*R. cyanomelas*) shows the strongly decurved bill typical of genus *Rhinopomastus*. Namibia.

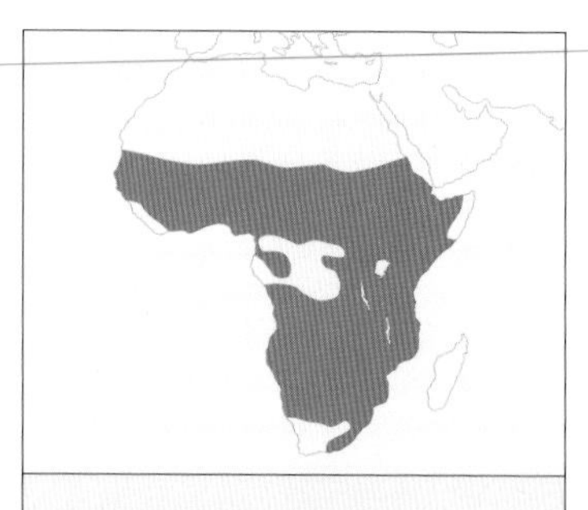

Number of genera 2
Number of species 8
Conservation Status Least Concern
Distribution Sub-Saharan Africa

HORNBILLS *Bucerotidae*

Wreathed Hornbill (*Rhyticeros undulatus*) has a variety of distinctive hornbill features, including a large, tough, decurved bill; a casque (albeit small); a long crest; and a gular pouch, which is yellow in males but blue in females. Borneo, Malaysia.

As their name suggests, hornbills are typified by their large, tough bill. Many species also have a large, horny structure called a casque on top of the bill, which is a unique feature of this family. Most hornbill species are similar in body structure. The main differences between them are in size (very few other bird families have such wide variations in size), plumage, and casque shapes and colors.

Structure Hornbills range from 12 to 47 in. (30–120 cm.) in length and from less than 3.5 oz. (100 g.) to more than 8.8 lb. (4 kg.) in weight. The family includes several smaller members, all of which lack the large casque above the bill, and two African species known as dwarf-hornbills. Sexual dimorphism is common mainly in size: males are larger than females, and have a larger and more colorful casque.

Hornbills have large or huge, broad, rounded wings and a long or very long tail. The legs are short, thick, and strong. The feet have broad soles and a syndactyl structure, in which the three forward toes are partially fused at the base.

The hornbills in genera *Aceros* and *Rhyticeros* are often referred to as the pouched hornbills, because they have a brightly colored, inflatable sac on the throat.

Plumage The brightest areas on most birds are the bill and casque, which vary from cream to banana yellow and black, to bright red and yellow. Some species have bare areas of blue, red, or yellow skin around the eyes or on the throat, which may be brighter in males. Most hornbills have boldly patterned black and white plumage. Some are browner or grayer, with white spots or pale fringes to the upperparts. Several have rufous or chestnut coloring on the head and neck. In several species, the tail is either very long or has elongated central feathers. The color may be entirely white, buffish,

Number of genera 13
Number of species 49
Conservation Status 2 species Critically Endangered; 2 species Endangered; 5 species Vulnerable
Distribution Africa, southern Asia, the Philippines, Indonesia, New Guinea, and Solomon Islands

The smallest hornbills belong to genus *Tockus*, like this pair of **Southern Yellow-billed Hornbills** (*T. leucomelas*). South Africa.

or black, or may be two-tone: black or dark gray in the center, with white outer feathers (or white tips), or mostly white with a broad black bar subterminally or at the tip.

The feathers of the head, neck, and body are fairly coarse and hair-like, while those on the belly and thighs are long and loose or fluffy. Several species have long head and face feathers that form a loose crest or that can be fluffed up, usually during displays, to form a halo-like crown. Silvery-cheeked Hornbill (*Bycanistes brevis*) has a huge head and neck frill, which is black with silvery tips and extends down to the sides of the breast.

Behavior Hornbills occur in pairs or family groups. Several of the larger species often live together in groups of up to 20, and some species have communal roosts. Hornbills are active foragers, spending long periods searching for food or feeding areas. Family groups often converge in areas of fruiting trees, occasionally forming flocks of hundreds of birds. However, each species has its own preferences, foraging heights, and techniques, so the species do not compete with each other and can live within the same area of forest.

Most hornbills fly with slow or labored, deep wingbeats, which make loud whooshing noises and are occasionally broken with short glides in a straight or slightly undulating direction. Some of the shorter-winged species fly with a rapid series of wingbeats and little gliding.

Voice Most hornbills are noisy and conspicuous birds, making frequent contact calls between pairs or family groups. Calls include distinctive single sharp notes or a series of cackles grunts, whistles, laughing, clanking, and howling noises.

Habitat The birds are found mostly in evergreen forests but also in open savanna woodlands and thornscrub. Most species occur between southeast Asia and New Guinea. Sulu Hornbill (*Anthracoceros montani*) and Walden's Hornbill (*Aceros waldeni*), both of which are endemic to small islands in the Philippines,

Like many hornbill species, **Indian Gray Hornbill** (*Ocyceros birostris*) is predominantly a fruit-eater; it prefers pipal (wild fig) and banyan fruits. India.

have declined rapidly in numbers and are now critically threatened, due to restricted ranges, habitat destruction, and hunting. Two other species are listed as Endangered and five others as Vulnerable, mainly due to continuing habitat destruction.

Movements Largely sedentary and territorial.

Diet Hornbills are omnivorous, but particular species take mainly fruit, insects, or small animals. The fruit-eating hornbills take a wide selection of fruits. Insects taken include beetles, cicadas, and butterflies. Other prey includes millipedes, earthworms, scorpions, small lizards and geckos, fish, frogs, birds and their eggs, squirrels, bats, and small poisonous snakes. Most of the fruit-eaters live in forests, while insect- and animal-eating species live in more open areas or edges of woodland. Fruit-eaters are entirely dependent on the availability of fruit, although some species also take a higher proportion of insects or animals in the breeding season.

An adult male **Knobbed Hornbill** (*Aceros cassidix*) brings food to his mate, who is sealed inside the cavity of the tree. Sulawesi, Indonesia.

Breeding Most species breed monogamously. Pairs bond for more than one season and possibly for life, and usually stay together year-round. Some species are known or thought to have helpers at the nest.

Hornbills naturally nest in holes in trees, sometimes with several species occupying the same tree. Where suitable trees are few, they may use earth banks or make use of artifical structures such as nestboxes. A pair may reuse the same nest site for several years.

Usually, the female chooses the nest site. With the male's help, she begins to close the entrance with a mixture of droppings, mud, and sticky food remains, which harden to a bricklike texture. Once the entrance is nearly complete, the female seals herself inside the nest-hole, leaving only a narrow slit through which the male passes food to her. This behavior, unique to hornbills, prevents the nest from being invaded by predators or by competitors for nesting space.

The clutch is normally one or two eggs in the larger species and up to four in the smaller ones. In smaller species, the female and young break out of the nest when the young are half-grown, at two or three weeks old, while in larger species they all remain within the nest until the young fledge. In the case of Silvery-cheeked Hornbill, this can mean that the female remains in the nest for a total of 130 days.

Taxonomy The hornbill family has often been thought to have the greatest resemblance to the toucans of the New World because of their large, colorful bills. However, recent research has found they are more closely linked, through the syndactyl foot structure, to hoopoes (family Upupidae), and to the wood hoopoes and scimitarbills (family Phoeniculidae).

GROUND HORNBILLS *Bucorvidae*

Northern Ground Hornbill has a larger and more tubular casque than that of the Southern species. Location unknown.

Imposing, long-legged African birds, ground hornbills resemble true hornbills, particularly in their pointed bill and casque. However, recent evidence suggests that they are distinct in their terrestrial existence. structure, and behavior. There are just two species.

Structure The birds stand up to 39 in. (100 cm.) tall. They have a large head and a massive, sharply pointed bill. Northern Ground Hornbill (*Bucorvus abyssinicus*) has a well developed, high, short casque, while Southern Ground Hornbill (*B. leadbeateri*) has only a prominent ridge. The eyes have long eyelashes, which are flattened to form a screen protecting the eyeball. Both species have broad areas of bare skin around the eyes and on the inflatable throat sac. They have a huge body; broad, rounded wings and a fairly short tail; long, thick legs; and toes with broad, strong soles and sharp claws.

Plumage Both species are almost entirely black, with white primaries. Male Southern Ground Hornbills have bright red facial and throat skin, while the females have a partially blue throat. Male Northern Ground Hornbills have a blue face patch and a red and blue throat patch, while in the females, the face and throat are entirely blue.

Behavior The birds usually occur in pairs or family groups of three, but occasionally in larger flocks; Southern Ground Hornbill breeds in cooperative groups of up to eight. They forage on the ground, digging, walking, or running after prey. They have been known to pursue eagles to take their prey, and may climb trees in pursuit of prey. Southern species also regularly roost in trees.

Voice Mainly silent, but in the breeding season, the birds inflate their air sacs to give a series of deep, booming *hoo-hoo-hoo-hoo* notes.

Habitat Fairly common and widespread in open woodlands, savanna and lightly wooded grasslands including edges of cultivated land.

Movements The birds are almost entirely sedentary. The southern species is highly territorial year-round.

Diet Mainly insects; also amphibians, small reptiles, mammals, and birds. Northern birds also take carrion, fruit, and seeds such as groundnuts.

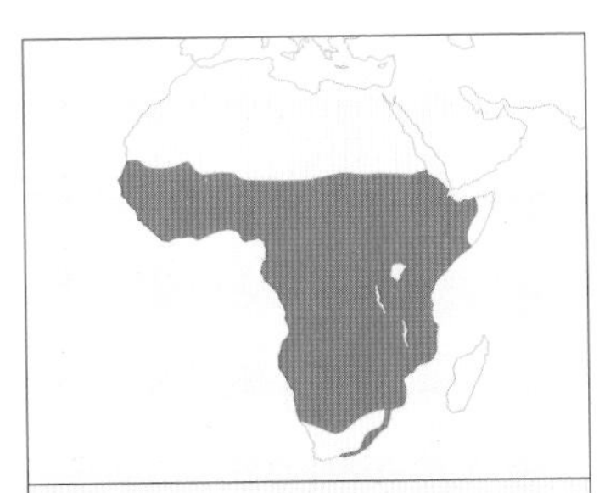

Number of genera 1
Number of species 2
Conservation Status Least Concern
Distribution Sub-Saharan Africa

Breeding Ground Hornbills are monogamous. In the southern species, the pair may be assisted by several adult and immature helpers. The nest site is a cavity in a tree or a hole in rocks, which the male lines with leaves and grass; unlike true hornbills, the birds do not seal the entrance. The female incubates the clutch of two eggs for up to 43 days. Chicks fledge at 80–90 days, but in the northern species the second chick rarely survives.

A **Southern Ground Hornbill** in flight. These birds fly with their bill and legs angled down, but they can reach speeds of up to 18 miles an hour (29 km/h.). South Africa.

BARBETS AND TOUCANS *Ramphastidae*

Ground-dwelling barbets, like this **Crested Barbet** (*Trachyphonus vaillantii*), have proportionately longer legs than most other members of the family. South Africa.

A diverse assemblage of primarily tropical species, the barbets (found in Africa, Asia, and the Americas) and the toucans (found only in the Neotropics) are notable for the complex taxonomic relationships within the family. Despite wide variations in shape and color, most of these birds are striking: toucans, especially, are instantly recognizable with their unique bill.

In **Prong-billed Barbet**, the hooked tip of the bill fits into the fork at the end of the lower mandible. Panama.

Structure The species vary widely in length, from the tiny tinkerbirds (genus *Pogoniulus*), at 3.5 in. (9 cm.), to Toco Toucan (*Ramphastos toco*), at 24 in. (60 cm.).

Barbets have a thick, heavy, pointed bill. In Toucan-Barbet (*Semnornis ramphastinus*) and Prong-billed Barbet (*S. frantzii*), the tip of the lower mandible is forked, forming 'prongs.' African ground barbets have a more slender bill, and several unrelated species have one or more prominent 'teeth' on the maxilla.

In toucans, the bill is huge and laterally compressed. The maxilla is keel-shaped and has serrated edges, and the decurved culmen ends in a hooked tip. Despite its size, the bill is extremely light. In some species, females have a shorter, deeper, straighter bill than males.

All species have a large head, a short neck, and a stout body. The tongue is proportionately long, and brush-tipped; in toucans, it also has fringed edges. The wings are short and rounded, while the tail may be short or long, and is graduated in some species. Toucans have an unusual articulation of the tail vertebrae, which allows roosting birds to bend the tail forward so that it rests on the back and extends over the bill (possibly to conceal it). The legs are short but strong. The feet are zygodactylous, with two toes facing forward and two behind.

Plumage In many species, the sexes are alike or differ only slightly, but some toucans (genera *Selenidera* and *Pteroglossus*) and New World barbets show marked sexual dichromatism. Toucans' plumage is more dense and lax than that of barbets. Juvenal plumage is typically softer and duller than that of adults. Molting has not been well studied, but most species have a complete postbreeding molt.

Most species are colorful and boldly patterned, often with contrasting chest bands, ear tufts, rump, or undertail coverts. Many barbets show spotting on the upperparts and streaking on the underparts.

Rictal bristles are prominent on many barbets but lacking in toucans. A few barbets have additional facial bristles on the crown or chin, and Fire-tufted Barbet (*Psilopogon pyrolophus*) has long red feathers at the base of the maxilla. Ground bar-

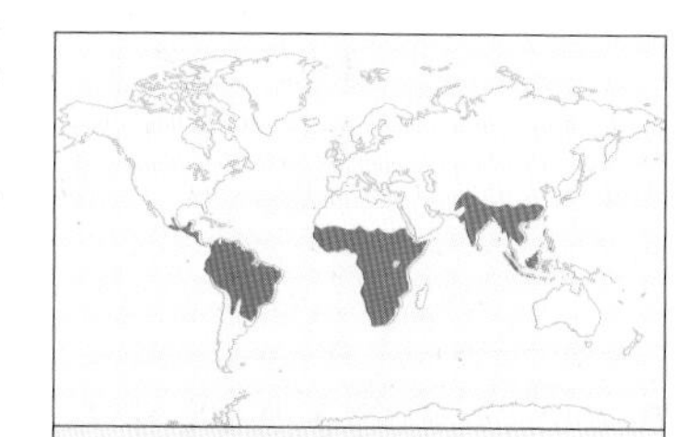

Number of genera 20
Number of species 120
Conservation Status 2 species Endangered, 1 species Vulnerable
Distribution Neotropics, sub-Saharan Africa, south Asia, Southeast Asia

A **Groove-billed Toucanet** (*Aulacorhynchus sulcatus*). The toucanets resemble toucans in shape but are smaller, with much more colorful body plumage. Venezuela.

bets have elongated crest feathers. Most unusual are the highly modified crown feathers of Curl-crested Araçari (*Pteroglossus beauharnaesii*), which have a texture like plastic.

Most toucans have colorful bare skin around the eye and the base of the bill. A few barbets also have bare skin around the eye; Naked-faced Barbet (*Gymnobucco calvus*) has a bald face and crown. The iris may be white, yellow, brown, red, or gray; most unique is the two-tone iris of certain toucanets and araçaris (genus *Selenidera*, some *Pteroglossus*), which has a dark horizontal bar.

In barbets, the bill is generally uniform or bicolored, in shades of ivory, gray, brown, yellow, red, and black. Many toucans, in contrast, have a spectacular, multicolored bill, sometimes with intricate patterning or with the serrations highlighted and appearing like teeth.

Behavior The birds roost in cavities, sometimes in pairs, families, or even small groups of unrelated individuals. Most forage singly or in pairs, but a few species are highly social, and several (especially toucans and *Semnornis* barbets) become more social outside the breeding season, foraging in small groups. New World barbets (such as genus *Capito*) join mixed-species foraging flocks, while some toucans join mixed-species flocks to follow army ant swarms. Barbets in genus *Stactolaema* use a hard surface as an 'anvil' to prepare hard insect food for their young. Toucan pairs will sometimes hunt as a team to catch vertebrate prey.

The flight is strong and, in many species, undulating; toucans often intersperse flapping with glides. Despite their short legs, most species, including the larger toucans, are very agile, able to cling vertically or hang from vegetation.

Voice Loud (especially some toucans), with a large repertoire of songs and calls—generally a monotonous series of hoots, growls, yelps, whistles, and trills. In many species, pairs duet, or groups counter-sing. Among barbets, the most complex vocalizations are those by some of the African genera *Lybius* and *Trachyphonus*; males and females have specific calls, which are synchronized and accompanied by displays. Nestlings make soft begging calls. Both barbets and toucans may also produce nonvocal sounds such as tapping or drumming on hard surfaces. Barbets may make bill-snapping sounds; toucans may clatter their bill or hit the bill against hard surfaces. *Ramphastos* toucans also make loud 'wing rustling' sounds using their two modified outer primaries.

Habitat A wide variety of tropical and subtropical habitats from open, arid woodland to humid primary forest, at elevations from sea level to 11,975 ft. (3,650 m.). Habitat loss or degradation pose the greatest threats for many species. Toucans are also threatened by hunting or by capture for the pet trade.

Movements Generally sedentary. Great Barbet (*Megalaima virens*) of Himalayas and some toucans undergo seasonal elevational movements, retreating to lower elevations during colder periods. Many undergo limited dispersal to follow local fruit supplies, or move to different habitats in the wet and dry seasons. Historically, large-scale irruptions of Toco Toucan (*R. toco*) and White-throated Toucan (*R. tucanus*) have occurred in northern Brazil and French Guiana, probably due to widespread fruit failure.

Diet Primarily fruit, but also invertebrates (insects and larvae, spiders, centipedes, and scorpions). The birds regurgitate large fruit pits, while smaller pits pass through their digestive system; in many areas, barbets and (especially) toucans are important seed-dispersing agents. Toucans may be opportunistic predators of small birds (including barbets

A **White-throated Toucan** (*R. tucanus*). Toucans are highly agile and very flexible. To sleep, a bird can rest its bill on its back and fold its tail right over to cover it. Guyana.

A male **D'Arnaud's Barbet** (*Trachyphonus darnaudii*) displaying. Courting pairs perch on tree branches facing each other, and lift and wag their tails, while bobbing up and down and singing a duet. Kenya.

and smaller toucan species) and their nestlings and eggs, using their huge bill to tear open or probe into nests. They may also take small vertebrates including roosting bats, reptiles, and amphibians. Nestlings are initially fed a diet of invertebrates (plus some small vertebrates in toucans), later mixed with fruit.

Breeding In most species, breeding coincides with the rainy season, when food resources are at their peak, but others can breed year-round. Most are monogamous. Pairs vigorously defend their territory and remain together through the year. At the nest, they show marked aggression to intruders from their own and other species. Courtship displays may be elaborate, and may include stereotyped postures associated with duets. Toucans may use their bill as a weapon during courtship fighting.

The colorful **Plate-billed Mountain Toucan** (*Andigena laminirostris*) lives in montane forests. It eats fruit from as many as 100 species of trees. Ecuador.

Some African barbets (such as some in genera *Trachyphonus* and *Stactolaema*), and some araçaris in genus *Pteroglossus*, nest in groups, in which nest-helpers assist in feeding nestlings; barbets' nest-helpers may also help with brooding chicks. Other barbets (*Gymnobucco* and *Calorhamphus*) form large colonies located in a single large tree. All are cavity nesters. In barbets, pairs use their powerful bills to excavate a fresh cavity in dead wood. Toucans tend to use natural cavities, or modify those made by barbets or woodpeckers; only rarely do they excavate new holes. Cavities made by smaller species are often stolen and enlarged by larger species. A few barbet species nest in fence posts or nest boxes. Some *Trachyphonus* barbets and *Ramphastos* toucans excavate or use burrows in dirt banks; some *Trachyphonus* also use termite mounds.

Depending on species, pairs raise one to four broods per season. The clutch is typically one to seven eggs in barbets, and one to six eggs in toucans; the eggs are incubated by both parents, for 12–19 days. The chicks hatch naked with eyes closed, and develop slowly; for example, toucans' eyes remain closed for two to four weeks. In barbets, the time taken to fledge varies from 17 days for tinkerbirds to 46 days for Toucan-Barbet. The young of some species, such as Coppersmith Barbet (*Megalaima haemacephala*), are virtually independent at fledging, while others receive extended parental

Coppersmith Barbet lives in forests and forest margins. Breeding pairs hollow out cavities in dead or dying branches to make their nest chambers. India.

care: in Black-collared Barbet (*Lybius torquatus*), the young are attended for more than five months. Toucan chicks fledge at 40–60 days old; little is known about the post-fledgling period, but the young are probably attended by their parents for two to eight weeks. African barbets may be parasitized by honeyguide adults or nestlings, which destroy the barbet's eggs and young, causing the loss of the barbet brood.

Taxonomy The toucans and barbets have historically been treated as two families: the Capitonidae (barbets) and the Ramphastidae (toucans). However, recent genetic analyses support uniting the two groups. In the classification method followed here, all the species are placed in family Ramphastidae, which is divided into four subfamilies: toucans and araçaris (Ramphastinae), New World barbets (Capitoninae), Asian barbets (Megalaiminae), and African barbets (Lybiinae).

HONEYGUIDES *Indicatoridae*

Although dull in appearance, honeyguides are well known and almost unique among birds because they feed on beeswax and honeycombs. Both their common name and their family name are derived from their habit of guiding humans and other animals, particularly honey-badgers, to bees' nests. The family is closely related to barbets and woodpeckers.

A **Lesser Honeyguide** (*I. minor*) feeds at an opened bees' nest in a rocky crevice. Some species rely on other animals to open nests inside rocks or trees. South Africa.

Structure Honeyguides are 4–8 in. (10–20 cm.) in length. They have a thicker skin than similarly sized passerines, for protection against insect stings. The head is small and the bill is short and stout. The wings are long and narrow, and the tail is slender. Usually, the outer tail feathers are slightly shorter, although Lyre-tailed Honeyguide (*Melichneutes robustus*) has outward-curving tips to the tail. All species have extremely good eyesight and a well developed sense of smell.

Plumage Mostly dull gray with greenish or olive-tinged upperparts and pale gray or whitish underparts. The wings and tail are dark gray, and the latter usually has prominent white panels in the outer feathers. Yellow-rumped Honeyguide (*Indicator xanthonotus*) has bright yellow patches on the face and rump.

Behavior Generally solitary, but up to 50 individuals of three or four species may gather at a bees' nest or honeycomb during a single day. Where more than one species occurs, there are dominance hierarchies, and within species immature birds are surprisingly dominant over adults. Some species habitually call and display to guide other animals, or humans, to the bees' nest. Once the other animal has opened the nest, the bird can reach the honey. Despite being fairly thick-skinned, honeyguides are not immune to bee-stings, and it is not unusual for them to be attacked by bee swarms; some have been found dead with more than 300 stings.

A **Greater Honeyguide** (*Indicator indicator*) with a piece of honeycomb. Most species eat beeswax. South Africa.

Voice The song of most species consists of a long, rapid trill which may be introduced by a shorter *wew* or catlike *miaow* note. Several African species have a repetitive song of two or three phrases, either musical or with the same notes repeated.

Habitat Tropical forests and woodlands. In the Himalayas, Yellow-rumped Honeyguide occurs in montane forests occasionally reaching the timberline.

Movements Mostly sedentary. Males invariably remain on or near their territory throughout the year; immatures and females wander further in search of new feeding areas.

Diet The birds eat the wax, eggs, and larvae of honeybees, and also collect wax from scale-insects. At other times they take aphids, beetles, ants, termites, and spiders.

Breeding Honeyguides are brood-parasites in the manner of cuckoos, laying their eggs and having their young raised principally by barbets, small woodpeckers, and occasionally flycatchers. Nestling honeyguides have a hooked bill with which they destroy the host's eggs or young.

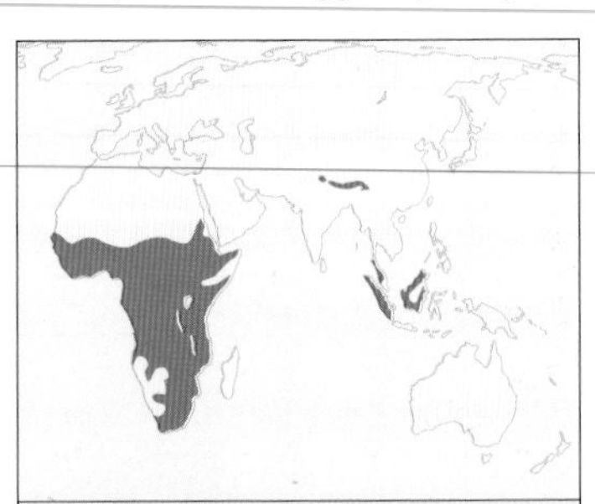

Number of genera 4
Number of species 17
Conservation Status 2 species Near Threatened
Distribution Africa, Himalayas to Southeast Asia

WOODPECKERS *Picidae*

A female **Green Woodpecker** (*Picus viridis*) brings food to the nest-hole for one of its growing chicks. Females typically lay between four and six eggs. U.K.

A large and familiar group of wood-chiseling species that are of fundamental importance in woodland and forest communities because of the nest cavities they construct. The birds are found in most temperate and tropical regions, but are conspicuously absent from the Australian and Papuan regions (they are among the taxa that do not cross Wallace's Line between Indonesia and the New Guinea region); woodpeckers are also absent from oceanic islands. The woodpecker family is usually divided into three subfamilies: one containing the wrynecks (genus *Jynx*, two cryptically patterned African and Eurasian species), another the piculets (*Picumnus*, *Sasia*, and *Nesoctites*, tiny short-tailed nuthatch-like woodpeckers, many quite range-restricted, found in the New World, African and Indo-Malayan tropics), and finally the widespread true woodpeckers. The last group is often subdivided into several tribes of related genera. Most woodpecker species retain healthy populations, and relatively few species are considered to be of conservation concern.

A male **Golden-fronted Woodpecker** (*Melanerpes aurifrons*) takes a drink from the Rio Grande, Texas, U.S.

Structure Great size range, from 4 in. (9 cm.) for the smallest piculets to 23 in. (60 cm.) in the Imperial Woodpecker (*Campephilus imperialis*). Weight range 0.25 oz. (7g.) to over 18 oz. (500g.). Sexes often differ slightly in size and structure; males are often slightly larger but in piculets they are subtly smaller.

Strong chisel-like bills are the hallmark of the family; stiff nasal feather tufts cover the base of the maxilla. In all true woodpeckers the tail feathers are stiffened, serving as support while climbing vertical surfaces; tails range from very short to moderately long. Tails are not stiffened in wrynecks and piculets. Toes are arranged with two pointing forward and two (only one in some species) to the rear, although functionally one or both rear toes are brought to the outer side while climbing. Wings are generally broad and somewhat rounded.

The skull is strong, with well-protected eye sockets; the base of the maxilla is hinged. These characters and the anatomy of the neck and jaw musculature protect the brain from injury during foraging and cavity excavation. Tongues can be exceptionally long, with the greatly elongated hyoid bones inserting over the top of the skull as far forward as the base of the bill or even around

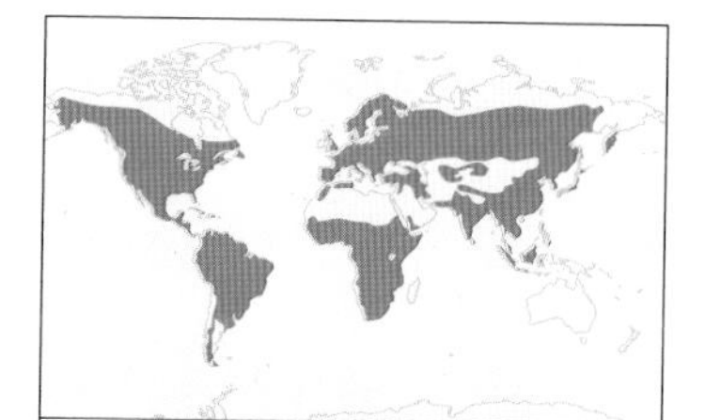

Number of genera 29
Number of species 210
Conservation Status 4 species Critically Endangered, 7 Species Vulnerable, 15 Species Near Threatened
Distribution Nearly worldwide, but absent from the Austro-New Guinea region and from oceanic islands

an eye; the tip of the tongue is slightly to strongly barbed in most species, but is brush-like in sap-sucking species.

Plumage The sexes are usually similar but often differ in the presence, or greater extent, of red on the head in the male—particularly on the crown, nape, and malar region. In a few species, notably Williamson's Sapsucker (*Sphyrapicus thyroideus*), there is strong sexual dichromatism. Juvenal plumage is usually distinguishable (the plumage is duller and more lax than in adults); definitive plumage is reached during the first year. Most familiar north temperate species are mainly pied or barred in black and white, with red patches on the crown or nape (usually only in males). However, plumage colors and patterns in the family overall are much more varied; many flickers and their relatives (genus *Colaptes*) and *Picus* species are green, yellow, or chestnut/red above and sometimes also on the underparts. In most species there is barring and spotting on the wings and underparts. A few woodpeckers are almost entirely black or (Great Slaty Woodpecker, *Mulleripicus pulverulentus*) gray. Many species (especially in *Dryocopus, Campephilus, Celeus* and several southeast Asian genera) are conspicuously crested. In Wrynecks (genus *Jynx*) the sexes are identical, and the birds are clad in cryptic grays and browns, almost suggesting nightjars (family Caprimulgidae) with their 'dead leaf' patterns. Piculets are mostly brown or olive above (frequently with white-spotted black crowns) and variously spotted, streaked, or barred below.

Chestnut-colored Woodpecker (*Celeus castaneus*) is a bird of moist tropical lowland forests in parts of Central America. It usually forages on the trunks and larger limbs of trees. Costa Rica.

Behavior Most woodpeckers forage at least in part by scaling tree trunks and limbs and using the strong bill to chisel into dead or living wood or flake away bark. Climbing behavior is aided by strong claws, toes splayed out forward and to the side, and stiffened tail feathers which provide support. Small species (and most notably the piculets) may be quite acrobatic on small branches, twigs, and vines. Some species forage mainly on the ground. Woodpeckers are generally solitary, or found in pairs or family groups. Some species occur in loose flocks or colonies; Acorn Woodpecker (*Melanerpes formicivorus*) and a few other species live in complex social groups. Flight is usually undulating, with bursts of wingbeats interspersed with short dips while the wings are folded. Some species have more continuous wingbeats; Red-headed (*M. erythrocephalus*) and especially Lewis's Woodpeckers (*M. lewis*) have regular crowlike wingbeats; these and some other species frequently hawk for insects in the air.

Voice Drumming (rapidly beating the bill against a resonant surface such as a dead branch) is the woodpecker equivalent of song, used in proclaiming territories and attracting or communicating with mates. Drum rolls may be of even cadence or may slow or accelerate slightly; sapsuckers and some other species have interrupted drums. Large woodpeckers in the genus *Campephilus* drum with a simple double knock. Vocalizations are varied, often with sharp, piercing, churring or crowing call notes; other calls

White-barred Piculets (*Picumnus cirratus*) live mostly in dry tropical and subtropical forests in South America, where they hammer vigorously, and audibly, to excavate small holes in their quest for insect prey. This woodpecker sometimes joins mixed-species feeding flocks. The red on its crown shows this bird to be a male. Brazil.

are given in longer series (whinnies, rattles, etc.). Softer calls typify interactions between pair members. Piculets give high, thin notes and trills. Begging young in nest cavities can be incessantly vocal.

Habitat In general, woodpeckers occur in woodlands and forests, from boreal and montane conifer forests to temperate hardwood forests, tropical lowland forests, and a variety of tree associations in between. Some thrive in parks, gardens, orchards, and other human-altered landscapes. Most woodpecker species make much use of dead or dying trees (snags) within their habitats. A few species occur in open savannas and even grasslands or deserts with only scattered woody plants. Trees are usually required for construction of nest cavities, but some species of open tropical environments excavate nests in termite mounds. Strong declines in a few

A female **Northen Flicker** (*Colaptes auratus*) showing characteristics of the yellow-shafted and red-shafted forms. Intergrades such as this are common in a broad zone on the Great Plains but may also be encountered anywhere within the species' range. U.S.

species (including four Critically Endangered and seven Vulnerable species) stem largely from deforestation and from forest management practices that alter the tree species composition and result in the loss of old-growth trees and snags. Some of the most highly threatened species are characterized either by very small overall ranges (such as Okinawa Woodpecker, *Sapheopipo noguchii*) or highly specialized habitat requirements (Red-cockaded Woodpecker, *Picoides borealis*).

Movements Most species are sedentary, or at most undergo limited annual movements (often altitudinal) that track food availability. A few northern temperate breeding species (most notably the North American sapsuckers) are strongly migratory, moving to lower latitude temperate or subtropical areas for fall and winter; Northern Wryneck (*Jynx torquilla*) is also strongly migratory. Some species that specialize on conifer seed crops or mast (for example, acorns from oaks) may undertake irregular irruptive movements outside the breeding season from areas with poor crops to more productive areas. Some of the pied woodpeckers, particularly the three-toed species, are somewhat irruptive and may colonize recently burned forests to exploit larvae of wood-boring beetles.

Diet Most species are largely insectivorous, many specializing on wood-boring beetle larvae and other grubs. Many species, including the flickers, the Green Woodpecker (*Picus viridis*) and related species, and the wrynecks are specialized ant-eaters (these species tend to have sticky tongues and enlarged salivary glands to neutralize the ants' acids). Termites are also important in the diet of many species. Others spend much of the year eating seeds, for example, acorns or pine seeds; some of these seed-eating species commonly store food in small caches (including many *Melanerpes* species and some pied woodpeckers) or even in large communal granaries (Acorn Woodpecker). Many more generalist feeders take a great variety of fruit and seeds and are sometimes pests in orchards. Some species (notably North American sapsuckers) eat living tissues from trees and tree sap that is harvested at small sap 'wells' drilled into the tree.

Breeding Generally in territorial, monogamous pairs, but some woodpeckers are loosely colonial. A few species breed in communal groups which may include young from previous breeding efforts. Nests are excavated in wood (often relatively soft wood found in dead or diseased snags or trunks, but also in living portions of trees); some species excavate into the ground or in termite mounds. Wrynecks use preexisting cavities rather than drilling their own. No nest material used, apart from wood chips and 'sawdust.' Eggs are white and unmarked. Clutch size varies; some tropical species lay two or three eggs, whereas temperate species may lay six, eight, or rarely even 12; in Acorn Woodpecker, at least, two or more females may lay eggs in a single nest. Incubation periods are relatively short, generally 10–14 days. The altricial young fledge in three to four weeks, but are attended by adults for weeks after fledging. Many small and medium-sized species suffer from the impacts of increasing populations of aggressive secondary cavity-nesters, such as European Starlings (*Sturnus vulgaris*) in North America.

Northern Wrynecks use their long tongue to extract ants from ant nests. This species is a long-distance migrant to and from most of its breeding range. The sexes are alike. U.K.

JACAMARS *Galbulidae*

The generally colorful, active jacamars are found only in the Neotropics. All species capture insect prey on the wing. Jacamars are most closely related to the puffbirds (family Bucconidae), although in the past various ornithologists placed them near the kingfishers (Alcedinidae). They bear a superficial physical and behavioral resemblance to the Old World bee-eaters (Meropidae), but the two families are unrelated.

Structure Jacamars range from 5.5 to 13.5 in. (14–34 cm.) in length. They are distinctive for their long, daggerlike bill. The wings are short and have 10 primaries, with the outermost notably reduced. Most species have long, rounded tails. The legs are short, with small, zygodactylous feet; the hind toe is absent in Three-toed Jacamar (*Jacamaralcyon tridactyla*), and much reduced in genus *Brachygalba*. The front toes are fused at the base (syndactylous), an adaptation seen in various species that excavate nesting tunnels.

Most jacamars, like this **Rufous-tailed Jacamar**, have a long bill, which enables them to catch large prey but avoid being struck in the face by flailing legs and wings. Brazil.

Plumage The sexes are similar; differences are limited mainly to throat coloration. Great Jacamar (*Jacamerops aureus*) and the 10 species in genus *Galbula* are generally iridescent green or bronzy on the upperparts, and have white, rufous, and sometimes iridescent green on the underparts. Chestnut Jacamar (*Galbalcyrhynchus purusianus*) and White-eared Jacamar (*G. leucotis*) are largely chestnut, while Three-toed Jacamar and the four species in genus *Brachygalba* are duller bronzy or brownish, some with whitish patches. Jacamars' contour feathers have a short aftershaft, a character shared with their closest relatives the puffbirds.

Behavior In general, jacamars are found singly or in pairs, but they will sometimes join mixed-species flocks. The birds typically perch upright with the tail hanging down and the bill held upward at an angle, much like bee-eaters or hummingbirds (family Trochilidae). Most species use open perches in the forest understory or at the forest margin, but Paradise Jacamar (*Galbula dea*) often chooses bare branches at the tops of trees. All species hunt by sitting quietly and watching for insect prey, then sallying out and catching it on the wing, in fast, acrobatic, twisting flights. To stun large insects or remove stings and venom, they beat their prey on a branch. They defend their territories by means of vocal displays; physical contact rarely occurs.

Number of genera 5
Number of species 18
Conservation Status 1 species Endangered, 1 species Vulnerable
Distribution Neotropics

Voice Most species have loud, high-pitched calls, sometimes given singly or more often in series; the calls may increase in tempo, sometimes to a trill, depending on the bird's level of excitement or agitation. Great Jacamar gives a long, eerie whistle, as well as other calls including a frequently heard meowing.

Habitats Jacamars are found mainly in lowland forests, where they have rather broad distributions, and prefer edges and clearings to the forest interior. Exceptions include Coppery-chested Jacamar (*Galbula pastazae*), which is found in subtropical woodland in Andean foothills, and Three-toed Jacamar, which prefers drier areas of the Atlantic forest in southeastern Brazil.

Three-toed Jacamar is listed as Endangered and Coppery-chested Jacamar is classed as Vulnerable. Two further species, Dusky-backed Jacamar (*Brachygalba salmoni*) and White-throated Jacamar (*B. albogularis*), are rather scarce, although they are not considered to be at risk.

Movements All species are sedentary within their ranges in the Neotropics. No significant movements are known, other than the dispersal of young and occasional short-distance wandering in adults.

A **Yellow-billed Jacamar** (*Galbula albirostris*) sits in a typical jacamar pose, with the bill angled upward. The syndactylous front toes can also be seen. Brazil.

A **Paradise Jacamar** with its prey. Jacamars readily learn which insects can be eaten; they reject unpalatable ones with vigorous head-shaking and bill-wiping.

Diet All species are insectivorous, with grasshoppers, butterflies, flies, beetles, dragonflies, and stinging insects included in their diet. Great Jacamar has also been known to take small vertebrates, such as lizards, from vegetation.

Breeding Most species nest as single, monogamous pairs; however, Three-toed Jacamar is more likely to nest colonially. In all species, rival males perform courtship displays involving vocalizations, perching side by side, flicking their wings, and pumping their tail. The birds nest in burrows, which pairs excavate using their specially adapted feet; they mainly nest in earthen banks, or sometimes use termite nests in trees. Some burrows may be used for roosting outside the breeding season. Most information on nesting is based on the only well-studied species, Rufous-tailed Jacamar (*Galbula ruficauda*). Typical clutches comprise two to four eggs, which are incubated for 18–26 days depending on species and conditions. The nestlings are altricial, and fledge after 20–26 days. Some species can be double-brooded.

PUFFBIRDS *Bucconidae*

A diverse family of smallish, stocky South American birds, puffbirds are found in Neotropical lowland forests. The family received its name from the birds' habit of fluffing out their feathers when perched, perhaps as a method of disguising their appearance from both predators and prey.

Puffbirds can be divided into several main groups. The 'typical' puffbirds include 15 species in four genera: *Notharchus*, *Bucco*, *Nystalus*, and *Hypnelus*. Similar, but with more streaky plumage, are the seven species in genus *Malacoptila*. The much smaller, monotypic Lanceolated Monklet (*Micromonacha lanceolata*) may be seen as intermediate between the *Malacoptila* birds and the nunlets. The six nunlets, in genus *Nonnula*, are the smallest and dullest group. The nunbirds, in contrast, are large puffbirds, comprising four species in genus *Monasa* and White-faced Nunbird (*Hapaloptila castanea*). The final species, Swallow-wing Puffbird (*Chelidoptera tenebrosa*), is the most distinctive in both shape and feeding style.

Structure Puffbirds range from 5 to 11.5 in. (13–29 cm.) in length. They are generally chunky in shape, with a large head, large eyes, and thick-based bill with a hooked tip. In most typical puffbirds, the tip of the upper mandible is forked and the tip of the lower one fits into this cleft; the purpose of this feature is not known, but it could enable the bird to take a firm grip on its prey. The birds also have short wings, a short to medium-length, rounded tail, and small, zygodactylous feet. Nunlets have a rather thinner bill and longer tail. Swallow-wing Puffbird is unique in having a smaller bill, longer wings, and a shorter tail.

Barred Puffbird (*Nystalus radiatus*) shows the typical puffbird plumage of muted colors but strong patterns. Ecuador.

Moustached Puffbird (*Malacoptila mystacalis*). The *Malacoptila* birds are sometimes called 'softwings' due to their loose plumage. Venezuela.

Plumage The coloration of typical puffbirds is a combination of earth-toned shades, often in bold, patchy patterns, and frequently includes a dark breast band or pale markings on the nape or head. A few species show entirely black-and-white plumage. Species in genus *Malacoptila* are similar to the typical puffbirds but are strongly streaked throughout with brown, buff, white, and black; in addition, they usually have patches of white on the breast and face, and sometimes show rufous areas. Lanceolated Monklet is uniform brown above and white below, with distinct black streaks. Nunlets are mostly brown, buff, and rufous with some gray, but have no bold patterning or patches. Nunbirds are larger and almost entirely black, with white patches on the face or shoulders; the exception is White-faced Nunbird, which is grayish-brown above, rufous below, and with white patches on the forehead and throat. Swallow-wing Puffbird has sooty gray upperparts, with a chestnut belly and a white rump.

All species have well-developed rictal bristles. Some species have additional bristles on the face and long feathers on the throat.

Behavior Generally secretive. Typical puffbirds tend to be found singly or in small family groups, while nunbirds are more visible in larger, boisterous, vocal groups.

Puffbirds' demeanor could be described as lethargic, except when in pursuit of prey. They sit motionless for long periods on open or hidden perches, keeping watch for insect prey. They capture their prey by using a 'sally-strike' method, which involves surprisingly quick and agile flights to the ground, tree trunks,

Number of genera 12
Number of species 55
Conservation Status 1 species Near Threatened
Distribution Neotropics

White-faced Nunbird is usually seen sitting in a hunched posture, ready to dart after prey; it also forages by using its bill to dig out insects from decaying wood.

leaves, or even in the air. Swallow-wing Puffbird is the only member of the family to capture insects from open perches, using aerial sallies.

Voice In keeping with their secretive nature, puffbirds vocalize very seldom: most often at dawn and dusk, but also at midday. In typical puffbirds, the main vocalizations consist of repeated, high-pitched whistles and trills, and in *Malacoptila* birds they comprise very high-pitched, sibilant notes. The nunbirds are the most vocal of the family: they sometimes give startlingly loud alarm calls, and groups use their loud and often melodious calls to bring their members together.

Habitats A variety of forested or wooded habitats, including lowland, foothills, open woodland, and savanna. White-faced Puffbird is the only species to occur in highlands.

No species is considered to be threatened, and only one, Sooty-capped Puffbird (*Bucco noanamae*), is listed as Near Threatened. However, several species are poorly known or have restricted ranges.

Movements Puffbirds are entirely sedentary within their range; no species has been recorded moving any significant distance beyond its home territory.

Diet Puffbirds take a variety of large insects including grasshoppers, butterflies, moths, caterpillars, cicadas, and beetles. They also eat other arthropods such as spiders, millipedes, centipedes, and scorpions, as well as small crabs and velvet worms (genus *Peripatus*). They occasionally take small vertebrates—most often lizards, but sometimes frogs and toads or even snakes. Some puffbirds include berries and other fruit in their diet.

Breeding Not well known. Puffbirds defend breeding territories mainly as monogamous pairs. Among the few species for which nesting habits are known, most typical puffbirds excavate cavities in termite nests in trees, while nunbirds, Swallow-wing Puffbird, and *Nystalus* puffbirds dig burrows on level or sloping ground. Nunlets and *Malacoptila* birds use both types of nest site.

Puffbirds are known to be only single-brooded, and lay clutches of two or three eggs. The eggs are small, rounded, and white (as is typical in cavity-nesting species). The incubation period for Swallow-wing Puffbird has been documented as 15 days. The young are altricial but are highly mobile; from a very young age, they are able to crawl to the entrance of the nest burrow to be fed. Known fledging periods range from 20 days in White-whiskered Puffbird (*Malacoptila panamensis*) to 30 days in the nunbirds.

Taxonomy The puffbirds show a superficial physical resemblance to certain of the kingfishers (family Alcedinidae), including the Asian forest kingfishers and the Australian kookaburras (genus *Dacelo*). The similarities extend to the birds' ecological niches and feeding habits, but the two families have never been considered to be closely related. The American barbets (subfamily Capitoninae) were formerly considered among their relatives, but current relationships are not well resolved. Puffbirds are currently placed in order Galbuliformes with only one other family, the jacamars (Galbulidae) as their closest relatives.

Swallow-wing Puffbird spends much more time in the air than the other puffbirds, typically making fast flights in pursuit of insects, or gliding to conserve energy when not actively hunting.

Spot-backed Puffbird (*Nystalus maculatus*) excavates its nest chambers in earth banks or even in flat ground. Brazil.

NEW ZEALAND WRENS *Acanthisittidae*

A female **Rifleman**. This species feeds by climbing trees in a spiraling path, probing the bark with its fine bill. New Zealand.

A tiny, ancient family, these passerines are endemic to New Zealand and its offshore islands. Up to seven species in five genera are recognized by scientists, of which only two, Rifleman (*Acanthisitta chloris*) and South Island Wren (*Xenicus gilviventris*), remain; the other species are now extinct, including Bush Wren (*X. longipes*), which was last reported on Kaimohu Island in 1972. The family has no close living relatives.

Structure Rifleman is the smallest of New Zealand's birds, at just 2.5–4 in. (7–10 cm.) in length and 0.18–0.25 oz. (5–7 g.) in weight. Both species are sexually dimorphic; unusually for passerines, females are larger than males. New Zealand Wrens have short, rounded wings and appear almost tailless. They are weak fliers; several of the extinct species were among the very few passerines to have become flightless. The birds have comparatively long legs and feet for their size. The female Rifleman, as well as being larger than the male, also has a slightly more upturned bill and a larger hind claw.

Plumage The male Rifleman has bright yellowish-green upperparts, while the female is streaked brown; both sexes have pale underparts. South Island Wrens are duller, with green and olive-brown upperparts, although the male has yellowish flanks. Both species have an obvious supercilium above the eye.

Behavior South Island Wren is a highly elusive bird, spending long periods foraging among rocks, crevices, and boulders in high mountain valleys. It also forages in swards of short alpine plants, including in snow-covered areas, and may even store food in rock crevices. When approached, the bird often bobs vigorously up and down, and it prefers to hop or run rather than fly. Rifleman feeds by gleaning insects from bark, foliage, and lichen as it works its way up trees, and rarely forages on the ground.

Voice Rifleman has a repeated *zipt* contact call, which is so high-pitched that some people cannot hear it. South Island Wren has two main vocalizations: a far-carrying three-note call and a 'whirring' note. Pairs occasionally duet.

Habitat The distribution of both extant species has shrunk considerably since the arrival of humans. Rifleman is still locally common in both native and exotic woodland, particularly in beech forest and on smaller offshore islands such as D'Urville Island and Stewart Island. South Island Wren is an alpine specialist, found from 4,000 to 8,000 ft. (900–2,400 m.). Listed as Vulnerable, its small, fragmented population is experiencing an ongoing decline due to nest predation by introduced stoats.

Movements Rifleman is sedentary and is territorial throughout the year. South Island Wren's seasonal movements still present a mystery: there is no evidence that birds move to a lower elevation during the harsh winter, as was formerly believed, but they often disappear from their usual territories at this time. It is possible that they may enter a state of torpor during the coldest part of the year, but this remains unproven.

Diet Both species take a wide variety of invertebrates, and occasionally fruit and seeds.

Breeding Both species form monogamous pairs to incubate the eggs and raise the young. Rifleman usually nests in tree holes. South Island Wren digs a hollow in a bank or among rocks, in which it builds a large nest of grasses, with a side entrance. Eggs are laid from September to December, and incubated for 19–21 days. The young are independent 28–35 days after fledging. Breeding pairs of Rifleman are often assisted by one or more 'helpers,' often unpaired males that may go on to breed with fledged female offspring from their 'adopted' family.

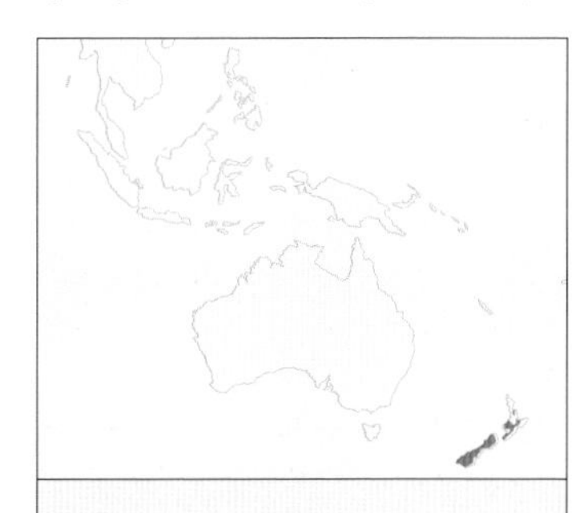

Number of genera 2
Number of species 2
Conservation Status 1 species Vulnerable
Distribution Northern and eastern Australia, southern New Guinea

BROADBILLS *Eurylaimidae*

Black-and-red Broadbill (*Cymbirhynchus macrorhynchos*) is a bird of moist lowland forests and mangrove forests. It typically builds its nest suspended from vegetation over standing water, presumably to minimize the threat of predation. Borneo, Malaysia.

The broadbills are small to medium-sized arboreal birds of tropical forests. They vary in size from the flycatcher-sized Rufous-sided Broadbill (*Smithornis rufolateralis*) to the roller-sized Dusky Broadbill (*Corydon sumatranus*).

Structure The species range from 4.5 to 11.25 in. (11.5–28.5 cm.) in length. Features common to all include a thickset body; a broad head with a wide gape and a broad, flat, hooked bill; and relatively large eyes. The African *Smithornis* species are relatively small, with stiff, twisted outer primaries, while the Asian *Calyptomena* broadbills are short-billed and short-tailed. Long-tailed Broadbill (*Psarisomus dalhousiae*) has a tail as long as its body.

Plumage Broadbill plumage coloration varies greatly. Sexual dichromatism is evident in some, but not all. Where males and females differ, the plumage of the former is brighter than the females. The three *Smithornis* species are relatively dull in color, with streaked brown, gray, and blackish upperparts, buffish underparts, and variable amounts of orange on the breast. The three *Calyptomena* species are mostly green, with variable amounts of black spotting and streaking, giving them excellent camouflage against the greens of foliage. Broadbills of the other genera vary greatly in color. Some are relatively dull: for example, Dusky Broadbill, which is mostly brown and buff, and Grauer's Broadbill, which is mostly green with some blue on the breast. Other species have bright, strikingly

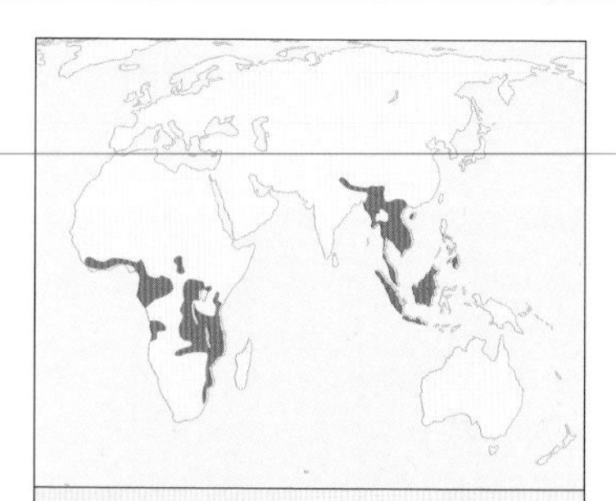

Number of genera 9
Number of species 14
Conservation Status 3 species Vulnerable
Distribution Parts of Africa south of the Sahara; parts of Southeast Asia

A female **Green Broadbill** (*Calyptomena viridis*). This species lives in forests up to 2,500 ft. (800 m.). Thailand.

patterned plumage. For example, Black-and-red Broadbill has bright maroon underparts, glossy black upperparts, and a blue and yellow bill. Black-and-yellow Broadbill (*Eurylaimus ochromalus*) has a black head, a pale blue bill, yellow irises, a peachy breast, a bold white collar, and black and yellow upperparts. The two species of wattled broadbills have blue eye wattles.

Behavior Mostly active early and late in the day, and some species are almost crepuscular. Broadbills often perch quietly in foliage with their neck drawn in, sometimes for long periods, occasionally making a short flight to take an insect or other prey before returning to the same perch. Some species are gregarious. For example, Long-tailed Broadbill often associates in groups of up to 15, and Dusky Broadbill often in groups of 10 to 20. The primarily insectivorous species readily join mixed-species feeding flocks.

Black-and-yellow Broadbill is notable for its blue bill and for its insectlike, quivering trill, a distinctive sound of Borneo's lowland forests. This species is heard far more often than it is seen. Borneo, Malaysia.

Voice All species use calls specific to alarm, foraging behavior, and the breeding season, but broadbills are often silent, and none is renowned for its song. African species produce mechanical sounds during display flights, and calls reminiscent of amplified frog croaks. Asian species are more vocal, producing a variety of whistles, squeaks, and screams. For example, Silver-breasted Broadbill (*Serilophus lunatus*) calls a mournful *ki-uu*, like a rusty hinge, and uses a contact call, *pri-iip*. Dusky Broadbill can be very vocal, producing six to eight screaming notes on an ascending scale: *ki-ky-ui, ky-ui, ky-ui*.

Habitat Mostly primary or secondary evergreen and deciduous tropical forest. Only African Broadbill (*Smithornis capensis*) lives extensively outside forest, in savanna woodland and scrub. Broadbills live from sea level to 8,350 ft. (2,550 m.). Some species are more specialist than others; for example, Black-and-red Broadbill prefers riparian forest.

Movements Sedentary apart from local movements in search of food.

Diet The *Calyptomena* species are specialist frugivores with a preference for figs, although some insects are taken in the breeding season. Other species eat mostly insects and small animals, including beetles, spiders, millipedes, snails, crustaceans, treefrogs, and lizards; they also eat fruits, and parties of broadbills may gather near fruiting trees. When hunting, broadbills either actively forage or adopt a sit-and-wait strategy, making a sally to take an insect or other invertebrate on a leaf.

Breeding Most species are probably monogamous, but lekking behavior observed in two species suggests polygyny in some. Some broadbills perform dramatic courtship displays. Male Green Broadbills spin around and around on a branch, with an open gape, occasionally touching the branch with their bill. The *Smithornis* species perform elliptical display flights, with the feathers of the lower back puffed up to show their white bases; as they fly, the twisted outer primaries vibrate, making a loud whistling sound. Other courtship displays involve bowing, crouching, frequently depressing the tail, and regurgitating food. The breeding season depends on locality.

Nest sites are usually camouflaged to reduce the risk of predation. Typically, nests are built from woven vegetation, suspended from a branch 10 to 33 ft. (3–10 m.) above the ground. Black-and-red Broadbill habitually builds its nest over water. Both sexes may cooperate in nest building, or the female may do all the work. Sometimes a nest is reused. Small clutches are typical, and females sometimes lay more than one clutch in a season. Two or three eggs are typical, with an incubation period of 17–18 days and fledging at 22–23 days. In three species, Long-tailed, Silver-breasted, and Dusky Broadbill, 'helpers' assist the parents with nest-building, and they may also help to incubate the eggs and feed the young. Post-fledging dependency is typical and may be long. For example, Banded Broadbill (*Eurylaimus javanicus*) parents are known to provide 70–80 percent of their offspring's food 13 weeks after fledging.

ASITIES *Philepittidae*

The asity family are small, forest-dwelling birds, endemic to Madagascar. They are classified in two genera: the asities (genus *Philepitta*) and the sunbird-asities (*Neodrepanis*). They are notable for the colorful wattles around the eyes of breeding males.

Structure *Philepitta* species are 5–6.5 in. (12.5–16.5 cm.) long, compared with 3.5–4.25 in. (9–10.5 cm.) in *Neodrepanis*. Males are slightly larger than females. *Philepitta* asities have a relatively short, slightly decurved bill, whereas that of the sunbird-asities is long, thin, and strongly decurved. *Philepitta* birds are plump and short-legged. All species have a very short tail with 12 feathers.

Plumage Asities are strongly sexually dichromatic. Male sunbird-asities have iridescent blue upperparts, yellow underparts, and green or blue wattles around the eyes; females are mostly green above and yellow below, and lack wattles. The adult male Velvet Asity (*Philepitta castanea*) is mostly black with green wattles, and breeding male Schlegel's Asity (*P. schlegeli*) has a black head with green and blue wattles, mostly green upperparts, and a yellow chest and belly. *Philepitta* females have olive-green upperparts and yellowish underparts, streaked with olive.

This female **Velvet Asity** has the plump body and stubby, slightly decurved bill typical of genus *Philepitta*. Madagascar.

In the breeding season, male **Common Sunbird-asity** has striking turquoise-blue wattles around the eyes. Madagascar.

Behavior The *Philepitta* species spend much of their time feeding alone in forest understory. When disturbed, they usually do not fly far. Sometimes they feed in the canopy, and they will also join mixed-species flocks on occasions. Sunbird-asities are more active feeders, rarely settling on a perch for very long, and behaving aggressively toward other birds and even humans.

Voice With the exception of Schlegel's Asity, songs and calls are squeaky and relatively quiet. Common Sunbird-asity (*Neodrepanis coruscans*) produces a series of rapidly repeated hisses, and Velvet Asity's calls have been likened to the squeaks of a teddy bear. The song of the male Schlegel's Asity is a series of whistles that first ascend and then descend the scale.

Habitat Schlegel's Asity inhabits seasonally dry rain forest and dry forest in western Madagascar. The other species live in rain forest, generally from 1,300 to 8,695 ft. (400–2,650 m.), with Velvet Asity more common at lower altitudes, and Yellow-bellied Sunbird-asity (*N. hypoxantha*) showing a preference for montane forest.

Movements Mostly sedentary. Common Sunbird-asity may make short-distance seasonal movements.

Diet *Philepitta* species feed mostly on the small fruits of understory plants, picked while the birds are perched or after a short hover. Sunbird-asities are nectar specialists, although they also take insects to feed their young.

Breeding Displays and nest-building start toward the end of the dry season, thus allowing the young to be fed during the rainy season, from late November to early May. Velvet Asity, the species studied in most detail, is polygynous. Males perform several displays within a small territory, to attract females; these displays include wing-flapping, gaping, inflating the wattles, and performing somersaults on a display perch. In this species, nest-building and incubation are the responsibility of the female alone. Displaying male Schlegel's Asities droop their wings, fluff-up their breast feathers, and gradually raise their tail. In all species, the nest is a hanging, spherical or pear-shaped structure constructed from grass, small leaves, bamboo strands, and moss, with an entrance hole and an overhanging 'porch.' Very little else is known of the family's breeding biology.

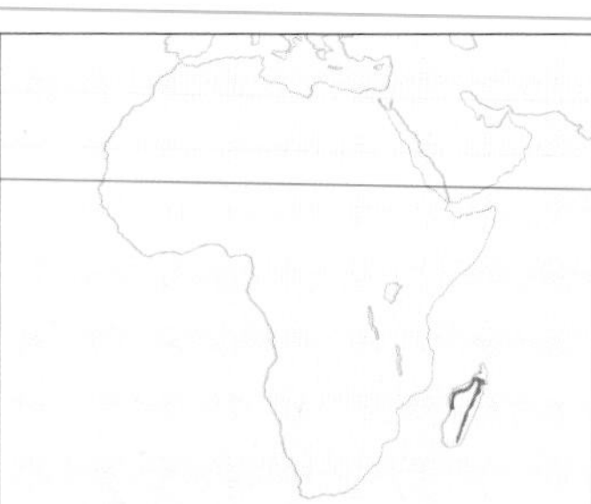

Number of genera 2
Number of species 4
Conservation Status 1 species Endangered, 1 species Near Threatened
Distribution Madagascar

SAPAYOA *Sapayoaidae*

The monotypic family Sapayoaidae contains only Sapayoa (*Sapayoa aenigma*), also known as Broad-billed Sapayoa: an inconspicuous and rarely seen bird that lives in the undergrowth of South American rain forests.

Structure Sapayoa is a smallish bird, about 6 in. (15 cm.) in length, with a distinctive broad, flat bill. It is superficially similar to Thrushlike Mourner (*Schiffornis turdinus*) or to a female manakin (family Pipridae), especially that of a subspecies of Green Manakin (*Chloropipo holochlora litae*). However, Sapayoa is smaller than Thrushlike Mourner and larger than most manakins, with a longer tail.

Plumage The coloration is an almost uniform olive, slightly darker on the wings and tail, and more yellowish on the throat and belly. There are no wing bars and no streaking. The legs and bill are gray. The eyes appear dark, with an orange-brown iris, and there is a pale, very indistinct eye ring. Adult males usually have a partially concealed (and difficult to observe) yellow coronal patch, a feature that is probably lacking in immature birds. Otherwise, the sexes are alike.

Behavior Sapayoa is a generally solitary and unobtrusive bird. To forage for food, it perches almost motionless, sometimes for long periods, on a branch, quietly watching for prey, before sallying to pick an insect from the air or to glean insects or fruit from the foliage of a nearby tree. At other times the bird will join a mixed feeding party of flycatchers, antwrens, or other forest-dwelling passerines as the group moves through the forest.

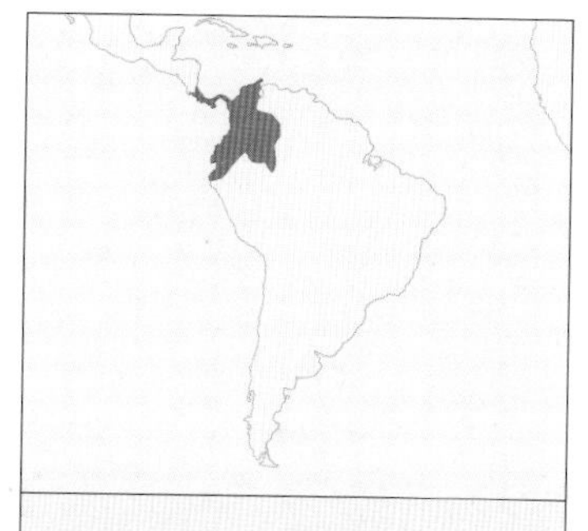

Number of genera 1
Number of species 1
Conservation Status Least Concern
Distribution Northwestern parts of South America

Voice A quiet trill and a louder *chip, ch-ch-ch* contact call.

Habitat The bird occupies the understory or middle level in humid lowland forest, at altitudes up to 3,400 ft. (1,100 m.), especially near small streams and in ravines.

Movements Sedentary.

Diet Mostly small invertebrates; the bird also takes fruits.

Breeding Few details are known, but breeding has been reported from February to April in the Choco region (which covers western Colombia, extreme eastern Panama, and northwest Ecuador).

Taxonomy The affinities of Sapayoa have long perplexed ornithologists. Once considered to be a species of manakin, it is now believed to be the only surviving New World relict of an ancestral group of broadbills (family Eurylamidae); the broadbills now live only in Asia and Africa. Some authorities, however, suggest that the species should be considered as a member of the broadbill family, or of the asity family (Philepittidae).

Sapayoa's inconspicuous, rather dull coloring helps the bird remain unnoticed as it quietly watches for prey in the forest.

PITTAS *Pittidae*

Hooded Pitta (*Pitta sordida*) shows a preference for primary riverine forest where it feeds on the ground. It forages among dead leaves for beetles, ants, termites, and their larvae, and earthworms. Borneo, Malaysia.

Small to medium-sized ground-dwelling birds of tropical forests, pittas are short-tailed, relatively long-legged birds that scour the forest floor for insects and other small prey. Most are very brightly colored, but they are shy and rarely seen.

Structure The largest species, Giant Pitta (*Pitta caerulea*), is up to 11.4 in. (29 cm.) long, while female Blue-banded (*P. arquata*) and Garnet Pittas (*P. granatina*) are only half this length. Males are usually slightly larger than females. The most obvious features of a pitta are its almost tailless appearance and its upright stance. Other features common to all pittas include a stout body; long, strong tarsi; long feet; a strong, slightly decurved bill; and shortish wings. Pittas are known to have a keen sense of smell as well as excellent eyesight and hearing.

Plumage Extremely varied and often very colorful: pittas include some of the most brightly colored of all birds. A majority of species have mostly green upperparts, and others feature a great deal of blue or brown on their upperparts. Many have a red belly. Some have a blue breast-band. A bold face pattern produced by black eye-stripes and paler supercilia are characteristic of several species. In 14 species, the sexes are virtually identical; seven others are almost identical; and in only nine species is there strong sexual dichromatism, with the males brighter than the females. Juveniles are mostly brownish. Most pittas have a black bill, while in a few species the bill is brown or ivory-colored. Leg coloration varies from pale pink to purplish pink, buff, or pale gray.

Behavior Pittas hop on the forest floor, turning over dead leaves with swift sideways movements of their bill. They are almost exclusively diurnal and feed most actively in the cooler parts of the day, in the early morning and late afternoon. In the hottest part of the day they may stand motionless on one leg. They are shy birds and when disturbed are more likely to hop away into dense cover rather than fly. Pittas usually roost above ground, on the branch of a tree. Strongly territorial even outside of the breeding season, pittas are usually solitary apart from during the breeding season.

Voice No songs are known, but all pittas call when alarmed and when

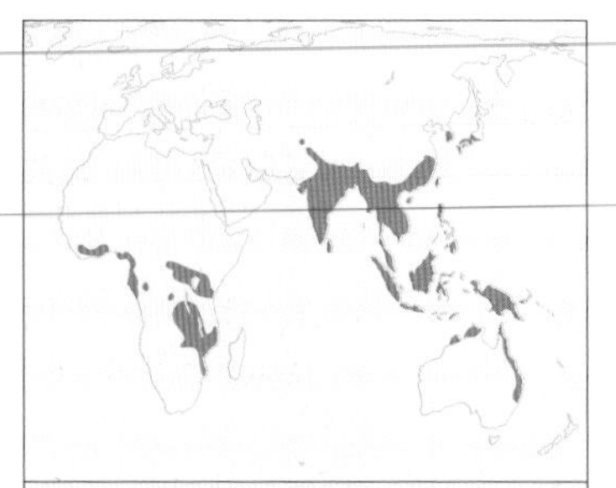

Number of genera 1
Number of species 30
Conservation Status 1 species Critically Endangered, 8 species Vulnerable
Distribution Tropical Africa south of the Sahara; South and Southeast Asia; Australasia

Indian Pitta breeds in the foothills of the Himalayas and winters as far south as southern India and Sri Lanka. Goa, India.

advertising their territory. Calls are monosyllabic or disyllabic and comprise fluty or whirring notes. Calls are most common in the early morning and the evening; Blue-winged Pitta (*P. moluccensis*) and Mangrove Pitta (*P. megarhyncha*) call on brightly moonlit nights. Male pittas make territorial calls more commonly than do females, and the call of one will often stimulate others to respond. The alarm calls of most species are similar: *keow* or *eeyow*. Territorial calls vary greatly. That of Eared Pitta (*P. phayrei*) has been likened to a yelping dog, while Blue-headed Pitta (*P. baudii*) is reminiscent of a crying baby, and Noisy Pitta (*P. versicolor*) has one call that sounds like *walk-to-work* and another that recalls a purring cat.

Habitat Most pittas live in moist tropical forest, but they may also occupy degraded forest and scrub. Ten species live mostly in lowland rain forest; others inhabit upland and montane forest up to 9,800 ft. (2,980 m.). Several show a liking for bamboo thickets; Mangrove Pitta has a strong preference for mangroves; Giant Pitta sometimes occupies marshy areas; and Indian Pitta (*P. brachyura*) may feed in open grassland or even on roadside verges.

Movements Most species are sedentary, although food shortages may induce local movements. Only four species are true migrants, including Indian Pitta, which moves from the Himalayan foothills to southern India and Sri Lanka.

Diet The most important element of most pittas' diet is earthworms, although these are not always available near the surface. Other food includes snails, which some species break on stone or wood anvils; termites, ants, and beetles; centipedes; and spiders. Less usual food includes small skinks, snakes, lizards, and frogs. For some species at least, plant matter is a minor part of the diet.

Breeding Pittas are monogamous, and copulation outside the pair bond is unknown. The breeding season is timed to coincide with the onset of the wet season, when earthworms and other invertebrates are plentiful and easier to find, so breeding varies from region to region. Little is known about courtship displays, although male African Pittas (*P. angolensis*) are known to jump up vertically from a branch and then parachute back to the branch, with their red belly feathers fluffed out; this is repeated several times a minute. Nests, which are fashioned by both parents, are loosely built, dome-shaped structures of dead leaves, sticks, moss, and grass, with an entrance off to one side. They may be situated on the ground, among tree roots, or between forked branches in a tree.

Clutch size varies from two to six eggs. Incubation, which lasts 14–18 days, is shared by the parents, as are feeding duties, which last for 15–17 days, until the young have fledged. Post-fledging dependency probably lasts for 10 days at most.

A male **Banded Pitta** (*P. guajana*). This species usually builds its nest 7–10 ft. (2–3 m.) above the ground. Both sexes incubate the clutch of three to five eggs. Malaysia.

MANAKINS *Pipridae*

A family of small forest birds, the manakins are known for their extreme sexual dichromatism, with males having highly colorful plumage, and for the males' elaborate courtship dances.

Structure The largest species are the tyrant-manakins in genus *Neopelma*, at up to 5.5 in. (14 cm.) long; in contrast, Blue-rumped Manakin (*Lepidothrix isidorei*) is just 3 in. (7.5 cm.) in length. All species have a stout body, a short tail, and a short bill that is very wide at the gape; this bill shape allows manakins to swallow food items that are large in proportion to the the birds' size.

Plumage Male manakins have some of the most striking and varied plumage of any birds. Many are mostly black, with brilliant patches of red, white, pale blue, or yellow on the head, rump, or elsewhere on the upperparts. Black Manakin (*Xenopipo atronitens*) and Jet Manakin (*X. unicolor*) are all black. Araripe Manakin (*Antilophia bokermanni*) is mainly white with a contrasting black tail and flight feathers and a bright red crest and nape. Golden-winged Manakin (*Masius chrysopterus*) is mainly black with a yellow crown; the feathers either side of the crown project backward to form two 'horns', and the feathers of the forecrown project forward.

A pair of **Striped Manakins** (*Machaeropterus regulus*), with the male at left and female at right, shows the dimorphism typical of the family. Males are very striking, but females are much more nondescript and can be difficult to identify to species. Brazil.

In some species, the males have unusual tail, flight, or head feathers. For example, in males of genus *Manacus*, the four outer primaries are stiff and slightly curved, producing a whirring sound when the birds fly. In Wire-tailed Manakin (*Pipra filicauda*), the shafts of the tail feathers end in filaments, which project beyond the end of the tail. In Long-tailed Manakin (*Chiroxiphia linearis*), the central tail feathers have extensions 4–6 in. (10–15 cm.) long—sometimes longer than the bird's head and body. Young males molt three distinct subadult plumages before acquiring full adult plumage in their fourth year.

Females do not share males' bright colors; they are often dull shades of green on the upperparts and yellowish on the underparts.

Behavior Foraging behavior is characterized by very quick aerial sallies from a perch to pluck fruits and seize insects. The birds sometimes join mixed-species flocks to feed on invertebrates disturbed by army ant swarms. Mixed-species social bathing has also been noted in some species. The most unusual aspect of males' behavior is the large proportion of daylight hours they spend at display sites (see feature box).

Voice Away from leks, both sexes are usually silent. At the display sites,

Number of genera 13
Number of species 48
Conservation Status 1 species Critically Endangered, 1 species Endangered, 2 species Near Threatened
Distribution Central America, northern South America

The male **Wire-tailed Manakin** uses his tail extensions for touching his mate rather than for visual display. Venezuela.

Club-winged Manakin (*Machaeropterus deliciosus*) makes a 'singing' sound like a violin's note by rubbing its modified secondaries together. Ecuador.

males make a variety of simple vocal calls. They also produce bizarre mechanical sounds in a variety of ways. In some species, air passing through their modified feathers makes a sound during flight. In others, males clap their wings over their back, or rapidly brush the tips of the primaries against the fanned tail.

Habitats Found from southeast Mexico to northeast Argentina in humid tropical forest, mostly below 3,300 ft. (1,000 m.). Some species prefer the interior of primary forest, while others favor forest margins and secondary growth. Black Manakin lives in gallery forest in mostly unforested areas, and Helmeted Manakin (*Antilophia galeata*) occupies gallery forest in cerrado. Habitat preference is driven by the need to find a constant supply of small fruits.

Movements Mostly sedentary, but some species make seasonal altitudinal movements.

Diet Mainly small fruits, supplemented by insects and spiders.

Breeding The most remarkable facet is the central role of the males' displays (see below). Manakins are polygamous; at leks, a small number of dominant males, sometimes just one bird, perform most of the matings. Males may not achieve such a position until they are several years old. Other aspects of breeding ecology are poorly known in all but a few species. Nests are small, located within a few feet of the ground, and built by the female alone. She also incubates alone, for 16–17 days in Golden-headed Manakin (*Pipra erythrocephala*) and 18–19 days in White-bearded Manakin (*Manacus manacus*). Fledging occurs at 13–15 days in Blue-crowned Manakin (*Lepidothrix coronata*), White-bearded Manakin, and Golden-collared Manakin (*M. manacus vitellinus*).

The rare **Black-capped Piprites** (*Piprites pileata*), endemic to southeastern Brazil, is traditionally classed with the manakins, but its placement continues to be disputed.

Amazing Show-offs: Manakins at the Lek

Male manakins have some of the most elaborate courtship displays seen in any passerines. Their life is dominated by the lek. Birds may attend their traditional sites for much of the day, sometimes perched motionless on a branch, but often displaying. Females, when ready to mate, visit males at leks; when they approach, the display activity becomes more impressive.

Male **Blue-backed Manakins** perform a 'cartwheel' display; as one bird moves forward, his partner jumps over him. Tobago.

Each species performs different displays, and there are often several 'movements' to a single display. Supporting males, presumably young birds waiting their turn to be dominant, often assist a lead male's choreography. For example, in the 'cartwheel' display of male Blue-backed Manakins (*Chiroxiphia pareola*), a dominant bird and an assistant take turns to leap backward over each other on a branch. The jumping gets faster and faster, giving the appearance of a rotating wheel, until the lead male's sharp call note brings it to a close.

When displaying to a female, Red-capped Manakin (*Pipra mentalis*) will first hop onto his favorite perch, then extend his thighs to show off his yellow feathering, or 'trousers.' Next, he flies quickly to another nearby perch, spins around to face the female, then flies back to the original perch, turning again. Each flight is accompanied by the whirring sound of his modified secondaries. The last part of the display involves the male holding his head down, lifting his tail, and moving backward along the perch with a series of short, rapid steps, giving the impression that he is gliding backward; finally, he fans his tail and raises his wings. ■

COTINGAS *Cotingidae*

Spangled Cotinga (*Cotinga cayana*) is typical of the blue cotingas, with iridescent turquoise plumage. Venezuela.

The cotingas are an amazingly diverse Neotropical family ranging from inconspicuous to spectacular birds, adapted for a wide variety of habitats and lifestyles. Members include the shimmering blue canopy dwellers; vibrant green fruiteaters; crowlike forest birds; glowing orange cocks-of-the-rock; plumed umbrellabirds; and the becards and tityras. Although the species look so dissimilar, the cotingas are united by structural characters such as skull and syrinx shape, toe structure, and tarsal scutellation (scaling on the lower legs), and behavioral features.

Structure Cotingas range from 3 to 20 in. (7–50 cm.) in length. Males are usually a little larger than females, but a few species exhibit reverse sexual dimorphism.

The bill, in general, is proportionately short. It varies in shape and size from wide and compressed to moderately long and crowlike, and may have a hooked tip. A few species have rictal bristles. Some fruiteaters have a very wide gape, enabling them to swallow large fruit. Plantcutters (genus *Phytotoma*) have a small, finchlike bill with a serrated maxilla, adapted for feeding on leaves, fruit, buds, and seeds.

Wing shapes also vary from long and pointed to short and rounded. The tail is typically short to medium-length, but Swallow-tailed Cotinga (*Phibalura flavirostris*) has a long, forked tail. The legs are short, and the feet large, with long, strong claws, especially in species that are adapted for gripping vertical perches.

Some species have bare skin on the head and throat, or wattles that can be inflated in courtship displays.

Plumage Sexes vary from similar to strongly dichromatic, males being more brightly colored than females. Plumage coloration ranges widely from uniform blue, black, gray, or white, to uniform above and streaked or barred below, to complexly patterned or garishly colored. Some species have violet and deep maroon coloring, which is found in few other birds. All but one of the fruiteaters (genus *Pipreola*) are vibrant green, patterned with yellow, some species with a contrasting black head or face and orange chest.

Certain species have glossy plumage due to structural modification of the feathers. The color shade may change with the angle of light, as in male blue cotingas (genus *Cotinga*), which can vary from deep blue to greenish blue. Male *Xipholena* cotingas' glossy burgundy to black feathers are hardened by heavy deposition of carotenoid pigments and modification of the feather barbs: the elongated greater wing coverts are pointed and, when the wings are folded, cascade decoratively over the otherwise contrasting white wing. The glistening red or orange feathers of fruitcrows (genera *Haematoderus*, *Querula*, and *Pyroderus*) are highly modified, with flattened barbs and no barbules. Capuchinbird (*Perissocephalus tricolor*) has curled orange undertail coverts, which it uses in courtship displays.

Males of some species have highly modified primaries—for example, curved or shortened—that create a

A male **Bare-necked Umbrellabird** (*Cephalopterus glabricollis*), with the red throat sac partially inflated. Peru.

Number of genera 34
Number of species 97
Conservation Status 1 species Critically Endangered, 6 species Endangered, 11 species Vulnerable
Distribution Central America, northern South America

sound as air rushes through them when the bird flies. Cocks-of-the-rock have very broad secondaries; in Guianan Cock-of-the-rock (*R. rupicola*), the outermost secondary has a long, filamentous orange fringe.

A few species have bushy, plush, or bristle-like crest feathers, or an elongated crest. Most unique are the upright crests of umbrellabirds and male cocks-of-the-rock. In cocks-of-the-rock, the crest is a fanned, comblike structure nearly obscuring the similarly colored bill. In umbrellabirds, the black crest curls forward, with dangling, hairlike tips resembling a fringed parasol.

In species with a bare head and throat, the skin may be brightly colored. Capuchinbird has a naked blue-gray head. Bare-necked Fruitcrow (*Gymnoderus foetidus*) has blue or bluish-white folds on the sides of the head and throat. Male Bare-necked Umbrellabird (*Cephalopterus glabricollis*) has an intense scarlet throat and upper breast, with a dangling, feather-tipped lappet of naked skin. Two other umbrellabird species have these feathered lappets; that on Long-wattled Umbrellabird (*C. penduliger*) can be up to 14 in. (35 cm.) long. Male Bare-throated Bellbird (*Procnias nudicollis*) has a blue-green throat with black bristles. Males of other bellbird species have dark, fleshy wattles above the bill or on the throat. In both umbrellabirds and bellbirds, the lappets or wattles can be inflated during displays.

Bill color varies from black to brownish-red, bright red (as in genera *Pipreola* and *Tijuca*), orange, or silvery-blue. Iris color ranges from dark brown to red, gray, white, or yellow. Leg color is typically black, gray, or brown, but is bright orange in many fruiteaters (*Pipreola*) or yellowish in cocks-of-the-rock.

Nestlings of most species that have been observed are covered by thick down, which may be cryptically colored or white (as in *Phibalura*). *Rupicola* species have sparse, long, downy feathers. Juvenal plumage resembles that of adults or females. There is a single complete molt, usually after breeding.

Barred Fruiteater (*Pipreola arcuata*) is named for the yellow and black barring on the underparts. Males have a black head, while in females the head is green.

Behavior Cotingas are arboreal. They are typically solitary or found in pairs or family groups, but several species regularly join mixed-species foraging flocks. Such flocks may gather at fruiting trees, but only plantcutters and Bare-necked Fruitcrow regularly form flocks.

Most species are generally quiet and perch in one spot for long periods. Fruit or insects may be plucked from branches or seized with the bill during sallies. Purple-tufts (genus *Iodopleura*) hawk flying insects.

Voice Songs are given by males, and cover an enormous range of simple sounds, from loud, piercing squeaks, quacks, and clangs, to deep, booming groans or mooing, and resonant, clear whistles. White Bellbird (*Procnias alba*) utters a loud, bell-like *ding*

Male **Andean Cock-of-the-rock** (*R. peruvianus*). The bird is instantly recognizable by its semicircular, upright crest, which almost conceals the bill. Peru.

ding. Screaming Piha (*Lipaugus vociferans*) gives an amazingly loud, resonant, three-note *squeeee screeeah*. Umbrellabirds produce a deep boom augmented by their throat sacs. Females are generally less vocal or silent (although female vocalizations are still unknown for some species). Both sexes of plantcutters give nasal rasps. The modified wing feathers of some species make whirring, rattling, buzzing, or whistling sounds, used in displays. For example, those of Purple-breasted Cotinga (*Cotinga cotinga*) rattle, and those of Black-necked Red Cotinga (*Phoenicircus nigricollis*) make a tinkling sound.

A male **Three-wattled Bellbird** (*Procnias tricarunculatus*) calling. The bird's bell-like sounds can be heard up to a quarter of a mile away. Costa Rica.

Habitat Coastal mangroves, desert scrub, open woodland, and humid tropical forest from sea level to 9,840 ft. (3,000 m.). Cotingas face serious threats from loss and degradation of their habitats.

Movements There is little information. Most species are presumably sedentary, but may commute long distances in search of fruit. Some undergo seasonal movements from higher to lower elevations or vice versa. Southernmost populations of Swallow-tailed Cotinga and plantcutters are partially migratory.

Diet Most species are primarily frugivorous. They swallow fruit whole, and regurgitate large seeds. Fruit-eating cotingas probably represent one of the most important seed-dispersal agents in their territories. Some species are more omnivorous and may take insects or even small vertebrates; Scaled Fruiteater (*Ampelioides tschudii*) also eats snails. Plantcutters eat leaves, buds, flowers, and grasses. The nestlings' diet has a higher percentage of insects.

Breeding Many aspects remain poorly known. Some genera (such as *Ampelion*, *Pipreola*, *Iodopleura*, and *Phytotoma*) are monogamous, with parental duties shared. Others are polygynous, with the female assuming all parental duties. Cooperative breeding has been documented for Purple-throated Fruitcrow (*Querula purpurata*), with three to eight adults attending a single nest.

Several of the polygynous cotingas (genera *Lipaugus*, *Rupicola*, *Procnias*, *Cephalopterus*, and *Pyroderus*) participate in courtship leks, where males engage in elaborate vocalization, displays, or dances. Some perform flight displays that include specialized wing sounds; for example, Scimitar-winged Piha (*Lipaugus uropygialis*) makes three 'puffs' during its falling spiral display. The dominant male at the lek performs the most copulations.

Nests range from very small to large. The structures may be open cups or platforms of loosely woven plant material, placed in a tree or shrub. Cocks-of-the-rock use mud and vegetation to form a bracket-shaped nest attached to a ledge or rock face. Clutches usually comprise one to four eggs (but one or two in *Rupicola* and *Pipreola*). Incubation typically takes 15–28 days, but may be protracted in some species. Fledging usually occurs at 28–33 days, but may take up to 48 days in *Rupicola* and, by contrast, average 17 days in plantcutters.

Taxonomy Some of the features common to cotingas overlap with those of the manakins (family Pipridae) and tyrant-flycatchers (Tyrannidae). Some taxa in the Cotingidae remain problematic. For example, genera *Laniocera*, *Tityra*, *Pachyramphus*, and *Xenopsaris* are sometimes placed in the Tyrannidae, and Kinglet Calyptura is sometimes left as *incertae sedis* ('of unknown position').

A male **Black-necked Red Cotinga** on a display perch. Peru.

TYRANT-FLYCATCHERS *Tyrannidae*

The strikingly colored **Lulu's Tody-tyrant** (*Poecilotriccus luluae*) is a newly described species that was only discovered in 2001. Peru.

The species that make up this huge New World family actually have no definitive characters uniting them. Tyrant-flycatchers are highly diverse both in appearance (from drab and cryptic to colorful and boldly patterned) and general behavior (secretive to conspicuous). They also inhabit a wide range of environmental niches; the richest diversity of species occurs in the South American tropics, as a result of specialization in microhabitat and foraging habits.

Structure The species range from tiny (genus *Myiornis* includes the smallest passerines) to medium-sized, and vary in length from 3 to 16 in. (6.5–40.5 cm.). In many cases, the body structure is adapted to the birds' lifestyle, particularly foraging strategy and size of prey taken.

The vast majority of tyrant-flycatchers have long rictal bristles and a slightly hooked tip to the maxilla, facilitating the capture of insects. Otherwise, bill shape and length vary dramatically, from small and narrow, to broad-based and triangular, to relatively long and stout with a strongly hooked maxilla tip; genus *Oncostoma* has a peculiar 'bent bill.'

Wing shape also varies according to foraging strategy and/or whether the species is migratory or sedentary. The wings are proportionately long and pointed in aerial hawkers and long-distance migrants, while they are short and rounded in sedentary, short-distance sallying species, or highly specialized, as in genus *Atalotriccus* (whose four outer primaries are very narrow and reduced in size) or genus *Pachyramphus* (in which primary nine is short and emarginated in males).

The tail is typically medium-length, with a square, slightly rounded, or strongly notched tip. However, in some species the tail is exceptionally short (as in *Myiornis* pygmy-tyrants, in which it is almost nonexistent), while in others it is extremely long and modified. For example, in Long-tailed Tyrant (*Colonia colonus*) the innermost rectrices, or tail feathers, are long and narrow, growing to 6 in. (15 cm.), and Fork-tailed Flycatcher (*Tyrannus savanna*) has outermost rectrices up to 10 in. (25 cm.) long. The tail is even more bizarre in two species of *Alectrurus*: Cock-tailed Tyrant (*A. tricolor*) has a plumed tail like a miniature rooster, while Strange-tailed Tyrant (*A. risora*) has long, twisted outer rectrices with a naked basal vane.

Most tyrant-flycatchers perch with a relatively upright posture,

Number of genera 98
Number of species 400
Conservation Status 2 species Critically Endangered, 7 species Endangered, 17 species Vulnerable
Distribution North, Central, and South America

and have fairly small legs and feet adapted for this purpose. The legs and feet are least well developed in more aerial species (such as Cliff Flycatcher, *Hirundinea ferruginea*), and are longer and better developed in largely terrestrial genera such as the ground-tyrants (*Muscisaxicola*). Short-tailed Field-Tyrant (*Muscigralla brevicauda*) has scutellation (scaling on the legs) extending above the tarsal joint.

Plumage Generally, the sexes are alike, although a few species are dimorphic, the male being more colorful. Most tyrant-flycatchers are cryptically colored in shades of green, brown, rufous, or gray, many with paler whitish or yellowish underparts. Plumage patterns range from uniformly drab to striking: some species have a contrastingly colored tail, cap, streaks, or bars; contrasting edges, tips, or webs of tail or wing feathers; a colorful coronal patch; or bold facial markings, chest bands, and rump patches. A few species are predominately black, black and white, or intensely colored with rufous, yellow, or red, as in the beautiful Many-colored Rush-Tyrant (*Tachuris rubrigastra*).

Many species have short, bushy crests, while some, such as tit-tyrants (genus *Anairetes*), have elongated crest feathers. Royal Flycatcher (*Onychorhynchus coronatus*) exposes an amazingly long and colorful, fanlike crest during displays. A few species have areas of colorful, bare, fleshy skin: Spectacled Tyrant (*Hymenops perspicillatus*) and some of the *Tityra* species have bare skin around the eyes, and male Strange-tailed Tyrant has a bright red throat during the breeding season. Iris colors range from dark brown to red, yellow, gray, or white. The bill is dark, often with a paler or more colorful base: for example, in many *Knipolegus* species it is a metallic silvery-blue. The legs and feet may range from black, brown, or gray to flesh-colored.

Juveniles have a distinctive, lax plumage that is similar to adult coloration but quickly replaced by a partial or complete molt. Adults have a single complete molt, usually following breeding.

Behavior Tyrant-flycatchers may be solitary, found in pairs, or gregarious, some with communal roosts. Several species join mixed-species foraging flocks.

Most species perch upright, while others are more horizontal. Some cock the tail upward, constantly dip or wag the tail, or flick or raise their wings. Most employ a 'search and sally' foraging strategy, sitting quietly in a strategic position to watch for prey. Methods of capture vary between species, from 'sentinel' aerial hawkers to those that sally-glean, hover-glean, or flush prey from within foliage. Others glean prey directly from foliage while perched, frequently changing perches to do so. Within genera, foraging behaviors are similar, or even stereotyped, such as the upward-striking behavior of tody-tyrants (*Platyrichini*). Terrestrial species employ similar capture methods: antpipits (*Corythopis*) walk after prey, in a fashion similar to pipits, and ground-tyrants hop and stop.

Voice In tyrant-flycatchers, voice is an important character used to define species limits. Calls are usually single or short series of whistles, chirps, clicks, weeps, peeps, or trills. Songs are longer, often incorporating a series of call notes. Most

An **Ochraceous Pewee** (*Contopus ochraceus*) watches alertly for prey, ready to dart into the air in pursuit. This species is rarely seen. Costa Rica.

species perform an even longer dawn song, often singing continuously and beginning before first light. A few species duet (such as those in genus *Suiriri*).

Habitat Very variable, from lowland desert, scrub, woodland, and forest to montane *puna* and *paramo*, and from sea level to around 18,370 ft. (5,600 m.) Habitat loss and degradation are the greatest threats facing the family.

Movements Species may be sedentary, dispersive, or short-distance or long-distance migrants.

Diet Mainly insects and their larvae. The birds may also take other arthropods, and small vertebrates such as reptiles, amphibians, and fish; the largest species sometimes take small birds and mammals. Most species take fruit seasonally; some are principally frugivorous.

Vermilion Flycatcher (*Pyrocephalus rubinus*), one of the most colorful tyrant-flycatchers, is also widespread, found from the southwestern U.S. to northern Chile. Location unknown.

Breeding Most species breed seasonally, with breeding times coinciding with peak prey or fruit abundance. The birds are generally monogamous and territorial. Many perform breeding displays that typically consist of stereotyped postures, movements, or aerial maneuvers; some perform mutual displays. The most elaborate displays occur in species with the strongest sexual dimorphism (of which some are, or are suspected to be, polygynous). Males of genus *Mionectes* gather at 'leks' to give courtship displays.

The nest is woven from vegetation and other materials. It varies in shape from a simple open cup to an elaborate pouch suspended in vegetation, a ball-shaped mass, or a pendulous structure with a side or bottom entrance. The nest may be located on or in the ground, on a ledge, in foliage, or in a natural or artificial cavity. Some species use the abandoned nests of other bird species. Piratic Flycatcher (*Legatus leucophaius*) steals recently constructed nests from a variety of species by harassing and evicting the owners. Typically, females are mainly responsible for nest-building, incubation, and brooding; males defend the territory, and both sexes feed the young. Clutches comprise two to six eggs, incubated for 12–23 days. The

Ornate Flycatcher (*Myiotriccus ornatus*) is recognizable by its yellow rump and the white patches above the bill. Ecuador.

Differing Migration Strategies among the Tyrant-flycatchers

Many of the tyrant-flycatchers breed in higher latitudes in both hemispheres, and have to migrate when insects and fruit diminish during the colder winter months. Most of the migratory species are 'complete migrants,' which means they completely evacuate their breeding range. Several northern and southern species are relatively short-distance partial migrants, vacating a portion of the breeding range and coexisting with more sedentary populations ouside the breeding season. Elevational migrations are also documented for some Andean species that move down slope.

Say's Phoebe (*Sayornis saya*) breeds as far north as Alaska and spends the winter in the southern U.S. and Mexico. Arizona, U.S.

North American breeders generally can be divided into two groups: those that winter from Mexico to Central America or just reach northern South America, and those that winter entirely in South America. The latter group includes some of the New World's longest-distance migrants. For example, Eastern Kingbird (*Tyrannus tyrannus*) breeds as far north as British Columbia and Newfoundland and winters as far south as southern Argentina. In the tropical latitudes, food resources tend to be more available year-round. As a result, these habitats support many sedentary tropical species. The abundance of food has led to the evolution of a wide range of microhabitat and foraging specializations in many of these species. The greater reliability and availability of resources in the tropics also allows wintering migrant species to coexist with the resident birds.

Birds that breed in the temperate and subtropical latitudes of South America also must contend with diminishing food resources during the austral winter. For this reason, they are also migratory, moving north toward the Equator to winter. A few of these species have separate resident and migratory populations that overlap their distributions during the austral winter. For example, two subspecies of Streaked Flycatcher (*Myiodynastes maculatus*) can be found together in Amazonia, where they are believed to partition resources by food preference (migrants consuming more fruit) and microhabitat. Although our knowledge remains incomplete regarding the movements of many South American species, some are also thought to be complete migrants; for example, Cinnamon-bellied Ground-Tyrant (*Muscisaxicola capistratus*) leaves breeding grounds in southern Chile and southern Argentina and moves as far north as southern Peru and western Bolivia. Although none of the austral migrants can rival the distance traveled by their northern counterparts, some species nonetheless make impressive journeys: for example, Small-billed Elaenia (*Elaenia parvirostris*) breeds as far south as Argentina and southern Brazil and moves north to northern Colombia and Trinidad.

As with many migratory species, navigational errors can lead to short- or long-distance vagrancy; well known examples include eastern North American species (such as Great Crested Flycatcher, *Myiarchus crinitus*) wandering to the west coast, or South American species (such as Fork-tailed Flycatcher, *T. savanna*) turning up in North America. ■

young are altricial. They hatch with thin, often translucent down, but juvenal plumage begins to appear within a few days. The chicks fledge at 12–28 days, but remain with their parents for further periods from two weeks to several months.

Taxonomy Certain problematic species and genera have caused several revisions and exchanges of taxa between the tyrant-flycatchers of family Tyrannidae, the cotingas (Cotingidae), the manakins (Pipridae), and other families (with most such revisions between the first two families). For example, the antpipits (genus *Corythopis*) have been variously classified as 'antbirds,' in their own family Corythopidae, and currently as tyrannids. Some authorities place genera *Laniocera*, *Tityra*, *Pachyramphus*, and *Xenopsaris* in this family, but others treat these birds as members of family Cotingidae (as has been done in this book). Genetic and vocal studies are ongoing to further resolve relationships.

ANTBIRDS *Thamnophilidae*

A large Neotropical family, the antbirds occur over a wide range of habitats, with the greatest diversity of species found in the rain forests of Amazonia. The number of species gradually decreases to the north and south, with only a few found in Mexico and south to northern Argentina.

The common names of various groups, such as 'antshrike,' 'antwren,' and 'antvireo,' often reflect the birds' similarity to other groups of birds, but the name 'antbird' itself derives from the close association of some species with army ant swarms. In fact, seven genera spend all their foraging time following ant swarms, which they rely on to flush insects and other arthropod prey.

Structure The birds range in length from 3 to 10 in. (8–25 cm.). Across the family, there are wide variations in overall body proportions, tail length, and posture (from upright, like tyrant flycatchers, to more horizontal, like vireos).

Most antbirds have a long, slender bill with a tomial tooth (reminiscent of falcons) and a hooked tip, but the length, depth, and hook vary, from robust with a prominent hook (antshrikes; most exaggerated in genera *Batara*, *Taraba*, *Frederickena*, and *Cymbilaimus*) to short and delicate with a slight hook (as in genus *Drymophila*). In three species (genera *Clytoctantes* and *Neoctantes*), the bill is stout with an upturned mandible, used to hammer or pry open prey. Several species have colorful bare skin around the eye or on the face, exaggerating the birds' big-eyed look; this feature is especially prominent on many of the ant-followers. A few species have stiff bristles on the face; for example, Spiny-faced Antshrike (*Xenornis setifrons*) has bristles below each eye. Two of the *Thamnomanes* antshrikes have bristles along the edge of the maxilla and lores.

Wings are generally short and rounded. Tail length varies from very short, in some *Myrmotherula* antwrens, to long and broad in large antshrikes, such as the aptly named Large-tailed Antshrike (*Mackenziaena leachii*). The legs and feet are strong, especially in the obligate ant-followers, which spend much of their time clinging to vertical stems; these species have well-developed leg muscles, rough soles on their feet, and long claws.

Number of genera 46
Number of species 206
Conservation Status 4 species Critically Endangered, 9 species Endangered, 13 species Vulnerable
Distribution Central and South America

Plumage The plumage is generally soft and fluffy, and some species have silky, elongated flank feathers. Most antbirds are cryptically colored in black, brown, rufous, and gray, often with paler underparts; a few are white or bright yellow below. Sexual dimorphism is fairly pronounced: males are predominantly black or gray, compared to brown, rufous or buff in females. Female plumage tends to be more geographically variable than that of males (heterogynism).

The body plumage may be uniform in color, or subtly or boldly patterned with spots, streaks, or barring; many species have contrasting cap, face, throat, or rump patches.

Ocellated Antbird (*Phaenostictus mcleannani*), found in the woodlands and lowland forests of Central America, is characterized by large eyes and blue facial skin. Panama.

The wings and tail are often contrastingly colored, highlighted with spots or bars. Several species have white patches, which may be either conspicuous or concealed (feather bases) on the back, scapulars, lesser coverts, or shoulder; these hidden patches are 'flashed' in threat or courtship displays. A few species, such as White-flanked Antwren (*Myrmotherula axillaris*), have white flank patches, which are often highlighted by wing-flicking behavior.

Many species have long, erectile crest feathers; this feature is most exaggerated in several antshrikes, such as Undulated Antshrike (*Frederickena unduligera*). White-plumed Antbird has modified feathers that form an erect, forked crest and spiky beard.

Iris color ranges from dark brown to reddish-brown, gray, yellow, white, or bright red (as in fire-eyes, genus *Pyriglena*). Bare skin around the eyes may be blue, green, yellow, or red; Male Bare-crowned Antbird (*Gymnocichla nudiceps*) has a bright blue face and forehead. The bill may be black to gray, often with a paler blue-gray lower mandible. Legs and feet are typically blue-gray, although some are brown to flesh-colored, pink (as in Silvered Antbird, *Sclateria naevia*), or bright orange (White-plumed Antbird, *Pithys albifrons*). Details of molting are poorly known. Juvenal plumage resembles that of females; some juveniles begin their molt to adult plumage during the post-fledging period.

Behavior Most species occur in pairs and defend territories of various sizes. Some species, such as Banded Antbird (*Dichrozona cincta*) are mainly solitary, but many others will join mixed-species foraging flocks passing through their territory, either occasionally (as in genus *Hylophylax*) or routinely (in the case of Plain-throated Antwren, *Myrmotherula hauxwelli*). Some species, such as *Thamnomanes* antbirds, are known to act as 'leaders' of such flocks, directing the flock's movements through the forest understory.

Seven genera (*Pithys*, *Gymnopithys*, *Pyriglena*, *Rhegmatorhina*, *Phlegopsis*, *Phaenostictus*, and *Skutchia*) are obligate followers of army ant swarms. These species constantly vocalize while keeping pace with the front of the swarm by jumping from stem to stem, and make sallies to capture fleeing prey flushed by the ants from vegetation and leaf litter. Unlike more sedentary antbirds that occur in pairs, several individuals of an ant-following species may be found at a swarm; these birds may cover long distances to find an active swarm. Other types of antbird, such as some *Myrmeciza* species, opportunistically attend swarms that move through or close to their territory.

Antbirds show stereotyped behaviors including various postures, wing-flicking, tail-wagging, quivering, and twitching. They may perform aggressive displays that involve raising the crest or groups of feathers to expose white or colorful areas of the plumage or face.

Voice Most species are highly vocal. Typical vocalizations are simple sounds given by both sexes, defined as loudsong (a long, stereotyped series of notes, typically metallic, nasal, or whistled in quality and ascending, descending, accelerating, or decelerating) or softsong (a quiet, hummed version of the loudsong). The birds also have a variety of calls, which include metallic, mewed, barked, buzzy, rattled, trilled, peeping, or chirring sounds; many calls are single notes, while others comprise multiple notes.

Habitat A variety of scrub, woodland, and forest, from sea level to about 7,875 ft. (2,400 m.). The complex Amazonian rain forest offers the greatest number of microhabitats; here, more closely related species coexist in relatively close proximity. Species are strongly partitioned by microhabitat, forest stratum, and foraging substrate. *Terenura* antwrens associate with mixed-species flocks in the canopy, and glean in

An immature **Long-tailed Antbird** (*Drymophila caudata*). Adults have more striking black-and-white plumage, with whiter underparts. Ecuador.

A male **Barred Antshrike** (*Thamnophilus doliatus*). Both sexes have a crest, which they raise in display or while singing. Panama.

branches, frequently investigating vine tangles. The 'checker-throated' *Myrmotherula* antwrens forage in dead leaf clusters as they travel with flocks in the understory. Ornate Antbird (*M. ornata*) joins flocks that pass through its territories in *Guadua* bamboo thickets. A few species, such as Banded Antbird, forage while walking on the forest floor. For all of the antbirds, habitat loss and degradation present the greatest threats.

Movements Generally sedentary, but some engage in local to moderate dispersal to locate army ants or between patchily distributed habitats.

Diet Insects and other arthropods; also snails, slugs, and small vertebrates such as lizards, snakes, and frogs. Many species are able to capture, kill, and eat prey that is large in proportion to themselves, typically by beating it against a substrate; ant-followers also vigorously shake their prey to dismember it. Exceptionally, Great Antshrike (*Taraba major*) is known to capture small rodents, tadpoles, and small fish. Some species take fruit. Indigestible material is regurgitated.

Breeding For many antbirds, breeding habits are poorly known. Most species are presumed to be monogamous and pair for life. Serial monogamy has been observed in White-plumed Antbird: the male remains with the first brood while the female produces a second with a different male. Both sexes participate in nest-building. Nests vary from a shallow, open, woven cup to a deeper purse, with the base supported by a branch, or slung by the rim from a fork in a branch. Some nests are placed in a tree trunk, or on or just above the ground; the latter may be domed or oven-shaped. Clutches comprise one or two eggs. Both sexes incubate the eggs and rear the chicks. The young are altricial and hatch naked with their eyes closed. Few data are available for incubation (14–20 days, where known), fledging (8–15 days), or duration of post-fledging care (a few weeks to three months).

Taxonomy Although diverse in size and appearance, the antbirds were originally allied as a family by shared morphological characters such as the hooked bill with tomial tooth, pattern of tarsal scutellation, fusion of toes, and sternal structure. Their relatedness is further supported by syringeal characters, and most recently by genetic analyses. All data support both family cohesion and separation from family Formicariidae ('ground antbirds:' the antthrushes and antpittas). Within the family, however, relationships between groups remain poorly resolved.

A female **Western Slaty Antshrike** (*Thamnophilus atrinucha*). This species typically forages in the understory of forest canopy. Panama.

GNATEATERS *Conopophagidae*

The gnateaters are a small family of species that live in the understory of Neotropical forests. Unobtrusive birds, they usually give away their presence only when they sing or call.

Structure The birds range from 4 to 6 in. (10–16 cm.) in length and from 0.7 to 1.5 oz. (20–40 g.) in weight. Almost all species are of similar length, apart from Black-bellied Gnateater (*Conopophaga melanogaster*), which is distinctly larger. All are relatively stocky. Their bill is flattened and slightly hooked at the tip. They have short, rounded wings, a short tail, and relatively long legs.

Plumage All species exhibit sexual dichromatism. In most species, the males have a black or gray breast, but Rufous Gnateater (*C. lineata*) has an orange-rufous breast. Chestnut-belted Gnateater (*C. aurita*) has a black throat and a rufous breast. The crown varies from black to rufous, and the belly may be white, pale rufous, dark brown, or gray. Females' coloration is generally not as bright, with rufous, gray, and brown tones dominant.

One of the most distinctive characteristics is the silvery or white plume behind each eye; this feature is most marked in males but is also present in females. It is seen in all species apart from most Black-cheeked Gnateaters (*C. melanops*). The plumes are usually obscured, but may be erected during territorial disputes and interactions between pair members, flaring out sideways as flashes of white.

Behavior The birds are not particularly shy, but are not often seen because they forage at low levels in relatively thick understory vegetation, usually no higher than 5 ft. (1.5 m.) above the ground. They forage in leaf-litter and low-level vegetation, and move around on the ground by hopping. While foraging, gnateaters usually flick their wings slightly, more so if alarmed.

Voice There are two main song types. Some species sing a series of two to eight irregularly spaced, high-pitched single notes: *greep ... greep ... greep*. Other species sing variations on a series of closely spaced notes that form a burst of chattering. All gnateaters give harsh or shrieking, single-note alarm or contact calls.

Habitat Tropical and subtropical forest from sea level to 8,000 ft. (2,450 m.). Black-bellied Gnateater favors the densest vegetation, such as bamboo thickets. Ash-throated Gnateater (*C. peruviana*) and Chestnut-belted Gnateater prefer terra firme forest in the Amazon Basin; Rufous Gnateater sometimes occupies secondary forest. Black-cheeked Gnateater lives in the Atlantic Forest in Brazil.

Movements Sedentary.

Diet Almost entirely arthropods, especially spiders, caterpillars, beetles, ants, and grasshoppers.

Breeding Probably monogamous. The birds remain in pairs year-round. Very little is known of the breeding ecology of most species.

The male **Black-cheeked Gnateater** is distinctive for his rufous crown and black mask; females lack these features. Brazil.

However, Rufous Gnateater is known to breed from August to January in southern Brazil, and Black-cheeked Gnateater's breeding season is between October and January. The birds build a small, shallow cup nest of rootlets and lichens, camouflaged with leaves and twigs. Nests are often less than 3 ft. (1 m.) above the ground. The typical clutch size is two eggs, and incubating birds usually allow very close approach by humans. Rufous Gnateater parents share incubation duties. One fledgling of this species was observed to be attended by a female for 45 days after it left the nest.

A **Rufous Gnateater** reaches up to take a caterpillar. The upright stance is typical of all species in this family.

Number of genera 1
Number of species 8
Conservation Status Least Concern
Distribution South America

TAPACULOS *Rhinocryptidae*

The tapaculos form a medium-sized New World family of small birds. Masters in the art of remaining unseen, they make up for their 'invisibility' with their loud, obvious, and diagnostic voices. The habitats that they are fond of, including thick bamboo tangles, often make them nearly impossible to find. In the majority of species this is compounded by their blackish or grayish colors, which help them blend in with their environment. Sometimes the small, grayish *Scytalopus* species are known as 'avian mice' for both their appearance and their habits, as they crawl under branches and run through tunnels of vegetation. The larger tapaculos of southern South America are more distinctive, sometimes colorful and boldly patterned. The colorful crescent-chests (genus *Melanopareia*) are considered by many not to be tapaculos, but instead to be more closely related to the antbirds (family Thamnophilidae).

Structure Tapaculos vary from 4–9 in. (10–23 cm.) in length. They are rotund birds with loose feathering on the flanks and tail coverts. The tail ranges from short and weak in *Scytalopus* species, to long in the crescent-chests; one feature all tapaculos share is that they commonly keep the tail stiffly cocked. They have strong legs, taken to the extreme in the huet-huets (genus *Pteroptochos*) and Moustached Turca (*P. megapodius*), which forage by digging and scratching on the ground. The family name Rhinocryptidae means 'hidden nose,' referring to a horny sheath that partially covers the nasal openings.

Chucao Tapaculo (*Scelorchilus rubecula*) forages alone or in pairs while walking on the ground. Chile.

Plumage The sexes resemble each other, with males being slightly brighter in some species. However, the two bristlefronts (genus *Merulaxis*) are clearly dimorphic, males having dark gray plumage, and females showing rusty or warm brown underparts. The great proportion of the family belong to the genus *Scytalopus*, and these are grayish or blackish birds, with varying amounts of rusty or brownish barring on the flanks. Several species have contrasting whitish eyebrows, foreheads, or crowns, and a group of Brazilian species is white on the underparts. Several of the larger species show rusty or cinnamon on parts of the throat, breast, belly, or flanks. Spotted Bamboo-wren (*Psilorhamphus guttatus*) and Ocellated Tapaculo (*Acropternis orthonyx*) are intricately spotted and uniquely patterned. The crescent-chests are cinnamon or buff below, sporting a black crescent on their chest; above, they vary but all show a dark mask and pale supercilium.

Behavior Tapaculos are generally terrestrial when foraging, although crescent-chests and Spotted Bamboo-wren may glean from leaves. Being highly terrestrial and usually living in dense habitats, tapaculos seldom fly. They avoid crossing areas without ground cover, and as such roads and rivers can be major barriers to movement. Tapaculos are highly territorial and stay on territories year-round.

Voice The loud and highly distinctive songs of tapaculos are key to

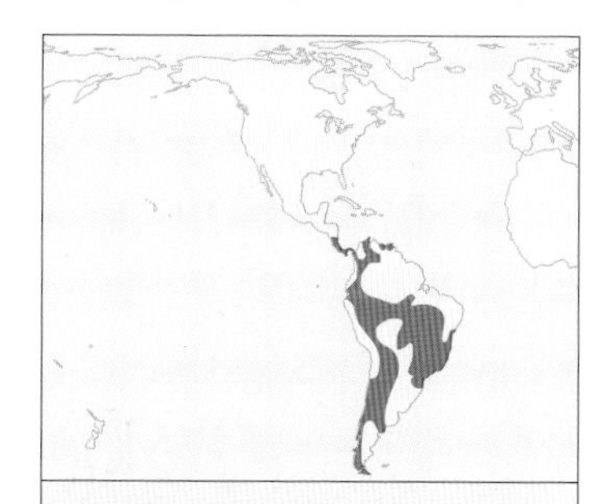

Number of genera 12
Number of species 55
Conservation Status 2 species Critically Endangered, 3 species Endangered, 1 species Vulnerable, 5 species Near Threatened
Distribution South and Central America

Although **Crested Gallito** (*Rhinocrypta lanceolata*) spends most of its time on the ground, males sometimes sing from exposed perches. Argentina.

their identification. Many species of *Scytalopus* are most easily differentiated by voice, and in fact recent taxonomic revisions have been based in great part on vocal differences. Songs are repetitive and characteristic, as well as being loud relative to the size of the birds. The quality of many tapaculo songs is distinctive, being mechanical and simple; they often have a sharp 'ticking' tone. In *Scytalopus* they range from two-syllable songs that are repeated incessantly to longer trills. The voices of the huet-huets and Moustached Turca are low-pitched and hooting, almost owl-like. Several species are known to duet.

Habitat Throughout South America, north to Costa Rica in Central America. Most *Scytalopus* species are found in moist montane forest, particularly where *Chusquea* bamboo is found. The birds also use thick ferns or other cover. Different *Scytalopus* species are sorted by elevation, with those occupying the highest elevations often in habitats above the tree line. In these cases, grass and fern cover, as well as large boulders, provide concealment. Tapaculos as a whole prefer dense habitats. In contrast, Sandy Gallito (*Teledromas fuscus*) lives in arid, shrubby areas, often along dry stream beds. Moustached Turca also occupies arid habitats, particularly those on slopes with large rocks, and it does not avoid foraging in the open.

Movements Mostly sedentary. Tapaculos are weak fliers and are not migratory. Most species are resident on their territory year-round. Long flights are nearly unknown in the family, making it a mystery how several species are common on Mocha Island, Chile, approximately 20 miles (32 km.) from the coast. Several southern temperate species range high enough in elevation that their breeding habitat is covered in snow for at least part of the winter. Some may move to lower elevations during the worst weather.

A juvenile **Magellanic Tapaculo** (*Scytalopus magellanicus*). Juveniles undergo one or more molts before attaining the gray-brown adult plumage. Tierra del Fuego, Argentina.

Diet Tapaculos feed mainly on ground-dwelling invertebrates, principally insects but also spiders. Species from southern temperate forests, such as Chucao Tapaculo (*Scelorchilus rubecula*), also take seeds and berries.

Breeding Very poorly known. Nests are known for fewer than half of all species of tapaculos. It appears that pairs are monogamous. Nests are often at the end of a tunnel dug in a bank. Ochre-flanked Tapaculo (*Eugralla paradoxa*) and Crested Gallito (*Rhynocrypta lanceolata*) nest above ground, making a globular nest hidden in dense vegetation. The huet-huets commonly nest in a cavity in a tree, often rather high up. The crescent-chests make a cup nest above ground. Eggs are rounded and unspotted white; they are more pointed and spotted in the crescent-chests. Clutch sizes tend to be two to three eggs, sometimes four.

Ocellated Tapaculo is one of the largest members of its family, at a little over 8 in. (21.5 cm.) long. It lives in dense Andean thickets, mostly above 8,200 ft. (2,500 m.). Ecuador.

ANTTHRUSHES AND ANTPITTAS *Formicariidae*

Shy terrestrial or semiterrestrial birds, antthrushes and antpittas live in dense undergrowth in the forests of Central and South America. The family is sometimes known as 'ground antbirds.'

Antthrushes, such as this **Striated Antthrush** (*Chamaeza nobilis*), derive their name from their thrushlike plumage, with brown upperparts and patterned underparts. Ecuador.

Structure The species range from 4 to 9.5 in. (10–24 cm.) in length. In most species, the sexes are alike or the males are slightly larger than the females; this characteristic is most pronounced in antpittas of genera *Pittasoma* and a few *Grallaricula* species. However, females are larger in some antthrushes (genera *Formicarius* and *Chamaeza*).

All species have a compact, rounded body; a sternum lacking a keel; short, rounded wings; a short tail; long legs; and large feet. Many have a proportionately large head and eyes. Antthrushes are medium-small, plump birds with a short, narrow bill, legs of medium length, a short tail, and a horizontal posture. *Formicarius* species cock their tail upward, a posture reminiscent of a rail. Antpittas are small to medium-sized, chunkier, with a shorter tail (some appear nearly tailless), upright posture, and very long legs. The bill varies from small and narrow to stout and heavy. Rictal bristles are present in most genera but absent in *Hylopezus* and *Myrmothera*.

Chestnut-crowned Antpitta (*Grallaria ruficapilla*) has the stout body, long legs, and big feet typical of antpittas. Ecuador.

Plumage Typically, the sexes are similar; however, males of two *Grallaricula* species have a rufous crown or ochraceous face lacking in females, and the sexes in *Pittasoma* species differ in throat or facial coloration. Plumage colors are generally dull earth tones. *Formicarius* antthrushes are dark gray to rufous, with a black face and throat and pale blue eye skin or eyelids, and a broad pale rufous band on the wing, which is revealed during displays. *Chamaeza* antthrushes are brown above, white or buffy below, and conspicuously scalloped, streaked, or barred. Antpittas may be uniform gray, brown, or rufous above, but generally with paler white, gray, or rufous underparts, which may be streaked, scalloped, or barred. Many species also have contrasting colors on the face, lores, crown, malar stripe, or breast. Iris color is typically brown, reddish-brown, or red; legs and feet may be blue-gray, brown, or flesh-colored. The bill is typically blackish, but in a few species it is pale, and in some there is a pale base to the mandible. Details of molt are poorly known. Juvenal plumage is similar to that of adults except in *Grallaria* and *Grallaricula*, which have a briefly held, fluffy juvenal plumage.

Behavior Most species are territorial and solitary or found in pairs, although *Formicarius* antthrushes have been seen roosting communally outside the breeding season. Many species are secretive and diffi-

Number of genera 7
Number of species 62
Conservation Status 4 species Endangered, 7 species Vulnerable
Distribution Central and South America

cult to observe. Some exceptions are the antthrushes, which are often seen on the forest floor, and certain antpittas living at high elevations; for example, Stripe-headed Antpitta (*Grallaria andicolus*) can sometimes be observed near the edges of more open *Polylepis* forest. Most species forage in leaf litter, but the smaller antpittas (genus *Grallaricula*) forage on low vegetation, clinging to and jumping between plant stems. Most of the family are thought to be poor fliers, and rarely do so, instead relying on their strong legs and feet: antthrushes walk or run, while antpittas hop.

Voice Songs are loud and ventriloquial, and include clear whistles or thrills, hoots, or popping sounds. They may be given as a simple single note, a two- or three-note series, or a long series (some up to 100 notes and lasting several seconds); the series may ascend or descend in pitch or volume, or accelerate or decelerate in pace. Most species sing from an elevated perch or the ground, usually at dawn or dusk. Calls are sharp or plaintive single or double notes, or shorter components of the song; *Formicarius* antthrushes sometimes call while in flight.

Habitat Mainly wet or humid forest habitats, but also dry or stunted forests, and isolated montane scrub in the *puna* and *páramo* zones, from sea level to 15,090 ft. (4,600 m.). Habitat loss and degradation pose the greatest threats to the family.

Movements Sedentary.

Diet Diverse: includes insects and their larvae, other arthropods, large snails, and earthworms, small vertebrates (frogs and snakes), and occasionally fruit.

Breeding Details are unknown or poorly studied in most species. The birds are presumably monogamous, some species found in pairs year-round. Breeding is likely seasonal.

A **Tawny Antpitta** (*Grallaria quitensis*) finds a worm. Location unknown.

The few described nests comprise a small cup of vegetation lined with finer materials and placed on or near the ground, on dead leaves or moss, or in the center of a plant or a tree cavity. Streak-chested Antpitta (*Hylopezus perspicillatus*) is known to use the old nests of other species. Clutches usually comprise two eggs, which may be blue-green and unmarked (genus *Grallaria*); spotted (*Myrmothera*); or white, light green, or buff to light brown with fine or coarse darker spots or blotches, possibly due to staining (as may be the case for genus *Formicarius*, which lay white eggs). The young are altricial; antpittas hatch naked with blackish-gray skin, while *Formicarius* have fluffy gray down. For the few species studied, both sexes incubate the eggs, for 16–20 days, and the young fledge 13–20 days later. To deter potential nest predators, Rusty-breasted Antpitta (*Grallaricula ferrugineipectus*) nestlings feign death and adults feign injury.

Taxonomy The family's separation from the 'typical antbirds' (family Thamnophilidae) is supported by syringeal characters as well as by DNA–DNA hybridization analysis. Genus *Pittasoma* is possibly more closely related to the *Conopophaga* genus of gnateaters, and may eventually be moved to the gnateater family (Conopophagidae), based on recent genetic evidence, sternal morphology, bill shape, and sexually dichromatic plumage unlike that of other antpittas.

A **Giant Antpitta** (*Grallaria gigantea*). These striking birds are not often seen. Having a limited range, they are at risk of decline due to habitat loss. Ecuador.

OVENBIRDS *Furnariidae*

The ovenbirds are a large neotropical family named for the dome-shaped nests, resembling clay ovens, that some species construct. They are diverse in appearance but similar in nest construction and the structure of the skull (mainly the nostrils and maxilla). Three subfamilies are traditionally recognized, based on morphology, plumage, similarity of nest materials and structures, and vocal or other behavior features: the Furnariinae (terrestrial), Synallaxinae (small-bodied, often with long, oddly shaped tails), and Philydorinae (more heavily built, forest-dwelling foliage-gleaners).

All races of **Pale-legged Hornero** (*Furnarius leucopus*), like this Pacific Hornero (*F. l. cinnamomeus*), have strong legs and feet, well adapted for walking and running. Peru.

Structure Ovenbirds range from 4 to 10 in. (10–26 cm.) in length. The sexes are similar; males are only slightly larger than females, with longer legs, wings, and tail. However, the family is diverse in body form and posture, and species can superficially resemble a variety of other bird groups: tits, warblers, larks, thrashers, jays, nuthatches, woodcreepers, and woodpeckers.

Tuftedcheeks, like this **Buffy Tuftedcheek** (*Pseudocolaptes lawrencii johnsoni*), occur mainly in wet mountain forests. Colombia.

The bill varies in shape from short, thin, and pointed, to long and slightly decurved for probing, or chisel-shaped with the tip of the mandible upturned and used for prying. Exceptionally, female tuftedcheeks (genus *Pseudocolaptes*) have a longer bill than males.

The wings are usually rounded or moderately pointed, but a few species have longer, strongly pointed wings. Tail shapes, lengths, and structures are extremely variable. The lark-like genus *Geositta* has a short, notched tail. Thistletails (genus *Schizoeaca*) and tit-spinetails (*Leptasthenura*) have a proportionately long, graduated tail. The tiny Des Murs's Wiretail (*Sylviorthorhynchus desmursii*) has only six rectrices, the central pair growing to 5.5 in. (14 cm.) long. Many species have pointed rectrices, some with exposed terminal shafts (as in Sharp-tailed Streamcreeper, *Lochmias nematura*). In treerunners (genus *Margarornis*), which use their tail woodpecker-style to brace against trees, the rectrices have sturdy vanes and exposed, decurved tips.

Plumage Generally the sexes are alike, but females are sometimes

Number of genera 55
Number of species 236
Conservation Status 3 species Critically Endangered, 7 species Endangered, 15 species Vulnerable
Distribution Central and South America

duller than males. All but one species are cryptically colored in shades of brown or rufous, or rarely gray, and most are paler below. Many species are strikingly patterned with white streaks or spots, or with conspicuously colored crown, throat, or cheek patches; some have bold wing bands or tail patterns. A few have a short crest. Orange-fronted Plushcrown (*Metopothrix aurantiaca*) is olive above and yellowish below, with a modified plush yellow-orange forecrown.

Iris color is typically brown, but ranges from red to yellow (in a few species). The bill is typically dark, but some species have a paler mandible. Legs and feet range from dark brown to gray; exceptionally, orange in Orange-fronted Plushcrown, and pink in Pink-legged Graveteiro (*Acrobatornis fonsecai*). Many species have distinct juvenal plumage. Molt details are poorly studied.

Behavior Most ovenbirds are arboreal, and a few are climbers, climbing up trees woodpecker-style; all have powerful legs and feet associated with acrobatic foraging maneuvers. Several genera are terrestrial and run well: some walk (such as *Furnarius*), some hop (such as *Asthenes*), and Sharp-tailed Streamcreeper will wade in shallow water.

The birds are mainly secretive and skulking. They are usually found in pairs or family groups, and many species defend territories year-round, sometimes using the nest as a roost outside the breeding season. Some are more conspicuous and gregarious, especially outside the breeding season, when they form large single-species flocks. Several species are regular participants in mixed-species foraging flocks.

Voice Loud, unmusical, or mechanical series of notes of variable length, often rising or falling in pitch, or accelerating or decelerating. Most songs are simple, many preceded or followed by a rapid trill or chatter. Leaftossers (genus *Sclerurus*) have a whistled song. Some genera (such as *Furnarius*) have a more complex series of notes, and pairs may duet. Vocalizations associated with other behaviors include notes given in song-flight displays (*Geositta*) and in aerial territorial displays (*Cinclodes*). Calls include single notes or short phrases or trills.

Habitat Ovenbirds occur in nearly every habitat, from rocky, marine intertidal shoreline to almost barren deserts, lowland forest, and *puna* (montane grassland), and at elevations from sea level to about 18,000 ft. (5,500 m.). Some species occupy a wide range of elevations and vegetation types, while others are extremely specialized on particular, patchily distributed plant species or microhabitats. For example, Peruvian Recurvebill (*Simoxenops ucayalae*) forages on *Guadua* bamboo, and Point-tailed Palmcreeper (*Berlepschia rikeri*) on groves of *Mauritia* palms. Many species coexisting within a single habitat partition resources by means of specialized foraging behavior, such as probing into dead leaves or epiphytes. Habitat loss is the most critical threat for ovenbirds as a whole.

Movements Most species are sedentary, although some engage in short-distance migrations or dispersal, including elevational movements after breeding or to escape harsh weather conditions. Otherwise, movements are generally poorly known and need further study. One species, Campo Miner (*Geositta*

A **Many-striped Canastero** (*Asthenes flammulata*) singing. The species has a soft, plaintive call like a cat's meow. Colombia.

A **Rufous Hornero** (*Furnarius rufus*) beside its finished nest. The nest is situated on a sturdy horizontal branch and is made of clay, mud, and straw, with a side entrance. Nests can weigh up to 10 lb. (4 kg.). Mato Grosso, Brazil.

poeciloptera), is nomadic, moving between recently burned grasslands.

Diet Primarily insects and other arthropods, some seeds, rarely fruit. In addition, Bar-winged Cinclodes (*Cinclodes fuscus*) takes mollusks; treehunters (genus *Thripadectes*) eat small lizards and frogs; and Brown Cachalote (*Pseudoseisura lophotes*) takes other birds' eggs.

Breeding Nearly every aspect of breeding biology is poorly studied for most species. The breeding period occurs during peak arthropod abundance, at the start of the wet season or during the austral spring and summer. The birds are presumably monogamous, and some at least are thought to mate for life.

Nest-building is an intensive project often begun months in advance of breeding. There are three basic types of nest, which can be categorized according to the materials and method of construction: the 'adobe oven,' made of mud and straw; a hanging purse of woven sticks; or a domed nest. The latter may incorporate specific softer materials such as *Sphagnum* moss, or be rigid and made primarily of large sticks; some are especially elaborate, with thatch used above the nest chamber, or with a roof or 'awning.' The floor may be lined with soft plant material. Some species nest in a tunnel dug into the ground, or in a natural cavity. In the nest chamber, they build a simple woven nest cup or sometimes a more elaborate, domed nest of fibrous materials.

Both sexes are thought to share parental duties; nest-helpers have been observed for a few species. Presumably, most species produce one clutch per year, but two broods have been documented for several species. Clutches comprise one to six eggs, with tropical species typically having smaller clutches. Incubation takes 14–22 days. The young are altricial and hatch with eyes closed, naked or with a very sparse covering of gray down. Fledging varies from 9–29 days; post-fledging care may continue for another month. Independent young of a few species are known to stay in their parents' territory for several more months.

Taxonomy The relationships among subfamilies, genera, and species are still matters of debate. Recent genetic analyses indicate that genera *Sclerurus* (leaftossers, six species) and *Geositta* (miners, 11 species) should be placed in a separate subfamily, Sclerurinae. Such studies also indicate that woodcreepers (family Dendrocolaptidae) are actually more closely related to ovenbirds than are leaftossers and miners, and should be merged as a subfamily into the Furnariidae, with the leaftossers and miners probably given familial status. Ovenbird genus *Xenops* may be more closely related to woodcreepers. Further study is needed to resolve these issues.

WOODCREEPERS *Dendrocolaptidae*

Found in neotropical forests, woodcreepers are a family of mainly arboreal birds. They look a little like some ovenbirds, but most species forage in the trees in a similar way to woodpeckers. Like woodpeckers, woodcreepers have evolved strong, pointed tails, which they use to brace against trees as they climb, and strong legs and feet to aid in hitching up tree trunks.

Structure Species range from 5 to 14 in. (13–36 cm.) in length. The sexes may be similar in size, or males may be slightly larger than females; genus *Deconychura* shows the greatest sexual size dimorphism. Bill shapes and lengths are highly variable, to suit different foraging strategies. They range from short and wedge-shaped (genus *Glyphorynchus*), or short and delicate (*Sittasomus*), to stout and powerful (*Xiphocolaptes*), very long and straight (*Nasica*), or very long, narrow, and strongly decurved (*Campylorhampus*). Most woodcreepers are similar in shape, with a relatively slender body and broad, rounded wings. The tail is long, and the rectrices have rigid shafts with exposed tips; in most species, they also have decurved ends, making it easier for the bird to brace itself against a trunk or branch. The feet have strongly curved claws, giving a secure grip on bark.

Plumage The sexes are alike. Body coloration is typically various shades of brown or rufous, with (often contrasting) rufous wings and tail. A few species are essentially uniform, but most are patterned on the head, chest, underparts, or upper back with distinctive, pale markings: streaks, teardrops, or bars. In most species, the iris is dark brown to red, but in a few it is gray or yellowish-brown. Bill colors range from ivory, blackish, or brownish, to gray or dull red, many with a darker culmen or maxilla. The legs and feet are typically blue-gray, gray, or brown. Juveniles are similar to adults. Limited evidence suggests that woodcreepers have a single annual molt.

Behavior Most species are arboreal, flying from tree to tree and using a variety of foraging strategies. Gleaners (such as *Xiphorhynchus* species) pick or probe food from bark. A few species have more specialized methods: for example, Buff-throated Woodcreeper (*X. guttatus*) searches dead leaves; Long-billed Woodcreeper (*Nasica longirostris*) probes into bromeliads; and Red-billed Scythebill (*Campylorhampus trochilirostris*) searches *Guadua* bamboo. A number of species regularly join mixed-species foraging flocks; some of these, such as genera *Sittasomus* and *Deconychura*, travel with flocks, sit on a vantage point such as a tree trunk to wait and watch for prey flushed by other flock members, then sally from this point to capture prey. Some genera, such as *Dendrocolaptes*, *Hylexetastes*, and *Dendrocincla*, are regular visitors to army ant

A **Wedge-billed Woodcreeper** (*Glyphorynchus spirurus*) balances with the tip of its tail as it climbs a tree. Ecuador.

swarms, and use a similar technique to capture prey flushed by the ants. Scimitar-billed Woodcreeper (*Drymornis bridgesii*) is the most terrestrial species, routinely feeding on the ground by probing in leaf litter or soft soil; it also runs well.

Voice The birds are most vocal at dawn or dusk. They have a variety of songs or calls, which consist of ascending or descending trills, rattles, or whistles. There is a great deal of individual variation, and many species have relatively large repertoires.

Number of genera 13
Number of species 50
Conservation Status 1 species Vulnerable
Distribution Central America, tropical South America

Habitat Forest and woodland from sea level to about 11,480 ft. (3,500 m.). In lowland tropical areas supporting multiple species, the birds coexist by occupying particular microhabitats or taking different food, or by interspecific aggression. Habitat loss or degradation are the greatest threats to woodcreepers.

Movements Most species are sedentary, but seasonal elevational movements are known for a few.

Diet Mainly arthropods: primarily insects, but also centipedes, millipedes, scorpions, and spiders. The birds also take crabs, snails, and occasionally small amphibians, snakes, and lizards; rarely, they take fruit or seeds. Great Rufous Woodcreeper (*Xiphocolaptes major*) has been documented eating a small bat.

Breeding Generally, breeding biology is poorly known. Most species breed (or are assumed to breed) annually, usually coinciding with seasonal peaks in food abundance. Most are monogamous, maintaining the pair bond year-round; both sexes contribute to nest-building, incubation, and feeding the young. Most *Dendrocincla* woodcreepers, however, are polygamous, and have no long-term pair bond; following mating, the female assumes all parental duties. Uniquely among woodcreepers, male Tyrannine Woodcreepers (*D. tyrannina*) gather to deliver long songs from 'leks' on top of ridges.

Nests are usually placed within 16–32 ft. (5–10 m.) of the ground, in a cavity. The birds may use a natural or woodpecker-excavated tree cavity, vine tangle or other dense vegetation, termite nest, or sometimes even a manmade structure. Clutches comprise one to three eggs, placed on a bed of vegetation or in a small lined cup. Incubation takes 14–21 days. The altricial young are initally naked but quickly grow a sparse, fluffy down; they fledge at 18–25 days. The young may remain with their parents for an extended period after fledging: a few months in smaller species, or up to a year in larger species, such as Red-billed Woodcreeper (*Hylexetastes perrotii*).

Taxonomy The woodcreepers are very closely allied to the Furnariidae (ovenbirds), with which they share syrinx morphology; recent genetic analyses suggest that they should be embedded as a subfamily within the Furnariidae. However, woodcreepers have a unique configuration of the ventral feather tracts, which, combined with their skull structure, ossified leg tendons, and foot morphology, sets them apart from some of the superficially similar ovenbirds. Further study is needed to resolve species limits, and may reveal additional cryptic species.

A **Strong-billed Woodcreeper** (*Xiphocolaptes promeropirhynchus*) with prey. This species forages on tree trunks and in bromeliads, and sometimes follows army ants. Ecuador.

LYREBIRDS *Menuridae*

A male **Superb Lyrebird** fans out and shakes his elegant tail. The lyre-shaped outermost feathers and delicate inner feathers can be clearly seen; also visible are the two ribbonlike feathers at the center of the tail. Tasmania, Australia.

This small family has only two species: Superb Lyrebird (*Menura novaehollandiae*) and Albert's Lyrebird (*M. alberti*). Both are ground-dwelling forest birds with a long, distinctive tail. Both are endemic to Australia. Superb Lyrebird occurs in Victoria and New South Wales, and in Tasmania (where it was introduced in 1934–45). Albert's Lyrebird is classified as Vulnerable: it is restricted to a small area of rain forest between Blackwall Range, New South Wales, and Mistake Range, Queensland.

Structure Lyrebirds are among the largest passerines in the world. Male Superb Lyrebirds are 31.5–38.5 in. (80–98 cm.) long, including a tail of about 21.6 in. (55 cm.). Females are slightly smaller, at 29–33 in. (74–84 cm.). Albert's Lyrebird is marginally smaller, with males reaching a maximum of 35 in. (90 cm.). Both species have long legs and powerful feet, adapted to their terrestrial lifestyle. However, their short, rounded wings are ill-suited for flying, so they can make only brief, cumbersome flights.

Plumage Both species have similar plumage, with chestnut to dark brown upperparts and rufous to buff underparts. Albert's Lyrebird has warmer-looking tones and appears darker, while Superb Lyrebird is markedly grayer in tone. The spectacular tail of the male Superb Lyrebird resembles the shape of a Greek lyre, hence the common name. The tail is black above and a contrasting pale silvery-gray below, and is held horizontally when the bird is not in display mode. It consists of 16 highly modified feathers: two outer lyrates patterned with chestnut, two ribbonlike medians, and 12 filamentaries. The tail feathers only become fully modified when the bird reaches sexual maturity at three to four years of age. The tail of the male Albert's Lyrebird has just 14 feathers, lacking the distinctive lyrate plumes of its showy congener. Females of both species have a simpler tail with a drooping tip.

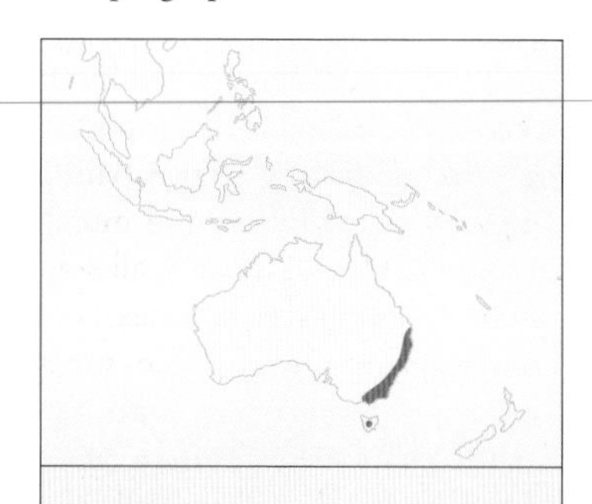

Number of genera 1
Number of species 2
Conservation Status 1 species Vulnerable
Distribution Australia

Behavior Lyrebirds are shy birds, usually first detected by voice, but are often seen crossing forest roads and occasionally entering bordering gardens. They escape danger by adopting a fluttering, bounding run. Although they spend most of the time on the ground, both species roost in tree branches to avoid predators. Small parties may form outside the breeding season, but males become highly territorial during autumn and winter. Lyrebirds' territories are highly variable in size: they have been recorded from 6 acres (2.5 ha.) to 86 acres (35 ha.), depending on the quality of the habitat and the population density. Studies of Albert's Lyrebird have revealed that its territories are widely spaced in optimal habitat, with a density of about 5 pairs per 0.3 square miles (1 km^2).

Voice Lyrebirds are among the most astonishing mimics in the natural world: they are unrivaled by any other bird in the variety and accuracy of the sounds they copy. Their own repetitious phrases are augmented by the songs and calls of many other birds—including the chattering sounds of entire flocks—along with a wide range of animal calls, human voices (including crying babies), musical instruments, sirens, gunshots, camera shutters, and machinery such as trains, car engines, and chainsaws. This remarkable vocal prowess is possible due to the lyrebirds' three pairs of highly developed syringeal muscles (the musculature used by all songbirds to produce complex vocal phrases). Females are also superb mimics, but are less vocal than males. Both species have a similar sharp *whisk whisk* alarm call, often given as the bird flees on foot.

Habitat Superb Lyrebird favors damp temperate and subtropical forest, but tolerates a wide altitudinal range from sea level to the snowline. Albert's Lyrebird, in contrast, requires rain forest, mainly above 985 ft. (300 m.), with Antarctic beech or a dense understory.

Movements Sedentary.

Diet The birds feed on invertebrates such as insects, mollusks, earthworms, and spiders. Occasionally, they eat seeds and small vertebrates such as frogs. They forage using their strong claws and feet to rake over soil, leaf litter, and bark in search of prey, leaving behind distinctive 'workings.'

Breeding The male Superb Lyrebird performs a spectacular courtship dance to woo receptive females, in which he inverts his tail and thrusts it forward over his head, so that it forms a shimmering silver fan. He first builds a bare, circular mound 3.25–6.5 ft. (1–2 m.) in diameter as a 'stage'; Albert's Lyrebird uses trampled platforms of vines and twigs or logs. Each male may have up to 15 display mounds within his territory. When a female appears, the male fans and raises his tail while singing constantly and slowly changing direction. The dance usually occurs at the climax of his display, and copulation follows soon after if the female is suitably impressed. Male Superb Lyrebirds are polygamous and play no part in raising the young. Females of both species build a large, domed nest with a side entrance. Nests are usually situated on or near ground level in vegetation, between rocks or buttress roots, or in a hollow log. Occasionally, however, Superb Lyrebirds have been found nesting up to 82 ft. (25 m.) above the forest floor, in the fork of a tree. The nest is constructed from twigs, bark, moss, ferns, and leaves, and the nest chamber is lined with a mass of soft flank feathers and small rootlets. Superb Lyrebirds breed from June to October, and lay a single heavily marked egg, variably colored gray to dark green. Albert's Lyrebirds nest from June to August, also producing just one egg per season, which is purplish with black and brown blotches and streaks. Incubation is a protracted process: the egg takes an average of 50 days to hatch, since the female leaves the nest for up to six hours each day to feed. Chicks take around 47 days to fledge, during which time they grow rapidly and gain weight until they are around 63 percent of the parent bird's weight. They remain with the female for up to eight months after leaving the nest.

A male **Superb Lyrebird** performs a display, throwing his tail right over his body. Victoria, Australia.

Taxonomy Lyrebirds were thought to belong to order Galliformes owing to their pheasantlike appearance, but they are now recognized as passerines. Fossils belonging to the family Menuridae have been found in Australia as far back as the Early Miocene (c.15 million years ago). Lyrebirds' closest living relatives are generally considered to be the scrubbirds (Atrichornithidae). Controversially, some experts have proposed an alternative close connection with the bowerbirds (Ptilonorhynchidae).

SCRUB-BIRDS *Atrichornithidae*

A male **Rufous Scrub-bird** (*Atrichornis rufescens*) displaying. Males of both species of scrub-birds display in a similar way to immature lyrebirds, with a fanned tail, drooping wings, and a quivering body. Australia.

Scrub-birds are extremely shy terrestrial birds. There are just two species: Noisy Scrub-bird (*Atrichornis clamosus*) and Rufous Scrub-bird. Both are rare, with restricted ranges. Remnants of an ancient family, their closest surviving relatives are the lyrebirds (family Menuridae) and possibly the bowerbirds (Ptilonorhynchidae) and Australasian treecreepers (Climacteridae).

Noisy Scrub-bird is so elusive it was believed extinct for many years until a tiny population was rediscovered in Western Australia in 1961.

Structure Noisy Scrub-bird is the larger, at 8.25–9 in. (21–23 cm.); Rufous Scrub-bird measures 6.25–7 in. (16–18 cm.). Males are slightly bigger than females. Both species have a strong, pointed, broad-based bill, stubby wings, and a longish tail, which is often cocked.

Plumage Finely-barred brown upperparts and pale underparts. Males have a blackish chest (more extensive in Rufous Scrub-bird) and a white throat. Females and juveniles are more uniform.

Behavior Scrub-birds are generally solitary birds. As poor fliers, they prefer to scurry around in thick cover, rarely emerging into the open. They forage in the leaf litter by using their strong feet and claws to rake over vegetation.

Voice In the breeding season, male Noisy Scrub-birds produce a variety of ringing and often ventriloquial calls in defense of their territory, which are so loud they may cause discomfort to the human ear at close range, and can be heard from several miles away. Females are usually silent, but occasionally Noisy Scrub-birds sing a duet with their mate in the breeding season. Rufous Scrub-birds are excellent mimics.

Habitat Both species are endemic to Australia. They prefer habitats with dense cover and thick leaf litter. Noisy Scrub-bird is dependent on semi-arid heathland with scrubby gullies. Rufous Scrub-bird occurs in temperate eucalyptus and Atlantic beech rain forest above 1,970 ft. (600 m.). The latter species is declining, with isolated populations in Queensland and New South Wales, and is considered Near Threatened. Noisy Scrub-bird, found only in a small area of southwestern Australia, is considered Vulnerable, although recent intensive conservation efforts have benefited this very rare species. It remains highly restricted in range, and is acutely sensitive to habitat change, particularly fire damage.

Movements Both species are sedentary, highly territorial, and tied to specific habitat requirements. Noisy Scrub-birds rely on corridors of closed vegetation to disperse, and, although they may break cover to cross roads, will not cross cleared land. This undoubtedly contributed to the fragmentation and decline of their population in the past.

Diet A variety of insects, spiders, and other invertebrates, as well as frogs and small lizards.

Breeding Noisy Scrub-birds breed from May to November, and Rufous Scrub-birds from September to December. Females build a domed nest close to the ground, lined with wood pulp, and with a side entrance. Noisy Scrub-bird lays one egg, while Rufous Scrub-bird lays two. Scrub-birds have a long incubation period for such small passerines, of 36–38 days. Females rear the young alone. Fledged offspring may remain with their mother for several months.

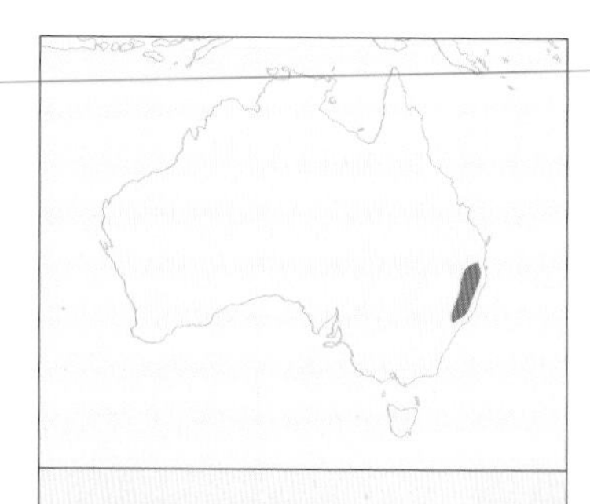

Number of genera 1
Number of species 2
Conservation Status 1 species Vulnerable, 1 species Near Threatened
Distribution Australia

BOWERBIRDS *Ptilonorhynchidae*

This family is famous for its elaborate courtship behavior and, in particular, for some species' spectacular bower constructions. Although bowerbirds have been widely studied, not all species are well known. The elusive Golden-fronted Bowerbird (*Amblyornis flavifrons*), confined to the remote Foja Mountains in Papua New Guinea, was originally described from skins obtained in 1895, but remained an enigma for almost a hundred years until it was rediscovered in 1981. The first photographs of the species were obtained as recently as 2005, during an exploration of the mountains conducted by an international team of scientists.

Included in the family are the two species of catbirds (genus *Ailuroedus*). These differ from other bowerbirds in that they are monogamous and do not build bowers.

Structure Bowerbirds are robust birds, 9–14.7 in. (23–37.5 cm.) in length; the largest species is Great Bowerbird (*Chlamydera nuchalis*) of northern Australia, at 12.6–14.7 in (32–37.5 cm.). Bowerbirds have a thick, stout bill that, in some species, is hook-tipped or equipped with notched cutting edges. Their feet are typical of passerines, with three slender toes in front and one behind to facilitate perching. Most species are strong fliers, with their wingtips held upswept on each downstroke in active flight.

A female or immature **Satin Bowerbird** (*Ptilonorhynchus violaceus*). In all bowerbirds, females and juveniles have similar coloration. Australia.

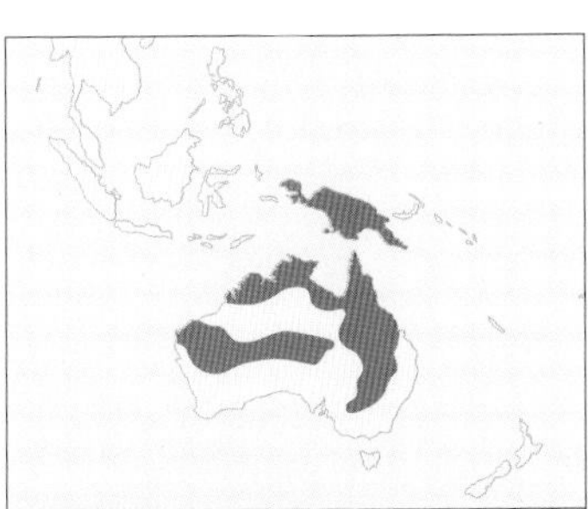

Number of genera 8
Number of species 18
Conservation Status 1 species Vulnerable, 1 species Near Threatened
Distribution New Guinea and nearby islands; Australia

Plumage The family shows a wide variety of plumages. Sexual dichromatism is common: some species have strikingly colored males that seem to glow in the forest gloom, whereas typically the females are drab and inconspicuous. For example, the unmistakable male Regent Bowerbird (*Sericulus chrysocephalus*) is a velvet-black bird with a golden crown, mantle, and wing feathers, yellow feet, and a striking yellow iris. By contrast, the female is olive-brown, scalloped and mottled with fawn. Several species in genera *Sericulus*, *Amblyornis*, and *Archboldia* have elaborate ruffs or manes extending from the hindcrown to the upper back. In several species in genus *Chlamydera*, both sexes share very similar, rather sober plumage, including a characteristic pale pink

The male **Regent Bowerbird**, with its golden crown and mantle, is one of the most striking of all bowerbirds. The male's eye color takes two years to develop fully, and the plumage color takes up to five years. Australia.

crest on the nape, which is erectile and both larger and more brightly colored in males. Green Catbird (*Ailuroedus crassirostris*) has bright green upperparts and a mottled brown head with streaked underparts. Juveniles of all bowerbird species tend to be similar in coloration to females.

Bowerbird species that build the most elaborate bowers have the least striking plumage, and show little sexual dimorphism, whereas species with boldly colored males tend to construct far less elaborate bowers. Scientists label this phenomenon the 'transfer effect'—suggesting that the master builders have transferred the advantages gained by exotic plumage to their bowers, which serve as their alternative 'shop window' to lure females seeking the fittest mate. A possible bonus of this strategy is that dull-colored males may be far less easy to spot by a hunting predator. However, the transfer effect in bowerbirds is still the subject of ongoing study, and the hypothesis has been questioned by some authors.

Behavior The most striking feature of bowerbirds' behavior is the construction of bowers: elaborate gardens or avenues in which to court potential mates (see box). Bower-building males devote a great percentage of their time to improving and maintaining these structures. Catbirds differ from the rest of the family: the males do not build bowers. In addition, catbirds are not polygamous, but instead form a pair bond with a single female and help her in rearing the young.

Voice Bowerbirds are highly vocal, producing a wide variety of whistles, churrs, chatters, rattles, croaks, and insectlike noises. Many are excellent mimics: for instance, Tooth-billed Bowerbird (*Scenopoeetes dentirostris*) faultlessly reproduces the songs and calls of many other birds, and Macgregor's Bowerbird (*Amblyornis macgregoriae*) has an eclectic repertoire that includes human voices, grunting pigs, and the sound of waterfalls. Some males have diagnostic calls that are given only at the bower site to attract females. Catbirds make a wailing cry, like a cat, from which they get their name.

Habitat Twelve species are endemic to New Guinea, and 10 species occur in mainland Australia. Most species are birds of humid forests, occurring in a range of subtropical and tropical woodlands from coastal lowlands to remote highlands. An exception is Western Bowerbird (*Chlamydera guttata*), which lives in open woodland in the Pilbara region of Western Australia and the continent's arid interior.

Three of the species endemic to New Guinea, all with a limited range, are currently a cause for concern for conservationists: Archbold's Bowerbird (*Archboldia papuensis*) is classed as Near Threatened due to ongoing habitat loss, and Fire-maned Bowerbird (*Sericulus bakeri*) is classed as Vulnerable, although this species may actually be more widely distributed than current research indicates.

Movements Many 'gardener' bowerbirds are sedentary, remaining loyal to one territory for life. However, avenue-building species such as Satin Bowerbird (*Ptilonorhynchus violaceus*) make seasonal movements. Outside the breeding season, feeding flocks of up to 50 or more birds are formed, consisting mainly of females and immature birds; although nomadic, such flocks tend to remain within a relatively small area. Some montane species, including Regent Bowerbird, move to lower elevations during the winter months.

Diet Bowerbirds are mainly frugivorous, eating a wide variety of berries and other fruits. Figs are a key food item, particularly favored by catbirds and Western Bowerbird. Many species also eat other plant material, such as shoots, stems, leaves, buds, flowers, and seeds. Tooth-billed Bowerbird relies heavily on leaves as a dietary staple during the winter months, using its thick bill to shred and pulp foliage.

Invertebrates are also taken, with different species demonstrating preferences; Golden Bowerbird (*Prionodura newtoniana*) often eats beetles, while Yellow-breasted Bowerbird (*Chlamydera lauterbachi*) shows a fondness for caterpillars and other soft-bodied insects. Some species eat small snakes, lizards, and tree frogs. The young of all bowerbirds are fed a variety of small animals; catbirds often steal other birds' nestlings to feed to their own chicks.

Breeding With the exception of catbirds, the males are polygamous and play no role in rearing the young; the females perform all of the parental duties, from nest-building and incubation to feeding the young. The nests are robust cups constructed from leaves, vines, twigs, and other plant materials, usually situated in the forks of tree branches or in dense cover. Most species lay one or two eggs per clutch, but catbirds occasionally produce three eggs. Incubation takes an average of 19–25 days, and the young fledge at 18–22 days. Most juveniles are independent by 60–80 days after fledging.

Taxonomy Bowerbirds were traditionally considered to be very closely related to the birds of paradise (family Paradisaeidae), but some authors now reject this classification and argue that DNA–DNA hybridization studies place them closer to the lyrebirds (family Menuridae). This new arrangement remains controversial, however, and further studies are needed to unravel the complex phylogenetic associations of the bowerbird family.

Master Builders: Bowerbirds and Their Creations

Male bowerbirds are widely renowned for building elaborate bowers at the center of their territory, in which they enact complex displays to attract as many females as possible. These diligent architects may spend as much as nine or 10 months of the year constantly building, enhancing, and rearranging their properties.

A male **Western Bowerbird** (*Chlamydera guttata*) adorns his avenue bower with green berries. Australia.

There are two types of bower, constructed by different branches of the family: avenues and gardens. The avenue-type bower is typical of genera *Sericulus*, *Chlamydera*, and *Ptilonorhynchus*. It is made from one or two arched rows of sticks or grasses forming a narrow avenue, at the center of which is a platform of twigs that the male uses as a stage. The second type of bower is the garden, which occurs in two forms: mat-type and maypole-type. Mat-type bowers are simple platforms of sticks and ferns, favored by the *Archboldia* bowerbirds of New Guinea. Maypole-type bowers are created by *Amblyornis* and *Prionodura* bowerbirds; these are dramatic single or double columns of sticks surrounding sapling trunks, joined at the base and equipped with a central displaying perch. Catbirds do not build a bower but defend a display territory near the forest floor.

Decoration is crucial to the success of a bower if a male is to outcompete his rival builders and entice females to mate with him. Even within a single species, each bird's bower is unique, and its decorations indicate the bird's personal preferences, along with his ability to procure new or unusual items from year to year. Bowerbirds use all manner of things to improve their creations, from pebbles, moss, snail shells, flowers, buds, and berries to man-made objects, particularly reflective items such as glass and plastic. One male Vogelkop Bowerbird (*Amblyornis inornata*) reportedly used batteries and a film casing taken from a visiting film crew to give his bower that 'extra something.' In some species, coveted items are often stolen by rival males.

Studies have shown that color preferences vary widely between bowerbird species. Often, color selection is related to the plumage, bill, and eye colors of their own species. Satin Bowerbirds, for example, have a strong attraction to blue objects. During displays, these objects are attacked as if they were an intruding male, to demonstrate the bower-owner's superiority as fathering material. Some species smear the interior of their bower with plant or berry juice to add extra color.

The male bowerbirds' displays to females are in themselves a remarkable and complex behavior that has been the subject of much study. Avenue-builders generally adopt a curious stiff-legged dance, while fanning their wings and seizing prized display objects. Maypole-builders such as Golden Bowerbird perform spectacular hovering displays together with bizarre, contorted posturing in which they use their wings, head, and tail.

Prospecting females flit from bower to bower, to inspect the quality of each structure and the caliber of its owner's display techniques. Often, the same star performers are chosen by a number of females, leaving underperforming rivals in the neighborhood without a mate. ■

AUSTRALASIAN TREECREEPERS *Climacteridae*

A small passerine family of two genera and seven species endemic to Australia and New Guinea. Australasian treecreepers are named for their habit of shuffling and hopping along tree trunks and branches, probing the bark for food. This activity has also earned them the erroneous common name 'woodpecker.' The family is not closely related to the treecreepers (family Certhiidae) of the northern hemisphere.

Structure Australasian Treecreepers are robust, medium-small birds, 5.5–7.5 in. (14–19 cm.) in length. Sexual dimorphism is subtle. All species have a similar structure, with a slightly decurved bill and large, strong-clawed feet.

Plumage Generally brown, but patterned with black, buff, or white streaks on the underparts. Rufous Treecreeper (*Climacteris rufus*) has warm reddish-buff coloring below. Red-browed Treecreeper (*C. erythrops*), White-browed Treecreeper (*C. affinis*), and Brown Treecreeper (*C. picumnus*) have brow markings, which are particularly prominent on males. White-throated Treecreeper (*Cormobates leucophaea*) has a scalloped crown and a white loral stripe, and the female has a pale orange spot on the neck. All species have a pale wing bar.

Behavior Usually seen climbing trees or 'hopping' along the underside of branches, Australasian treecreepers feed by working their way up from the base of a tree to the upper branches, then dropping to the base of an adjacent tree and repeating the process. Some species also descend to the ground to forage in rotting timber and leaf litter. Several species are gregarious, forming feeding parties and nesting cooperatively.

White-throated Treecreeper has scalloped patterning on the crown, and a white throat and chest. Australia.

Voice Social species have a variety of contact notes to keep in touch as they forage, such as high-pitched whistles, trills, chatters, and rattles. White-throated Treecreepers' complex vocalizations vary markedly from those of the other Australasian species, and include a repetitive, high-pitched piping.

Habitat Temperate, tropical, and subtropical woodland. Two species occur at higher altitudes: Red-browed Treecreeper in the subalpine zone of eastern Australia, up to about 4,920 ft. (1,500 m.), and Papuan Treecreeper (*Cormobates placens*) in the montane forests of north, west, and southeast New Guinea. None are currently causes for concern among conservationists, although Red-browed, White-browed, and Papuan Treecreepers are uncommon over much of their respective ranges.

Movements All are sedentary.

Diet Insects and other invertebrates are taken from the surface and underside of bark. Ants are a favorite food, but some species occasionally consume plant material such as seeds. Several species also hunt on the ground, foraging in leaf-litter and on fallen timber.

Breeding Most Australian species breed between August and January, although Black-tailed Treecreeper (*Climacteris melanurus*) nests from March to May and September to November. Several species are cooperative breeders.

Nests are made of grasses or bark, often reinforced with animal dung, and usually lined with feathers, grass down, or fur. They are concealed in tree hollows 10–50 ft. (3–15 m.) off the ground. Most species lay two or three eggs and incubate them for 16–18 days; however, in White-throated Treecreeper the incubation period is 22–23 days. In species that nest cooperatively, the young are fed by parents and two or more 'nest-helpers,' often single males.

Little is known about the breeding biology of Papuan Treecreeper.

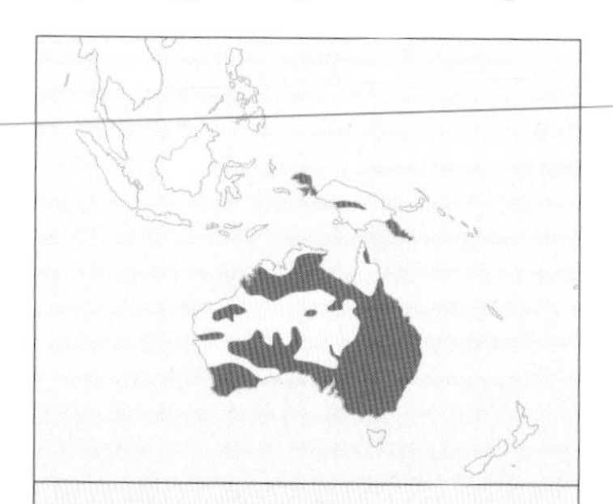

Number of genera 2
Number of species 7
Conservation Status Least Concern
Distribution 6 species occur in Australia and 1 species is endemic to Papua New Guinea

AUSTRALASIAN WRENS *Maluridae*

A family of small, colorful, long-tailed birds that spend much of their lives on or near the ground. They resemble the true wrens (family Troglodytidae) in their shape and their habit of cocking their long tail. There are three distinct groups: fairy-wrens (three genera; 15 species), emu-wrens (one genus; three species), and grasswrens (one genus; 10 species).

Structure The smallest family members are the emu-wrens—some of the tiniest Australian birds, at 5–7.5 in. (12–19 cm.) long, including the tail. The largest is White-throated Grasswren (*Amytornis woodwardi*), at 8–8.5 in. (20–22 cm.). Fairy-wrens and emu-wrens have the short, fine bill typical of insectivorous birds, while grasswrens have a slightly thicker bill. The body is dumpy, but the legs and tail are long.

Plumage Sexual dichromatism is common, and at its most dramatic in the fairy-wrens. Male fairy-wrens in breeding plumage are striking birds. Many have blue, violet, and black plumage, while Red-backed Fairy-wren (*Malurus melanocephalus*) is jet black or chocolate-brown with a fiery red or orange-red 'saddle.' In contrast, female fairy-wrens are mainly muted brown, with pale underparts and, in some species, a washed-out blue tail. Males also have a nonbreeding plumage that closely resembles that of females and juveniles. Emu-wrens are mainly streaky and rufous, with males having a pale blue throat and breast. Grasswrens are streaked rufous, brown, black, and white, and several species have black moustachial stripes.

Behavior Fairy-wrens and emu-wrens are highly social birds. They live in family groups comprising a dominant male, a female, and various

Male fairy-wrens, like this **Splendid Fairy-wren**, are at their most spectacular when in breeding plumage. Australia.

nonbreeding and immature birds. Grasswrens and emu-wrens are shy and hide in thick vegetation when an intruder approaches. All family members typically scurry like rodents to escape from danger or lure a predator away from their nest.

Voice Fairy-wrens have complex, melodic songs. Splendid Fairy-wren (*M. splendens*) and Superb Fairy-wren (*M. cyaneus*) also give a song-like call when they encounter predatory birds such as butcherbirds. Experts are still uncertain of the true meaning of this call, as it does not appear to serve the same function as an alarm call.

Habitat Fairy-wrens occur throughout Australia, where they are found in every major habitat; four additional species are endemic to New Guinea. Most grasswrens have localized distribution and inhabit remote rocky or arid areas of coastal or interior Australia, usually in association with spinifex grasses. Mallee Emu-Wren (*Stipiturus mallee*) is confined to a small, fragmented area of Victoria and South Australia and is listed as Vulnerable, as is White-throated Grasswren. Carpentarian Grasswren (*A. dorotheae*) also has a highly restricted range in eastern Northern Territory and northwestern Queensland, and is listed as Near Threatened due to habitat loss.

Movements Most species are sedentary, but White-winged Fairy-wren (*M. leucopterus*) and Splendid Fairy-wren are partially nomadic within their extensive ranges, as is the localized Gray Grasswren (*A. barbatus*).

Diet The birds are largely insectivorous, but take a variety of small invertebrates, as well as earthworms. Seeds may also be eaten, particularly by grasswrens.

Breeding Most species breed from July to March, with regional variations. The nests are domed structures made from grass, bark, spiders' webs, fur, and moss, and usually sited low in vegetation. Fairy-wrens are often polygamous. They are co-operative breeders, with several other group members helping the parents feed the young. Most fairy-wrens have several broods a year, with two to four young each; the young fledge in 10–12 days. Grasswrens and emu-wrens have one brood of two or three young.

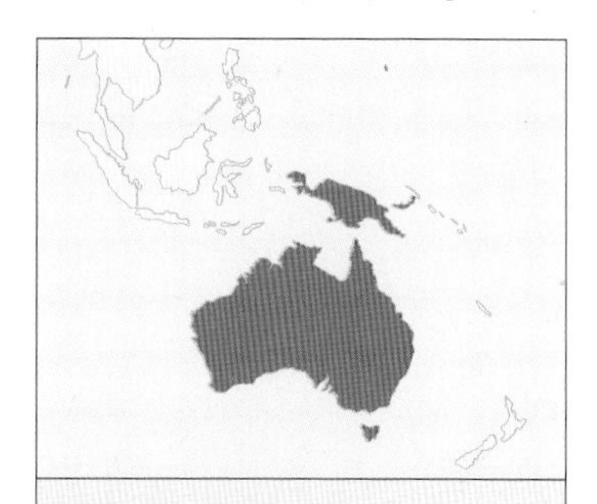

Number of genera 5
Number of species 28
Conservation Status 1 species Near Threatened, 2 species Vulnerable
Distribution Australia and New Guinea

HONEYEATERS *Meliphagidae*

A **Blue-faced Honeyeater** (*Entomyzon cyanotis*) takes nectar from an *Aloe vera* plant. Queensland, Australia.

Aptly named, honeyeaters are often seen probing flowering plants for the nectar that forms the major part of their diet. This large, diverse family of passerines is found in Australasia, Indonesia, and islands in the Pacific.

Structure Honeyeaters vary greatly in size, ranging from 3 to 20 in. (7–50 cm.) in length and 0.25 to 7.0 oz. (7–200 g.) in weight. All of the genera have a slender body, pointed wings, and short, sturdy legs with sharp-clawed feet.

The most remarkable feature that all honeyeaters share is their tongue, which is partially tubelike and split, with a brushlike tip superbly adapted for extracting nectar. Honeyeaters also have specially adapted kidneys that allow them to process maximum nutritional benefit from this food source.

The bill varies in shape and size depending on the species' feeding adaptations, from small and straight or slightly decurved, to thick and robust, or slender and markedly decurved. Other anatomical variations include ear-tufts, gape skin, wattles, bearded plumes, and warty facial skin. Male friarbirds (genera *Melitograis* and *Philemon*) have a knobbly casque just above the base of the bill, and bare black skin on the head. Crow-honeyeater (*Gymnomyza aubryana*), an endangered species endemic to New Caledonia, is so named because it resembles a crow, although with more rounded wings and a longer neck and tail.

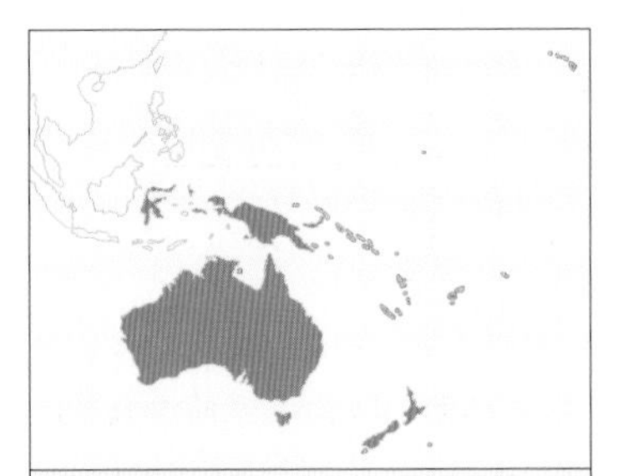

Number of genera 44
Number of species 174
Conservation Status 4 species Endangered, 5 species Vulnerable, 5 species Near Threatened
Distribution Australasia, Indonesia, Pacific islands

Plumage Sexual dichromatism is comparatively rare and most notable in species with brightly marked males and drab females, such as Red-headed Honeyeater (*Myzomela erythrocephala*) and Western Spinebill (*Acanthorhynchus superciliosus*). Several genera, such as the miners (genus *Manorina*) and the *Meliphaga* honeyeaters, have a number of lookalike species.

There are a variety of plumage colors, but most honeyeaters are olive, brown, or green, with some smaller species having white or yellow underparts. A few species are striking black-and-white birds, such as Banded Honeyeater (*Certhionyx pectoralis*) and male Pied Honeyeater (*C. variegatus*), while others have cheek patches or strong head patterns, such as birds in genus *Melithreptus*. Many show bare areas of colored skin around the eyes and face; the colors of bare parts and adornments often change as the bird ages or enters the breeding season. Most juvenal plumages are broadly similar to those of adults, though more drably colored.

Male friarbirds, like this **Noisy Friarbird** (*Philemon corniculatus*), are unusual among honeyeaters in having a bony bump (casque) on their bill. Australia.

Behavior To glean nectar, a honeyeater darts its tongue into a flower repeatedly, while the bird closes its bill and uses the upper mandible to squeeze out any remaining liquid from the flower. The birds also forage for insects by gleaning them from leaves and branches, or by hawking from the air. Larger species, such as Strong-billed Honeyeater (*Melithreptus validirostris*), use their robust bill to probe bark for food.

Outside the breeding season, honeyeaters are often very conspicuous birds: noisy, active, social, and aggressive. Many form family parties or loose flocks that may include several different species; larger flocks occur when some species are on migration. Mixed-species aggregations occur where nectar supplies are plentiful, although inevitably the larger, more aggressive species usually dominate the richest niches. A number of honeyeater species have also adapted well to living alongside humans in towns and gardens.

Bell Miner (*Manorina melanophrys*) is widely known for its musical call, which consists of a bell-like *tink* sound. Australia.

Voice Honeyeaters have a wide variety of songs and calls, ranging from sweet, mellifluous songs in some smaller species to harsh coughing and rasping calls in the larger members of the family. O-os (genus *Moho*) derived their name from their distinctive *owow, owow-ow* calls, while the rare Mao (*Gymnomyza samoensis*) of the Samoan Islands produces loud, wailing calls, primarily at dawn and dusk.

Habitat A widespread and successful family, honeyeaters have adapted to many habitat niches. The majority of species occur in Australia, New Guinea, and Indonesia, but others are found in New Zealand, Wallacea, Polynesia, Micronesia, and other Pacific islands eastward to Hawaii. Just one honeyeater occurs outside the Australasian region, west of Wallace's Line (which marks the western border of Australasia): a subspecies of Brown Honeyeater (*Lichmera indistincta*), known as Indonesian Honeyeater (*L. i. limbata*), which is confined to Bali.

Some six species are known to be extinct, including some of the o-os and the poorly known Kioea of Hawaii, which vanished in the mid- to late 19th century.

Movements Much remains unknown about honeyeaters' movements, though most species are nomadic at least locally, due to seasonal changes and fluctuations in food supply, with many species following the availability of favorite food plants. Nomadism tends to be most pronounced in species of arid habitats and in some rain forests where favored foods are highly seasonal; for this reason, the birds are sometimes called 'blossom nomads.' Some species show great variations in local abundance from year to year, and specialists of more arid habitats tend to travel greater distances and have less predictable movements. Some locally nomadic or migratory species form large flocks when on the move. Certain species, such as Yellow-faced Honeyeater (*Lichenostomus chrysops*) and White-naped Honeyeater (*Melithreptus lunatus*), travel in mixed flocks; in coastal eastern Australia, these may number many thousands during peak migration.

Diet Although nectar is a primary food for virtually all honeyeaters, most supplement their diet with regular supplies of fruit, manna (sugary plant secretions), lerps (honeydew), and invertebrates such as insects, caterpillars, and spiders. The very largest species will also take the eggs and chicks of other birds, and occasionally they capture small vertebrates such as lizards.

Most species usually take insects in the breeding season, to provide much needed protein for egg production and nestlings. They may also turn to this food source at the end of the day, when nectar supplies have been depleted. In some species, such as New Zealand Bellbird (*Anthornis melanura*), it has been noted that females consume less nectar than males, but this is believed to be due to the more aggressive males monopolizing local supplies of flowering plants. A few honeyeaters take fruit in much greater quantities: for example, Painted Honeyeater (*Grantiella picta*) specializes on mistletoe berries.

Breeding The birds employ a variety of breeding strategies, including monogamy, polygamy, and cooperative nesting in sometimes socially complex colonies. Striped Honeyeater (*Plectorhyncha lanceolata*) often nests in association with Gray Butcherbirds (*Cracticus torquatus*; family Cracticidae), perhaps as protection from predators. Breeding seasons vary between species and by region, but some Australian honeyeaters have a long season that may span six months, though most breed from late winter through spring.

A **New Holland Honeyeater** (*Phylidonyris novaehollandiae*) displaying. Breeding males spend a lot of time defending their nest and territory. Tasmania, Australia.

Nest sites too, are varied: from hollows and low-level shrubs to the top branches of mature trees, with the nest usually suspended or placed in a fork. The nests are constructed from a variety of materials, such as plant matter, fur and hair, cobwebs, feathers, and man-made materials.

Clutch sizes range from one to four eggs, incubated primarily by the female. Incubation lasts for an average of 12–17 days. The young are fed by both parents; cooperative breeders such as the miners receive aid from 'helpers.' The young fledge at 11–20 days, but fledging may take 30 days or more in some of the hollow-nesting species.

Taxonomy Recent DNA studies have led some authorities to remove the Australasian chats (genus *Epthianura*) from the family and add Bonin Honeyeater (*Apalopteron*; family Zosteropidae).

Honeyeaters and Flowering Plants

Honeyeaters play a major part in pollinating many types of flowering plants, trees, and bushes; in doing so, they fulfil a similar convergent evolutionary role to the hummingbirds of the New World. In Australia, some honeyeaters have a particularly close mutual association with flora such as *Eucalyptus*, *Banksia*, *Epacris*, and *Proteacae*, collecting dustings of pollen on their heads from one flower and fertilizing another when they visit, seeking more nectar. Other honeyeaters act as important agents of dispersal for seeds, notably mistletoe and acacias.

Macleay's Honeyeater (*Xanthotis macleayanus*) plays a major role in pollinating rain forest flowers. Queensland, Australia.

The true extent of this widespread symbiotic partnership remains unknown, but a number of honeyeaters have evolved specially modified bills that allow them to feed on the nectar of particular flowers. For instance, spinebills are able to access nectar in longer, tubular-shaped flowers that their shorter-billed congeners would be unable to reach. However, the association is not all favorable—some species of honeyeater have developed a liking for cultivated fruit, and this has caused them to be viewed as pests in certain agricultural and suburban areas. ■

BRISTLEBIRDS *Dasyornithidae*

Rufous Bristlebird is so named for the rich, reddish-brown coloration of the head, wings, and rump.

A tiny family endemic to Australia, the Dasyornithidae comprises three similar species of ground-living birds: Western Bristlebird (*Dasyornis longirostris*), Eastern Bristlebird (*D. brachypterus*), and Rufous Bristlebird (*D. broadbenti*). The birds are named for the bristles at the base of their bill. They resemble scrub-birds (family Atrichornithidae) in their structure, behavior, and habitat preference, although the two families are not related.

Structure Bristlebirds are robust, long-legged, thrush-sized passerines, 7.8–10.6 in. (20–27 cm.) in length. They have four or more stiff, forward-curving rictal bristles. All species are similar in appearance and sexes are generally alike, although adult male Eastern Bristlebirds are, on average, larger than females. Western Bristlebird is smaller than its relatives, with a marginally shorter tail and bill.

Plumage All are somber-colored brown, olive, and gray birds. Rufous Bristlebird has a rusty crown and cheeks with pale eyebrows and lores, which give it a hooded appearance; its throat and breast are pale with darker scaly markings, and the wings, rump, and tail are darker shades of cinnamon and brown. Eastern and Western Bristlebirds are both duller overall but have more contrasting rust-colored wings; the latter species also has subtle scalloped markings.

Behavior Bristlebirds are very elusive, spending most of their life on the ground in thick vegetation. To forage, they use the bill to sweep aside leaf litter and poke into the surface layer of soil. Like scrub-birds, they prefer to run from danger; if they do take to the air, their flights are short and low. When anxious, they often fan or cock their long tail.

Voice Bristlebirds have distinctive, ringing vocalizations, which are usually the only clue to their presence. The males have far-carrying, silvery songs, and pairs sometimes duet. The birds also produce a variety of chattering and whipcrack-like calls. Alarm calls are short and sharp.

Habitat The three species are widely separated in range, but all are specialists of dense coastal or montane heathland, woodland, and scrub. Western Bristlebird is confined to a small coastal area of southwestern Australia, while Eastern Bristlebird is found in fragmented pockets in the southeast. Rufous Bristlebird occurs between South Australia and Victoria. Eastern Bristlebird is listed as Endangered and the Western species as Vulnerable. While Rufous Bristlebird is not presently at risk, the subspecies *litoralis*, endemic to Western Australia, was last recorded in 1906 and is considered extinct.

Movements All three species are sedentary. However, recent conservation programs reintroducing Eastern Bristlebirds to suitable areas of habitat have revealed that translocated individuals disperse over much greater areas than in their native habitat; one male traveled more than 2.5 miles (4 km.) from the point of release. While this shows bristlebirds can move over a reasonably wide area in unbroken tracts of suitable habitat, individuals still seem to prefer remaining close to other bristlebirds and are reluctant to colonize new areas.

Diet All species are insectivorous.

Breeding Much remains unknown about the birds' behavior, but it is thought that pairs remain monogamous for life and dwell in permanent territories. Bristlebirds breed from August to October. Their nests are large domed or globular structures made from grasses, leaves, and twigs, sited in low vegetation. Two whitish to pinkish eggs with speckled markings are laid per brood.

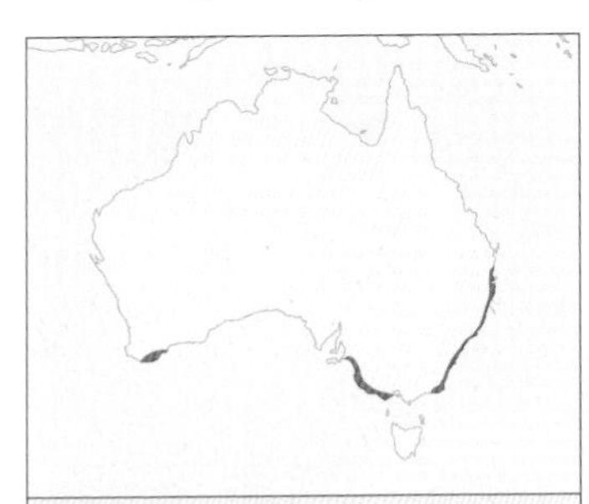

Number of genera 1
Number of species 3
Conservation Status 1 species Endangered, 1 species Vulnerable
Distribution Australia

PARDALOTES *Pardalotidae*

Forty-spotted Pardalote (*Pardalotus quadragintus*) is the rarest species in the family. Tasmania.

The tiny family Pardalotidae consists of a single genus, with four species. The birds are endemic to Australia, where they are widespread. The name 'pardalote' is derived from a Greek word meaning 'spotted.'

Structure Small passerines, 3–4.5 in. (8–12 cm.) in length, with a stubby bill, a short tail, and strong legs.

Plumage Often boldly marked. Spotted Pardalote (*Pardalotus punctatus*) is also known as 'Diamondbird' due to its colorful, jewel-encrusted appearance, with black skullcap and wings studded with white spots. Striated Pardalote (*P. striatus*) varies between its six subspecies, but all are unspotted with a striking black or streaked crown, a prominent eye-stripe, and white wing stripes. Red-browed Pardalote (*P. rubricatus*) has paler underparts, a plainer back, a spotted crown and striped wings. Males are more boldly marked in all species except Forty-spotted Pardalote, which is the plainest of the family, with olive-green upperparts, pale underparts, and spotting that is restricted to the wings. In all pardalotes, juveniles have duller, less marked plumage than adults.

Striated Pardalotes (*P. striatus*) vary widely in plumage, but all have a white brow and yellow spot in front of the eye. Tasmania.

Behavior Pardalotes live in pairs or small family groups, but some species form large flocks after breeding. They are energetic and acrobatic foragers, moving through the tree-tops in search of insects.

Voice Paired birds constantly make soft, whistling contact calls to each other; call and response are so rapid that an observer may find it hard to locate the birds, because the calls seem to issue from two different directions simultaneously. The birds also make a clicking sound with their bill as they feed.

Habitat Mostly eucalypt and white gum forests, as well as a variety of scrubby habitats. Also rain forest, mangroves, parks, and backyards. Striated Pardalote is found in most habitats except the western deserts. Forty-spotted Pardalote is now confined to a few colonies in the coastal white gum forest of southeast Tasmania. One of Australia's rarest birds, and in decline mainly due to habitat loss and fragmentation, it is listed as Endangered.

Movements Spotted and Striated Pardalotes are both migratory, often forming very large flocks during dispersal. In the fall, they move from moist montane forests to interior and northern coastal regions; the Tasmanian population of Spotted Pardalote moves as far north as Queensland. Forty-spotted Pardalote is more sedentary, but makes small-scale movements within Tasmania during the winter.

Diet Pardalotes are foliage-gleaners, feeding on arthropods such as insects and spiders. In particular, they feed on lerps, a sugary casing produced by sap-sucking psyllid bugs. It is believed that pardalotes may play a significant role in controlling lerp infestations in eucalypt forests.

Breeding Striated and Forty-spotted Pardalotes are colonial nesters; cooperative breeding may also occur. Breeding takes place broadly between June and February, depending on the species and region.

Pardalote nests are cups or domes of dry grasses and other vegetation. Nest sites vary between species: in tree hollows, burrows in the ground, or holes excavated in earth banks up to 39 in. (1 m.) deep. Clutches average two to five eggs, incubated for 19–23 days. Both parents incubate the eggs and rear the young, which fledge at 21–25 days.

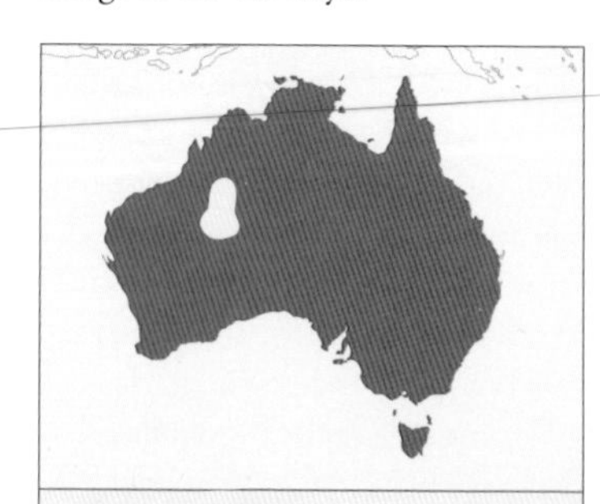

Number of genera 1
Number of species 4
Conservation Status 1 species Endangered
Distribution Australia

THORNBILLS *Acanthizidae*

The thornbills and their allies (acanthizids) are a diverse Australo-Papuan family of small passerines. The larger genera include the widespread thornbills (genus *Acanthiza*) and gerygones (*Gerygone*), plus scrubwrens (*Sericornis*), mouse warblers (*Crateroscelis*), heathwrens and fieldwrens (*Calamanthus*), Redthroat and Speckled Warbler (*Pyrrholaemus*), and whitefaces (*Aphelocephala*). Pilotbird (*Pycnoptilus floccosus*), Rock Warbler (*Origma solitaria*), Fernwren (*Oreoscopus gutturalis*), Scrubtit (*Acanthornis magna*), and Weebill (*Smicrornis brevirostris*) are all placed in monotypic genera. Most species are similar in structure and there is little sexual dimorphism, particularly among the gerygones and thornbills. Many are superficially similar to the Old World warblers, family Sylviidae, (hence the name 'fairy-warblers' formerly given to gerygones) but are related neither to this family nor to the New World warblers of family Parulidae.

Buff-rumped Thornbill (*Acanthiza reguloides*) can be identified by its buff-colored rump and the white irides. The birds are often seen feeding in acacias. Australia.

The similarities between species makes the acanthizids notoriously difficult to separate in the field, but attention to details such as head pattern and eye color, in conjunction with calls, behavior, and habitat, usually make identification possible.

Structure Acanthizids are dainty birds with a delicate structure and a short, thin bill. The bulky, long-tailed Pilotbird is the largest species, reaching 6.5 in. (17 cm.) in length. Whitefaces are also larger, dumpier birds, with a thicker, blunter bill. By contrast, Weebill is the smallest Australian bird, at 3–3.5 in. (8–9 cm.) and has, as the name suggests, a very small, short bill.

Plumage The most common coloration is olive-brown above and cream, buff, or gray below. Some species have patches of contrasting color on the rump. Many species have 12 tail feathers with a dark subterminal band, and ground-feeding species usually have pale eyes.

Whitefaces have distinctive face patterns, chestnut flanks, and rufous, brown, and gray upperparts; the male Banded Whiteface (*Aphelocephala nigricincta*) has a diagnostic black breastband. Heathwrens and fieldwrens have streaked underparts and, like scrubwrens, have a prominent white supercilium. The endangered Yellowhead (*Mohoua ochrocephala*) of New Zealand has lemon-yellow underparts and head (which are brighter in the male), and the tip of its tail often appears spiny when the feathers have become worn.

Behavior Hyperactive and acrobatic foliage-gleaners, most thornbills are highly visible, familiar birds. Many are social, nesting cooperatively and forming mixed-species feeding flocks outside the breeding season. Weebills often associate with flocks of other thornbills, which are joined by other passerines such as fantails, fairy-wrens, and honeyeaters.

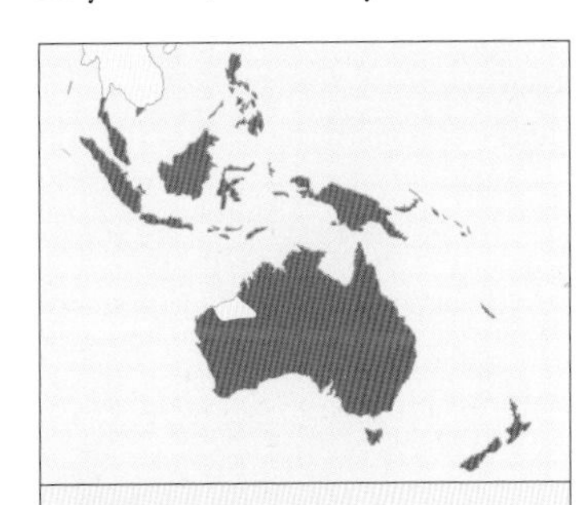

Number of genera 14
Number of species 60
Conservation Status 2 species Endangered, 1 species Vulnerable, 2 species Near Threatened
Distribution Australia, New Zealand, Southeast Asia, Melanesia

Gerygones feed on the outer foliage of trees and bushes, often hovering in the open before snapping at insects in the air or on nearby leaves. Other species, such as Fernwren, Speckled Warbler (*Pyrrholaemus sagittatus*), and the whitefaces feed almost exclusively on the ground. Pilotbird gains its name from its habit of accompanying foraging lyrebirds and snatching exposed invertebrates whenever the larger bird disturbs the soil.

Voice Gerygones have sweet, silvery songs, usually on a descending scale. Whitefaces have distinctive, plaintive, bell-like tinkling calls. Some species are accomplished mimics, notably Redthroat (*P. brunneus*) and Speckled Warbler. Feeding flocks are usually vocal, with high-pitched chittering and insectlike calls. Alarm calls are harsh scolding and chattering notes.

Habitat The family occupies a broad range of habitats in Australia, New Zealand, Indonesia, New Guinea, and Melanesia. Golden-bellied Gerygone (*G. sulphurea*) is the only representative found in South-East Asia (to Malaysia, Thailand, and the Philippines). Many species are woodland, mangrove, and rain forest specialists, but several, such as the whitefaces, have adapted to saltmarsh and open, arid habitats.

Most species are common within their range, but four are at risk. Yellowhead is classified as Endangered, as is one subspecies, Biak Gerygone (a subspecies of Large-billed Gerygone, *G. magnirostris*). Lord Howe Gerygone (*G. insularis*) became extinct around 1936, probably due to nest predation by introduced rats.

Movements Most species are sedentary, but Weebill, several thornbills, and the whitefaces are locally nomadic outside the breeding season.

Diet Largely insectivorous: the birds take a wide variety of small arthropods, including bugs, flies, and beetles, as well as larvae and spiders. Whitefaces also take seeds and fruit in season, using their stouter bill to deal with hard-cased food.

Breeding Many acanthizids are cooperative breeders, forming loose territorial 'families.' Their nests are large domes with a spout-shaped side entrance, hung from a branch; Yellow-rumped Thornbill (*Acanthiza chrysorrhoa*) adds an internal chamber above the nest cup, but the purpose of this structure remains unknown. The female usually builds the nest and incubates the eggs, but the male and several helpers may feed her and the young at the nest. Clutches average three eggs, and incubation lasts for around 19 days. The young fledge at about 17 days.

Curiously, Weebill has been documented as nesting in association with a communal-nesting spider (*Badumna candida*) in southwestern New South Wales. It is thought that the birds use the spiders' webs as protection for their nest.

Some acanthizids are hosts for smaller parasitic cuckoos (family Cuculidae), such as Fan-tailed Cuckoo (*Cacomantis flabelliformis*) and bronze-cuckoos (genus *Chrysococcyx*).

A **Yellow-throated Scrubwren** (*Sericornis citreogularis*) forages on the forest floor. One of the brightest acanthizids, males have a black face mask, whitish-yellow supercilium, and bright yellow throat. Females have no face mask, and a paler brow and throat. Australia.

AUSTRALASIAN BABBLERS *Pomatostomidae*

The Australasian babblers are strikingly marked, vocal birds comprising five species in two genera (*Pomatostomus* and *Garritornis*). All are similar in appearance. Three of the species have several distinct subspecies.

Structure Most of the family are 7–10 in. (18–25 cm.) in length; Gray-crowned Babbler (*P. temporalis*) is the largest Australian species, reaching 12 in. (30 cm.). They are robust birds with a heavy, decurved bill, a long tail, and strong legs and feet.

Plumage Long, pale superciliary markings ('eyebrows') and a dark band across the eye give the birds a characteristic masked appearance. All have gray-brown upperparts, with a paler throat and underparts, and white outer tail feathers. Chestnut-crowned Babbler (*P. ruficeps*) has a more contrasting crown and white wing bars; Gray-crowned Babbler has pale orange-buff wing patches, which are clearly visible in flight. Adult Gray-crowned Babbler has a pale yellow iris, which the bird gains by its third calendar year.

Behavior Highly social, Australasian babblers live in territorial family groups and flocks comprising up to 20 individuals. They forage, dust-bathe, preen, and roost together. Territorial disputes with neighboring groups occur frequently and are noisy and often protracted, with occasional fighting. Flight is low and rapid, interspersed with glides. The birds also run with a bouncing gait.

Number of genera 2
Number of species 5
Conservation Status Least Concern
Distribution Australia, New Guinea, Indonesia

Gray-crowned Babbler populations are rapidly declining due to human activities such as agriculture and timber removal, leading to habitat degradation. Australia.

Voice Foraging groups call loudly throughout the day to keep in touch and warn others of predators or incursions by neighboring groups. Gray-crowned Babbler's popular name, 'Yahoo,' is due to the loud, repetitive calls given by duetting pairs. Various other common names, such as 'Happy Family,' 'Chatterer,' and 'Barking-bird,' are also derived from the birds' vocalizations.

Habitat Three species are endemic to Australia, living in *Eucalyptus* forests and dry scrublands, while Isidore's Rufous Babbler (*G. isidorei*) is endemic to the lowland forest areas of New Guinea and Indonesia. Gray-crowned Babbler has a wide distribution across both regions.

Movements Hall's Babbler (*P. halli*) is locally nomadic but generally Australasian babblers are sedentary; groups remain loyal to a territory of 2.5–123 acres (1–50 ha.), which is defended year-round. Reluctance to disperse across unsuitable habitat has been cited as a factor in the decline of the eastern race of Gray-crowned Babbler.

Diet Arthropods, small vertebrates, fruits, and seeds, taken from trees such as eucalypts, acacia, and mallee, or from the ground.

Breeding Australian species breed from June to November, but maintain a number of nests year-round, which also serve as roost-sites. Nests are large, dome-shaped stick structures, conspicuously located in shrubs or on tree branches up to 20 ft. (6 m.) high. Usually one to six eggs are laid, depending on species; females incubate alone but are fed by the male and several helpers. The chicks are fed by all other members of the group.

Taxonomy The Australasian babblers were formerly included among the Old World babblers (family Timaliidae), but DNA studies have revealed this association to be erroneous, despite similar habits and morphological traits (which have given rise to the family's alternative name, 'pseudo-babblers').

White-browed Babbler (*P. superciliosus*) is mainly found in dry woodlands, mallee, and saltbush, as well as near water courses. Australia.

LOGRUNNERS *Orthonychidae*

A tiny Australo-Papuan family of rain forest passerines comprising three species, in genus *Orthonyx*: Australian Logrunner (*O. temminckii*), Papuan Logrunner (*O. novaeguineae*), and Chowchilla (*O. spaldingii*).

Structure Logrunners are 7–8 in. (17–20 cm.) in length. The bulky, thrushlike Chowchilla is larger, at 10.25–11 in. (26–28 cm.). All are robust birds with a sturdy bill and strong legs and feet.

Plumage Chowchillas are predominantly dark brown above, with a blackish head and a contrasting blue-gray eye ring. Logrunners are patterned rufous-olive, gray, and mottled black. All of the family are sexually dichromatic: males have a white throat, while females have a rusty to rufous throat and upper breast. All have tail-feather shafts that extend as spinelike protrusions beyond the feather-vanes; the birds use them as a brace while foraging.

Behavior Usually seen in pairs or small groups on the rain forest floor, where they forage noisily in leaf litter. An unusual skeletal adaptation allows the bird to thrust debris aside with sideways strokes of the leg, rather than back to front (which is more usual in ground-foraging passerines). The bill is also used to toss leaves aside; together, these movements create characteristic small, circular depressions on the forest floor. Studies have indicated that these birds may play a significant role in the seed dispersal and germination rates of rain forest trees.

Voice All species are highly vocal. Chowchillas are particularly noisy around dawn and dusk, when small groups call loudly to each other with the onomatopoeic call that gives the species its name. They also produce a range of other complex vocalizations, including harsh-sounding notes and mimicry. Papuan Logrunner is a shy species, and its song of four to six descending notes is often the only clue to its presence.

An **Australian Logrunner**. Both sexes have mottled upperparts and gray and black wing bars, but this bird's rufous breast indicates that it is a female. Australia.

Habitat All species are endemic to relatively small areas, although with widely separated ranges. Australian Logrunner is restricted to moist lowland forest along the eastern coast of Australia. Papuan Logrunner is found in Papua New Guinea and Indonesia, where it inhabits subtropical or tropical montane forest. Chowchilla is endemic to the rain forests of northeastern Queensland.

Movements All species are sedentary, defending local territories throughout the year.

Diet Mainly invertebrates such as earthworms and insects, but also small vertebrates.

Breeding Both Australian species breed from April to October. They build bulky, dome-shaped nests of sticks, moss, vegetation, and roots, with side entrances. Nests are placed on or near the ground, often on logs or tree stumps or in vegetation. Australian and Papuan Logrunners lay two eggs per clutch, Chowchilla usually just one. In both Australian species, the female builds the nest and incubates the eggs, while the male provides food for her and the young. Incubation takes 25 days, and the young fledge in 18 days.

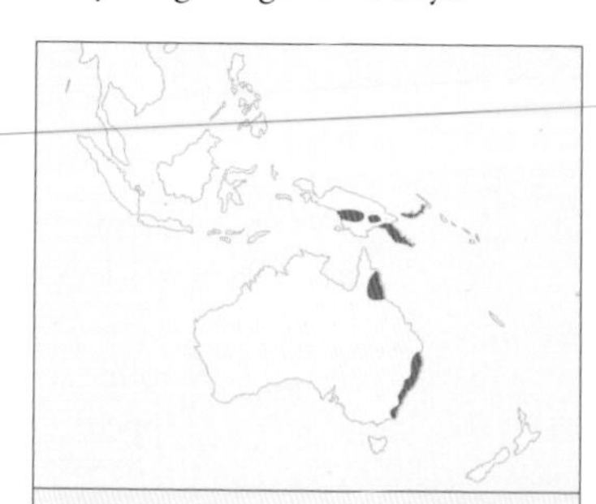

Number of genera 1
Number of species 3
Conservation Status Least Concern
Distribution Australia, Papua New Guinea, Indonesia

SATIN BIRDS *Cnemophilidae*

Yellow-breasted Cnemophilus, like all satin birds, has a short bill and a wide gape; for this reason, satin birds are also called 'wide-gaped birds of paradise.' New Guinea.

A small Old World family, the satin birds were previously considered part of the larger family of birds of paradise (Paradisaeidae), but they lack the latter's elaborate plumes. Satin birds are now regarded as being of older lineage, and in structure and appearance bridge the gap between the birds of paradise and the bowerbirds (family Ptilonorhynchidae).

Structure These compact birds are 7–10 in. (18–25 cm.) in length. They have a large, rounded head and a fairly short, weak bill. The male Yellow-breasted Cnemophilus (*Loboparadisea sericea*) has a large, tubelike wattle covering the nostrils. The wings are short and rounded, and the tail is square-tipped or slightly rounded.

Plumage All species are sexually dichromatic and take several years to reach maturity. Crested Cnemophilus (*Cnemophilus macgregorii*) is the brightest, with a reddish-orange or golden-yellow crest and upperparts and blackish-brown underparts. The male Loria's Cnemophilus (*C. loriae*) is velvet-black with a purple-bluish forehead and iridescent tertials. Females of both species are almost entirely olive-brown or olive-green. Yellow-breasted Cnemophilus has a dark brown face and crown, grading into warmer brown upperparts and pale yellow underparts; the male's wattle is pale blue-green. Females are duller or paler, with no wattle.

Behavior Satin birds may be found alone, in pairs, or occasionally in groups of up to 10 in fruiting trees; Yellow-breasted Cnemophilus is usually more solitary and sluggish than the other two species. The birds forage at all levels in forest trees, mostly on moss-covered branches in lower levels. They are occasionally inactive and inconspicuous for long periods.

Voice Most calls are given by males, as either contact or alarm. They include loud, musical, bell-like notes, harsh, rasping notes, muffled barks, and a series of rising, grating notes.

Habitat Entirely restricted to New Guinea, where the birds occur in montane forests, secondary growth, and shrubberies. Two species are common and widespread; the third, Yellow-breasted Cnemophilus, is locally common but considered to be Near Threatened due to habitat degradation and logging.

Movements Sedentary.

Diet All species mostly eat small fruits and berries, but occasionally take small invertebrates.

Breeding Not well-known for any species. All are polygamous. The male Loria's Cnemophilus has a courtship display in which it hangs upside-down from a high branch, with wings partially open, while giving a clicking noise. The male Crested Cnemophilus appears to have a large territory or display area, which it patrols regularly at the start of the breeding season. The nest is a large dome of moss, ferns, and orchids, with a front entrance of sticks and fronds; it is placed in vegetation on a rock face, or in a mossy and epiphyte-covered tree-trunk. The single egg is incubated by the female for about 26 days. The nestling is also cared for solely by the female.

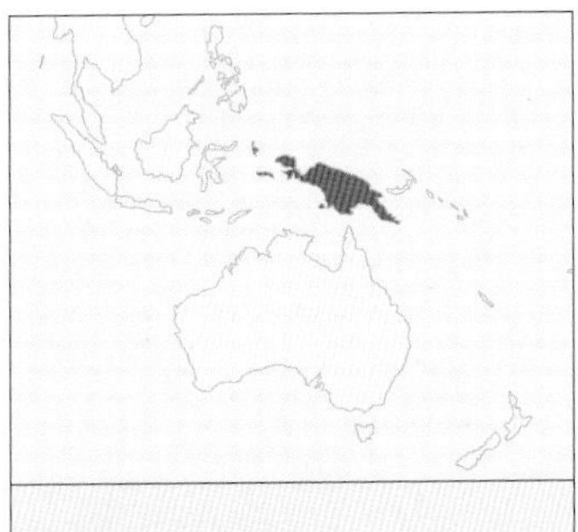

Number of genera 2
Number of species 3
Conservation Status 1 species Near Threatened
Distribution New Guinea

BERRYPECKERS *Melanocharitidae*

This Old World family comprises two groups: the berrypeckers and the longbills. The birds resemble flowerpeckers (family Dicaeidae), but the berrypeckers' tongue is less specifically adapted for collecting nectar. Several of the smaller berrypeckers also resemble small or short-tailed honeyeaters that have made the transition to fruit-eating. The longbills more closely resemble sunbirds, with long, decurved bills.

A **Crested Berrypecker**. The long black crest is usually laid flat on the head, but is raised almost vertical when the bird is displaying or alarmed. New Guinea.

Structure Most of the berrypeckers and longbills are 4–6 in. (10–15 cm.) in length, apart from two exceptions: the tiny Pygmy Longbill (*Oedistoma pygmaeum*), at about 3 in. (7 cm.), and the largest species, Crested Berrypecker (*Paramythia montium*), at 8 in. (21 cm.). Berrypeckers have a fairly small, rounded head. The bill is short but strong, broad-based, and tapering or pointed. The wings are fairly short and pointed, and the tail is medium-length. Longbills are small, short-tailed, and long-billed: in two species, the bill is longer than the head.

Plumage Most berrypeckers are sexually dichromatic, except for Crested Berrypecker. They range from glossy bluish-black to bluish-green in males and white to brown in most females, and are boldly spotted with white. Male Tit-Berrypeckers (*Oreocharis arfaki*) have a pattern very like that of a true tit (family Paridae). Crested Berrypecker has a broad, white, flaring supercilium; a black crest; a black chin and throat; deep blue on the sides of the neck and the underparts; and a bright yellow undertail. Longbills are mostly dull yellowish-green with brighter yellow eye ring and underparts.

Behavior Most berrypeckers and longbills live singly or in pairs. They forage by hovering and picking fruit and insects from foliage, either low down in undergrowth or high in the canopy of fruiting trees. Longbills are extremely shy; they are usually seen actively foraging and nervously flicking their wings, or in rapid flight. Tit-Berrypeckers and Crested Berrypeckers are more social, often occurring together in loose flocks.

Voice Berrypeckers give a series of rapid, high-pitched whistles or down-slurred, spluttering notes, as well as some scratchy notes and long or nasal wheezing. Songs are mostly a series of repeated twittering notes, but are not well known. Longbills have a short, sharp, disyllabic note.

Habitat Endemic to the forests of New Guinea and its satellite islands. In eastern New Guinea, certain species occupy different elevations in the same area: Black Berrypecker (*Melanocharis nigra*) is a lowland bird; Lemon-breasted Berrypecker (*M. longicauda*) is found higher up, in mid-mountain forest; and Fan-tailed Berrypecker (*M. versteri*) occurs in montane forest above 4,500 ft. (1,400 m.). No species is considered to be at risk, although Obscure Berrypecker (*M. arfakiana*) remains rare and very little known, and may be threatened by logging and land clearance in its habitat.

Movements Sedentary.

Diet Both subfamilies take small fruits, especially berries and soft figs, and insects; longbills also sip nectar from flowers, trunks and vines.

Breeding Berrypeckers' nests are a neat cup-shape made of plant fibers, fern fronds, lichens, and animal hair, usually slung from a slender branch up to 7 ft. (2.2 m.) from the ground. The clutch of one or two eggs is incubated solely by the female, although both partners feed the young. The longbills' nesting and egg-laying habits are poorly known.

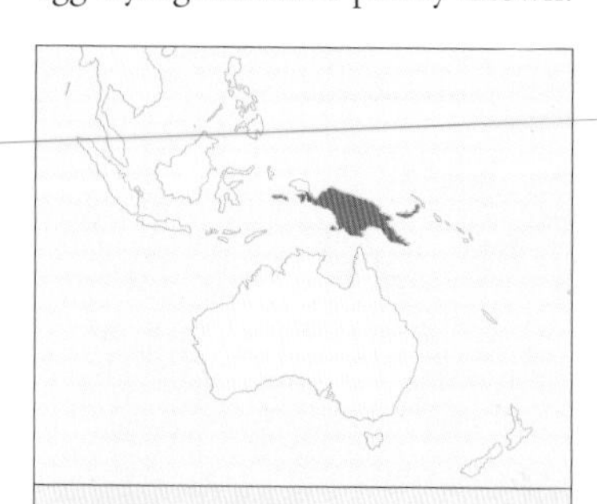

Number of genera 6
Number of species 12
Conservation Status Least Concern
Distribution New Guinea and surrounding islands

WATTLED CROWS *Callaeatidae*

A tiny family endemic to New Zealand, the wattled crows comprise two species in two monotypic genera: Kokako (*Callaeas cinereus*) and Saddleback (*Philesturnus carunculatus*). They are unrelated to the true crows (Corvidae) of the northern hemisphere.

A third species, Huia (*Heteralocha acutirostris*), once existed but is probably extinct; it was last seen alive in 1907. All three, including Huia, were widely distributed up to the end of the 19th century, but populations crashed dramatically due to the rapid spread of introduced predators: mainly stoats, cats, and rats. Huia was also hunted by both Maoris and Europeans, who used its feathers as adornments.

Structure The species differ widely in size, from 9 to 19 in. (23–48 cm.). They also vary in shape, and in bill size and structure. However, both have fleshy wattles at the base of the bill. Saddleback is the smaller bird, with a slender, slightly decurved bill. Kokako is larger and more robust, with a short, thick, arched bill, and long legs and tail. Both species have fairly short, rounded wings, and as a result are weak fliers.

Kokako chicks hatch with pink wattles; the blue color comes with age. New Zealand.

Plumage Saddleback is mainly blue-black with a broad, bright chestnut 'saddle' across the back and wing coverts, and bright orange-red wattles. Kokako is bluish-gray, grading to paler on the undertail, with a small black facial mask. North Island adults have blue wattles.

Behavior Saddlebacks forage in dead and rotting wood, often feeding on the forest floor or stripping bark and searching crevices and epiphytes for insects. Kokakos leap around in the branches, and are most often found near berry bushes and fruiting trees.

Voice Saddlebacks have rich, fluting, melodic songs, with 'dialects' that vary between groups, including those in fairly close proximity to one another. Groups of males often indulge in singing against each other, and individual birds within groups can be told apart by their specific songs. Kokako's song is a slow series of loud, mournful organ-like notes.

Habitat Subtropical beech or *Nothofagus* (tropical beech) forest. Kokako is considered to be Endangered (the South Island race *cinerea* has not been seen since 1967, and is currently presumed to be extinct). Saddleback is listed as Near Threatened.

Movements Sedentary.

Diet Saddleback is mainly insectivorous; Kokako is mainly a fruit-eater.

Breeding Pair bonds are strong and enduring. Saddlebacks are known to pair for life and remain together year-round. Their nests are shallow platforms, while Kokako's nests are more bulky; both are made mostly of twigs, leaves, moss, and ferns, and usually placed in a hollow tree up to 33 ft. (10 m.) from the ground. Clutches comprise two to four eggs, incubated by the female for 18–25 days. Both parents attend the young, which fledge at 28–31 days.

Number of genera 2
Number of species 2
Conservation Status 1 species Endangered, 1 species Near Threatened
Distribution New Zealand

Saddleback occurs on small islands off North Island. Once in severe decline due to predators, the species was saved by being relocated to predator-free islands. New Zealand.

WEDGEBILLS AND WHIPBIRDS *Eupetidae*

A widely disparate family of terrestrial birds, the Eupetidae include the wedgebills and whipbirds, plus the jewel-babblers (genus *Ptilorrhoa*) and Malay Rail-Babbler (*Eupetes macrocerus*).

Structure The jewel-babblers and Malay Rail-Babbler are thrushlike birds, 8–12 in. (20–30 cm.) long, with a long, tapering or fan-shaped tail. They have short, rounded wings and long legs and toes. Whipbirds and wedgebills are slightly smaller. In all species, the bill is fairly short and straight. In the whipbirds and wedgebills it is shorter, deeper, and wedge-shaped, while in the rail-babbler and jewel-babblers it is longer and more slender. Malay Rail-Babbler has inflatable sacs on the neck.

Plumage Whipbirds are dark olive-green, with black on the head or face and broad white malar stripes. Wedgebills are almost entirely gray-brown, paler below, with large white tips to the tail. Both whipbirds and wedgebills, with the exception of Papuan Whipbird (*Androphobus viridis*), have a crest, which is more prominent and slightly forward-pointing in the wedgebills. Malay Rail-Babbler is mostly warm brown, with a rufous throat, breast, and tail; the head has a broad white supercilium, black from the lores to the sides of the neck, and bluish lower ear-coverts. The jewel-babblers are deep blue or brown, or mixtures of both colors, with a broad white throat and a black face.

Spotted Jewel-babbler (*Ptilorrhoa leucosticta*) has a blue eye ring and white spots on the wing-coverts. Papua New Guinea.

Behavior The birds may be solitary or found in pairs. Most species are shy or extremely retiring, and are more often heard than seen. All are terrestrial, foraging in ground vegetation and leaf-litter; the whipbirds and wedgebills also forage in low undergrowth. Jewel-babblers and Malay Rail-Babbler walk with a pigeonlike back-and-forth motion of the head, while holding the body horizontally and the tail low or partly spread.

Voice Some species are highly vocal. Whipbirds give loud calls that carry for up to half a mile (800 m.). Wedgebills and whipbirds also perform song-duets, in which the partners contribute different phrases as part of a complete song. Jewel-babblers have a series of whistles and bell-like notes. Malay Rail-Babbler has a thin, rising monotone whistle, given with the neck sac inflated. Its alarm note is a froglike *tok* or *goink*.

Habitat The birds occur in a variety of habitats, from tropical rain forests to temperate woodlands, heath, and low, arid scrub. None is currently thought to be at risk, but two species are listed as Near Threatened due to continuing clearance and fragmentation of their habitat.

Movements Sedentary.

Diet Mainly seeds, shoots, fruit, and small insects; occasionally frogs and small lizards are taken.

Breeding The breeding habits of many species are poorly known or unknown. The best known are the Australian species. Eastern and Western Whipbirds defend their breeding territory year-round. They make their nests mostly from plant fibers, fern fronds, twigs, and grass, and place them on the ground or in dead vegetation. Chiming Wedgebill (*Psophodes occidentalis*) builds its nest up to 15 ft. (4.5 m.) above the ground, in the fork of a dense shrub. All lay two or three eggs. In whipbirds, the eggs are incubated by the female for 16–25 days. The young leave the nest after 17–29 days.

Taxonomy Blue-capped Ifrit (*Ifrita kowaldi*) is sometimes included in the family, but this placing is not universally accepted.

Whipbirds, like this **Eastern Whipbird** (*Psophodes olivaceus*), raise their crest when excited. Australia.

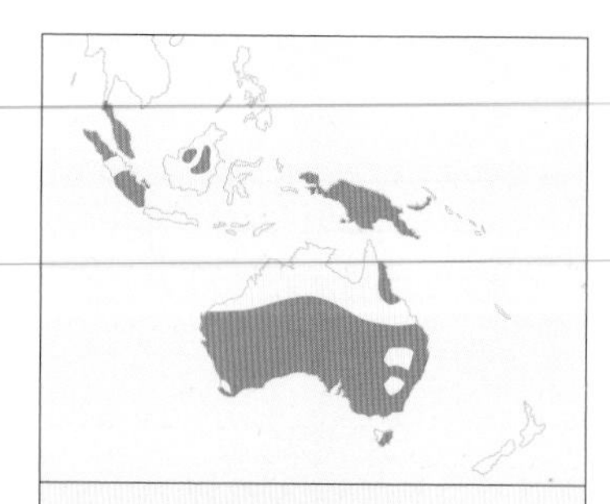

Number of genera 4
Number of species 10
Conservation Status 1 species Endangered, 1 species Near Threatened
Distribution Southern Thailand, Indonesia, New Guinea, Australia

QUAIL-THRUSHES *Cinclosomatidae*

Quail-thrushes, like this male **Chestnut-backed Quail-thrush**, have a boldly striped head, with white lines over the brows and cheeks. Australia.

A small family of ground-dwelling birds that are found in forest undergrowth. They look a little like thrushes and behave in some respects like quail.

Structure Quail-thrushes are 7–11 in. (17–28 cm.) in length and thrush-like in shape, with a straight, pointed bill and a dumpy body. The wings are fairly short and rounded, except in Chestnut-backed Quail-thrush (*Cinclosoma castanotum*), which has more pointed wingtips. The birds have strong legs and toes.

Plumage The body coloration is a complex pattern of browns to orange or chestnut, gray, black, and white, broken up with white spots and blackish patches. It may closely resemble the soil color of the birds' habitat. The underparts are often more boldly marked with blackish lines and crescents, and on most species the lower breast and belly are white. The females are less brightly colored than the males. The tail is long, with a fan-shaped tip; all species have broad white tips to the outer tail feathers, usually most visible when the bird is in flight.

Behavior Quail-thrushes are generally shy. They spend most of the time foraging on the ground, turning over undergrowth and leaf litter, and probing the soil beneath. While walking, they have a dovelike action, with the head moving forward and back. The 'quail' element in their name relates to their secretive nature and quail-like response to predators: when disturbed, they freeze and then burst into flight at the last moment, with a flurry of wings. After a dashing flight, they land at speed and run into cover. At least three Australian species perform distraction displays, feigning injury with wings and tail spread, to lure predators away from their nest or young.

Voice The calls of most species comprise a very thin, high-pitched *seep* or *see see see*, or, in the New Guinea species, an insectlike *tk*. The songs are soft, fluting, tremulous, and far-carrying; the New Guinea species has a low-pitched whistle followed by a rising *ooo-whit*.

Habitat In Australia, the birds are mainly found in the undergrowth of dry or semi-arid 'sclerophyll' forest, woodlands, and scrub: this habitat comprises acacia and eucalyptus trees on sandy soil and stony plains. In New Guinea, the birds are found in monsoon forests and rain forests.

Movements The birds are generally sedentary or locally nomadic within their breeding area.

Diet Quail-thrushes mainly eat a variety of insects and their larvae. They may also take small frogs and lizards.

Breeding The breeding seasons are largely governed by climatic conditions. At times, quail-thrushes breed year-round, but in severe droughts they may not breed at all. The nest is a loose, shallow cup of grass, twigs, and bark located under ground-level plants. The clutch of two (occasionally three) eggs is usually incubated by the female, for 11–14 days. The young leave the nest at about 14 days old, before they can fly.

Taxonomy Some authorities consider that the New Guinea jewel-babblers (genus *Ptilorrhoa*) should be included with the quail-thrushes in an enlarged family. Some also think that the distinctive race *alisteri* of Cinnamon Quail-thrush (*C. cinnamomeum*) should be regarded as a separate species: Nullarbor Quail-thrush.

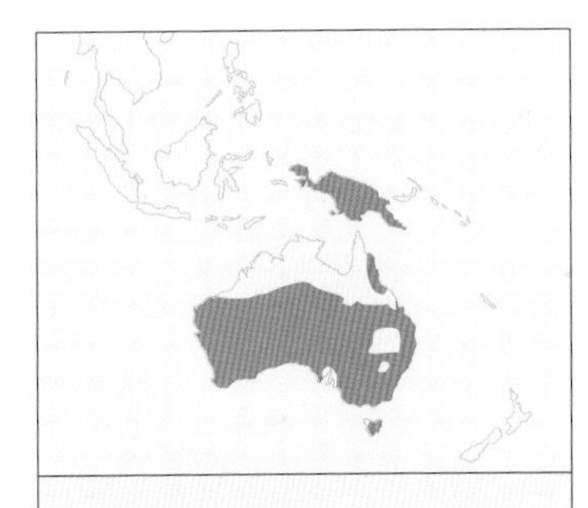

Number of genera 1
Number of species 5
Conservation Status Least Concern
Distribution Australia, New Guinea

WATTLE-EYES AND BATISES *Platysteiridae*

Small, mostly black and white African flycatchers with pale eyes and bright eye rings or red wattles over the eyes. The larger species resemble shrikes—a possible indication of their closest relatives.

Structure The birds are 3–6 in. (8–16 cm.) in length. The large, broad head is rounded in most species and with a short crest in one. The bill is broad and flat (substantially larger in genera *Megabyas* and *Bias*) and hooked at the tip. The wings are short and pointed, and the legs fairly short. Two genera are short-tailed and one comparatively long-tailed.

Several species have prominent, fleshy eye rings, or hornlike wattles above each eye; White-spotted Wattle-eye (*Dyaphorophyia tonsa*) has both. In all species, these features are present only in adults; they serve in communication between pairs, and greatly expand in size when the bird is excited or alarmed.

Plumage In wattle-eyes, the head and upperparts are black or gray and white, brown, or reddish-brown. The underparts are white, except in Yellow-bellied Wattle-eye (*D. concreta*), which has yellowish or chestnut coloring. All batises have pale gray to dark gray or black crown and upperparts; a black face mask; white on the chin and sides of the neck; a broad black, rufous, or orange breastband; a white wing flash; and a dark gray or black tail with white edges. Females are slightly duller or browner. All species have yellowish, pale orange, or reddish eyes.

All species, like this **Chinspot Batis** (*Batis molitor*), build open cup nests of fine plant stems and roots, moss, bark strips, fungi, and lichens, bound with spiders' webs. Botswana.

Jameson's Wattle-eye (*D. jamesoni*) has a vivid blue eye ring. Location unknown.

Behavior The birds may be solitary, found in pairs, or, particularly in the batises, occur in mixed-species foraging flocks outside the breeding season. They forage between trees or in the canopy of tall trees, and catch insects in flight or by gleaning from foliage; they usually seize their prey with a distinct snap of the bill.

Voice Highly vocal, the birds mostly make piping calls, whistles, churrs, buzzes, and trills.

Habitat Lowland to montane forests; woodlands, including edges and open savanna woodlands with *Acacia* thickets; also mangroves.

Movements Largely sedentary, but several of the species that breed at higher altitudes move to lower levels outside the breeding season.

Diet Mainly flying insects: principally flies and mosquitoes, but also lacewings, moths, butterflies, large grasshoppers, dragonflies, and mantises. The birds may also take insect eggs and pupae from within foliage, and take small lizards.

Breeding Pairs mate for at least one breeding season, and some species are territorial throughout the year. They lay between one and three eggs, which are incubated for 17–19 days. The chicks fledge after 21–23 days. They are fed by the adults for a further three to five months, and usually stay with the family group until the next breeding season.

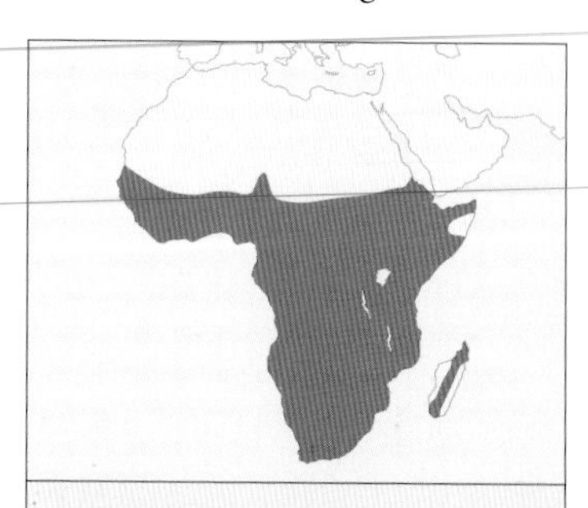

Number of genera 6
Number of species 28
Conservation Status 1 species Endangered
Distribution Sub-Saharan Africa, Madagascar

HELMET-SHRIKES AND PUFFBACKS *Malaconotidae*

White Helmet-shrike (*Prionops plumatus*) has the stiff, up-curved lores and crest typical of its genus. It is also recognizable by the yellow eye wattle. South Africa.

A large family of African birds, which includes the helmet-shrikes (genus *Prionops*), bush shrikes (genera *Malaconotus* and *Chlorophoneus*), tchagras (*Tchagra*), puffbacks (*Dryoscopus*), and boubous and gonoleks (*Laniarius*). Some authorities also include the Asian woodshrikes (*Tephrodornis*) and philentomas (*Philentoma*) in this family, as has been done here.

Gray-headed Bush Shrike (*Malaconotus blanchoti*) is called the 'ghost-bird' due to its secretive ways and eerie call. Tanzania.

Helmet-shrikes derive their name from the crests seen in many species. Within each genus, all species are similar in structure, clearly resembling the true shrikes (family Laniidae), but they vary widely in plumage. In most species, the sexes are alike or very similar.

Structure The birds are 6–10 in. (15–25 cm.) in length. They have a broad head and a short, stout bill with a notched and hooked tip; the bill is smallest or weakest in the puffbacks and helmet-shrikes. All species have broad wattles around the eyes. The body is slender. The wings are fairly short and pointed or rounded. The tail is long, with 12 feathers and a square or rounded tip, or graduated in the tchagras. The birds also have fairly long, strong legs and feet with sharp claws.

Plumage Helmet-shrikes in genus *Prionops* are most notable for the stiff, upward-curving feathers on the lores, forehead, and crown. In two species, these form a prominent crest. Yellow-crested Helmet-shrike (*P. alberti*) has a golden-yellow crest and crown. The five subspecies of White Helmet-shrike (*P. plumatus*) have a crest varying from short to more pronounced. Several other groups (mainly the bush-shrikes) have long, stiff or bristle-like feathers on the back of the neck.

Coloration varies within and between genera. Helmet-shrikes are mainly black-and-white, although four species are almost entirely black, with only a white undertail. Two species have a pale gray crown and are buffish or chestnut from belly to undertail. Puffbacks are also mostly black above and white below. The males have large rump patches of loose or fluffy white feathers, which are raised or fluffed out in display; in three species, the the females are browner, with white in the wings. Tchagras are mostly brown and chestnut, with a black head (or stripes across the head) and black tail, and paler below. Boubous are mostly black above and white or yellow below, with a broad white band across the closed wing; four species are entirely black. The brightest groups are the gonoleks and the bush shrikes. Gonoleks are mostly black above and deep orange or crimson on the underparts, and two species also have a bright golden crown.

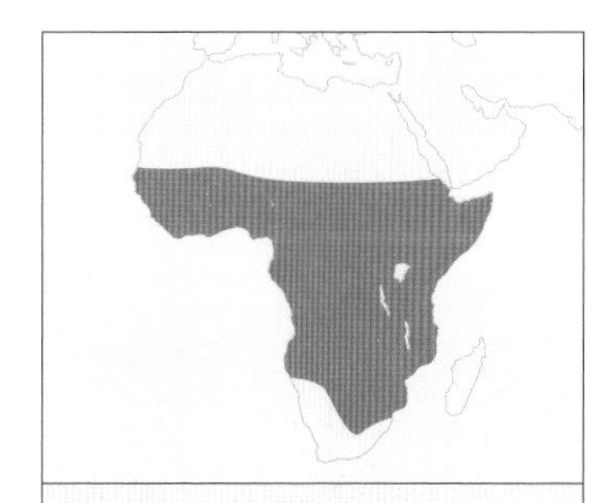

Number of genera 12
Number of species 56
Conservation Status 6 species Endangered, 2 species Vulnerable
Distribution Sub-Saharan Africa

Four of the bush shrikes have deep green upperparts, sulfur-yellow underparts, a black tail, a black facial mask (continuing into a broad breastband in one species), and crimson throat patches or a red, yellow, or black forehead. Many-colored Bush Shrike (*Chlorophoneus multicolor*) has several morphs across the range, in which the breast varies from buff to orange, scarlet, or black. Similarly, Black-fronted Bush Shrike (*C. nigrifrons*) has an orange breast over most of its range, but morphs with a buff, red, or black breast also occur. The two philentomas are blue-gray; Maroon-breasted Philentoma (*Philentoma velata*) has black from face to breast, and Rufous-winged Philentoma (*P. pyrrhoptera*) has bright rufous wings and tail.

Behavior Most species occur in pairs or small family parties. Helmet-shrikes are more social: they are generally seen in groups of 12 to 20 birds, and occasionally in mixed-species foraging flocks. Most of the tchagras and bush shrikes are highly territorial and rarely travel long distances. Helmet-shrikes forage over a wide area and are usually more arboreal than the other groups. While foraging, the birds often tilt their head to one side, as if to locate concealed prey by sound; they also forage along branches high into the canopy of trees.

Voice The birds have a variety of notes, including soft, mellow whistles, flutelike phrases, dry rasping, and nasal, chattering rattles, as well as nonvocal bill-snapping. The songs of several species, including helmet-shrikes, boubous, and bush-shrikes, are melodious and complex, with rich, rolling, fluted, bell-like phrases. The songs are also ventriloquial or mournful in quality. Bush shrikes and helmet-shrikes often duet, with two or more birds contributing to the song. Tchagras have a series of rising and falling, slow, or drawn-out whistles, often beginning with slow churring notes.

Habitats The birds are widely distributed in sub-Saharan Africa. They occur in lightly wooded savanna and open woodlands with undergrowth, rain forests (mostly in the lowlands, but one species is montane), swamp woodlands, mangroves, and papyrus beds. Several species are very poorly-known, and several others have very restricted ranges. One, Bulo Burti Bush Shrike (*Laniarius liberatus*), is known only from one individual caught and examined for DNA in central Somalia in 1988; it is assumed to have a small, diminishing population in an area threatened by uncontrolled tree-cutting and land clearance. Five other species are considered Endangered, and two more Vulnerable, for the same reasons.

Movements Mostly sedentary, although some birds make short-distance movements.

Diet All species feed mostly on insects and their larvae, including mantises up to 2 in. (5 cm.) long; they also take geckos, some seeds, and small fruits. The larger-billed species of boubou and bush shrike may take tree frogs, young chameleons, birds' eggs, and nestlings.

Breeding Most species are monogamous, solitary, and territorial. The helmet-shrikes, however, are cooperative breeders, with group members helping build the nest, incubate the eggs, and care for the nestlings.

Nests are mostly woven of bark or vine strips, twigs, plant fibers, grass, and lichens, bound together with spiders' webs, and placed in a fork or along a horizontal branch of a tree or shrub. The clutch of three to five eggs is incubated for 12–15 days by the female or pair, and nestlings remain in the nest for up to 20 days. After fledging, the young stay with the parents for up to four months, or longer in the helmet-shrikes.

Taxonomy There is considerable disagreement over the grouping used here. Several authorities consider the bush-shrikes, with the boubous, as a separate family, and the helmet-shrikes as more distantly related, or all as subfamilies of the true shrikes. Others place them closer to the vangas (family Vangidae) and the wattle-eyes and batises (Platysteridae). In the present grouping, the position and affinities of Brubru (*Nilaus afer*) are unclear. This arrangement also requires further consideration of the two wood-shrikes in genus *Tephrodornis*, and of the philentomas.

A male **Black-backed Puffback** (*Dryoscopus cubla*) feeding young. The bird's breeding behavior includes a courtship display in which the male puffs out his back feathers. Zambia.

BOATBILLS *Machaerirhynchidae*

The boatbills are a tiny family restricted to New Guinea and northern Australia. They derive their name from their large, flat bill. The two species are similar in structure, including in bill shape, and differ only in plumage.

Structure The birds are 4–6 in. (10–15 cm.) in length. They have a broad head and a flat, broad bill with a keel-like hooked tip to the upper mandible. The body is small and compact, with short wings and a long tail. The legs and toes are fairly short and thick.

Plumage Yellow-breasted Boatbill (*Machaerirhynchus flaviventer*) has black from the crown and upperparts to the wings and tail; bright yellow on the forehead and supercilium (although the latter is white in northern birds); a black face; a white throat; and rich yellow underparts. Black-breasted Boatbill (*M. nigripectus*) is dark olive from mantle to rump, and has broad black lores, with the rest of the face and the underparts deep or bright yellow except for a large black breast patch. Both species have a pure white double wing bar on the covert tips and fine white lines on the edges of the inner flight feathers. Females of both species are dull olive-green above, rather than black, and have paler underparts with fine spots or bars.

Behavior The birds may forage alone, in pairs, or as part of a mixed-species foraging flock with other insectivores including warblers, fantails, and honeyeaters. They feed mainly in the middle to upper levels of forest trees, gleaning insects from foliage, branches, and vines. They may sometimes sally to take flies and small wasps in mid-flight, before landing to eat them. They also spend long periods sitting quietly, and occasionally hold the tail cocked up in a wrenlike posture.

A male **Yellow-breasted Boatbill**. The distinctive bill is more than half the length of the head, with a central ridge. Boatbills also have long rictal bristles. Queensland, Australia.

Voice Mostly a series of rapid buzzing notes, soft whistles ending in a trill, or repeated *chee-chip* notes.

Habitat Boatbills are found mainly in lowland rain forest, thick or mature secondary woodlands, and montane forest edges. Yellow-breasted Boatbill is a lowland bird, while Black-breasted Boatbill is found at higher altitudes. Both species are generally common, and neither is considered to be at risk.

Movements Mostly sedentary, but Yellow-breasted Boatbill may disperse to more open forest after breeding.

Diet Insectivorous.

Breeding The breeding season is August to March, and occasionally to June in New Guinea. The nest is a small, shallow, fragile basket or saucer (neater or more substantial in Black-breasted birds). It is made from vine tendrils and plant fibers, bound with cobwebs, and placed up to 65 ft. (20 m.) from the ground in a fork of a tall tree. The clutch of two or three eggs is incubated by both parents for at least 14 days, and both parents care for the nestlings.

Taxonomy The boatbills were previously considered to be closely related to the monarch flycatchers (family Monarchidae) on the basis of wing and tail plumage and behavior (including their habit of vibrating the tail when perched). However, recent molecular work has revealed that they are only distantly related; their closest affinities are uncertain.

Number of genera 1
Number of species 2
Conservation Status Least Concern
Distribution New Guinea, Australia

VANGAS *Vangidae*

Chabert's Vanga (*Leptopterus chabert*), unusually for its family, is sometimes found in degraded woodlands far from primary forest. Madagascar.

A little-known family of arboreal birds restricted to Madagascar (and the Comoros for one species). The different species have varying bill shapes, which are the result of evolving in isolation to fill numerous niches and adapting to particular habitats and food.

Structure Vangas are 5–13 in. (13–32 cm.) in length. They generally resemble shrikes. However, the bill shapes are highly diverse: they may be short, pointed, and titlike; longer and flycatcher-like; larger and shrike-like; or very stout or flattened. Helmet Vanga (*Euryceros prevostii*) has a prominent, laterally flattened, casque-like bill reminiscent of a hornbill, while Sickle-billed Vanga (*Falculea palliata*) has a long, fine, scythe-shaped bill. Three species have a powerful, chisel-shaped bill.

Plumage Varied and complex combinations of black, gray, brown, olive-green, chestnut, or white. In most species, males are brighter than females. In some, males have a black hood. The most colorful species is Blue Vanga (*Cyanolanius madagascarinus*), its ultramarine blue head, blue wings, and blue and black tail contrasting with white underparts; the steel-blue bill is emphasized by blackish lores and pale blue eyes.

Behavior Usually seen in small to large groups and frequently in mixed-species foraging flocks (often solely with other vangas). The birds forage by gleaning or picking insects from the ground or from tree foliage, probing in holes, or stripping bark. They also catch prey by sallying in loops, or by aerial flycatching with swooping flights like those of woodswallows (family Artamidae). Large prey are held by one foot against a perch and then torn apart and eaten in a shrike-like fashion.

Voice Mainly whistles, dry churrs, hisses, chatters, chortles, *chuck* or *whip* notes, cries, and rattles, some of which are prolonged, sustained, and penetrating. Some species have a rich variety of loud, musical, ringing notes. Songs are varied whistling phrases either rising or falling with a rolling cadence, or unmusical warbles. Pairs of some species call or duet together, while songs of other species have a ventriloquial quality.

Habitat Primary rain forest and deciduous forest, also arid xeric 'spiny' forest; some species also occur in plantations and euphorbia scrub.

Movements Entirely sedentary.

Diet Principally insects and spiders; also frogs, lizards, snails, mouse lemurs, young birds, and fruit.

Breeding Most vangas breed between August and January. Nesting and egg-laying habits are not well known for any species. The nest is a cup or bowl placed along or in the fork of a branch, woven from plant fibers, roots, twigs, and moss, and sometimes bound with spiders' webs. Clutches comprise one to four eggs. 'Nest-helpers,' either related or unrelated, may aid in incubating the eggs and feeding and guarding the nestlings while the parents are foraging.

Number of genera 15
Number of species 22
Conservation Status 5 species Vulnerable
Distribution Endemic to Madagascar and the Comoros

A male **Helmet Vanga** on its nest of twigs. Northern Madagascar.

BUTCHERBIRDS *Cracticidae*

The Cracticidae comprise butcherbirds (genus *Cracticus*), currawongs (*Strepera*), and Australian Magpie (*Gymnorhina tibicen*). Butcherbirds are so called from the way that they wedge their prey into a tree-trunk or fence-post, then kill and dismember it. Currawongs are named for their distinctive call. The Australian Magpie is named after the unrelated northern hemisphere bird.

Structure The birds are 10–20 in. (25–50 cm.) in length. All have a large, powerful, hook-tipped bill; long, pointed wings; and strong legs. Butcherbirds are the smallest of the family; they have a broad, flat-crowned head, a steel-blue bill with a black tip, and a long tail. Australian Magpies have a similarly colored bill, but lack the long tail. Currawongs have bright yellow eyes, a black bill, a more slender head and neck, and a long tail.

Plumage Mostly black and white or gray, with white in the wings and tail. Black Butcherbird (*Cracticus quoyi*) is entirely black except for juveniles of the race *rufescens*, which are deep rufous. Pied Butcherbird (*C. nigrogularis*) has a black hood, mostly black upperparts, white underparts, and a white collar on the back of the neck. The northern race of Gray Butcherbird (*C. torquatus*) has a silver-gray mantle and back. Currawongs are also black and white, or gray, black, and white. Australian Magpie is mostly black with white or pale gray in the wings and upperparts.

Behavior The birds live in pairs or groups during the breeding season, but are found in larger flocks when not breeding. Australian Magpies live in social groups of up to 20 birds. They are strongly territorial, with territories ranging from 5–44.5 acres (2–18 ha.) depending on the size of the social unit dominated by a single male. All species forage by hunting or scavenging.

Young **Pied Butcherbirds** are paler in color than adults. Australia.

Voice Butcherbirds and Australian Magpies have a range of clear, high-pitched, mellow, piping songs, often given by two or three birds alternately. Currawongs make calls that sound like their name, including *curra-awok-awok-curra* or *kar-week*; they also produce loud, wailing cries and ringing notes.

Habitat Forests, woodlands, heaths, scrub areas, edges of cultivated land, orchards, rubbish dumps, mangroves, beaches, suburban streets, and gardens. Australian Magpies occur in most open areas with trees.

Movements Sedentary, partly nomadic, or migrating from higher to lower altitudes in winter.

Diet Butcherbirds are rapacious hunters, taking large insects, small birds, mammals, and reptiles; they also cache food remains in holes in trees. Currawongs forage mostly in trees and on the ground, and often take birds' eggs and nestlings. Around houses and picnic sites, they quickly learn to accept meaty morsels. Magpies are mostly opportunistic ground foragers.

Breeding The birds breed from August to December, with pairs occupying small territories. Australian Magpies will fiercely defend their nests from all intruders, including humans and domestic animals.

The nests are large, untidy baskets of sticks and grass situated in trees, up to 20 ft. (6 m.) from the ground. Butcherbirds lay three to five eggs, which are incubated for about 25 days. The chicks fledge at about 28 days. The female is solely responsible for incubating the eggs, but both parents feed the young, which remain dependent on their parents and other adults in the family group for some time after fledging. The young birds remain with their parents throughout their first year, and help the adults rear the young that hatch in the following breeding season.

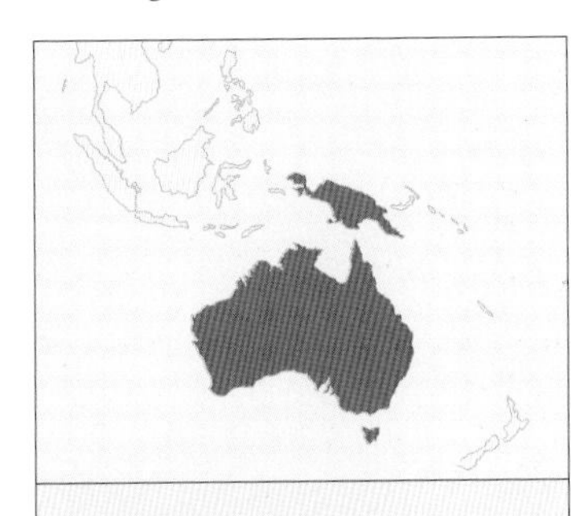

Number of genera 4
Number of species 13
Conservation Status Least Concern
Distribution New Guinea and Australia

WOODSWALLOWS *Artamidae*

Dusky Woodswallow (*Artamus cyanopterus*) is named for its smoky brown and gray plumage. Tasmania, Australia.

Stocky, gregarious birds, woodswallows are agile fliers, like their namesakes, but are not closely related to true swallows (family Hirundinidae). They are also called wood-shrikes, but are not related to the shrikes (Laniidae) either.

Structure The birds are 5 to 9 in. (12–23 cm.) in length. They have a rounded head; a stout bill, which is wide at the base and pointed; long, triangular wings, reaching to the tip of the short, square-tipped tail when at rest; and short legs and feet.

A tightly-packed group of **Black-faced Woodswallows** settles down to roost on a tree trunk. Australia.

Plumage Body coloration varies from gray to brown, with paler or whiter underparts. The birds usually have blackish facial markings. In Masked Woodswallow (*Artamus personatus*) and White-browed Woodswallow (*A. superciliosus*), the females are duller and less well marked than males. Woodswallows also have fine, waxy powder-down, which they spread over their plumage when preening; rarely found in passerines, this feature is shared only with the cotingas of South America.

Behavior The birds perch in prominent positions on treetops or roadside wires, from where they swoop out with graceful glides and spirals, often soaring up to some height in pursuit of prey. Woodswallows are highly social: flocks often number more than 100, and may comprise two or more species. When perched, they sit close together and often preen other members of the group. Roosting flocks pack together tightly on the trunks and canopy branches of forest trees.

Voice Mostly a series of sharp or high-pitched chirps, musical notes, harsh scolds, and rolling notes. The songs are animated and include mimicry of other birds' songs. Birds of prey are greeted and mobbed with loud, aggressive calls. Little Woodswallow (*A. minor*) is generally less vocal than other species, giving only a chattering contact note.

Habitat The birds live in tall, live or dead trees at the margins of forests and open woodlands; along roads, rivers, lakes, and coasts; and also in open, treeless areas of saltbush and semidesert scrub. Some species, particularly Black-faced Woodswallow (*A. cinereus*), are well adapted for life in the harsh conditions of the arid Australian interior.

Movements Mostly sedentary, but several species are nomadic outside the breeding season. White-breasted Woodswallow (*A. leucorynchus*) migrates from north to southeast Australia to breed. White-browed Woodswallow migrates in small numbers to coastal east and southeast Australia, but is occasionally irruptive in larger numbers.

Diet High-flying insects such as bees, wasps, cicadas, grasshoppers, and dragonflies. The birds also glean nectar from blossoms while in flight.

Breeding Woodswallows tend to breed at times of heavy rainfall, when food is abundant. During courtship, the male offers food to his mate, and the pair display to each other, fluttering their wings and rotating their spread tails. Their flimsy nests are made of twigs, lined with grass and plant stems, and sited in low trees, shrubs, or rocky crevices. The females lay two to four eggs, which they incubate for 12–16 days. The chicks fledge in 14–20 days. Both parents share parental duties, feeding the young for up to a month after fledging. In some species, other group members also help defend the nest and care for the young.

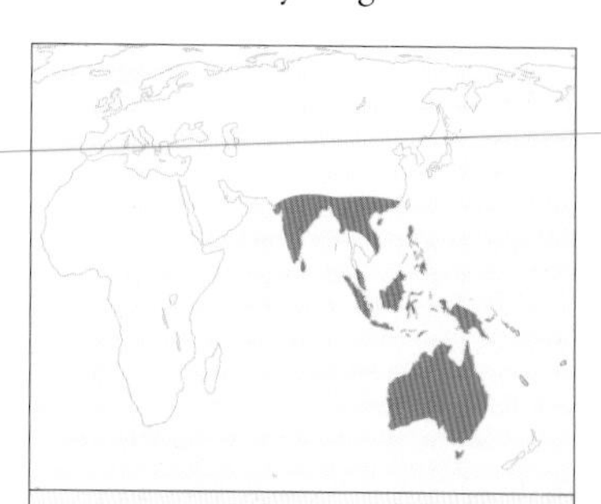

Number of genera 1
Number of species 10
Conservation Status Least Concern
Distribution South and Southeast Asia to Australasia

IORAS *Aegithinidae*

A small family of Asian passerines. Formerly considered within the larger group of leafbirds (family Chloropseidae), the ioras are now regarded as a distinct family on their own.

Structure Ioras are 5 to 7 in. (12–17 cm.) in length. Most species are similar in size and shape, at 5–6 in. (12–15 cm.) in length, but Great Iora (*Aegithina lafresnayei*) is larger. They have a short, stout bill with a fine, hooked tip to the upper mandible; fairly short, rounded wings; a fairly short, square-tipped tail; and short legs and toes.

Plumage All species are fairly similar in plumage and all are sexually dichromatic. They are mostly either bright green (brighter in males than in females) to greenish-yellow or black on the upperparts, and bright yellow below. The wings and tail are black with a prominent white double wing bar to the tips of the wing coverts (except in Great Iora). Marshall's Iora (*A. nigrolutea*) has broad white wingtips and a yellowish hindcollar. Males of the nominate race of Great Iora have black crown and upperparts. Male Common Iora (*A. tiphia*) molts in black on the head and upperparts in the summer.

Behavior The birds may forage alone, in pairs, or as part of foraging flocks with other small passerines. They actively search and glean the foliage for insects. Due to their small size and quick, warbler-like actions, they can reach to the outermost leaves and twigs; they occasionally hover briefly to catch prey.

Voice Ioras are fairly vocal. Pairs and birds in foraging flocks often utter distinctive, strident whistling contact calls. Common Iora has a variety of calls, including a long, high-pitched, down-slurred whistle, *weeeeeeeeeeeeeee-tuu*, ending on a lower pitch, and a short, harsh chatter. The songs of all species are fairly loud and melodious, and include imitations of other birds' songs.

A female **Common Iora** gleans prey from foliage. These small birds can balance even on the tips of branches or twigs, sometimes hanging upside-down as they forage. Thailand.

Habitat Evergreen deciduous forest, woodlands, acacia scrub, and mangroves. Common Iora is widespread across parts of Asia from northern India to the Philippines, with considerable differences between populations. In contrast, Green Iora (*A. viridissima*), found in lowland primary forest, is currently considered Near Threatened due to the rapid, extensive clearance of its habitat through logging and conversion of land to agriculture; its survival may depend on its adaptation to secondary growth areas.

Movements Mostly sedentary.

Diet Insects and their larvae.

Breeding Male Common and Great Ioras have elaborate chasing and courtship flight displays in which they spring into the air, hover, and then 'parachute' down with feathers fluffed out so they resemble fluffy green balls. Ioras' nests are small, compact cups of bark strips and grass, bound together with cobwebs and strongly fixed into the fork of a horizontal branch or outlying twig. Clutches comprise two or three eggs. Incubation, performed by both parents, takes about 14 days. Common Ioras may tend one chick each.

Number of genera 1
Number of species 4
Conservation Status 1 species Near Threatened
Distribution South and Southeast Asia

BRISTLEHEAD *Pityriasidae*

An enigmatic bird, Bristlehead (*Pityriasis gymnocephala*) is the only member of its family, Pityriasidae. Generally elusive, and more often heard than seen, it lives high up in dense lowland forest on the island of Borneo (and is also known as Bornean Bristlehead). The bird is notable for its striking red and black plumage, and for the bristles on its crown, from which it gets is name.

Structure About 10 in. (25 cm.) in length, Bristlehead looks similar to a crow. It has a broad head; a large, sharply pointed, hook-tipped bill; fairly short, rounded wings; and a short, square-tipped tail.

Plumage Mostly black, with bright red on the face, sides of the neck, throat, and thighs. The head is bare, with yellow bristles on the crown and nape, and gray ear-coverts. In flight, a small white patch is visible on the wings. The sexes are alike, but the female also has red spots or a patch on the flanks. The bill is black, and the legs and feet pink.

Behavior Bristleheads spend long periods foraging, in groups of up to 10 birds, in the canopy and upper levels of tall trees, where they search the foliage for large insects. They are surprisingly agile in negotiating the upper branches and outer foliage, and climbing vertical lianas, but also make heavy, ungainly sideways leaps and bounds along branches. On capturing an insect, they hold it in the bill and strike it against a branch before removing the hard wing-cases and legs. Flocks rarely spend long in one tree, moving on quickly if a preliminary inspection fails to reveal potential prey. Occasionally, however, the flock stops and individuals peer around, seemingly uncertain of which direction to take. The birds often forage in mixed-species flocks with other large insect-eaters such as babblers, trogons, drongos, malkohas, and woodpeckers.

Despite their striking plumage, **Bristleheads** are often hard to see as they move around in the topmost levels of forest trees. Borneo.

Voice Flock members keep in contact almost continually, with a nasal, whining call, a frequently given *wit-wit-peeoo*, and a crowlike chatter.

Habitat Primary and secondary forest, including partly logged and cut-over forest, below 3,280 ft. (1,000 m); also *Acacia* groves and forest edges, and sometimes mangroves. The species is considered to be near-threatened due to the rapid and continuing rate of forest loss.

Movements Mostly sedentary, although foraging flocks are known to cover large areas in a single day. In parts of its small range, Bristlehead is known to occur or to be more numerous only seasonally; this suggests that some birds move longer distances, possibly to higher areas.

Diet Mainly crickets, cicadas, beetles, cockroaches, and spiders, as well as moth caterpillars; the bird also takes small, plum-sized fruits.

Breeding Very little is known. It has been suggested that the birds build cup-shaped nests, like members of family Cracticidae, and that they are cooperative breeders, with other adults helping feed the young.

Taxonomy The family was formerly considered closely related to shrikes (family Laniidae), but is now thought to be closer to the Australian magpies, butcherbirds, and currawongs (family Cracticidae).

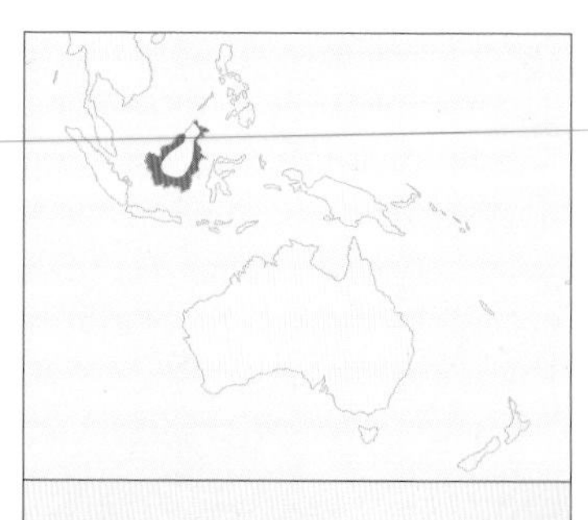

Number of genera 1
Number of species 1
Conservation Status Near Threatened
Distribution Borneo

CUCKOO-SHRIKES *Campephagidae*

The cuckoo-shrikes are a large family of Old World birds. They are named for their similarities to both cuckoos (family Cuculidae) and shrikes (family Laniidae). Despite their name, however, they are unrelated to either of these birds. Many have a powerful bill like that of a shrike and closely resemble cuckoos in their plumages. Their alternative names are 'gray-birds,' for their color, and 'caterpillar-birds,' due to several species' preference for eating large, hairy caterpillars.

The family consists of two distinct groups: the cuckoo-shrikes (genera *Campephaga*, *Coracina*, *Lobotos*, and *Campochaera*); and the trillers (*Lalage*), flycatcher-shrikes (*Hemipus*), and minivets (*Pericrocotus*). The cuckoo-shrikes are slightly larger than the other groups and usually gray in color; the minivets are smaller but much more colorful.

Structure The species range widely in length, from 5 to 15 in. (12–38 cm.), and in weight, from 0.2–0.4 oz. (6–12 g.) in Small Minivet (*Pericrocotus cinnamomeus*) to about 6 oz. (180 g.) in Melanesian Cuckoo-Shrike (*Coracina caledonica*). Bill shapes vary from broad-based, stout, and notched and hooked at the tip in cuckoo-shrikes, to short and rather slender in trillers, and flatter and very broadly based in the flycatcher-shrikes. The two species of wattled cuckoo-shrikes (genus *Lobotos*) have prominent wattles below the eyes. The birds generally have a slender body; fairly long, pointed wings with 10 primaries; and a long tail. In the cuckoo-shrikes, the tail is mostly square, rounded, or graduated at the tip, although Ground Cuckoo-shrike (*C. maxima*) has a deeply forked tail. In the minivets and some of the trillers, the tail is long and slender, while in the flycatcher-shrikes it is fairly short, narrow, and square-tipped. The legs are short, and slightly thicker in the trillers. In contrast, the mainly terrestrial Ground Cuckoo-shrike has long, well-developed legs and toes.

Black-faced Cuckoo-shrike (*Coracina novaehollandiae*) is sometimes called the 'shufflewing' for its habit of shuffling its wings when it lands. Australia.

Plumage In some species, the sexes are alike, while other species are strongly dichromatic. The plumage is mostly dense and fairly soft. However, the cuckoo-shrikes in genus *Coracina* also have rictal bristles. These birds, and the minivets, also have spine-like bristles on the lower back and rump, which the birds can raise in displays.

Most cuckoo-shrikes are entirely gray or black, or are mixtures of both colors, with a black head or face, or with white on the lower back and rump or the underparts. Several species also have barred underparts and pale or yellow eyes. The range of plumages also includes deep azure blue to blackish-blue in a few species. The females are either the same as the males, or vary from brown and gray to bright rufous on the upperparts and underparts. The most brightly colored cuckoo-shrike is the monotypic Golden Cuckoo-shrike (*Campochaera sloetti*), which has a golden-yellow body, gray crown, black from face to breast, and white in the wing. The African cuckoo-shrikes in genus *Campephaga* are also generally brighter in color: the males are all black or have red or yellow shoulder patches, while females are variably olive to greenish-yellow with a gray head and bright yellow underparts, or paler with heavily barred underparts. The two wattled cuckoo-shrikes have black heads, bright orange wattles, and yellow or orange-yellow underparts.

Trillers are mainly black above and white below, with large white wing patches on the coverts or the edges of the flight feathers. The most colorful is Rufous-bellied Triller (*Lalage aurea*). Several other species have warm buff, finely

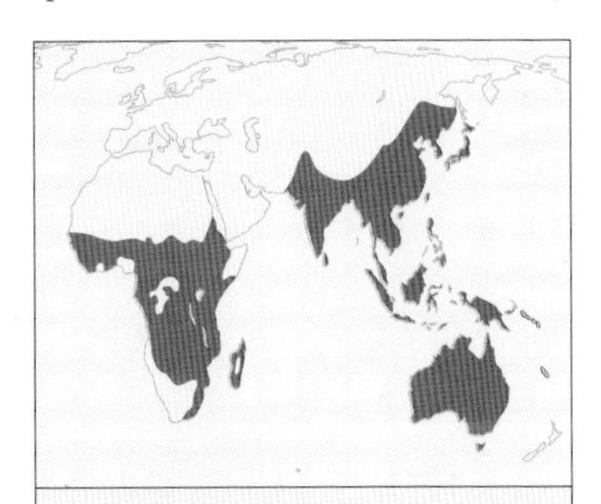

Number of genera 7
Number of species 81
Conservation Status 1 species Endangered, 3 species Vulnerable, 9 species Near Threatened
Distribution Africa, southern and eastern Asia, Philippines, Indonesia, New Guinea, Australia, and islands in the Indian and Pacific Oceans

barred underparts. The females of some species are paler brown and lack the white in the wing.

The minivets are the most colorful group: most of the males are a mixture of striking red and black, while the females are grayer, with yellow or orange underparts and wing patches. Several species, however (including those in which the sexes are alike), are black or gray above and white below.

The flycatcher-shrikes have black caps and upperparts, a white lower face and rump patch, and off-white to sandy-buff underparts. Bar-winged Flycatcher-shrike (*Hemipus picatus*) also has a broad white slash across the wing from the lesser coverts to the edges of the tertials.

Behavior Most species are almost entirely arboreal. They may occur singly, in pairs, or in small groups of family parties, although migrant and wintering Black-faced Cuckoo-shrikes (*Coracina novaehollandiae*) often number more than 100 birds together. All species regularly occur in mixed-species foraging flocks outside the breeding season.

Most *Coracina* cuckoo-shrikes are easily seen as they actively forage in the tops of trees or perch on the outermost parts of the foliage, but some of the African *Campephaga* species move more slowly and unobtrusively through the lower levels of trees and dense undergrowth. Ground Cuckoo-shrike forages solely on the ground. Trillers, either alone or in pairs, often spend long periods in a single tree or group of trees, rigorously searching and making short forays among the branches. Minivets are usually found in flocks, which constantly call to each other as they move from tree to tree. The exception is Rosy Minivet (*Pericrocotus roseus*), which more usually occurs in pairs and is generally a slower, more sluggish forager. Both species of flycatcher-shrike are commonly found in mixed-species feeding flocks year-round, although Bar-winged Flycatcher-shrike can also be found in pairs; the latter may feed in a similar manner to a flycatcher, pursuing insects in the air.

Voice Most cuckoo-shrikes are silent while foraging, but make contact and territorial calls. These are a variety of loud or high-pitched trills, churrs, whistles, and chattering notes, which may develop into a longer song often containing some scratchy notes. Nine of the *Coracina* cuckoo-shrikes are known as 'cicada-birds' due to their loud, mechanical buzzing songs, reminiscent of the sounds made by large bugs and cicadas. Trillers have a variety of simple songs of repeated piping whistles or musical notes, with trills, chatters, and chuckles. Minivets have comparatively thin or high-pitched twitters or whistled trilling notes.

Habitats All species principally occur in various types of forest habitat, from primary rain forest to *Eucalyptus*, moss, and swamp forests, as well as secondary growth areas and occasionally logged or degraded forest. None is critically endangered, but one, Réunion Cuckoo-Shrike (*Coracina newtoni*), is listed as Endangered as it now occurs in only two small areas of the island, and further destruction of its habitat or invasion by exotic plants or predators such as rats and cats may reduce its chances

The male **Scarlet Minivet** (*Pericrocotus flammeus*) has brilliant orange-red underparts, tail, and wing patches. Females and juveniles are gray above and yellow below. Nepal.

A **White-bellied Cuckoo-shrike** (*Coracina papuensis*) with a frog. Cuckoo-shrikes mainly forage for insects among tree branches and foliage, but some may take prey from the air or the ground. New Guinea.

of survival. Three other species are listed as Vulnerable, due mainly to habitat clearance; a further nine are listed as Near Threatened.

Movements Most species are entirely sedentary. However, Black-winged Cuckoo-shrike, which breeds north to central and eastern China, and two minivet species that breed in China, eastern Russia, Korea, and Japan, move south to winter in southern China, Southeast Asia, Sumatra, and Borneo. Shorter-distance movements are undertaken by cuckoo-shrikes that breed at high altitudes and winter at lower levels. Some of the Black-faced Cuckoo-shrikes that breed throughout Australia move north to winter in New Guinea and parts of Indonesia.

Diet Mostly insects and their larvae, especially caterpillars; also fruit (mostly figs but also berries), nectar, some seeds, and buds. Some species, particularly the minivets, flycatcher-shrikes, and several of the trillers, are predominantly insectivorous, while others are mainly fruit-eaters and take insects only occasionally. As well as a wide range of invertebrate prey, the birds may take snails and earthworms; the larger species of cuckoo-shrike also prey on geckos, small lizards, and frogs.

Breeding Most members of the family are monogamous, territorial, and generally solitary, although several species are cooperative breeders, forming groups of adults and young, and with nests placed close to each other. Several species have impressive courtship chases, flights, and spiraling displays. Nests are often small: many are only large enough to contain the eggs, and are later dwarfed by the incubating adult and growing nestlings. They are often flimsily built from twigs, plant fibers, and grass, camouflaged with moss and lichens, and bound together and to a supporting branch by spiders' webs. The nests are usually placed in a fork or along a horizontal branch. The clutch of one to five eggs is incubated by the female in some species and by both partners in others; incubation takes about 14 days in the trillers, and up to 27 days in some *Coracina* cuckoo-shrikes. Fledging periods similarly vary from about 12 to 25 days.

Taxonomy The two species of wood-shrike in genus *Tephrodornis* are sometimes placed here, but in this book have been grouped with the helmet-shrikes (family Malaconotidae). Fruithunter (*Chlamydochaera jefferyi*) of Borneo was also previously considered to be part of the Campephagidae, but DNA studies have shown this species to have closer links to the cochoas in family Turdidae (thrushes).

SITTELLAS *Neosittidae*

A male **Varied Sittella** climbs along a tree branch. Like nuthatches, sittellas can easily climb downward as well as upward. Australia.

The sittellas are a tiny family of arboreal birds found in Australia and New Guinea. They are usually grouped as three species in one genus (*Daphoenositta*). However, Varied Sittella (*D. chrysoptera*) has several races, each of which has been considered as a full species; it has also been seen as conspecific with Papuan Sittella (*D. papuensis*).

Structure The birds are 4–4.5 in. (10–14 cm.) in length, and weigh 0.3–0.7 oz. (8–20 g.). The head appears slightly large for the bird's size. The bill is short and daggerlike. The body is plump, but with long, broad wings, and the tail is short with a square tip. The legs are short, but the toes and claws are fairly long.

Plumage The sexes differ. Varied Sittella and Papuan Sittella both have a blackish crown and gray back; the face, wings, and tail finely streaked darker and black; a broad chestnut wing panel; and whitish underparts with fine streaks. Varied Sittella has been divided into five or more races, several of which are distinct in markings, with a black cap or head, white forehead and wing panel, or unstreaked underparts. Papuan Sittella males have a gray head and browner back, while females are white on the head and breast. Black Sittella (*D. miranda*) is almost all black except for a bright pink face and tail. The legs are yellowish or orange-yellow.

Behavior Sittellas are usually found in small flocks or family groups. They forage in the canopy of trees, in a similar way to nuthatches (family Sittidae), examining the tops and undersides of branches and tree trunks in search of insects.

Voice The contact call is a frequently repeated *chip* or a drawn-out, whistled or squeaky *pee pee tout*, usually given before the flock moves on.

Habitat Black and Papuan Sittellas occur in high-altitude forests, while Varied Sittella occurs in a wide variety of eucalypt woodland and forest, preferring rough-backed trees. None are considered to be threatened, but Black Sittella is uncommon to rare.

Movements Sedentary.

Diet Mostly beetles and larvae.

Breeding The birds are monogamous. Most pairs are usually assisted by nest-helpers, probably juveniles from previous broods. The nest is a deep cup of plant fibers and down, lichen, fur or feathers, and bark strips, bound with cobwebs and placed 30–130 ft. (9–40 m.) from the ground in an upright fork of a tree. The clutch of one to five eggs is incubated by the female for 20 days. The female, and then the chicks, are fed by the male and helpers. The chicks fledge at about 18 days.

Taxonomy Sittellas were thought to be closely linked to the nuthatches of the northern hemisphere, or to the Australasian babblers (family Pomatostomidae) and the Australasian treecreepers (Climacteridae). Recent DNA analysis indicates closer affinities with the whistlers (Pachycephalidae) and the berrypeckers (Melanocharitidae).

Number of genera 1
Number of species 3
Conservation Status Least Concern
Distribution New Guinea, Australia

SHRIKE-TITS *Falcunculidae*

A small Old World family, restricted to New Guinea and Australia, the Falcunculidae comprise the Eastern, Western, and Northern species of shrike-tits (genus *Falcunculus*) and the monotypic Wattled Ploughbill (*Eulacestoma nigropectus*). The two genera are linked by their bill shape and foraging behavior. They differ in plumage, Wattled Ploughbill being rather drab, whereas the shrike-tits are brighter, with a distinctive head pattern like that of a 'true' tit (family Paridae).

A male **Eastern Shrike-tit** at the nest. The birds often build their nests in eucalpyts; they may remove some of the leaves and twigs around the nest for easy access. Australia.

Structure The birds are 5–7.5 in. (12.5–19 cm.) in length. Their most distinctive feature is the wedge-shaped bill, which is short, deep, and laterally compressed, with a finely hooked tip. The wings are short and pointed. The tail is medium-length and square-tipped, or slightly notched in the shrike-tits. The male Wattled Ploughbill has large, round wattles on either side of the throat.

Plumage Wattled Ploughbill is mostly a dull yellowish-olive, with a brighter or golden-yellow forehead, face, and throat. The male's wattles are bright pink, and the wings and tail are black. The shrike-tits have a striking black-and-white face and a black, shaggy crest. The chin and throat are black in males but olive-green in females. The upperparts vary from yellowish-olive to green or brownish, and the underparts are yellow or white. In all members of the family, the bill is black.

Behavior The birds forage alone or in small family groups; ploughbills often join mixed-species foraging flocks. All species use their specially shaped bill to strip bark, open hard seed-cases and plant galls, or dig into dead wood; they also glean insects from branches. In addition, shrike-tits use twigs as tools, holding them in the bill and inserting them into holes and crevices to catch insects.

Voice Calls include thin, rising and falling whistles. Shrike-tits also have a mournful three-note call.

Habitat Shrike-tits occur in forests, including montane forests with bamboo, and in dry eucalypt and sclerophyll woodlands. In Australia, Eastern Shrike-tit (*F. frontatus*) occasionally occurs in parks and gardens. Northern Shrike-tit (*F. whitei*) has a small and declining population of fewer than 2,500 individuals in a heavily fragmented range, and is considered to be Endangered.

Movements Mostly sedentary, but some of the shrike-tits make local or seasonal movements away from their breeding areas.

Diet Mostly invertebrates, including large cicadas and their larvae.

Breeding The breeding habits are well-known only for shrike-tits, which are territorial year-round. The nest is a deep cup of bark strips, grasses, moss, and lichens, bound with spiders' webs and placed in a fork of a tree up to 100 ft. (30 m.) from the ground. The clutch of two or three eggs is incubated mostly by the female, for 18–20 days, and the nestlings are fed by both adults for 15–17 days. Fledglings may remain with the parents for 3–6 months. Shrike-tits' nests are often parasitized by several species of cuckoos.

Taxonomy Often classed within the family of whistlers (Pachycephalidae). The three populations of shrike-tits are also sometimes considered as conspecific with Eastern Shrike-tit.

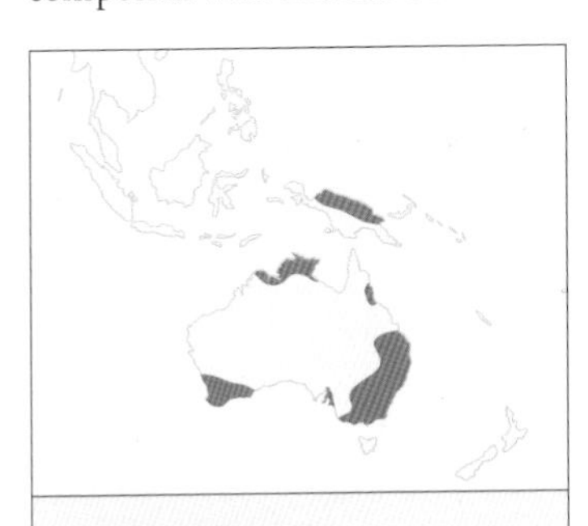

Number of genera 2
Number of species 4
Conservation Status 1 species Endangered
Distribution Australia, New Guinea

WHISTLERS *Pachycephalidae*

A male **Rufous Whistler** (*Pachycephala rufiventris*). This bird is common throughout Australia; it naturally occurs in forests and scrub, but is often seen in backyards. Australia.

Whistlers are mainly Australasian birds, many of which are renowned for their rich, melodious songs. Another distinctive feature is their relatively large, rounded head; one of their earlier alternative names was 'thickheads' (from which the Latin family name, Pachycephalidae, is derived).

Structure The birds are 5–8 in. (12–20 cm.) in length, and weigh 0.5–1.5 oz. (13–44 g). Obvious features are the fairly large, rounded head, short, thick neck, and short, stubby, thick bill. The wings are short and broad, with 10 primaries; in most species the wings are rounded, but in two (including the most migratory), the wing-tips are pointed. Tails vary in length, being longest in the larger species; they have 12 feathers and are square-ended or slightly notched. The legs and feet are strong, especially in the larger species.

Plumage The sexes are similar in some species, but in most the males are more brightly colored; females of all species are generally duller than the males. Several species have bold black-and-white head markings.

The family shows a variety of plumages, and can be divided into species that are rather dull or drab and those that are much brighter, with golden-yellow underparts. In many of the uniformly drab species, their color is reflected in their name, such as Olive Whistler (*Pachycephala olivacea*), Gray Whistler (*P. simplex*), and Brown-backed Whistler (*P. modesta*). The more brightly colored species have a contrastingly dark head and face, and a large white throat patch bordered by a broad black band. Within this range there are birds with gray heads, brown to maroon backs, yellow and white-speckled throats, and orange-rufous underparts. Rufous-naped Whistler (*Aleadryas rufinucha*) has a gray head, a bright rufous center to the nape, and yellowish-olive upperparts. Red-lored Whistler (*P. rufogularis*) has light rufous-orange from lores to underparts, interrupted by a gray breastband. The small Goldenface (*Pachycare flavogriseum*) recalls the patterns of some New World warblers of genus *Dendroica*, with gray upperparts, broad white tips to the tertials, and golden-yellow face and underparts separated by a broad black band over the crown and sides of the neck. The different subspecies of Golden Whistler (*Pachycephala pectoralis*) vary from olive-green to black or black and yellow on the upperparts and wings, with a yellow throat patch and a broad black breastband (or no breastband at all in one subspecies).

Behavior Whistlers forage in trees, either alone, in pairs, or as part of a mixed-species flock. They feed in the foliage and branches from mid-height to canopy level, although one or two species, including Rufous-naped and Olive Whistlers, feed at lower levels and in the understory. Their main food is insects and spiders, which they glean from foliage, either by sitting on a branch and scanning the leaves above or by picking prey off the outer edges of leaves while hovering or in flight.

Voice Songs are loud or powerful series of clear, melodious whistles. The different species' songs vary in

Number of genera 6
Number of species 41
Conservation Status 2 species Near Threatened
Distribution Southeast Asia, Australasia, islands in Pacific Ocean

Goldenface, found in moist tropical or subtropical forest, is the smallest member of the family, at just 5 in. (12.5 cm.). Papua New Guinea.

speed, volume, and delivery: some are sweet and haunting or pensive, whereas others are softer, shorter, or more complex, with differing phrases. Songs may also conclude with sharp or whipcrack-like notes.

Habitats Most whistlers occur in forests, from rain forest to drier lowland woodland, mallee, and coastal and arid scrub; one species, Mangrove Golden Whistler (*P. melanura*), lives solely in mangroves. The greatest center of diversity is New Guinea, with 15 species, followed by Australia with eight. None is currently at risk, but Red-lored Whistler and Tongan Whistler (*P. jacquinoti*) have declining populations and are considered Near Threatened due to habitat loss and introduced predators. One other species, Bare-throated Whistler (*P. nudigula*), is potentially threatened by trapping for the cage-bird trade.

Movements Most species are entirely sedentary. However, some populations of Rufous Whistler (*P. rufiventris*) in eastern Australia make complex (and poorly understood) movements, mainly northward, in August and September, returning to the same areas between February and May. Some Golden Whistlers move to more open habitats at lower altitudes outside the breeding season, but many are resident year-round.

Diet Insects and spiders; some species also take fruit and small crabs.

Breeding Monogamous and territorial. Most breed in individual pairs, but nest-helpers have been recorded in some species. Most whistlers breed from July to January, but Gray Whistler breeds throughout the year in the northern part of its range. Nests are woven cups of grass, lichens, plant fibers, and bark strips usually placed in a tangle of vines or a fork of a tree or shrub. The clutch is typically two or three eggs, incubated by both parents for 15–16 days. The nestlings are fed in the nest until they fledge at 10–12 days, and for some time after they leave.

Taxonomy Whistlers were previously considered to be part of the Old World flycatcher family (Muscicapidae), but recent molecular evidence has found no link. Their closest relatives are in the family Colluricinclidae: the shrike-thrushes (genus *Colluricincla*) and the New Guinea pitohuis (*Pitohui*). There is widespread disagreement over the composition of the whistler family. In particular, Golden Whistler is considered by some taxonomists to comprise 55 to 70 subspecies, many of which differ greatly from the nominate species. Other authorities propose raising six groups to full species status, but more detailed research is needed; the situation is complicated by interbreeding between subspecies in parts of the range.

A female **Golden Whistler** feeding young. The birds raise one brood per season; both parents build the nest and care for the young. Australia.

SHRIKES *Laniidae*

A widely distributed family of predatory passerines. Most shrikes are boldly colored birds of open country. All except four species belong to genus *Lanius*: the members of this genus (named for the Latin word for butcher) are noted for their habit of impaling their prey on thorns.

Structure The species range from 6 to 20 in. (15–50 cm.) in length, and the sexes are similar in size. Chinese Gray Shrike (*Lanius sphenocercus*) is the largest member of genus *Lanius*, at 12 in. (30 cm.) long and up to 3.5 oz. (100 g.) in weight. The longest species (by virtue of its long tail) is Magpie Shrike (*Urolestes melanoleucus*), at 16–20 in. (40–50 cm.). The family is characterized by a relatively large head and a stout, hooked bill with a tomial 'tooth' similar to that of falcons, allowing the capture and killing of large prey. Compared to similarly sized passerines, the legs are strong with sharp claws.

Plumage The shrikes are characterized by a black 'mask' over the eyes, and by large white wing spots. Body plumage varies widely in coloration, but many species have boldly contrasting patterns of black, brown, gray, and white. Juveniles are similar to adults but often with extensive barring. Many shrikes are thought to undergo a complete postbreeding molt starting with the first fall, although at least some species retain primaries and/or primary coverts until their second year.

Behavior Shrikes are sit-and-wait predators, usually scanning their surroundings from high perches and then dropping to the ground to catch prey. Some species spend relatively more time fly-catching or gleaning prey from foliage. It has been suggested that shrikes flash the white spots on their wings to startle and flush prey, in a similar fashion to mockingbirds.

Lanius shrikes are noted for their habit of impaling prey on thorns or wedging it in forked branches. This practice allows the shrikes to eat larger prey by pulling off pieces. Food may also be secured in this fashion to be stored for later consumption by the birds themselves or by their mate or offspring, or to allow poisons in toxic prey to degrade over time.

Voice Most species are quite vocal, with extensive mimicry common. In some languages, the birds are named

A **Loggerhead Shrike** (*L. ludovicianus*) watches for prey. The bird's hook-tipped bill is well adapted for dismembering prey such as mice and large insects. Wyoming, U.S.

In **Isabelline Shrike** (*L. isabellinus*), the head and upperparts are sandy-colored and the rump and tail reddish. U.A.E.

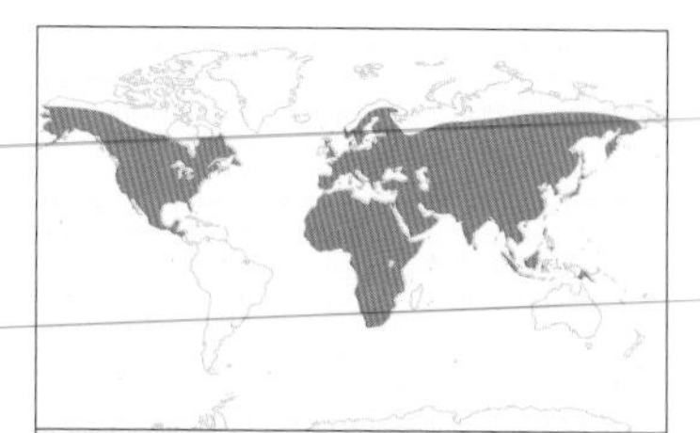

Number of genera 4
Number of species 30
Conservation Status 1 species Critically Endangered, 1 species Near Threatened
Distribution Africa, Europe, Asia, North America; also islands of East and Southeast Asia, reaching as far south as New Guinea

for the volume and variety of their calls. The Chinese name for the shrike means 'hundred tongues,' reflecting the widely varied repertoire of several Asian species. As well as varied and warbling calls, many species have a harsh alarm call, described as screeching, grating, or shrieking. This may have given rise to the common name, 'shrike,' which comes from the same root as the English word 'shriek.'

Habitat Shrikes are generally birds of open country, including savanna, woodlands, tundra or steppe, sagebrush, and scrublands. Many species are abundant in agricultural areas that contain a mixture of short grass and trees and shrubs for nesting. A few occur in dense forests.

Some species and subspecies are thought to be declining due to the loss or modification of their habitat, including the conversion of grasslands to cultivated fields.

Movements The shrike family includes resident, partly or altitudinally migrant, and long-distance migrant species.

Smaller species that occur in temperate climates are typically long-distance migrants, wintering in areas with abundant invertebrates. In some species, populations that occur at high latitudes or elevations migrate into areas at lower latitudes or elevations, where they may overlap with resident populations in winter. Such species may exhibit strong territoriality on the wintering grounds. For example, Bull-headed Shrike (*L. bucephalus*) calls most frequently in fall to defend its territory; in Japan, it is seen as a symbol of autumn in the same way that cherry blossoms symbolize spring.

Some African species are apparently nomadic. For example, individuals of the serially polyandrous Loggerhead Shrike may disperse several miles between nest sites.

A **Long-tailed Shrike** (*L. schach*) with its prey. The bird might either consume the insect immediately or impale it on a thorn and eat it later. India.

Diet Invertebrates such as grasshoppers and large beetles form the major part of the diet for most species. However, in larger species such as Northern Shrike (*L. excubitor*), more of the diet may comprise small mammals such as mice, and small birds, especially in winter. Some species are known to eat fruit. Like hawks and owls, shrikes regurgitate pellets of indigestible material (chitin, bones, and fur).

Red-backed Shrike (*L. collurio*) breeds in Europe and Russia, and migrates via the eastern Mediterranean and Arabia to tropical Africa. Poland.

Breeding Because this family spans so many climate zones, their breeding seasons are highly varied. In the tropics, breeding may coincide with the beginning of the rainy season, when insects are most abundant. In temperate zones, shrikes are often some of the earliest-breeding passerines. Most species breed in pairs, and are monogamous or serially polyandrous, but some are cooperative breeders. The nests are usually open cups, located in shrubs or trees. Clutch size ranges from three to eight eggs, which are incubated for 13–17 days. The nestling period lasts for 13–21 days, longer than is typical for similarly-sized passerines. Although incubation is carried out primarily or entirely by the female, males often provide food for their mate and the chicks. The young usually remain under the care of their parents for several weeks after leaving the nest.

Shrikes may raise one to three broods per season. In the cooperatively breeding Yellow-billed Shrike (*Corvinella corvina*), 14-week-old young have been seen feeding the fledged siblings of a later brood.

VIREOS *Vireonidae*

Yellow-throated Vireo (*Vireo flavifrons*) is typical of the Vireonidae in its size and greenish upperparts, although no other species has the distinctive yellow throat.

A family of mainly small, forest-dwelling passerines. Most of the North American species are classified in genus *Vireo*, while greenlets, found in tropical areas of South America, belong to genus *Hylophilus*. The two other genera are the shrike-vireos (*Vireolanius*) and the peppershrikes (*Cyclarhis*), both of which are found in the Neotropics. The majority of the family resemble warblers in their general appearance. Some of the most common forest-living New World birds belong in this family, such as Red-eyed Vireo (*Vireo olivaceus*) in the eastern U.S.

White-eyed Vireo might be hard to see in vegetation, but its loud, sharp song is unmistakable. Ohio, U.S.

Structure Most of the vireos, greenlets, and peppershrikes are rather small, at 4–6 in. (10–15 cm.), while the shrike-vireos are larger, at 5.5–7 in. (13–17.5 cm.). The birds tend to be 'chunky' in appearance, with a head that looks rather large. All members of the family share a similarly shaped, slightly hook-tipped bill, which is thicker than that of a warbler; shrike-vireos and peppershrikes have a heavier bill. Highly migratory species have longer and more pointed wings than resident, nonmigratory species.

Plumage Most vireos and greenlets have rather dull plumage, usually greenish on the upperparts and whitish or yellowish below. A number of vireo species have pale wing bars, while others, and most of the greenlets, have plain wings. Some species also have white eye rings or 'spectacles.' One, Hutton's Vireo (*V. huttoni*), looks almost identical to Ruby-crowned Kinglet (*Regulus calendula*, family Reguliidae), but can be distinguished by its bill shape and more sluggish behavior.

Slaty Vireo (*V. brevipennis*), endemic to Mexico, is uniquely colored, being slaty-gray and green. Another Mexican species, Chestnut-sided Shrike-Vireo (*Vireolanius melitophrys*), is strikingly patterned green above and white below, with a bright chestnut breastband and sides and a bright yellow eyebrow.

Some members of the family have pale or reddish irides, while a majority have dark eyes.

Behavior Vireos are rather inconspicuous, usually staying high off the ground or keeping well hidden, but are usually easy to detect when singing. Many species move rather sluggishly, and often sing from the same perch for long periods. Most vireos are highly territorial. In winter, the birds often join mixed-species flocks along with warblers and kinglets. Greenlets are often integral members of resident mixed-species flocks, either in the canopy, as in Dusky-capped Greenlet (*Hylophilus hypoxanthus*) of Amazonia, or in the understory, as with Tawny-crowned Greenlet (*H. ochraceiceps*) of Central and South America. Both shrike-vireos and peppershrikes, and in particular Green Shrike-Vireo (*Vireolanius pulchellus*), are canopy species that sing incessantly, but are notoriously difficult to see.

Voice Vireos are very vocal, with many species more easily identified by song than by plumage. Typical vireo songs comprise repetitive phrases of squeaky and burry notes, often sung over and over. Red-eyed Vireos have been known to sing

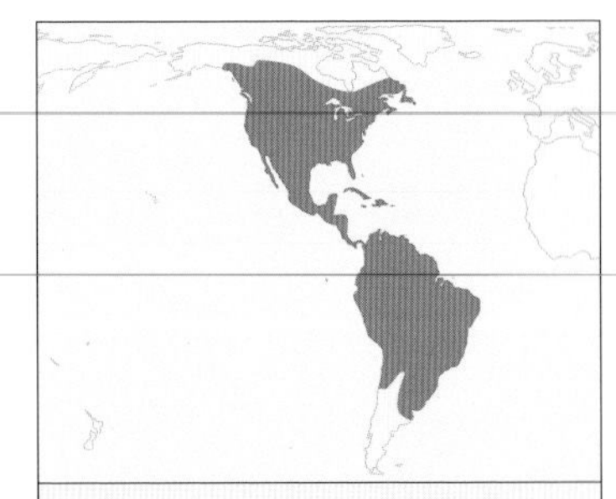

Number of genera 4
Number of species 52
Conservation Status 1 species Endangered, 2 species Vulnerable
Distribution North, Central, and South America

The peppershrikes of the Neotropics, like this **Rufous-browed Peppershrike** (*Cyclarhis gujanensis*), are larger than the true vireos, with a thicker bill. Trinidad.

every three seconds for hours on end. All vireo species have calls or 'scold' notes, which are used more commonly outside the breeding season. Greenlets' vocalizations are very similar to those of vireos.

Habitat Many species occur in forest canopy or subcanopy, while a few, such as White-eyed Vireo (*Vireo griseus*) and Bell's Vireo (*V. bellii*), live in dense, low vegetation. Some species have quite specialized habitat requirements, such as Black-whiskered Vireo (*V. atiloquus*), which breeds almost exclusively in mangroves.

Movements Most of the North American vireos are highly migratory, vacating their breeding grounds and wintering mainly in the southern U.S. and Central America. Red-eyed Vireo has the longest migration, with the majority wintering in Amazonia. Gray Vireo (*V. vicinior*) perhaps has the shortest, breeding from southern California to west Texas and wintering mainly in northwest Mexico. A few species, such as Philadelphia Vireo (*V. philadelphicus*), are typically trans-Gulf migrants (traveling across the Gulf of Mexico), while many of the Blue-headed Vireos (*V. solitarius*) winter either in southern Florida or in the Caribbean. Virtually all of the Central and South American vireos and greenlets are resident and sedentary.

Diet All members of the vireo family are mainly insectivorous, although some, in particular the shrike-vireos, are known to eat small fruits or berries occasionally.

Breeding Vireos and greenlets build small cup nests out of grasses and other plant material. Normally, they lay two to five eggs, and both male and female usually incubate and feed the young. Little is known about the breeding activity of the shrike-vireos, which live almost exclusively in the canopy of very tall forest. Some tropical species time their breeding to coincide with either the wet or the dry season.

A **Plumbeous Vireo** (*V. plumbeus*) feeds its young at the nest. The cup-shaped nest is suspended from a fork in a small tree branch. California, U.S.

OLD WORLD ORIOLES *Oriolidae*

A male **Eurasian Golden Oriole** (*Oriolus oriolus*) at the nest. Orioles' nests, made from vegetation such as grass, bark, and moss, are deep, open, cup-shaped structures, often slung like a hammock in the fork between two branches. Poland.

A small Old World family of mainly colorful passerines best known for their rich, flutelike calls and melodic songs. Although having no close affinities to other families, the orioles are usually considered to be close relations of the drongos and crows. Despite some physical similarities, they are unrelated to the New World orioles (family Icteridae). The name 'oriole' is derived from the Latin *aureolus*, meaning 'yellow' or 'golden,' although several of the eastern Asian and Australian species have neither of these colors in their plumage. Figbirds (genus *Sphecotheres*), part of the same family as orioles, are very similar in size and plumage, but are slightly more stocky and have a patch of bare skin around the eyes.

Structure Most species are medium to large passerines, ranging from 8 to 12 in. (20–30 cm.) in length. They are sexually dimorphic. The birds have a slender head and body. The bill is strong, robust, and slightly decurved in orioles and shorter, more stout, and hook-tipped in figbirds. In some species, the wings are long and pointed, with 10 primaries, while in island-living species they are short and rounded. The tail is of medium length, usually with a square or slightly rounded tip and 12 feathers. The legs are of medium length, with strong feet.

Plumage African, European, and Asian species are predominantly bright golden yellow and black on the wings and tail. Several have an entirely black head or a broad black mask that runs across the eyes and meets at the nape. One species, Green-headed Oriole (*Oriolus chlorocephalus*), has a dark green head and upperparts; females and immatures are often duller or greener and have streaked or striped underparts. Species living in New Guinea and Australia are olive-green or brown and are heavily streaked on the the upperparts and underparts. In contrast to most species, Maroon Oriole (*O. traillii*) has a dark maroon body, with a pale maroon tail and a blue-gray bill. Black Oriole (*O. hosii*) is all black except for chestnut

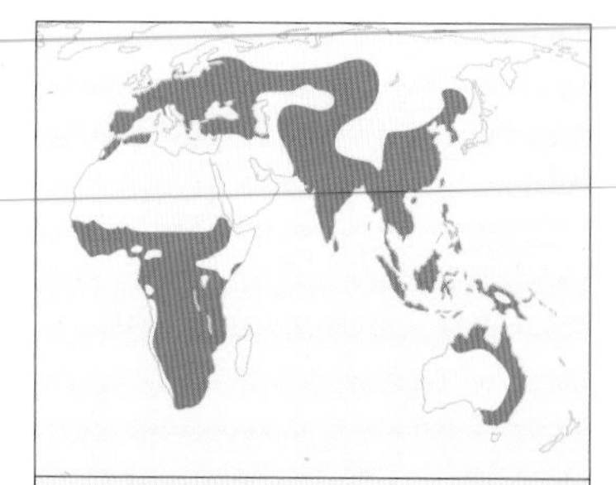

Number of genera 2
Number of species 29
Conservation Status 1 species Endangered, 2 species Vulnerable
Distribution Europe, Africa, Asia to Indonesia, Philippines, Australia

undertail-coverts. The most striking species is the scarce Silver Oriole (*O. mellianus*) of China, with a silver-white body, black wings, and maroon tail. The figbirds are similar to most orioles, with bright olive-green upperparts and yellow underparts, and a black head and face; however, they also have prominent bare patches of skin around the eyes.

Behavior Most are solitary or occur in pairs or family groups, although Eurasian Golden Orioles can form loose flocks of 50 on spring passage. Figbirds are generally more social than the other Old World orioles and often collect in noisy flocks of up to 30. Orioles forage in trees, often in the canopy—where, despite their bright plumage, they are often difficult to see and are usually detected by the frequently given call. They occasionally forage on the ground for insects and fallen fruit. The birds often travel some distances daily to find fruiting trees; some are regular members of mixed-species flocks that forage in woodlands and forest edges. In parts of the range, their foraging habits bring them into conflict with commercial growers of figs and cherries. Orioles' flight is strong and gently undulating, with deep wingbeats.

Voice Most of the African and European orioles have a distinctly recognizable vocalization: a clear, rising, flutelike, liquid series of *ou-weely-weeo* calls or variations. The sound is usually given as a contact note between pairs or as part of a longer song, and occasionally as a single *olio*. Asian species have a liquid, whistling *phu-phi-uu*, while Indonesian and Australian species have a mellow, rolling or bubbling *olly, olly-ole* or a disyllabic *oo-loo*. Some species, including Olive-backed Oriole (*O. sagittatus*), have a scratchy subsong that includes mimicry of other species. Other notes include a far-carrying, harsh or catlike growling *skaaa* or *riaaow*. Figbirds have a loud and less melodious, descending *see-kew, see-kew*, regularly repeated, and a weak, rustling, or metallic trill.

Habitat Orioles are predominantly woodland birds, with habitats ranging from rain forest to forest edges, mangroves, edges of cultivated land, orchards, copses, and parklands. Some of the Australian figbirds regularly occur in parks and gardens, including in the center of large cities; the family has its greatest density of species in Africa and in southern Asia to Indonesia.

Movements Most oriole species and all figbirds are sedentary. Eurasian Golden Oriole breeds across Europe to central Asia and migrates through northeast Africa and across the Sahara to winter in central and southern Africa. African Golden Oriole (*O. auratus*) is mostly resident, but outside the breeding season wanders to more arid areas. Black-naped Oriole (*O. chinensis*) breeds over much of eastern Asia and moves west to winter in southern China, Bangladesh, and western India. Two other Asiatic orioles make short-distance movements to winter in Myanmar (Burma) and India.

A female **Eurasian Golden Oriole** calling. The loud, melodious, fluting notes are given by both sexes. Oman.

Green Figbird (*Sphecotheres vieilloti vieilloti*) has a bare face, which is red in males but gray in females. Queensland, Australia.

Diet Orioles are mainly insect-eaters, taking ants, bugs, beetles, cicadas, crickets, flies, and caterpillars, but supplement their diet with fruit, mainly in late summer and early fall. They probably specialize on large, hairy caterpillars avoided by other species: they hold these in the bill and rub them vigorously against a branch to remove hairs and toxic gut contents. Figbirds feed mainly on figs, as well as ripe berries, cherries, grapes, and currants. Several orioles, including Eastern Black-headed Oriole (*O. larvatus*), also take pollen and nectar from aloe plants.

Breeding Oriole nests are cup-shaped structures, woven with grass, bark strips, moss, and lichens and slung in the fork of a branch. Figbird nests are shallow, flimsy structures made mostly of twigs; in some nests, the eggs can be seen through the base. The clutch of two to four eggs is incubated by the female for 13–15 days. The young are blind when they hatch, and fledge after 2–3 weeks; they are fed by both parents, and sometimes also by male helpers. In several species, egg-laying and nesting behavior is unknown.

SHRIKE-THRUSHES *Colluricinclidae*

Mainly Australasian birds, shrike-thrushes resemble thrushes but actually have no affinities to either the thrushes (family Turdidae) or the shrikes (Laniidae). The family includes the pitohuis (genus *Pitohui*) and Crested Bellbird (*Oreoica gutturalis*). One species, Sangihe Shrike-thrush (*Colluricincla sanghirensis*), was rediscovered after almost 100 years, in 1985, at a single site; there are probably fewer than 100 individuals of this species remaining.

Structure Shrike-thrushes range from 6.5 to 11 in. (16.5–28.5 cm.) in length. They are very similar in shape to thrushes in genus *Turdus*, except that they have a slightly shorter, stouter, broad-based bill, and shorter legs. Two species have short, erectile crests. One pitohui has thick legs and feet adapted to ground-living.

Plumage Most species are sexually dimorphic, and range in color from rufous to brown or gray. The shrike-thrushes have a fairly restricted range of drab grays to browns, with rufous predominating in two shrike-thrushes and several of the pitohuis. The most boldly colored species are Hooded Pitohui (*P. dichrous*), which has a bright rufous-chestnut body, and Variable Pitohui (*P. kirhocephalus*), showing a variety of gray or black coloring. Crested Bellbird has a bold black-and-white face and a black breastband and crest.

The pitohuis are unique in that at least three species have toxic feathers. They are thought to derive the toxin from their diet, as a defense against predators such as snakes and hawks.

Behavior The birds mostly occur in pairs or small family groups at the end of the breeding season. They are fairly shy, but in parts of the range some species become curious and tolerant of human presence, and will respond to imitated whistles. They forage in trees and on the ground, searching the vegetation for fruit and insects. Gray Shrike-thrush is one of the few species known to use tools, occasionally poking small sticks into crevices in trees or behind bark to flush out insects. Several of the shrike-thushes are more terrestrial than others. At least one species of pitohui occasionally joins mixed-species foraging flocks.

Gray Shrike-thrush (*C. harmonica*) is mostly gray, with brownish upperparts and paler underparts. Australia.

Voice Most of the shrike-thrushes and pitohuis have loud, rich, musical songs, often with a flutelike quality. Songs comprise drawn-out notes and clear whistles, or may include upslurred and downslurred notes mixed with scratchy phrases. Two of the finest singers are Gray Shrike-thrush and Crested Bellbird, the latter known for its remarkable ringing, ventriloquial song.

Habitat Most species occur in tropical forests and secondary growth areas. Sandstone Shrike-thrush (*C. woodwardi*) occurs more widely in scattered trees and low shrubs, on sandstone hills and gorges.

Movements Mostly sedentary.

Diet Mostly seeds, buds, fruit, and invertebrates including millipedes, snails, and worms. Shrike-thrushes (and possibly pitohuis) take frogs, lizards, baby birds, and small mammals such as pygmy possums. They kill and eat prey by striking it on a branch or the ground, or dismembering it while holding it in one foot.

Breeding Most species are monogamous. Their nests are open cups of grass, leaves, bark strips, and twigs, located in the fork of a tree, a crevice in a cliff, or in a cave. Both sexes share parental duties. The clutch of two to four eggs is incubated for 14–21 days, and the young fledge in 13–17 days. At least two pitohuis are known to have nest-helpers to assist with defense and feeding the young. Most species have two broods; in some years the number may rise to four or five.

Number of genera 3
Number of species 14
Conservation Status 1 species Critically Endangered
Distribution Australia and New Guinea; two species restricted to single islands in Indonesia

DRONGOS *Dicruridae*

A widely distributed, mostly Asian passerine family, drongos are often easy to see as they perch and hunt in the open.

Structure The birds are 7–25 in. (18–64 cm.) long, including the tail. The bill is usually short and stout, slightly curved and with a hooked tip. At its base are prominent rictal bristles; in Shining Drongo (*Dicrurus atripennis*), these extend as stiff feathers onto the forehead. Drongos have a slender body, with long, pointed wings and a fairly long tail.

Plumage Coloration is mostly black or blackish, with a green to bluish-purple sheen. One species, Ashy Drongo (*D. leucophaeus*), is paler or predominately grayer, with a white face in one race. White-bellied Drongo (*D. caerulescens*) is named for its white underparts. Shining

Ashy Drongo has paler plumage than other species. India.

Common Drongo (*D. adsimilis*) is also known as Fork-tailed Drongo due to the shape of its tail. Location unknown.

Drongo has head, upperparts and breast of a glossy, deep blue-green, while Spangled Drongo (*D. bracteatus*) has glossy green spots on the head, neck, and upper breast, and an iridescent green gloss to the wings and tail. In some species, the crown and the back of the head are finely crested. The tail is square or forked. The outermost feathers may be twisted outward, as in Spangled Drongo, or finely elongated but reduced to shafts except at the tip, as in Lesser Racket-tailed Drongo (*D. remifer*) and Greater Racket-tailed Drongo (*D. paradiseus*).

Behavior Drongos forage in or at the edges of trees. Several species perch in the open, often on telephone wires and fenceposts, and watch for insects, then either dash down and take them on the ground or pursue them in darting, twisting flight. Drongos may also join mixed-species foraging flocks, and gather in small numbers at termite hatches.

Voice Mostly rasping, buzzing, rattling, or chattering notes, given singly or in a series. Some species have varied or discordant songs, including jumbles of whistles, squeaks, and creaking notes, and imitations of other birds' calls and songs.

Habitat Primary and secondary forest, in thickets, glades, and forest edges; also open woodland with savanna, parkland, cultivated land with scattered trees, and villages.

Movements Most tropical species are sedentary. Some Black, Ashy, and Spangled Drongos migrate from the Himalayas and eastern China to winter in the plains of India and Sri Lanka and in Southeast Asia. Some Australian-breeding Spangled Drongos winter in New Guinea.

Diet Mostly insects, including beetles, mantises, moths, butterflies, cicadas, and dragonflies. Larger species occasionally take small birds.

Breeding Drongos nest in the outer foliage of trees. The nest is a shallow cup of plant fibers and vines, slung in the fork of a branch. Clutches comprise three or four eggs, incubated for up to 17 days. Both parents incubate the eggs and care for the young; they will aggressively defend the nest from intruders (including humans) who approach it.

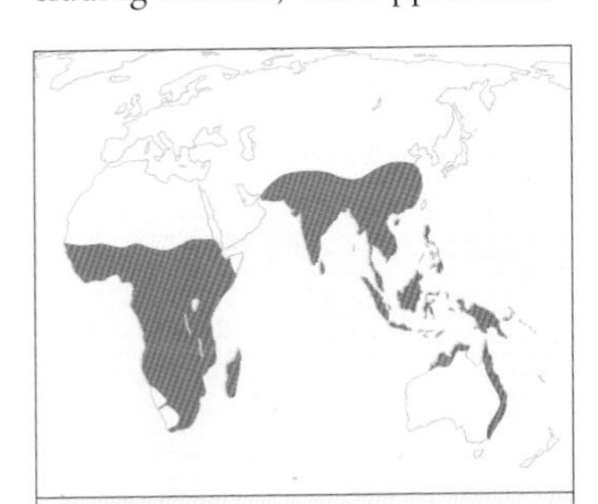

Number of genera 2
Number of species 22
Conservation Status 2 species Endangered; 3 species Near-Threatened
Distribution Africa; Asia to New Guinea; Australia

FANTAILS *Rhipiduridae*

A **Gray Fantail** (*Rhipidura albiscapa*) adopts the characteristic fantail posture, with spread tail feathers and drooping wings. Australia.

A family of Asian and Australasian flycatchers in a single genus, *Rhipidura*. All species are extremely similar, with a short bill and a small, compact body, usually held horizontally. The birds are named for their long tail, which they often raise, spread, and excitably wave from side to side.

Structure In most fantails, the sexes are alike or differ only slightly, but in some species, females are smaller than males. The birds are 4.5–8 in. (11.5–21 cm.) in length, and weigh 0.2–0.9 oz. (6–25 g.) The bill is short, flat, broad-based, and pointed with a small hook at the tip. The birds have a number of prominent rictal bristles, which in some species may be as long as the bill. The wings vary from short and rounded to longer and tapering, with 10 primaries and nine secondaries. The tail accounts for more than half the birds' total length, and is graduated, fan-shaped, or rounded. The legs and feet are fairly short and delicate, except in the largest species, Willie-wagtail (*R. leucophrys*).

Plumage The family is divided into species that are mainly gray, black, brown, or rufous, and species that are a mixture of colors. Many species have distinct white supercilia, throat patches, or moustachial streaks. Some have a dark breastband, often of heavy dark spots, or a dark breast with pale or whitish spots. In most species, the tail is uniform or two-tone, often with broad white or pale buff tips, or with pale or white in the outer feathers. Blue Fantail (*R. superciliaris*) is almost entirely dull blue, with a bright or silvery-blue forehead and supercilium. The smallest species, Yellow-bellied Fantail (*R. hypoxantha*), is mostly grayish-olive, with a black face patch, a broad, bright yellow supercilium, and yellow underparts; by contrast, Willie-wagtail is mostly black-and-white with an all-black tail.

Behavior Fantails may occur alone, in pairs, or as part of large mixed-species foraging flocks. Active and restless birds, even when perched most fantails twist and turn from side to side, and stiffly raise, fan, or flick their spread tail; they also keep the tail spread while in flight. They forage in foliage and along branches; some also sally after passing insects, often in aerobatic, looping flights of 165 ft. (50 m.) or more. Several of the common species are tame and inquisitive and often fly toward human observers, usually in search of insects that have been flushed. At least three species accompany and forage from the backs of cattle,

Yellow-bellied Fantail (*R. hypoxantha*) is notable for being the smallest species as well as one of the most colorful. China.

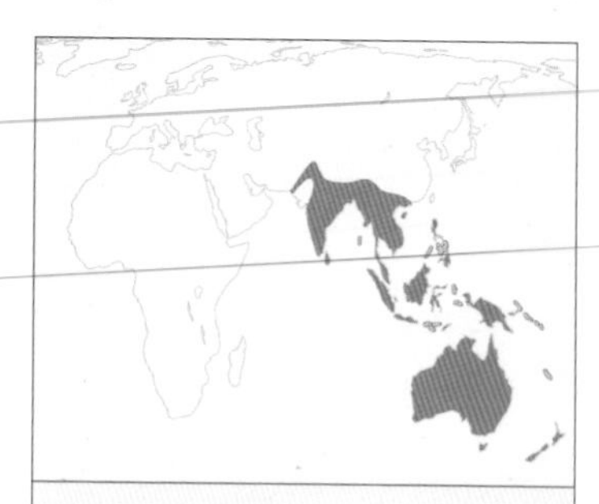

Number of genera 1
Number of species 43
Conservation Status 2 species Vulnerable, 5 species Near Threatened
Distribution South and Southeast Asia, Australia, New Zealand, Pacific Ocean islands to Fiji and Samoa

dashing out to glean insects from an animal's path. Willie-wagtails have also been recorded using kangaroos as look-out posts. Other species, particularly Northern Fantail (*R. rufiventris*) and White-winged Fantail (*R. cockerelli*), are less active feeders, preferring to spend long periods perching upright and watching before flying out to catch an insect.

Voice The calls are mostly single or repeated series of high-pitched squeaks, including *chip*, *swit*, or *sweet* notes, and harsher scolding in alarm. Fantails are often one of the early singers in a dawn chorus. Their songs are rapid, simple, repeated, high-pitched phrases. Willie-wagtail's song is three strong, musical notes, often given as *sweet pretty creature* and endlessly repeated.

Habitat The birds occur in a range of forest habitats, from primary rain forests to wet or monsoon forests, dry forest margins, and mangroves. The highest density of species is found in the forests of New Guinea. Five species are found in southern Asia, and a further five in Australia. Most are fairly common, but two, Manus Fantail (*R. semirubra*) and Malaita Fantail (*R. malaitae*), both endemic to small islands, are considered Vulnerable due to very small populations that appear to have declined or been unrecorded in recent years. Five other fantails are regarded as Near Threatened due to large areas of their habitat being destroyed or degraded through logging.

Movements Very few species make movements of any distance. Populations of Yellow-bellied Fantail and White-throated Fantail (*R. albicollis*) breeding at high altitudes make short-distance movements to lower foothills and valleys. In Australia, northern-breeding populations of Rufous Fantail (*R. rufifrons*) move northward to southern New Guinea from August to mid-November; similarly, Gray Fantails from the southwest move northward outside the breeding season.

A pair of **Willie-wagtails** mobbing a crow. These birds fiercely defend their territory from intruders; however, they are quite tame around humans. Queensland, Australia.

Diet Insects.

Breeding Courtship displays involve pairs raising and showing off their white supercilia by turning the head rapidly from side to side, and bobbing and jumping over each other; some species have courtship flights and singing. Females are usually more territorial than males, and actively repel intruders. The nest is built by both partners, on a horizontal branch up to 50 ft. (15 m.) from the ground. It is a small, deep cup of grass and fine plant fibers, bound together and fixed to the branch with cobwebs. Several species include a dangling tailpiece below the nest. The clutch of two to four eggs is incubated by both parents for 12–14 days. The nestlings usually fledge at about 12 days, or, in Willie-wagtails, 14–15 days.

Taxonomy The fantails are closely related to the monarch flycatchers (family Monarchidae); both are closer to the drongos (Dicruridae) than to the true flycatchers (Muscicapidae).

A **Rufous Fantail** at the nest. As with other fantails, both parents build the nest, incubate the eggs, and feed and care for the young. Location unknown.

MONARCHS *Monarchidae*

A large family of flycatchers, the monarchs are found across the Old World, from sub-Saharan Africa to islands in the Pacific. Most species are similar in structure; the main variations are the head tufts of the *Monarcha* species and paradise-flycatchers, and the long tails of the paradise-flycatchers.

Structure Small to medium-sized, slender passerines, most monarch flycatchers range from 3.5 to 8 in. (9–21 cm.) in length. The exceptions are the paradise-flycatchers, at 20 in. (50 cm.), including their long tail. The bill is generally sturdy, and short to medium-length, but is disproportionately large in the shrike-bills, and notched and hook-tipped in some of the other species. The wings are mainly short and rounded or blunt. The tail is usually short to medium length, and square-tipped or slightly rounded. However, in the crested-flycatchers the tail is long and graduated, while in the paradise-flycatchers it is very long. The legs are short or slender, with strong, sharp-clawed feet.

Plumage Many species have striking plumage, in a variety of colors. The monarch flycatchers range from pale to deep blue; females may be either similar in color or grayer, or have a brown back and wings. Most male paradise-flycatchers have a black head and orange-rufous upperparts and tail, but some are entirely or mainly red, rufous, or grayer, and in three species the male is mostly or entirely black. Several species have white morphs in which the orange-rufous areas are replaced with white; Madagascan Paradise-flycatcher (*Terpsiphone mutata*) has longer-tailed rufous, white, and black-backed white morphs. The monarch flycatchers in genera *Pomarea*, *Metabolus*, *Monarcha*, and *Arses* are predominantly black and white, and several of the last two genera have distinctive black and white head patterns or black face masks. Variations include all gray, gray and buff, or rufous. The male of one species is almost entirely white, and another is bright gold and black. The *Myiagra* flycatchers are predominantly black or gray and white with a black head; the underparts may be white or warmer rufous. The shrikebills are mainly rich rufous-brown, and two species have black face patches. The crested-flycatchers are pale blue or gray and white. The *Erythrocercus* flycatchers are mainly yellow, but

Spectacled Monarch (*Monarcha trivirgatus*) is named for the black band that extends across the eyes; it also has a black patch on the throat. Queensland, Australia.

White-eared Monarch (*Monarcha leucotis*) is restricted to Queensland and northern New South Wales. Queensland, Australia.

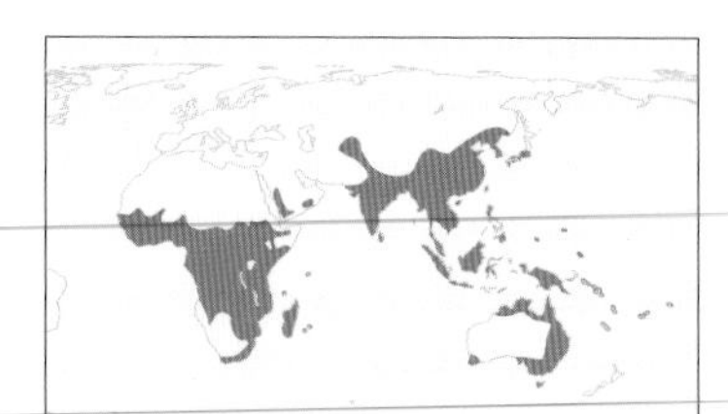

Number of genera 17
Number of species 95
Conservation Status
5 species Critically Endangered, 7 species Endangered, 6 species Vulnerable, 12 species Near Threatened
Distribution Africa, southern Asia, Indonesia, Australasia, several islands in the Pacific ocean

Chestnut-capped Flycatcher (*E. mccallii*) has a chestnut head and tail and gray upperparts.

The *Hypothymis* Monarchs and Cerulean Paradise-flycatcher (*Eutrichomyias rowleyi*) have prominent head tufts or crests. The *Myiagra* and *Elminia* flycatchers have smaller erectile crests, while those in genus *Arses* have fluffy or erectile nape feathers and prominent broad eye rings and wattles. Narrow eye rings are also found in the Monarch, *Myiagra*, and Paradise-flycatchers.

Behavior Monarch-flycatchers may be found either singly, in pairs, or in flocks. All are arboreal and predominantly insect-eaters. They actively forage by gleaning from foliage, hovering, and pursuing insects in flight. Several of the monarchs regularly join mixed-species foraging flocks.

Several species have a threat display in which they droop their wings and fan their tail. The African crested-flycatchers and *Erythrocercus* species often adopt a posture with the tail raised.

Voice All species are fairly vocal. Songs, mostly given by males, comprise a series of high-pitched, musical, fluting whistles or trills and harsh rasping notes. The paradise-flycatchers have a contact call of a querulous or rising *zweet* or *zweeoy*.

Habitat Mainly forests, but monarchs are also found in mangroves, savanna woodlands, thickets, and wooded edges of cultivated land.

The family includes many threatened species, including four that are critically close to extinction. Cerulean Paradise-flycatcher is thought to number less than 20 individuals on the Indonesian island of Sangihe, where almost all suitable habitat has been cleared for agriculture. Fatuhiva Monarch (*Pomarea whitneyi*), of the Marquesas Islands in the Pacific, has declined rapidly following the introduction of rats and invasive plants. All species in genus *Pomarea* have suffered greatly through man-induced threats: five subspecies have recently become extinct, two species are currently regarded as Critical, two are Endangered, and another is Vulnerable. The subspecies *freycineti* of Oceanic Flycatcher (*Myiagra oceanica*) on Guam declined rapidly and ultimately to extinction in 1983 following predation from the introduced brown tree-snake.

Leaden Flycatchers (*Myiagra rubecula*) often perch on exposed branches to give their calls. Australia.

Movements All tropical species are sedentary or make short-distance movements. Truly migratory species include several of the Australian monarchs, which breed along the eastern coast south to the Bass Strait and move north in winter, one species traveling to New Guinea. The longest-distance migrant is Japanese Paradise-flycatcher (*Terpsiphone atrocaudata*), which breeds throughout Japan and Korea and migrates through eastern China and southeast Asia to winter in Malaysia, Sumatra, and the Philippines.

Diet Mainly small insects (such as grasshoppers, dragonflies, and flying ants) and spiders. The shrikebills take scorpions and stick-insects. Several species take small crustaceans; some take small fruits and seeds.

Breeding Most species are monogamous. Males give courtship displays in which they may fan the tail, bow, and raise the crest. In many species, both sexes share parental duties; in some, the parents are assisted by other adults. Nests are small and cup-shaped, made of plant fibers, moss, bark, and spiders' webs, and sited in the fork of a branch. Clutches comprise two to four eggs, which are incubated for 12–18 days. The chicks leave the nest after 10–18 days, but may stay with the parents for a short time afterward.

Taxonomy The monarch flycatchers are usually considered most closely related to the flycatchers in family Muscicapidae, but their true affinities are not well established. The families may have evolved convergently. The five *Elminia* and three *Erythrocercus* species are here included in the Monarchidae, but some authorities dispute this placement.

A male **Madagascan Paradise-flycatcher** (rufous morph) at the nest. Madagascar.

CROWS AND JAYS *Corvidae*

Diverse in appearance and color, this group of birds, collectively known as corvids, is familiar almost worldwide. Their bold, intelligent behavior has led the birds to figure in human cultures through the ages: one species in particular, Common Raven (*Corvus corax*), has historically been revered or feared.

A **Pied Crow** (*Corvus albus*) with a dead and skinned chick. Pied Crow shares some features, such as throat hackles, with Common Raven. Kenya.

This charismatic group includes the largest passerines. They share various physical characters, including a strong bill, nostrils covered by feathers (in most species), and sturdy feet. The family Corvidae can be subdivided into several morphologically similar, but not necessarily closely related, types: crows and ravens, choughs, nutcrackers, jays, magpies, treepies, and ground jays.

Structure The species range in length from 7.5 to 25 in. (19–64 cm.). The sexes may be similar, or males may be larger than females (especially in terms of bill size).

The bill is typically strong, enabling the bird to tear or pummel its prey. The shape is stout and short to long, ending in a short hook or point. The culmen is curved, exceptionally so in some species: Thick-billed Raven (*C. crassirostris*) has a high, arched culmen. Choughs (genus *Pyrrhocorax*) and ground jays (*Podoces*) have a slender, curving bill. Many have a pouch under the tongue, for carrying food.

Common Raven was traditionally feared because it ate carrion from battlefields, but was respected for its intelligence. U.S.

Body shapes vary from thickset to slender. The wings may be long and pointed, as in many *Corvus* species, or short and rounded, as in tropical jays. The birds have sturdy legs and feet, used to hold down food as well as to move around on the ground. The legs are prominently scutellated on the front and smooth along the rear. The tail is typically medium to very long, but very short in Fan-tailed Raven (*C. rhipidurus*).

Plumage The flight feathers are typically stiff, but some jays have more lax flight feathers as well as lax, soft, fluffy body plumage. The plumage is subtly to highly glossy.

The sexes are generally alike. Most crows (genus *Corvus*) and choughs are predominately black or bicolored. Nutcrackers (*Nucifraga*), treepies, and ground jays tend to be cryptically colored in black, browns, grays, and white. Jays and magpies are more colorful, many with blues, violets, greens, or pinks. Some have intricately patterned plumage, with bars or spots.

Most species have forward-facing feathers, modified as bristles, that cover the nostrils. However, these are absent in three unrelated species: Rook (*C. frugilegus*), Gray Crow (*C. tristis*), and Pinyon Jay (*Gymnorhinus*

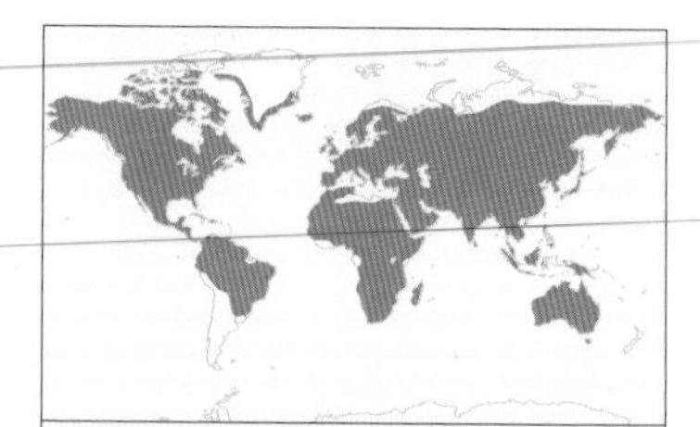

Number of genera 24
Number of species 117
Conservation Status 1 species Critically Endangered, 4 species Endangered, 8 species Vulnerable
Distribution Worldwide apart from Arctic, Antarctic, and southern South America

cyanocephalus). Many *Corvus* species also have a 'hackle' of distinctly shaped feathers on the head or throat; for example, Common Raven has lanceolate feathers on its throat. Many have long crown or forehead feathers forming a tuft or bushy crest; in a few, the crest is very long and pointed, or curled. In some species, the bristles over the nostrils are an extension of the bushy or plush crest; in treepies (genus *Crypsirina*), these feathers are velvety.

The tail is usually slightly rounded at the tip, but in some corvids it is more graduated, while in others it is wedge-shaped. Several genera (*Pica, Calocitta, Urocissa*) have a very long, graduated tail. Some treepies (*Dendrocitta* and *Crypsirina*) have modified central rectrices that broaden at the tip; in addition, *Crypsirina* species have 10 instead of the usual 12 rectrices. Piapiac (*Ptilostomus afer*) has long, pointed rectrices. Most unusual is the tail of Ratchet-tailed Treepie (*Temnurus temnurus*): the rectrices are jagged and elongated, with outward-pointing edges.

Most species have a distinct, although briefly held, juvenal plumage. Adults have a single complete post-breeding molt.

Short-tailed Green Magpies (*Cissa thalassina*) often hunt with their mate; they both stalk their prey, then the male kills it. Java.

Clark's Nutcracker (*Nucifraga columbiana*) lives in montane pine forests. It feeds on pine nuts, which it carries in a pouch under its tongue. California, U.S.

Many species have a fleshy eye ring. Some have bare eye or facial skin; this is most extensive on Rook and Gray Crow, while Stresemann's Bush Crow (*Zavattoriornis stresemanni*) has blue skin around the eye. In some species, soft part coloration is very different in juveniles.

The bill is black in most species, but in others it is ivory, gray, yellow, or red. The iris is typically dark brown, but in some corvids it is white, pale blue, red or yellow. The legs and feet are usually black or blackish-gray, but are red in green magpies (genus *Cissa*) and choughs.

Behavior Gregarious birds, corvids are typically found in pairs or family groups, and many occur in flocks. Some species defend temporary feeding territories, while others maintain territories year-round. Many roost communally, some with other corvids: for example, Rooks with Eurasian Jackdaws (*Corvus monedula*). Pairs or groups vigorously mob predators. Most social species communicate by means of complex displays or postures and vocalizations.

Corvids primarily forage in vegetation or on the ground; a few species perform aerial sallies, while others use their bill to probe and dig for concealed food. Several species use tools to get at hard food items such as shelled prey. Strategies include dropping the items so they break on impact with ground, using the bill to pound items against an 'anvil,' or fashioning tools from sticks, to probe into crevices. Food may also be dunked in water to soften it. All corvids typically carry food in their sublingual pouch, mouth, or throat. Food caching behavior is widespread in the family.

The birds maintain their feathers by bathing and dust-bathing, and by 'anting'—deliberately coming into contact with ants, possibly so that the acid secreted by the ants will kill parasites. Flight is typically strong, often interspersed with glides; many *Corvus* species soar. All corvids walk or hop, and some run very fast; most ground jays escape danger by running with a unique upright posture.

Voice Corvids are loudly vocal. They have a large variety of calls for different situations, such as alarm, location, assembly, but their songs are poorly known. The birds generally produce harsh or eerie sounds including grating, rattling, croaking, and caws, but also melodic whistles and metallic notes; some can mimic other birds or sounds.

A **Blue Jay** (*Cyanocitta cristata*) 'anting.' In this behavior, the bird lies on top of ants, or even rubs the insects over its feathers, as shown here.

Many corvids cache hundreds of nuts and seeds for later consumption, which makes them important in propagating trees. This **Eurasian Jay** is burying acorns. U.K.

Habitats Found in virtually all habitats, including desert, grasslands, woodland, forests, alpine zone, as well as urban areas, and from sea level to about 26,250 ft. (8,000 m.).
Movements Generally sedentary, but the northernmost populations of some northern-hemisphere species, such as Blue Jay (*Cyanocitta cristata*) and Eurasian Jay (*Garrulus glandarius*), undertake short- to fairly long-distance migrations. Some species, especially montane birds, make seasonal elevational movements or undergo periodic larger-scale irruptions to track food resources.
Diet Omnivorous. Most have a diverse diet: individuals take a wide assortment of plant material (such as nuts, seeds, fruit, tender leaves and shoots), invertebrates (especially insects and crustaceans, also snails and shellfish), and small vertebrates (fish, reptiles, amphibians, birds, and mammals) including eggs, young, or carrion. Nutcrackers and Pinyon Jay specialize more on conifer seeds. The larger crows and ravens are accomplished predators, able to take a variety of vertebrate prey; pairs often hunt as a team.
Breeding Most corvids are believed to be monogamous, and some pair for life. Pairs defend an exclusive territory, while social species defend a group territory. The birds may nest either singly, within the group territory, or colonially.

The nest, built by both of the pair, is typically a bulky, cuplike structure made of sticks and plant material, with a softer lining. It may be placed in a tree or shrub; on a rock fall, cliff ledge, or cave; or on a utility pole or other manmade structure. Exceptionally, Lidth's Jay (*Garrulus lidthi*) nests in cavities in trees; magpies (genus *Pica*) add a canopied top to their nest; and Stresemann's Bush Crow builds a globular nest.

Clutches comprise two to nine eggs, incubated by the female. Incubation takes 14–25 days, and in most species it begins after the first egg is laid. The male feeds the female during this time, and feeds the female and small young during the brooding period. Both parents feed larger young. In some social species, such as *Aphelocoma* jays, they may be assisted by nest-helpers. Most species raise one brood in a season, but many will nest again if the eggs or young are lost.

Nestlings hatch naked with eyes closed. Some acquire sparse down; nutcrackers have denser down. Fledging occurs in 17–42 days (the largest species taking longest). Depending on the species, post-fledging parental care can last as little as 1–2 weeks or as long as 6–9 months (in some ravens and nutcrackers). The young of many species remain in their parents' breeding territory until the next breeding season.
Taxonomy Although the corvids are well defined, some species are of uncertain position. Crested Jay (*Platylophus galericulatus*) may be more closely related to helmet-shrikes (family Malaconotidae) or shrikes (Laniidae). The affinities of Azure-winged Magpie (*Cyanopica cyanus*) and the ground jays (genus *Podoces*) also remain unresolved. Hume's Groundpecker (*Pseudopodoces humilis*), currently considered the smallest corvid, may be a tit (Paridae).

Rufous Treepie (*Dendrocitta vagabunda*) is mainly arboreal. It takes fruits, invertebrates, and other birds' eggs and young, and is highly agile at climbing through foliage. India.

AUSTRALIAN MUDNESTERS *Corcoracidae*

The Corcoracidae is a tiny family comprising just two genera, each with a single species. The species differ in size, shape and plumage, but are united mainly by their practice of building mud nests. Apostlebird (*Struthidea cinerea*) is named for its habit of living in groups, which were traditionally thought always to number 12 birds. White-winged Chough (*Corcorax melanorhamphos*) is named for its resemblance to the choughs in the northern-hemisphere family Corvidae, but it is not related to them.

Structure The birds are 12–18 in. (30–47 cm.) in length. Apostlebird is the smaller species, and has a short, robust, finchlike bill. White-winged Chough is larger, with a long, slender, arched, crowlike bill. The wings are fairly short and rounded, and the tail is long. Both species have long, strong legs and feet.

Plumage The plumage is soft and fluffy. The sexes are alike in color. Apostlebird is mostly gray, with paler streaks on the crown, back, and breast, brown wings, and a black tail. White-winged Chough is sooty-black with extensive white bases to the primaries and bright red eyes. Both species have a black bill and black legs.

White-winged Chough has pure white wing patches, which are usually seen while the bird is in flight or when the wings are spread. Queensland, Australia.

Groups of **Apostlebirds** are often seen playing together as well as foraging and socializing. Queensland, Australia.

Behavior Social and highly gregarious, the birds spend their lives in groups of 6 to 20. Where food is plentiful, hundreds may be found. Groups of White-winged Choughs (and possibly of Apostlebirds) appear to consist of a dominant male and one or more females, with the young from previous seasons (White-winged Choughs take up to four years to reach maturity). Both species forage on the ground, walking, hopping, or running. Apostlebirds have a habit of flicking their tail upward and then letting it subside. White-winged Choughs walk with a swaggering back-and-forth motion of the head, and carry their tail horizontally.

Voice Apostlebird has a harsh, scratchy call, which has been likened to the sound of tearing sandpaper, and a rough, nasal *git-out*. White-winged Chough has a more mellow, descending whistle.

Habitat Both species are endemic to eastern Australia, where they are fairly commonly seen in woodlands, dry, open forests, roadside trees, parks, and orchards. White-winged Chough was once thought to be threatened by the replacement of its natural habitat with new plantations of exotic pines, but, almost alone among Australian birds, it appears to have adapted well to an environment with non-native trees.

Movements Sedentary.

Diet Both species feed on seeds and insects. Apostlebirds mainly take seeds; White-winged Choughs dig into leaf-litter to search for beetles and worms.

Breeding All group members help build the nest and rear the young. The birds build large, bowl-shaped nests of grass, bark, mud, and manure, plastered onto a horizontal branch. They start with a flat platform and build up the structure in stages, each stage being troweled into place with the sides of the bill before being left to harden. Clutches range from two to nine eggs, often with more than one female laying in the same nest. Incubation takes 18–19 days, and nestlings take up to 29 days to leave the nest.

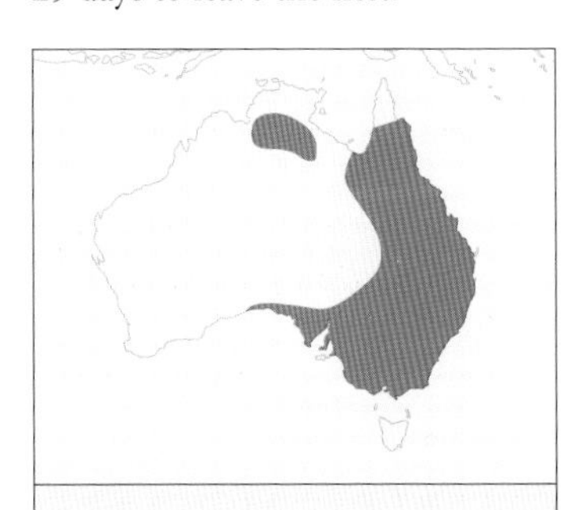

Number of genera 2
Number of species 2
Conservation Status Least Concern
Distribution Eastern Australia

BIRDS OF PARADISE *Paradisaeidae*

Among the most remarkable and enigmatic bird families in the world, the birds of paradise are famous for their spectacular plumage and bizarre courtship displays. Many inhabit isolated montane forests, where the absence of predators makes it safe for them to be highly flamboyant, and marked separation of species across a comparatively narrow geographic range has enabled the family to develop its astonishing diversity.

Polygamy and sexual dimorphism are common; in about two-thirds of all species, the males are larger and more colorful than the females. Exceptions include the monotypic Paradise-Crow (*Lycocorax pyrrhopterus*), both members of genus *Paradigalla*, and the manucodes, many of which are monogamous and exhibit far less marked sexual dimorphism.

Structure There is a wide variation in lengths across the family, from the tiny King Bird-of-Paradise (*Cicinnurus regius*), at 6 in. (15 cm.), to Black Sicklebill (*Epimachus fastuosus*), at 44 in. (110 cm.). However, most species are similar in shape to the related family Corvidae (crows and jays). General characteristics include strong feet adapted for perching, and a strong, stout, pointed bill.

A few genera, such as *Drepanornis*, *Epimachus*, *Ptiloris*, and *Seleucidis*, have longer, more specialized bill shapes; this adaptation may be linked to their more insectivorous feeding habits. Manucodes, uniquely among passerines, have a looped, elongated trachea, which is particularly well-developed in males (more than 20 times longer than in a bird of comparative size) and gives them a deep, resonant voice.

Plumage The plumage forms and colors that have earned birds of paradise their name are truly dazzling. Among the elaborate adornments sported by males are ornate head plumes, crests, beards, wattles, long tail streamers, and iridescent breast shields. Females, however, tend to be much more drab in color.

Wilson's Bird-of-Paradise (*Diphyllodes respublica*) has his whole body adorned, from his vivid blue crown to his curly tail, which shines silver in the light. Indonesia.

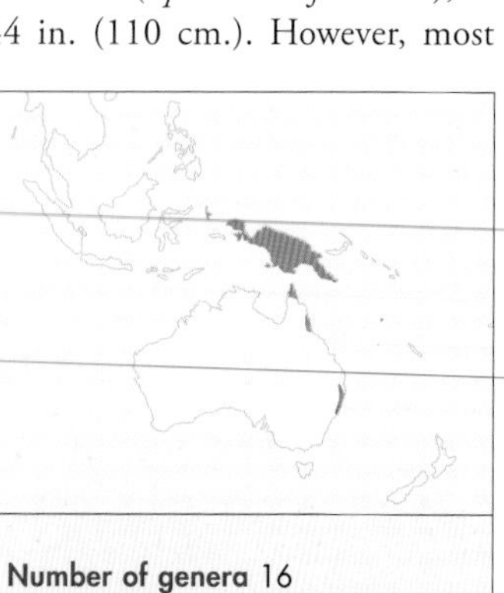

Number of genera 16
Number of species 40
Conservation Status 3 species Vulnerable, 7 species Near Threatened
Distribution New Guinea, eastern Indonesian islands, northeastern Australia

Lesser Bird-of-Paradise (*Paradisaea minor*) was traditionally hunted for its extravagant white flank plumes; however, it is not now considered to be threatened. New Guinea.

Twelve-wired Bird-of-Paradise (*Seleucidis melanoleucus*) uses his tail wires to brush the female during courtship. New Guinea.

Ribbon-tailed Astrapia (*Astrapia mayeri*) has proportionally the longest tail streamers of any bird species. The black and yellow King of Saxony Bird-of-Paradise (*Pteridophora alberti*) has two huge pale blue head plumes scalloped with black, which tremble when it calls.

The male Blue Bird-of-Paradise (*Paradisaea rudolphi*) has bright turquoise wings and striking cinnamon and violet-blue flank plumes, which it fans out as it hangs upside-down during its courtship displays.

The elusive Carola's Parotia (*Parotia carolae*) has six drooping, wirelike head plumes, bright yellow eyes, and an iridescent golden breast shield scalloped with black.

Emperor Bird-of-Paradise (*Paradisaea guilielmi*) is yellow and chestnut with a dark emerald throat, face and crown, white ornamental flank plumes, and two very long tail wires. It has been over-hunted for its beautiful plumage in some areas, and is now among the species considered Near Threatened.

The little-known Splendid Astrapia (*Astrapia splendidissima*) is predominantly black above and dark green below, with an iridescent yellow-green mantle, a bluish-green throat, and black-tipped white tail.

Perhaps most stunning of all is Wilson's Bird-of-Paradise (*Diphyllodes respublica*), with its crimson upperparts, lemon-yellow mantle, naked blue head, velvet green-black underparts, and two curlicued violet tail feathers. In the low light of dense forest, the bird appears to glow with unbelievably vivid color.

In contrast, the manucodes (genus *Manucodia*) are more soberly plumaged, primarily in black, although they have glossy breast and neck feathers; Glossy-mantled Manucode (*M. ater*) has purple, green, and blue plumage.

Behavior Elaborate courtship rituals, which vary between genera, are a key aspect of behavior in the Paradisaeidae. In these displays, the males perform a variety of amazing movements to show off their plumage to best advantage.

When not displaying, many male birds of paradise can be highly elusive, and the cryptically colored females and juveniles are even harder to observe. Exceptions include the more gregarious species of lowland rainforest, which often join mixed flocks with species such as drongos and babblers, foraging in the trees at heights of 65–100 ft. (20–30 m.).

Voice Birds of paradise have as wide a range of vocalizations as they do of plumage forms. Many have raucous crowlike calls. In addition, males make a range of bizarre sounds, particularly during courtship rituals. Brown Sicklebill (*Epimachus meyeri*) produces a deafening machine-gun-like rattle. King of Saxony Bird-of-Paradise makes a harsh crackling sound that is often likened to radio static. Superb Bird-of-Paradise (*Lophorina superba*) makes a loud, urgent snapping sound as it hops back and forth trying to impress a prospective mate.

Manucodes use their modified trachea to produce deep, loud, resonant calls that carry farther in the forest. Scientists think this adaptation may help members of roving flocks to stay in touch.

The male **Red Bird-of-Paradise** (*Paradisaea rubra*) takes at least six years to attain his full breeding plumage, including the red plumes and long tail wires. Indonesia.

Habitats Most species are endemic to New Guinea, but birds of paradise are also found on eastern Indonesian islands, including the Moluccan Islands and Aru Island, on Torres Strait Island, and in northeastern Australia.

The family is found in a range of forested habitats, including tropical and subtropical rain forest, lowland sago swamps, mangroves, and subalpine *Podocarpus* (coniferous) forest. The greatest number and diversity of species occur in the pristine forests of New Guinea; Berlepsch's Parotia (*Parotia carolae berlepschi*) was rediscovered there in 2005, more than 100 years since it was first described.

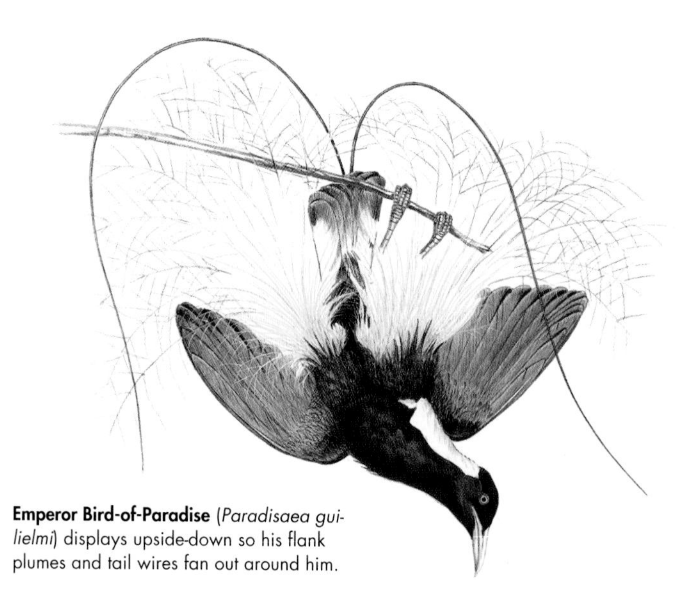

Emperor Bird-of-Paradise (*Paradisaea guilielmi*) displays upside-down so his flank plumes and tail wires fan out around him.

The different species occupy a wide range of elevational zones, from just under 10,000 ft. (about 3,000 m.) to near sea level. Many have altitude-restricted and physically isolated ranges: for example, the sicklebills and astrapias occur at 6,560–9,840 ft. (2,000–3,000 m.), while the parotias, Superb Bird-of-Paradise, and Black-billed Sicklebill (*Drepanornis albertisi*) are found at 3,280 to 6,560 ft. (1,000–2,000 m.). Other species, such as King Bird-of-Paradise, Pale-billed Sicklebill (*D. bruijnii*), Twelve-wired Bird-of-Paradise, and the manucodes, inhabit lowland forest below 1,640 ft. (500 m.). Several species are at risk, mainly due to loss of their restricted habitat.

Movements Many species are sedentary and tied to limited ranges; some spend their entire lives in small home territories. Manucodes are a notable exception: semi-nomadic and non-territorial, they roam the forest in response to the availability of their favorite foods, such as figs.

Diet Most species feed mainly on fruit, berries, and invertebrates (mainly insects, larvae, spiders and worms). Seeds, nectar, and small vertebrates such as lizards and frogs may also be taken. A degree of specialization occurs in a few species, notably the largely insectivorous riflebirds (genus *Ptiloris*) and sicklebills (*Epimachus*), which use their strong, decurved bills to probe for prey beneath bark and in decaying wood, in a similar manner to the northern-hemisphere treecreepers (family Certhiidae) and the woodpeckers (Picidae).

Breeding Although the males of polygynous species are famous for their displays, females are responsible for parental care. In monogamous species such as the manucodes, however, the pair share parental duties.

In most species, females build the nest and incubate the eggs alone. Nesting material may include twigs, stems, leaves, vine tendrils, ferns, and, occasionally, decorative items such as snakeskin. Species differ in preferred nest style and location: manucodes and parotias are among those that nest on branches, while Victoria's Riflebird (*Ptilotis victoriae*) and Ribbon-tailed Astrapia prefer dense vines or the tops of tree ferns. The nest shape is generally either a bulky, shallow cup or a suspended basket or cup. Unusually, King Bird-of-Paradise is a hole-nester.

Most species lay one or two eggs. Incubation takes 14–26 days, with the known average for most species being 18–19 days. The young are altricial and are fed by their mother (or both parents in monogamous species). They fledge in 16–30 days.

Females usually reach sexual maturity at two years, but the males of some more spectacular species may take up to five years to attain full breeding plumage (a characteristic known as sexual bimaturism).

Hybridization has been recorded often among the polygynous species, which suggests that they are all closely related despite their dramatically different plumage.

The female **Brown Sicklebill**, although much duller than the male, is distinctive for her red cap and blue eyes. New Guinea.

AUSTRALASIAN ROBINS *Petroicidae*

Australasian robins get their name from the reddish underparts of some species, such as this **Scarlet Robin** (*Petroica boodang*). Tasmania, Australia.

The Australasian robins are a family of flycatchers sometimes divided into three subfamilies: the pink-fronted robins (Petroicinae), thrushlike robins (Drymodinae), and yellow-fronted robins (Eopsaltriinae). Within each subfamily, most species have a similar structure, but there is a wide variation in plumage between species.

Although named 'robins' by early European colonists in Australia, because they resemble European Robin (*Erithacus rubecula*), the family has no close affinities to the 'true' robins in family Muscicapidae or the North American robins in family Turdidae.

Structure Lengths range from 4 to 9 in. (10–23 cm.). The birds have a large, rounded head with large or prominent eyes (pale or whitish-gray in one species). The bill is thin, straight, and short to medium-length, with a strongly hooked tip in some species. The body is round and stocky. The wings have 10 primaries and are mostly short and rounded, but slightly more pointed in aerial flycatching species. The tail varies from short to proportionately long (in the *Drymodes* scrub robins), with 12 feathers and usually a square tip (rounded in one species). The legs are moderately long, especially in the terrestrial scrub robins.

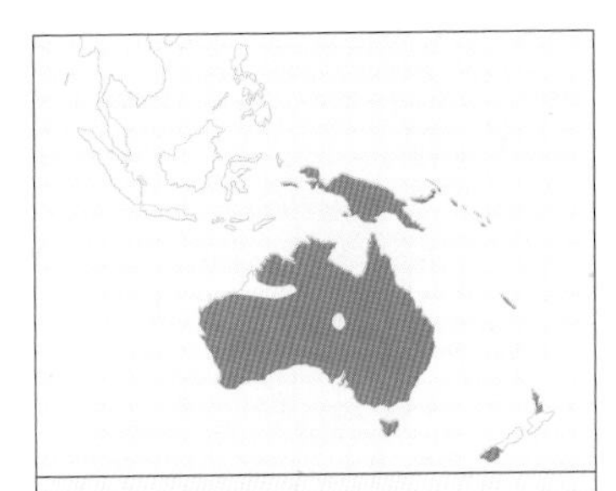

Number of genera 13
Number of species 45
Conservation Status 1 species Endangered
Distribution Australia, New Guinea, New Zealand, southwest Pacific

Plumage Sexual dichromatism is common in genera *Petroica* and *Eugerygone*; otherwise, the sexes are alike. The *Petroica* robins have a black head and face, strongly defined upperparts, a pink to rose-red breast, and underparts that are usually white or yellowish. Most members of this genus also have a variably sized white spot on the forehead, although one, Red-capped Robin (*P. goodenovii*), has a scarlet spot. Most also have a white patch on the wing coverts or the edges of the tertials. Black Robin (*P. traversi*), however, is entirely black.

The other Australasian robins are mainly deep olive-green above and yellow below, some with a white face and dusky breast patch. The *Microeca* species are mainly olive-green above and bright yellow below, although Jacky Winter (*M. fascinans*) is mostly gray-brown with a whitish eyebrow line, whitish underparts, and white outer tail feathers. Other species are gray or black and white. Torrent Flycatcher (*Monachella muelleriana*) has a black cap, black and gray upperparts, and white face and underparts. The ground-robins

A **Jacky Winter** (*Microeca fascinans*) feeding its offspring. Juveniles are paler than adults, with grayish-white streaking and mottling. Australia.

have a brown or black head and upperparts, a broad white eyebrow line, white wing patches, and whitish underparts. The scrub robins recall their similarly named but unrelated African counterparts, and are brown or dusky-brown with black stripes through the eyes.

Behavior The birds are usually solitary or found in pairs, but some species form small family groups, and several occur in larger flocks outside the breeding season.

Most robins have an upright stance and repeatedly flick their wings and tail. Several species, including Dusky Robin (*Melanodryas vittata*), Eastern Yellow Robin (*Eopsaltria australis*), and New Zealand Robin (*Petroica australis*), are known for their tameness, often approaching people and remaining nearby for some time. Others, mostly the New Guinea robins, are shy and secretive birds living in dense undergrowth. Most species forage in a similar way to a chat or a flycatcher, by watching and waiting, often apparently motionless while they survey the ground around them, and then pouncing. Several species also catch insects in flight. The ground robins and scrub robins forage at ground level, in the leaf-litter. New Zealand Robins, and possibly several other species, store food in tree-stumps, holes and crevices, or mossy logs.

A **Hooded Robin** (*Melanodryas cucullata*) sits quietly on a wire fence as it watches for prey. Australia.

Voice Mostly a series of short, high-pitched, staccato whistles, trills, or rising and falling, piping notes. Contact notes include a short, abrupt *chuck*, and harsh, aggressive scolding is given as an alarm call. Eastern Yellow Robin is noted for its clear, bell-like song. Australasian robins are some of the first birds to sing before dawn and some of the last to cease at dusk.

Habitat Australasian robins occupy a wide range of habitats, from dry, open savannas to rocky slopes, cliffs, high-altitude forests above 8,250 ft. (2,500 m.), and alpine grassland and shrubberies. Torrent Flycatcher frequents rocks and boulders by fast-moving hill and mountain streams.

One species, Black Robin, is currently considered endangered. The population in its native range of the Chatham Islands suffered drastic decline due to the introduction of rats and cats, and by the early 1980s only five individuals remained. All were relocated to a rat-free island and intensively managed and protected. The population increased, and in early 2000 was slightly in excess of 250 birds. Several species in New Guinea have extremely restricted ranges for which there is no information on population.

Movements Most are largely sedentary, or make short dispersal movements, usually of less than 6 miles (10 km.). In Australia, part of the population of Flame Robin (*Petroica phoenicea*) is resident in the same areas of the southeast year-round, while others move farther north to open-country areas in southeastern Queensland and west to Victoria; these often occur in flocks of up to 15 when on passage.

Diet Mostly small insects, including beetles, weevils, ants, centipedes, and small wasps; also earthworms, spiders, mollusks, grass seeds, and berries. Mangrove Robin (*Peneoenanthe pulverulenta*) takes small crabs.

Breeding Most species are monogamous, and the pair bond may last for more than one breeding season. Cooperative breeding or nest-helpers are known in several of the robins. Most species build a small, saucer-shaped nest of grass, plant fibers, lichens, bark strips, and spiders' webs, placed in the fork of a branch; ground-robins and scrub robins build a larger, cup-shape nest on the ground. The clutch of two or three eggs is incubated for 14–20 days, by the female in most species or both partners in the scrub robins. Both parents feed the young, which leave the nest after 12–17 days.

Pale-yellow Robin (*Tregellasia capito*) exists in two separate populations, in New South Wales and Queensland. Australia.

BALD CROWS *Picathartidae*

The family of bald crows contains just the single genus *Picathartes*, and two species: Gray-necked Picathartes (*Picathartes oreas*) and White-necked Picathartes (*P. gymnocephalus*). Their alternative name is rockfowl.

Structure The birds are 13–15 in. (33–38 cm.) in length. The sexes are alike, although male Gray-necked birds are larger than females. Bald crows have a fairly small head and a large, crowlike bill, which is slightly decurved in Gray-necked Picathartes. Both species have rounded wings, a long, broad tail, and long legs.

Plumage The most obvious feature, which gives these birds their name, is that the head and nape are bare, apart from a few short crest feathers. The skin is highly colored. In White-necked Picathartes, it is bright yellow with a black hindcrown and nape. In Gray-necked Picathartes, it is powder-blue on the forecrown and carmine red on the hindcrown and nape, with blackish, triangular patches on the sides of the face. White-necked Picathartes has dark gray upperparts and tail, while in the Gray-necked species, the neck and upperparts are pale bluish-gray. The neck and belly are white, sometimes with a yellow tinge.

Behavior The birds are usually seen in pairs or small groups, foraging on the forest floor or in low undergrowth. They often follow army ant columns to take insects flushed by the ants. When disturbed, they quickly vanish into the forest. Although generally shy and elusive, bald crows can become surprisingly tame and confiding at breeding sites.

Voice Usually silent. However, when alarmed or disturbed, White-necked birds make a loud, drawn-out *kaa* sound; Gray-necked birds make a slightly softer or more sibilant alarm call. At pre-roost or breeding gatherings, flocks make a churring noise, together with hisses or growls.

A **White-necked Picathartes**. Both species of bald crow are instantly recognizable due to their bare, brightly colored head and striking plumage.

Habitat Lowland rain forest and secondary forest on hills or mountains; also scrubby edges of farmland. Bald crows need cliffs, caves, or boulders and streams for nesting. Both species are at risk from habitat destruction and fragmentation.

Movements Largely sedentary; makes local movements when not breeding.

Diet Bald crows mostly eat beetles, termites, and ants; they may also take millipedes, earthworms, and occasionally small frogs and lizards.

Breeding Both species breed twice yearly, in small colonies of up to 40 pairs. The nest is a large cup of dried mud mixed with leaves and twigs and is placed on the wall or roof of a cave, under overhanging rocks, or on a cliff ledge. Clutches comprise one to three eggs, which are incubated by both parents for 17–28 days. The young remain in the nest for 24–25 days. Breeding success rates for White-necked Picathartes have declined in recent years, possibly due to natural causes.

Taxonomy Bald crows were formerly considered to have affinities with crows—hence the English name—but they may be more closely related to babblers or rockjumpers.

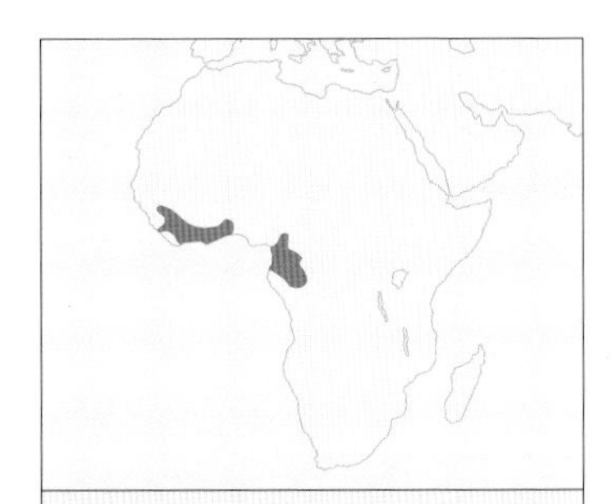

Number of genera 1
Number of species 2
Conservation Status Both species Vulnerable
Distribution West and Central Africa

WAXWINGS *Bombycillidae*

A pair of **Cedar Waxwings** feeds on a china berry tree. Waxwings are sometimes seen passing food from one bird to another, or down a line of birds. Texas, U.S

The waxwings comprise three similar species: Cedar Waxwing (*Bombycilla cedrorum*), Bohemian Waxwing (*B. garrulus*), and Japanese Waxwing (*B. japonica*). Their name comes from the bright tips to the secondaries, which look like drops of wax.

Structure Waxwings are stocky birds 6–9 in. (15–23 cm.) in length. The head is small, and the bill is short but broad at the base. The birds have a short neck, plump body, and short legs. The wings are long and pointed, often held down by the tail.

Plumage All species have soft, thick plumage of milk-chocolate brown upperparts, paler underparts, and gray on the rump and base of the tail. The head has a prominent, backward-pointing crest. A black band runs from the lower forehead over the eyes and beneath the crest; this continues as a large patch on the chin and throat. Japanese and Bohemian Waxwings have a chestnut tinge around the face. Two species have a yellowish belly; in addition, Japanese Waxwing has reddish undertail-coverts, while those of Bohemian Waxwing are deep chestnut. The wings are black (with white patches in Bohemian Waxwing), with fine red tips to the secondaries. The tail is short and square; the tip is bright red in Japanese Waxwing, and yellow in the other two species.

Behavior Waxwings gather and forage in large flocks. They appear suddenly in areas where there are plenty of fruiting trees; they can strip a tree of fruit or berries in just a few days.

Voice The call is a high-pitched, ringing trill. The song is a warbled series of wheezing notes, twitters, and coarser notes, often with a prolonged version of the call included.

Habitat Boreal and temperate forests: mainly coniferous, but also birch and mixed. In winter, waxwings occupy a wider range of habitats, including parks, farms, roadsides, and gardens.

Movements Waxwings are migratory, or are partial migrants, traveling within or south of their breeding range. In years when the food supply in their native range is poor, waxwings are highly irruptive.

Diet Mainly high-sugar fruits and berries, such as bilberry, rowan, bramble, juniper, cotoneaster, rosehips, and hawthorn. When fruit is less easy to find, they may take blossoms from trees including pears and apples, as well as insects and larvae.

Breeding Waxwings breed in small or loose colonies. The timing is largely determined by the availability of fruit for feeding young. They build a nest of twigs, plant fibers, and grasses, usually on a horizontal tree branch. The clutch of three to seven eggs is laid from late June to early August, and is incubated by the female for 11–13 days. The young leave the nest after 14–18 days.

Taxonomy The waxwings' nearest relatives may be the silky-flycatchers and Phainopepla (*Phainopepla nitens*) of the southwestern U.S. and Central America. Some authorities include these birds with the waxwings, as a subfamily (Ptilogonatinae). Gray Hypocolius (*Hypocolius ampelinus*) of southwestern Asia is also considered to be closely related to the waxwings and is also sometimes included as a subfamily (Hypocoliinae).

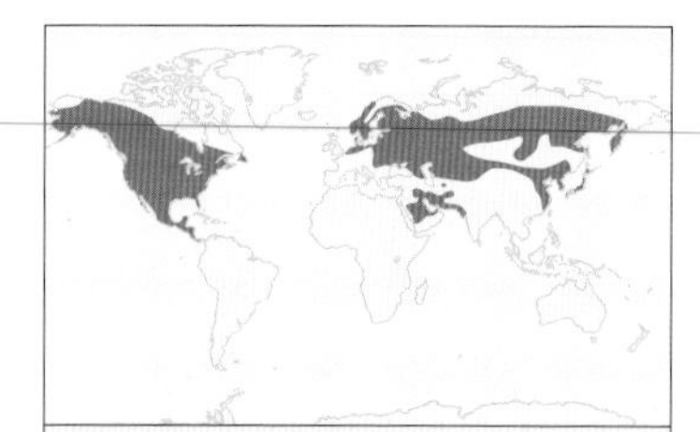

Number of genera 1
Number of species 3
Conservation Status 1 species Near Threatened
Distribution Holarctic to northern Neotropical areas of North and Central America, Europe, and Asia

PALMCHAT *Dulidae*

Previously grouped with the waxwings (family Bombycillidae), Palmchat (*Dulus dominicus*) is now classed in its own distinct family. A noisy, gregarious bird, it is found on the Caribbean island of Hispaniola and the offshore island of Gonâve. It is the national bird of the Dominican Republic.

Structure The bird is 8 in. (20 cm.) in length. It has a broad head with a large, stubby, laterally compressed bill; rounded wings; a fairly long tail with a notched tip; and strong legs and feet, with long toes and claws.

Plumage The sexes are alike: both have dark brown on the head and nape, slightly brighter brown on the mantle, back, and wings, and olive-green on the rump and tail, and broadly on the edges of the flight feathers. The underparts are buffish-white, boldly streaked with brown. Palmchats also have bright red eyes and a pale cream bill. Immatures are similar to adults but have a dark brown throat and neck and a buffish-brown rump.

Behavior This highly social bird lives in small groups of 4–10 pairs, each pair apparently living independently of the others. Palmchat actively forages with quick, agile movements in trees, shrubs, and bushes. It flies rather like a European Starling (*Sturnus vulgaris*), in a direct or slightly undulating manner, but rarely covers long distances.

Voice Fairly noisy and vocal, with a variety of short, harsh or melodious whistles, twitters, scraping or gargling notes, and loose, warbled phrases; seems to have no true song.

Habitat Palmchat is endemic to Hispaniola and the adjacent islands of Gonâve and Saona, where it is common in a range of open habitats, from gardens and scrub to woodland edges. It is most numerous in areas of lowland palm trees, but may occur at up to 6,500 ft (2,000 m.).

Two **Palmchats** perch on a palm frond. The birds usually nest and roost in palm trees, but at higher altitudes they may occupy conifers. Dominican Republic.

Movements Sedentary.

Diet The bird prefers the small, berrylike fruits of palms, but may take buds and flowers, and possibly pollen and nectar from other trees; it may also take small insects.

Breeding Palmchats build large, bulky communal nests, which are occupied by up to 50 pairs. The nests are made mostly of sticks, and are often more than 3 ft. (1 m.) across. They are placed high up, usually in the crown of a palm. Within the nest, each pair has its own entrance and nest chamber, although occasionally more than one female lays in the same chamber. The clutch of two to four eggs is incubated by both parents for about 15 days. The nestlings fledge after 32 days, but remain with their parents for some time after leaving the nest.

Number of genera 1
Number of species 1
Conservation Status Least Concern
Distribution Caribbean

TITS AND CHICKADEES *Paridae*

A **Blue Tit** (*Parus caeruleus*) in flight. With its striking blue, green, yellow, and white plumage, Blue Tit is unusually colorful among tit species. U.K.

A characterful family of small, acrobatic, arboreal birds, the 'true' tits are found across most of Eurasia, Africa, and North America through Mexico, with the highest diversity of species in the Old World. In English, the tit was first named the 'titmouse,' a derivation of Old English or Icelandic words meaning 'very small bird.' The name was later shortened to 'tit.' In the New World, similar-looking species were named 'chickadees,' for their call, and other, more 'mouselike' species were given the name 'titmouse.'

Structure Tits range from 4 to 8.5 in. (10–22 cm.) in length. The sexes are generally alike, and most species are relatively similar in size. At the extreme ends of the range are two atypical monotypic genera inhabiting eastern Asia: the smallest is the kinglet-like Yellow-browed Tit (*Sylviparus modestus*) at 4 in. (10 cm.), while the largest is Sultan Tit (*Melanochlora sultanea*), at 8.5 in. (22 cm.). Most members of the family are also similar in their overall proportions. Tits are generally small and compact. The bill is short and stubby, although rather more delicate in some species; in certain species or populations, males and females differ in bill size and shape. The wings are short to medium-length and rounded, with 10 primaries and nine secondaries (including three tertials). Most species have a square or notched tail of medium length; White-browed Tit (*Parus superciliosus*) has the longest tail, at nearly the same length as the body. The feet and legs are strong and of medium length, enabling the birds to hang upside-down from branches or hold food against a perch.

Plumage The feather structure is soft and dense. Siberian Tit (*P. cinctus*) has an increased density of barbs and barbules on its feathers, which provides greater than normal insulation against the cold: night temperatures in its taiga habitat may drop to -76°F (-60°C). The nostrils are partially covered by bristles.

The sexes are generally alike in color, or the females may sometimes be duller; however, a few species, such as Yellow-bellied Tit (*P. venustulus*), show more pronounced sexual dichromatism.

Most are somber shades of black, brown, gray, or white. Many, however, have bold patterning, which may include a darker, often glossy, crown or cap, a stripe down the throat or breast, or a streaked crest or face. The patterning is often further highlighted by a pale to white cheek or eyebrow. Pale patches on the nape and bars on the wing coverts are characteristic of many species, and some also have white- or paler-edged feathers or patterns on the back, wings, or tail. A few species are more colorful, with rufous, yellow, green, or blue in their plumage. Several have a crest; this feature is longest and most prominent in Sultan Tit, which is also strikingly black and yellow.

The iris is typically dark brown to reddish-brown. In Gray-crested Tit (*P. dichrous*) and Dusky Tit (*P. funereus*) it is crimson. A few African species—Rufous-bellied Tit (*P. rufiventris*), Southern Black Tit (*P. niger*), and Northern Black Tit (*P. leucomelas*)—have a pale yellow or

Number of genera 3
Number of species 54
Conservation Status 1 species Vulnerable
Distribution North and Central America, Europe, Asia, Africa

whitish iris. The bill is usually black or with a paler base to the mandible. The feet and legs are blue-gray.

Most tits have a single, full post-breeding molt; a few are thought to have an additional partial molt prior to the breeding season.

Behavior Tits are active, highly social, and generally tame around humans. Their social behavior is complex and varies according to species, geography, or resources. The birds roost in cavities or crevices. Pairs are typically territorial during the breeding season; afterward, they form the nucleus of a family flock, expanding their range to overlap with other family groups. Alternatively, they form mixed-species flocks with other tit species, nuthatches, finches, kinglets, and woodpeckers.

Food is obtained by gleaning or probing beneath the substrate. Smaller species may perform acrobatic maneuvers, especially hanging upside-down, dangling by one leg, hovering, and flycatching. Tits also 'hammer' prey items to kill or open them, while grasping the prey against a perch. Certain species are known to use tools, held in the bill: for example, Great Tit (*P. major*) uses pine needles to extract insects from crevices. Simple problem-solving has also been documented. Highly resourceful, many species quickly adapt to use birdfeeders. Food-caching, both short-term and long-term, is characteristic behavior in several groups of species; however, success in remembering locations of stored resources and retrieving food varies greatly by species. The survival of many species through the winter is dependent on long-term storage of seeds or invertebrates: Siberian Tit is estimated to store 500,000 items during a season. Great Tit is known to monitor the activities of other caching species to find and steal food from their caches.

Black-capped Chickadee (*P. atricapillus*) is a widely known and well-loved species in North America, and has successfully adapted to living around humans. U.S.

Voice Tits are highly vocal. The calls are usually series of single or multiple notes, which may be whistled or burry; they are given during flight and courtship feeding. Softer notes are used for staying in touch with other flock members. Warning calls include the *chick a dee dee dee* notes characteristic of many species. Males have a loud, repetitive, multi-noted

Several tit species have a distinctive, upright crest; one example is **Gray-crested Tit**, found in the Himalayas and eastern Asia. China.

A **Siberian Tit** forages in winter. The birds survive severe cold by having thick plumage, and by becoming torpid at night to conserve energy. Finland.

song, typically a composite of other vocalizations given by both sexes. Females disturbed at the nest may hiss. Nestlings are silent, or give a multi-note *tsee-see-see* begging call.

Habitat Preferred habitats include a wide range of humid or dry, coniferous or deciduous forests and woodlands. Tits also occur in arid scrub, savannah, marsh edges, bamboo groves with trees or cavities for nesting, and rural or suburban yards. They are found at elevations from sea level to 15,000 ft. (4,575 m.). Habitat loss and degradation are the greatest threats to most species.

Movements Generally sedentary. Any movements are generally associated with tracking food. Other movements may include dispersal of young, elevational movements to avoid severe weather, and irruptions outside the birds' normal range.

Diet Insects and other invertebrates; also seeds and fruit. Females take food rich in calcium (such as snails) during egg production. The young are fed a diet of insects, predominately small caterpillars.

Breeding The breeding season is timed to coincide with maximum food resources. The birds are monogamous, and many species pair for life. Courtship behavior varies, and may include stereotyped postures; all tits engage in courtship feeding.

All species are hole-nesters: they may use cavities excavated in soft, rotting wood, those excavated by other species (such as woodpeckers), natural crevices, or manmade structures. White-browed Tit will nest in rodent burrows. In most species, the female builds the nest by piling a pad of soft mosses or fine grasses in the cavity; the nest may also incorporate feathers, snakeskin, or bark strips, and be lined with hair.

Most species raise a single brood, but second and third clutches are common for some, such as Coal Tit (*P. ater*). Clutches range from two to 14 eggs; birds breeding at higher latitudes produce the largest clutches. Incubation (where known) is performed by the female, and takes 12–18 days (longest in species that inhabit the coldest climates). It generally begins once the clutch is complete, although some species begin a few days beforehand, which results in asynchronous hatching. The young are altricial, hatching naked and with closed eyes. Fledging occurs in 16–25 days depending on species. Fledglings may be cared for by the male, both parents, or helpers. The time taken to reach independence varies greatly between species, from as little as seven days for Blue Tit to more than 80 days in Varied Tit (*P. varius*).

Taxonomy Due to their similarities of plumage and morphology, nearly all members of the family are placed in a single genus (*Parus*). However, recent genetic research suggests partitioning *Parus* into multiple genera or subgenera: *Poecile* (chickadees), *Periparus* (coal tits), *Lophophanes* (Old World crested tits), *Baeolophus* (New World titmice), *Cyanistes* (blue tits), and perhaps also *Pardaliparus* (sexually dichromatic Indonesian taxa), *Melaniparus* (African 'black tits'), *Macholophus* (Yellow Tit of Taiwan), and *Sittiparus* (Varied Tit and Asian tits with white foreheads). The name *Parus* would be restricted to the 'great tits.' These divisions correlate with plumage appearance and some behavioral characters (notably food caching).

Bone and genetic analyses also support inclusion of the unusual Hume's Groundpecker or Ground-Tit (*Pseudopodoces humilis*), a terrestrial species with a long, curved bill. It is currently classed as a 'ground-jay' (family Corvidae). More study is needed to resolve the phylogeny and delineate species and generic limits.

A brood of young **Great Tits**. Juvenal plumage is typically duller and less elaborate than that of adults. Location unknown.

PENDULINE TITS *Remizidae*

The penduline tits are very small birds with delicate bills. They are named for the distinctive hanging nests that most species construct.

Structure The birds are 3–4 in. (8–11 cm.) in length, and weigh just under to just over 0.3 oz. (7–9 g.). They have a short, fine, conical or needle-like bill. The wings are short, and the tail is short to medium in length, with a square tip. The legs and feet are slender but moderately strong.

Plumage The sexes are dichromatic, males being brighter than females. Species in genus *Anthoscopus* have pale to deep green or gray upperparts and whitish to buff or yellow underparts. Penduline Tit (*Remiz pendulinus*) has a white, brown, or black head, and two subspecies have a broad black mask. Tit-Hylia (*Pholidornis rushiae*) has a streaked grayish head and breast and bright yellow rump and underparts. Fire-capped Tit (*Cephalopyrus flammiceps*) is small and greenish; males in breeding plumage are reddish-orange on the forehead and throat. Verdin (*Auriparus flaviceps*) is mostly pale gray with a yellowish head and face.

Verdin has red shoulder patches as well as a yellow face. California, U.S.

Behavior Most Old World species are gregarious: they usually occur in foraging groups of up to 20, and, rarely, in mixed-species flocks. Verdin and Tit-Hylia are less social and usually occur in pairs. All species actively forage for insects in undergrowth and low scrub. Penduline Tit of Europe and Asia also hangs upside down while taking seeds from reeds and bullrushes.

Voice The birds are often fairly vocal in flocks. Contact calls are a variety of thin or high-pitched wheezes or whistled notes. Songs are a short, simple repetition of the call notes.

Habitat Verdin, the sole New World species, occurs in the southwest US and northern Mexico. The Old World species are spread across Europe, Asia, and Africa. Most species occupy a range of woodland habitats, from open woods to montane forest and rain forest. Verdin occurs in dry, bushy areas and Penduline Tit is also found in reeds and the margins of marshy areas. No species is currently considered to be at risk.

Movements Most species are sedentary. Penduline Tit is partly migratory, northern-breeding populations moving south to winter in warmer temperate areas. Fire-capped Tit, which breeds in the Himalayas, is an altitudinal migrant, wintering at lower levels south to central India.

Diet Mainly insectivorous; Penduline Tit also takes seeds.

Breeding Several Old World species are cooperative breeders, with nest-helpers, probably young from previous broods, assisting the parents. In most species, the nests are purse- or flask-shaped structures with a short, tunnel-like side entrance, woven by the male. They are made from plant fibers and grasses and covered with plant down or animal fur. Verdins make a nest of thorn twigs woven into a large, strong ball and often reused for several years. Fire-capped Tits nest in holes in trees.

A male **Penduline Tit** at the nest. The nests are often coated with down from willows, kapoks, poplars, or reed-mace. Poland.

Taxonomy The family is closely related to the true tits in family Paridae. The two main genera, *Remiz* and *Anthoscopus*, are very closely related, but the affinities of the other genera are less clear. The warbler- or finch-like Fire-capped Tit is often considered to belong to the Paridae.

Number of genera 5
Number of species 10
Conservation Status Least Concern
Distribution Parts of North and Central America, sub-Saharan Africa, southern Europe, and parts of Asia

SWALLOWS AND MARTINS *Hirundinidae*

An easily identifiable group of accomplished aerial-foraging songbirds. The greatest diversity of species is in Africa (29 breeding species) and Central and South America (19 breeding species). The names 'swallow' and 'martin' are used fairly interchangeably, with the square-tailed species generally being referred to as martins and the fork-tailed species as swallows.

Structure The members of this family range from 4 to 9.5 in. (10–24 cm.) in length. They all have similar body structures and generally unlike any other passerines. Sexual dimorphism is primarily limited to martins in the genus *Progne*, with moderate to minimal sexual dimorphism in the other genera.

The bill is small and compressed, with a wide gape and sometimes with rictal bristles. The feathers of the lores are directed forward; this adaptation gives shade to the eyes and helps protect them from flying insects. The structure of the syrinx (vocal apparatus) is very different from that of other passerine families. Swallows have long, pointed wings with 10 primaries (the tenth being extremely reduced). The birds also have a short neck, short legs, and a square to deeply forked tail. The toes are fused at the base in burrowing species of swallows. Two species of river-martins have unusually large heads and more colorful eyes, legs, and bills than other swallows.

Plumage Variable, but with much consistency within individual genera. Often with iridescent blue or green above, and dark, white, or rufous below. Many species show a pale rump or a dark breast band, and a few Old World species are streaked below. Species occurring in rocky or dry habitats are often brownish or grayish.

Behavior Swallows forage aerially for food, and all species are very skilled fliers. Many species will even drink on the wing, swooping down to water at a shallow angle.

Some populations of **Red-rumped Swallow** (*Cecropis daurica*) spend the summer months in southern Europe but winter in Africa. Other populations are sedentary or make only localized movements within Africa. Spain.

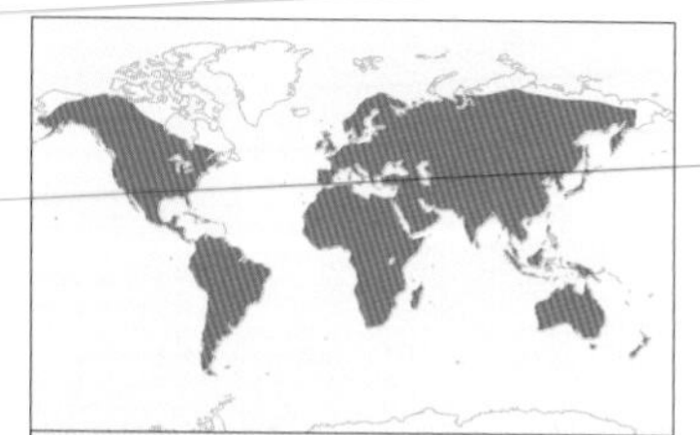

Number of genera 20
Number of species 84
Conservation Status 1 species Critically Endangered, 8 species Vulnerable
Distribution Virtually worldwide, except for Antarctic, some oceanic islands, and very high northern latitudes

Shortages of insects cause **Lesser Striped Swallow** (*Cecropis abyssinica*) to vacate parts of its range in the dry season. Breeding is timed to coincide with plentiful food. South Africa.

Voice The songs of most swallows consist of series of simple twitters or buzzes. Calls are not well documented in most species. Martins in genus *Progne* are known to sing dawn songs, consisting of varied churring and chortling notes.

Habitat Generally found in open areas, sometimes near water and sometimes along forest margins or in clearings. Many species, including all the North American-breeding species, have adapted well to the influence of humans on their environment. In fact, some species no longer nest in natural situations, preferring to nest on man-made structures or in nest boxes.

Movements Many temperate species undergo long-distance migrations, some among the longest of any passerines. Barn Swallows (*Hirundo rustica*), for example, that breed in northern Europe may spend the northern hemisphere winter in South Africa, and those that breed in northern Canada may winter as far south as central Chile and northern Argentina. Some southern hemisphere swallows are short-distance austral migrants. Some species in Africa make irregular, or even nomadic, movements.

Diet All species are insectivorous year-round, with prey being caught on the wing. Tree Swallow (*Tachycineta bicolor*) is unique as it is able to feed on wax myrtle berries in winter, allowing it to winter farther north than most species. Greater Striped Swallow (*C. cucullata*) and Lesser Striped Swallow also occasionally feed on fruit.

Breeding Breeding is seasonal, typically occurring during peak insect activity. Displays are not well documented. Males often sing at the nest site to attract females. Aerial fights sometimes occur, which can result in serious injury. The birds generally do not defend breeding or feeding territories, particularly the colonially nesting species. Male martins in genus *Progne* circle over their nest sites at dawn while vocalizing to defend the site and attract mates.

The nests are built of mud, sometimes with vegetation mixed in, and attached to vertical faces of natural and man-made structures. In some cases, nests may be built within cavities excavated by other species. Some species dig their own nest burrows. The rare Montane Blue Swallow (*H. atrocaerulea*) from southern Africa is the only species that nests below ground, using potholes or Aardvark burrows in open grassland. Pairs are generally monogamous, with both sexes participating in nest-building, incubation, and feeding of young. Many species nest colonially. They lay four to five eggs, rarely up to eight per clutch in temperate species, and two to three in tropical species. Many species have two or even three broods. The incubation period is 11–20 days, and the altricial young fledge in 17–30 days.

An adult **Barn Swallow** feeds its chicks. Typically, four or five eggs are laid in a cup-shaped nest. After hatching, the chicks remain in the nest for 18–23 days before they are able to fly. They are fed by both parents. U.K.

A **Northern House Martin** (*Delichon urbicum*) collects damp mud for nest construction. Hundreds of mud pellets are attached to the wall of a house and crafted into a cup shape. When dry, the cup is lined before the eggs are laid. U.K.

LONG-TAILED TITS *Aegithalidae*

Long-tailed Tit (*Aegithalos caudatus*) is easily recognized by its very long, slender tail, which contrasts with its round body. Scotland, U.K.

A widely distributed group of small birds with long, often graduated tails. Traditionally, the long-tailed tits were classed in the family of 'true' tits (Paridae), but differences in structure, molt, and nest-building have led to their recognition as a distinct family.

Structure Average length is 3–5 in. (9–14 cm.), about half of which is the tail. The birds have a stout, conical bill; short, rounded wings; and small legs and feet.

Plumage Long-tailed Tit shows wide variation between populations, being divided into 18 races including white-headed and dark-crowned birds. The head and face are often distinctly marked, with pale to warm brown coloring from the center to the crown, and black stripes on the sides of the crown, or a broad mask through the eyes. Sooty Tit (*Aegithalos fuliginosus*) has an entirely dark crown and a pale face. Several species have brightly colored eye rings. The upperparts are gray to brown. The underparts are white tinged with gray or pink, or with a warmer or brownish tinge; two species have a large black throat patch, and a third has a broad brown breastband. The tail is usually dark, but in some species the outer feathers are white.

Behavior The birds are highly social and usually occur in groups of up to 30. They actively forage together in fairly fast-moving flocks; outside the breeding season, they also join mixed-species flocks.

Voice Contact calls between flock members are usually a single note or a short series of sharp or sibilant *psip* or *tsee* notes. The birds may also make a more drawn out and lisping *si-si-si-sississip*; other notes include a dry *chup* or *tup*, soft clicking or twittering notes, and a dry rattle. The song comprises a longer series of twitters, trills, or chatters and short warbling phrases.

Habitat Coniferous, deciduous, mixed, and evergreen woodlands with shrubby edges, tangles, and undergrowth; open woodlands and bushes; heaths with scattered trees; hedges; and edges of cultivated land.

Movements Largely sedentary. However, some species, such as Bushtit (*Psaltriparus minimus*), disperse away from breeding areas in winter. Northern-breeding species move southward in winter, mostly within the breeding range.

Diet Mostly small insects, seeds, and fruit, and occasionally flower buds. The birds pick their prey from foliage, often while hanging from the ends of twigs, or may sometimes hover briefly to take insects in flight. Long-tailed tits have also been recorded drinking tree sap.

Breeding The ball-shaped nest is made of lichens, moss, feathers, and cobwebs, with a side entrance hole, and placed low in a bush or tree. Clutches comprise eight to 12 eggs, incubated by the female for 12–18 days. Nestlings are fed by their parents and by helpers from the group, and continue to be fed for about two weeks after leaving the nest.

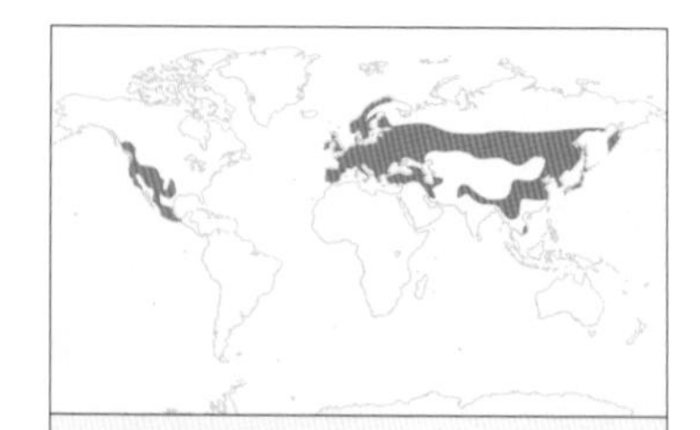

Number of genera 4
Number of species 11
Conservation Status Least Concern
Distribution North and Central America, Europe, mainland Asia, Indonesia

LARKS *Alaudidae*

Larks are renowned for singing while in flight, as this **Bimaculated Lark** (*Melanocorypha bimaculata*) is doing. Armenia.

Smallish, cryptically colored, mainly terrestrial birds, larks are widely known for the sweet, beautiful songs typical of many species. Most species are similar in structure and there is perhaps less plumage variation than in many other passerine families. Sexual dimorphism is limited primarily to size, with males of many species larger than females, and some subtle differences in bill size and shape.

Larks occur on all continents except Antarctica, although only two species breed in the New World, and one of these (Eurasian Skylark, *Alauda arvensis*), was introduced. The widespread Horned Lark (*Eremophila alpestris*) may be one of the most abundant passerines in the world, with 42 subspecies occurring across North and South America, Eurasia, and North Africa. In contrast, many species have highly restricted ranges and habitats.

Horned Lark is distinctive for its black facial mask and the 'horns' of feathers sticking up from the head. U.S.

Structure Lengths range from 5 to 9 in. (12–23 cm.). Bill shapes vary: in many species the bill is thin and fairly straight, but in others it is long and decurved or thick and conical. The wings have nine visible primaries, the tenth being vestigial or reduced; they also have unusually long tertials, often completely covering the primaries on the folded wing. The legs are fairly long, and the hind toe usually has a long, straight claw.

Plumage Most species are cryptically colored in dull browns, rufous, buff, black, and white, matching the soils where they occur. The plumage may be nondescript and uniform, but many species have streaked patterns. A few larks are more boldly marked, especially the sparrow-larks (genus *Eremopteryx*). Black Lark (*Melanocorypha yeltoniensis*) and Black-eared Sparrow-Lark (*E. australis*) are mostly black. Many species have bold patterns of black and white in the wings and tail, which are visible only in flight and thought to function as a distraction to predators.

Behavior Larks occur singly or in pairs in the breeding season, with many species becoming more social outside the breeding season, sometimes forming large flocks that often include other species. The birds forage on the ground by picking at the soil or at low vegetation, running in pursuit of prey, or digging in the earth. Larks often perform display flights, which often involve singing together with exaggerated wing-flapping or extensive climbing or diving. Some species, especially those in genus *Mirafra*, are best known for these wing-flapping displays, in which the wings actually strike each other, creating distinctive sounds.

Voice Larks typically have high-pitched songs. In many cases, the songs are complex and varied, but some are repetitious and monotonous. Songs may also include imitations of other bird species. Some larks have short songs, while others, notably Eurasian Skylark, have long, highly melodious songs. Larks sing while displaying, and even while being pursued by a predator or chasing off rivals. Flight and contact calls are typically short, sometimes high-pitched, buzzy, or sweet.

Habitats Larks occur in a wide range of habitats, most often open terrain, arid or semi-arid grasslands, and shrublands. The family reaches its greatest diversity in Africa, where 80

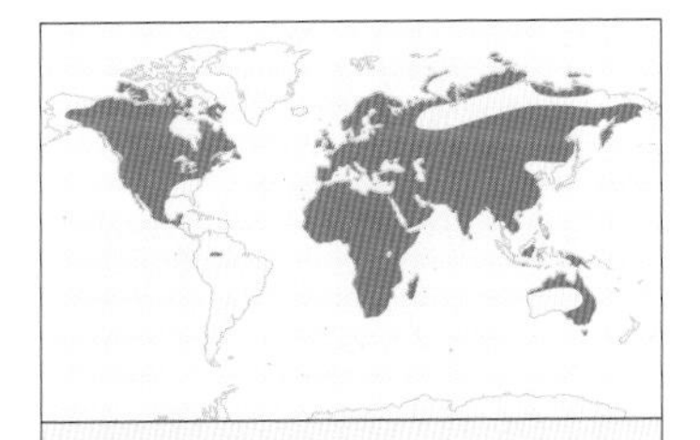

Number of genera 19
Number of species 92
Conservation Status 2 species Critically Endangered, 2 species Endangered, 4 species Vulnerable
Distribution Worldwide apart from Antarctica

Red-winged Bushlark (*Mirafra hypermetra*) often sings from the ground or from low shrubs as well as in flight. Kenya.

percent of the species occur. The lowest diversities are seen in South America, with a single isolated breeding population of Horned Lark (*E. a. peregrina*) in the Andes near Bogotà, Colombia; naturally, single species also occur in North America (Horned Lark), Australia (Horsfield's Bushlark, *Mirafra javanica*), and Madagascar (the endemic Madagascan Bushlark, *M. hova*). Two species, Rudd's Lark (*Heteromirafra ruddi*) and Raso Island Lark (*Alauda razae*), are Critically Endangered, two more are Endangered, and four are Vulnerable. Larks' habitat may be threatened by human agricultural activities, as well as urbanization, although some species prosper in agricultural areas.

Movements In arid areas of Africa, most species are entirely sedentary, although some of the specialized seed-eating species are nomadic, with flocks following rainfall in search of food. Eurasian and Nearctic species are partially or fully migratory. Migration often occurs across a broad front, with concentrations mainly along mountain ranges and coastlines.

Diet Species with short, stout bills tend to feed primarily on seeds, whereas species with longer, thinner bills tend to feed primarily on invertebrates, although most species feed on both. Green vegetation and small fruits are a small component of the diet in some species. Most larks feed their young exclusively on invertebrates, with a small number of specialized seed-eaters feeding their young mainly on seeds.

Breeding Monogamous and territorial, larks breed in isolated pairs or loose colonies. The majority of species that nest in arid regions breed in the rainy season, when food is most abundant. Nests are built on the ground by the females; most species build a cup made of dead grass, while some build more complex domed structures. Clutch sizes vary from one to eight eggs. The number of clutches per season is not known for most members of the family, but those species that have been studied raise from one to four broods per year. Incubation is performed by the female alone, and takes 9–16 days. The young are altricial. The fledging period is rather short, at 8–14 days, although the post-fledging period is unusually long, lasting up to a month.

Taxonomy Historically, larks have been thought to be the most primitive of the oscine passerines, despite their complex songs. Recent DNA evidence places them in a more advanced position, close to the Old World sparrows (family Passeridae), while further genetic studies have placed them closer to the Old World warblers (Sylviidae), bulbuls (Pycnonotidae), and swallows (Hirundinidae), among other families.

Sykes's Lark (*Galerida deva*) belongs to a genus whose members all have long, spiky, erectile crests. Gujarat, India.

Eurasian Skylark males perform display flights in which they climb and hover at more than 165 ft. (50 m.), while their song can be heard over great distances. U.K.

CISTICOLAS *Cisticolidae*

Golden-headed Cisticola (*Cisticola exilis*) males in breeding plumage have an orange-gold head with a crest that they raise when calling. Queensland, Australia.

The cisticolas and allied birds, sometimes collectively referred to as cisticolids, are small, often longish-tailed warblers, mostly found in dry grassland. This large family includes the cisticolas (genus *Cisticola*), prinias (genera *Prinia* and *Schistolais*), apalises (*Apalis*), wren-warblers (*Calamonastes*), and camaropteras (*Camaroptera*). Previously considered to be part of the larger family of Old World warblers (family Sylviidae), the cisticolids are now regarded as being sufficiently different to be recognized as a separate family.

The monotypic **Socotra Warbler** (*Incana incana*), endemic to Socotra, occurs in dry shrubland. Socotra Island, Yemen.

Structure The birds range from 3.5 to 8 in. (9–20 cm.) in length, with a corresponding weight range of 0.2–1.13 oz. (5–32 g.). Most species are at the smaller end of the range, and weigh less than 0.3 oz. (10 g.). Males are generally larger than females: in some *Cisticola* species, males outweigh females by 20 percent or more, and exceptionally in one or two species, the males are 50 percent heavier. Bill shapes vary from fine or slender to deeper and slightly arched; the bill is longest in Oriole-Warbler (*Hypergerus atriceps*). The body is slender and the wings are short and rounded, with 10 primaries and six secondaries. Tail shapes differ widely, from very short and rounded or square-tipped in several genera to extremely long and graduated in many others. Leg length also varies, from short to fairly long. The legs' structure largely depends on the species and lifestyle, being slender and weak in most arboreal species, and thicker in the ground-living birds. In all species, the claws are fairly short.

Plumage The sexes are alike in most species but differ in some of the *Apalis* warblers. Cisticolas are largely divisible into species with plain upperparts and those in which the upperparts are heavily streaked (although some of the streaked species have the mantle, back, and rump plain). Head colors vary from gray to buff, rufous, or golden-brown; the underparts are usually white or buffish; and tails are usually plain brown, with a subterminal black band and white tips.

Prinias are more varied, and include streaked and plain species ranging from gray to rufous, blue, olive, and yellow. Their underparts are either plain or streaked (some heavily streaked). One species, Banded Prinia (*Prinia bairdii*), has finely barred underparts and is very similar to several of the wren-warblers in genus *Calamonastes*.

Apalises are the most colorful, with upperparts ranging from bright green (contrasting with a blackish head in several) to gray or black-and-white, and underparts that are whitish to deep golden-yellow. They characteristically have a white throat and a broad breastband, but in some species the throat is all black. The most distinctive species is the monotypic Oriole-Warbler, with a jet black head and nape, and a breast that is finely fringed with silver edges, contrasting with the yellowish color in the rest of the plumage.

Behavior Most species are usually found alone or in pairs, or occasionally in small family parties toward the end of the breeding season. Several species of cisticolas, prinias, and scrub-warblers may regularly occur

Number of genera 21
Number of species 110
Conservation Status 1 species Critically Endangered, 3 species Endangered, 4 species Vulnerable, 3 species Near Threatened
Distribution Southern Europe, Africa, Asia, Australia

A **Hunter's Cisticola** (*C. hunteri*) takes a bath in the rain. Outside the breeding season, this species is more often heard than seen. Kenya.

in larger flocks of up to 50 birds when not breeding. Cisticolas and prinias usually favor lower levels of grass, scrub, and tangled vegetation, whereas apalises are more arboreal and forage in trees. Although all species can be fairly elusive and skulking, they often give away their presence by their contact calls, especially in the vicinity of a predator, or by their constant activity.

Voice Cisticolas and prinias have a range of short, sharp call notes and trills, including high-pitched rattles and warbling phrases. Several species are named for their calls: examples include Rattling Cisticola (*C. chiniana*), Chirping Cisticola (*C. pipiens*), Wailing Cisticola (*C. lais*), and Churring Cisticola (*C. njombe*). Apalises have shorter, softer, single call notes, which are repeated at frequent intervals.

The songs of cisticolas are often a series of musical or whistling notes with scratchy or harsh notes intermixed. They are given from a prominent perch, or during bouncing or elevated, circling song-flights. Wing-snapping Cisticola (*C. ayresii*) has a distinctive aerial song-flight accompanied by clicking notes and sharp cracks of the wings.

Habitat The birds occur in a range of habitats, from dry, grassy plains to scrub, reedy swamps, saltmarshes, light woodlands, and the margins of rain forests. One species is listed as Critically Endangered. Taita Apalis (*A. thoracica fuscigularis*), a subspecies of Bar-throated Apalis (*A. thoracica*), lives in fragmented areas of forest in southeast Kenya totaling 0.5 square miles (1.5 sq. km.) and is under continuing threat from habitat reduction; the entire population is considered to number fewer than 250 birds. Several other species are also at risk, mostly as a result of habitat clearance for agriculture.

A male **Yellow-breasted Apalis** (*Apalis flavida*). This striking species, with its gray head and yellow upperparts, is among the most common of the apalises. Kenya.

Movements Most species are sedentary or make only local dispersive movements outside the breeding season. Zitting Cisticola is a summer-breeding visitor to parts of central Europe, northern China, Korea, and northern Japan.

Diet Most species take small insects and their larvae, including bugs and grasshoppers. Larger items such as cicadas, dragonflies, and small lizards have also been recorded.

Breeding Most species are monogamous and solitary when breeding. Two, Zitting Cisticola and Golden-headed Cisticola (*C. exilis*), are polygamous, the latter often breeding year-round in parts of its range. Several species are more social when breeding, and defend common territories throughout the year; they may also be cooperative breeders.

The nest, usually built by the female, is an oval ball with a side entrance, or a deep cup of plant fibers and down, leaves, grass, moss, and cobwebs. It may be placed either low down in vegetation or up to 65 ft. (20 m.) from the ground. The clutch of two to five eggs is incubated, mostly by the female, for about 12 days. The nestlings are fed by both parents (except in the polygamous species) for up to 20 days before they disperse.

Taxonomy The exact limits and number of genera that comprise the family are still widely disputed between taxonomists. Several genera, including the Madagascar jerys (*Neomixis*), tailorbirds (*Orthotomus* and *Artisornis*), and *Poliolais*, are often placed either with the Old World warblers (family Sylviidae), as has been done in this book, or with the babblers (family Timaliidae). The monotypic White-browed Chinese Warbler (*Rhopophilus pekinensis*) is considered by some authorities, from recent DNA analysis, to have closer affinities with the babblers.

BULBULS *Pycnonotidae*

White-spectacled Bulbul (*Pycnonotus xanthopygos*), the most common species in the Middle East, is recognized by its black head, white eye rings, and fluting song. Oman.

An extensive Old World family, the bulbuls are found across Africa, Asia, and the Middle East; the name 'bulbul' is of Arabic derivation and possibly originated in imitation of the calls. The family is almost equally divided, in terms of species, between Africa and Asia. Only two genera span both continents; of these, the species in genus *Pycnonotus* are the most widespread and well known. The bulbuls also include the finchbills (genus *Spizixos*), the greenbuls and brownbuls (*Phyllastrephus* and six other genera), and bristlebills (*Bleda*). Some authorities also consider the nicators (*Nicator*) and Black-collared Bulbul (*Neolestes torquatus*) to belong to this family.

Shelley's Greenbul (*Andropadus masukuensis*) is a shy resident of dense Central African forest. Tanzania.

Structure The birds range from 5 to 11.5 in. (13–29 cm.) in length. The sexes are usually alike in size, although in some greenbuls the females are smaller than the males.

One of the defining features of 'true' bulbuls is the oval nostrils, which have a thin sheet of bone or cartilaginous tissue covering the rear part of the opening; this feature is present in all species except those of genera *Neolestes* and *Nicator*. The bill is short or slender; the upper mandible may be slightly decurved or ridged, and hooked at the tip. In the finchbills, the bill is shorter and more blunt-tipped. All species have well-developed bristles around the gape. This is particularly true of the bristlebills; in Common Bristlebill (*Bleda syndactylus*), the bristles may be as long as the bill.

Bulbuls have a fairly short neck and a slender body. The short, rounded wings have 10 primaries, the outermost of which is usually the shortest. The tail is moderately long, with 12 feathers, and mostly square-tipped but rounded or notched in some species. The legs and feet are short and fairly weak.

Plumage The sexes generally are alike, or females are occasionally duller than males. The plumage of most species is long and fluffy, especially on the back and rump; several species have a crest that may be loose and floppy, or erect and pointed.

Coloration varies from black or gray to dull brown or green, to bright yellow, usually with darker upperparts. Several species have brighter or white patches on the face, vent, and underparts, or are barred or streaked darker or whiter. A number of the *Pycnonotus* bulbuls have a black head or face, while several of the *Criniger* bulbuls have a white throat, which they puff out while calling or singing. In some of the duller brown species, the tail is warmer or more rufous. In most, the bill and legs are dark, but in the *Hypsipetes* bulbuls they are bright red. The usually plain-colored greenbuls and brownbuls are almost indistinguishable from one another.

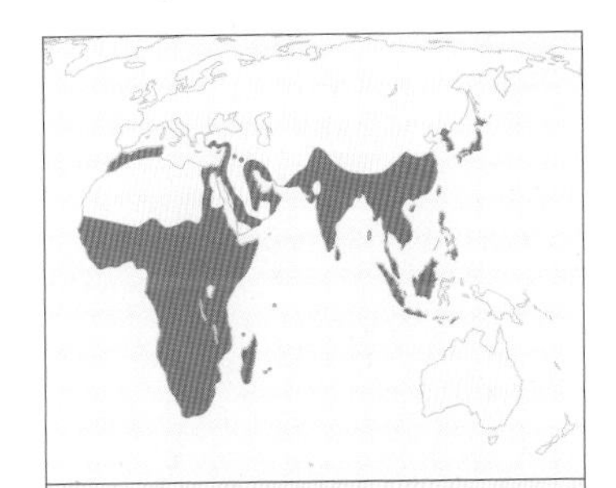

Number of genera 24
Number of species 122
Conservation Status 1 species Critically Endangered, 3 species Endangered, 11 species Near Threatened
Distribution Africa, Middle East, Asia

Perhaps the most striking crested species is **Red-whiskered Bulbul** (*Pycnonotus jocosus*), native to tropical Asia but now introduced to Australia and the U.S. Thailand.

The *Baeopogon* greenbuls closely mimic the plumage pattern of the honeyguides (family Indicatoridae), with gray faces and underparts and broad white outer tail feathers.

Behavior The species vary from social, conspicuous, and easily seen to shy, solitary, and generally skulking or elusive. Most open-country bulbuls are noisy, active, and gregarious, and forage in bushes, trees, and on the ground. Forest species are only seen fleetingly, foraging in tangled foliage or as part of mixed-species flocks that often follow ant-swarms for the insects they disturb. Several species feed while clinging to bark and lichens. Most forage in regular territories, but several of the *Hypsipetes* bulbuls cover large areas.

Voice Many bulbuls have distinctive calls, which often give the best clue to the birds' identity, particularly in the case of the greenbuls and brownbuls. Open-country species are generally more vocal, with a wide range of single- and double-whistle calls, as well as nasal and husky chattering notes. Songs are usually longer or repeated versions of the call. Few have extensive vocal repertoires, except for the nicators, which have loud or explosive, musical, varied songs similar in quality to those of Common Nightingale (*Luscinia megarhynchos*) or robin-chats (genus *Cossypha*).

Habitat The birds occupy a wide range of forest and woodland habitats, including semidesert scrub and thickets in dry savanna; several species also occur around villages and edges of cultivated land. Some bulbuls are common and widely spread across parts of the Middle East, Asia, and Africa; in Asia, bulbuls extend north to Japan and south to Indonesia. In contrast, Liberian Greenbul (*Phyllastrephus leucolepis*) is listed as Critically Endangered: it has only been seen in two small areas at the time of its discovery in 1985, and occurs in fragmented forest where uncontrolled clearance for agriculture is continuing. Three species are regarded as Endangered, and 11 others, many of which have very small ranges, as Vulnerable, mostly due to habitat degradation and loss.

Movements Most species are sedentary, or make only short-distance altitudinal movements. However, northern-breeding populations of Brown-eared Bulbuls (*Microscelis amaurotis*) move up to 1,000 miles (1,600 km.) south to winter in southern Japan and Korea.

Diet The diet consists mostly (or, in some cases, exclusively) of fruit or insects. Bulbuls may also take seeds, buds, nuts, nectar, and pollen; a few take bee larvae, wax, and birds' eggs.

Breeding Some species have dancing and fluttering courtship displays. Others, including Yellow-whiskered Greenbul (*Andropadus latirostris*), which is normally solitary, gather at leks. Most species form individual pairs, but several are cooperative nesters. Both partners build the nest, incubate the eggs, and rear the young. The cup-shaped nests are made of plant fibers, grass, and cobwebs, and placed in a fork or between upright twigs up to 30 ft. (9 m.) from the ground; nicators secure their nests to a branch using a fungus. The clutch of one to five eggs is incubated for 11–14 days, and the nestlings fledge at 11–18 days.

Taxonomy The bulbuls have been considered close allies of several families, including the cuckoo-shrikes (family Campephagidae), drongos (Dicruridae), Old World orioles (Oriolidae), and starlings (Sturnidae). Recent DNA evidence has shown a close link to the African grass warblers in family Cisticolidae. On structural differences, the nicators may not be closely related; however, they are included here.

Brown-eared Bulbul (*Microscelis amaurotis*) mainly eats fruit, seeds, and insects, but this bird has been feeding on *Camellia* flowers, and has pollen on its face. Japan.

OLD WORLD WARBLERS *Sylviidae*

Although small, elusive, and often inconspicuous, these warblers are fairly common across the Old World, with the greatest diversity of species found in Africa. There are four main subfamilies: the Megalurinae, or grassbirds (such as genus *Megalurus*); Acrocephalinae, or reed warblers (including *Acrocephalus* and *Hippolais*); Sylviinae, or scrub warblers (including *Sylvia*); and Phylloscopinae, or leaf-warblers (including *Phylloscopus*, *Eremomela*, and *Sylvietta*). Some authorities also include the tailorbirds (genera *Orthotomus* and *Artisornis*) and White-tailed Warbler (*Poliolais lopesi*) within the Old World warblers, as has been done here. Within these groups, most species are similar in structure, although often very different in plumage.

Hume's Leaf Warbler (*Phylloscopus humei*) is typical of most Old World warblers in having a compact body, fine bill, and subtle coloring. Sweden.

Structure Small, often slender passerines, warblers range from 4 to 11 in. (10–28 cm.) in length. Bill shapes vary from short and decurved to longer and finely pointed. The wings are short and rounded in sedentary species, and longer and pointed in migratory species. The species in genus *Megalurus* also have a vestigial claw at the bend of the wing. All species have 10 primaries, although the outermost is short to tiny. Tail lengths vary widely, from short in the stubtails (genus *Urosphena*), or apparently absent or invisible in the tesias, to medium or comparatively long in most warblers, and longest in the *Megalurus* grassbirds. The tail shape may be square, rounded, or graduated at the tip. Most warblers have 12 tail feathers, but some genera, including *Cettia*, have only 10. The legs and feet are proportionate to the size of the birds. In the ground-dwelling tesias, they are strong and stout. The legs in all Old World warblers are well adapted for their function—whether running along the ground (genus *Locustella*), grasping grass blades or reeds (genera *Cettia*, *Bradypterus*, *Acrocephalus*, and *Megalurus*), climbing up trees (*Sylvietta*), or moving along branches and through foliage in trees and bushes (*Sylvia* and *Phylloscopus*).

A male **Greater Whitethroat** (*Sylvia communis*). Males differ from females in having a gray head and a whiter throat. Sweden.

Plumage Sexual dichromatism is common in some genera and rare in others. Coloration varies between genera and species: most of these warblers are dull, brownish colors, but many have distinct supercilia (eyebrow lines) and other markings.

All grassbirds have buffish brown or darker head and upperparts, with a paler supercilium and darker or

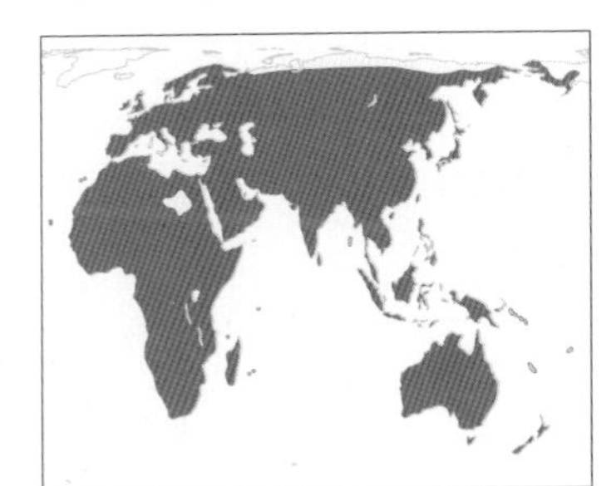

Number of genera 51
Number of species 279
Conservation Status 3 species Critically Endangered, 9 species Endangered, 26 species Vulnerable
Distribution Europe, Asia, Africa, Australasia, and Pacific islands

Melodious Warbler (*Hippolais polyglotta*) is recognizable by the yellowish underparts, plain lores, and brownish upperparts. Spain.

blackish streaks; the underparts are mostly pale buff to whitish. Most tesias are dull olive-green to brown, but several of the species have a distinct head pattern. Chestnut-headed Tesia (*Oligura castaneocoronata*) is particularly colorful, with a chestnut head and face.

The stubtails are warm brown to deep rufous-brown with very long, creamy-buff supercilia and broad black eyestripes. The bush-warblers are similar in color to the stubtails; in addition, some species have a warm or bright rufous crown and buffish to yellowish underparts. The *Bradypterus* warblers are darker in color than the bush-warblers, and usually have shorter or less prominent supercilia. Most species have dark underparts, and several have spots on the breast.

The *Acrocephalus* and *Locustella* warblers are various shades of brown, mostly with short, pale supercilia and pale to whitish underparts. However, several species, such as Sedge Warbler (*A. schoenobaenus*) and Aquatic Warbler (*A. paludicola*), have strong markings or stripes on the head and upperparts. Most *Locustella* warblers have well-marked undertail coverts, Lanceolated Warbler (*L. lanceolata*) being the most heavily streaked of the genus.

The *Hippolais* warblers have pale green to gray on the head and upperparts and white or yellowish underparts. Leaf-warblers (*Phylloscopus*) are mainly green to grayish or yellowish, with variably marked bright head patterns and wing bars and paler rumps, and whitish to yellow underparts. The *Seicercus* warblers are brighter versions of the *Phylloscopus* species, with a dark-streaked crown and pale yellow eye rings; some of these birds also have a brown head or a black face.

The *Sylvia* warblers have gray to bluish or brown upperparts, and many species have a gray to blackish head and face. The upperparts of Desert Warbler (*S. nana*) are sandy-buff, and Dartford Warbler (*S. undata*) has wine-red underparts. The short-tailed crombecs in genus *Sylvietta* vary from gray to olive, several also having orange-buff on the head, face, or underparts.

The tailorbirds are gray and green, often with warmer brown or black on the head and face and white or yellow underparts.

Behavior Most Old World warblers are solitary outside the breeding season. The exceptions are migratory species, which often feed in large or loose flocks when on passage, and tropical species, sometimes seen in pairs or in loose or mixed-species foraging flocks.

Several genera, including *Tesia*, *Cettia*, *Bradypterus*, *Locustella*, and most of the grassbirds, are secretive, skulking on the ground or in thick vegetation and usually most noticeable when singing. *Acrocephalus* warblers may be secretive, but several of the wetland species are more visible, especially when feeding at the margins of reedbeds, in ditches, or among floating vegetation. *Sylvia* warblers are most active in spring and summer.

Warblers actively forage in grassland vegetation, shrubs, bushes, and trees. The arboreal genera *Hippolais*, *Sylvietta*, *Eremomela*, and *Phylloscopus* spend long periods foraging in the foliage of trees and bushes, often well-camouflaged by their plumage. *Sylvietta* warblers feed in a similar manner to nuthatches (family Sittidae), climbing the trunks and branches of trees and inspecting the

Spectacled Warbler (*Sylvia conspicillata*) is named for the males' white eye rings. Canary Islands.

undersides for insects. *Phylloscopus* warblers may also hover briefly while picking insects from leaves. In addition, most species, particularly the *Phylloscopus*, *Seicercus*, and *Abroscopus* warblers, catch insects in flight; the last two genera are often referred to as 'flycatcher-warblers.'

Voice The songs of some species can be highly complex, while those of others are more simple and repetitive. The range covers the insect-like buzzes and trills of *Locustella* warblers, scratchy trills and warbles in the *Sylvia* warblers, and more musical songs in the *Cettia* and *Acrocephalus* warblers. Furthermore, in each family there are usually distinctive singers. Among the *Sylvia* warblers, the songs of Blackcap (*S. atricapilla*) and Garden Warbler (*S. borin*) are very similar, except that the range of the former is more accomplished, with a rich and varied song often mistaken for that of a Common Nightingale (*Luscinia megarhynchos*, family Muscicapidae), while that of the latter is slightly more scratchy. Songs of the *Tesia* and *Cettia* warblers often begin with a sudden or explosive start. Several of the *Acrocephalus* warblers are extremely accomplished mimics: for example, both Paddyfield Warbler (*A. agricola*) and Marsh Warbler (*A. palustris*) have few or no songs of their own but have a repertoire comprising perfect imitations of the songs of at least 20 other species. Songs can aid identification of very similar species, as in Europe, where there are several species of *Phylloscopus* warblers that look very much alike but each has a distinctive song.

Habitat Old World warblers have evolved and spread to occupy almost every type of terrestrial habitat, from desert edges and oases to grasslands, reedbeds, scrub, hedgerows, and scattered woodlands and the undergrowth of extensive forests. They are also found from sea level to montane scrub above the tree-line, at more than 16,500 ft. (5,000 m.), in the Himalayas. The species are widely distributed across the Old World, as far as Australia, New Zealand, and Pacific islands including Hawaii; one species, Arctic Warbler (*Phylloscopus borealis*), also breeds in west and southwest Alaska.

Arctic Warbler (*Phylloscopus borealis*) has a loud but rather monotonous song, which includes sharp *tzick* sounds. Sweden.

Several species, including those with a restricted habitat and range, such as Aquatic Warbler, are declining due to habitat destruction. A high proportion of island-endemic species have been adversely affected by the introduction of alien mammals, snakes, insects, and plants. In contrast, Hume's Reed Warbler (*A. orinus*) was rediscovered in Thailand in March 2006—the first sighting since its discovery in 1867.

Movements Most tropical species are entirely sedentary or make only short-distance nomadic movements. In contrast, northern-breeding species in Europe and across Russia, mostly the *Phylloscopus*, *Acrocephalus*, *Locustella*, and *Sylvia* warblers, are migratory, often moving long distances to wintering areas. These species include several that breed around the Mediterranean, although Dartford Warbler, which breeds north to the southern U.K., is usually sedentary unless winters become severe.

Several of the *Phylloscopus* warblers undergo particularly long migrations. All Willow Warblers (*P. trochilus*), which breed from Ireland to eastern Siberia, move south or

Orphean Warbler (*Sylvia hortensis*), one of the largest of all Old World warblers, breeds around the Mediterranean but winters in sub-Saharan Africa. Spain.

Seychelles Brush Warbler (*Acrocephalus sechellensis*) almost died out by 1960, due to destruction of the mangroves and palms in which it lives, but it has been saved by breeding programs. Seychelles Islands.

southwest to winter in Africa. Many Common Chiffchaffs (*P. collybita*) do the same, although in mild winters small numbers often stay north in central and northern Europe. Arctic Warblers, which breed from northern Scandinavia to Alaska, winter in southeast Asia, those from Alaska often moving through Japan and islands in the northern Pacific; those from the far northwest of the range have a journey of about 8,100 miles (13,000 km.).

Diet Most species are insectivorous, living on a wide range of terrestrial and aquatic insects and larvae. Large caterpillars and grubs (up to several times the length of the bird's bill) are beaten against a branch to remove hairs and toxic contents before being eaten. Larger species, such as Great Reed Warbler (*Acrocephalus arundinaceus*), also take fish fry, leeches, small amphibians such as tadpoles and newts, and lizards. Some *Sylvia* and *Acrocephalus* warblers also take seeds and fruits: mainly the berries of thornbushes, acacias, and ivy, but also larger fruits including cherries, figs, plums, and olives. The grassbirds frequently take seeds of grasses and sedges. Several species, mainly the smaller *Phylloscopus* warblers, are known to sip nectar from blossoms, gum from acacias, and tree sap.

Breeding Most species are monogamous, remaining in pairs for the duration of the breeding season. However, most *Cettia*, several of the *Acrocephalus* warblers, some *Phylloscopus*, and at least one of the *Sylvia* warblers are known to be polygamous, males mating with two or three females. All species are territorial; single pairs breed within their territory, including in loosely colonial species such as Barred Warbler (*S. nisoria*). Territories are held and defended by males singing vigorously, or even by aggressive behavior and brief fights.

Nests vary according to genera and locations. Some species, such as Willow Warbler, build a domed or ball-shaped structure with side entrance, on or near the ground. Others build an open cup fixed between fronds or branches or suspended from tall grass stems and reeds (as in reed-warblers' nests). Tailorbirds' nests are placed within a purse of two or more leaves, stitched together by the birds. Nests may be made from a wide variety of materials, including grasses, sedges, plant fibers and down, twigs, flowerheads, leaves, moss, lichens, vine or bark strips or feathers, and often bound together with cobwebs. Clutches vary from two or three eggs in tropical species that have more than one brood a year to four to seven eggs in the single-brooded northern species. In most species, eggs are incubated for 12–14 days, by the female. The young are fed by both parents for up to a further two weeks.

Taxonomy Evidence from DNA analysis indicates that the *Sylvia* warblers are more closely allied to the babblers (family Timaliidae) than to other warblers; however, this position is not universally accepted. The tailorbirds (*Orthotomus*, 11 species, and *Artisornis*, two species) and White-tailed Warbler (*Poliolais lopesi*) are included here in the Sylviidae, but some authorities are unsure of these groups' affinity.

A **Tawny Grassbird** (*Megalurus timoriensis*) feeds its young. The nest, made of rough grasses and sedges, blends in well with the surrounding grass. Australia.

BABBLERS AND PARROTBILLS *Timaliidae*

Black-lored Babbler (*Turdoides sharpei*) is one of the 'true' babblers, with a thrushlike body, a robust bill, and strong legs. Kenya.

This huge family of small or medium-sized birds, most of which are found in the Old World, includes a wide range of groups. In addition to the 'true' babblers, such as genus *Turdoides*, there are the parrotbills, laughing-thrushes, thrush-babblers (genus *Illadopsis*), scimitar babblers, wren-babblers, tit-babblers, shrike-babblers, barwings, minlas, fulvettas, crocias, sibias, yuhinas, and Wren-Tit (*Chamaea fasciata*).

Most parrotbills are classified in genus *Paradoxornis*: the Latin name, 'paradox bird,' refers to their odd, pushed-in, parrotlike face. The laughing-thrushes are well-known, noisy members of various Asian habitats, and are the most species-rich group within the babblers.

Brown-winged Parrotbill (*Paradoxornis brunneus*) shows the tiny, stubby bill that is typical of this group. Yunnan, China.

Structure Babblers come in a variety of shapes and sizes. They range in length from the jaylike laughing-thrushes, from 6 to 14 in. (15–35 cm.), to the small, long-tailed parrotbills, at about 7 in. (18 cm.), and titlike or even wrenlike species such as Wren-Tit, at only 6 in. (15 cm.).

The family is not well defined, particularly as many of its groups are not closely related. The typical babblers are thrush-sized but with a longer and often decurved bill, and a long tail. Otherwise, the bills vary from roughly thrushlike, to long and sickle-shaped in the scimitar-babblers, elongated in the yuhinas, or short, thick, and parrotlike in the parrotbills. The wings are typically short and rounded, with 10 primaries. The legs and feet are strong, especially in the ground-dwelling species. Tail lengths vary between groups; on average the group is long-tailed, but the wren-babblers have a very short tail.

Plumage Babblers tend to show loose, fluffy plumage, and in most species the sexes are alike. The groups differ in coloration and patterning. A large proportion are dull and brownish or olive, often with markings on the head and throat.

The typical *Turdoides* babblers are brownish, often with loose streaking or barring; many species have distinctive scaled patterns on the head and underparts, and most have white, yellow, or reddish eyes.

A superb variety of colors and patterns exists just in the laughing-thrushes. Some species have striking plumage, such as White-crested Laughing-thrush (*Garrulax leucolophus*), with its chestnut body, tail, and wings, contrasting white head, and black mask. Other bold markings include white throat patches, white or yellow underparts, white eyebrows, white cheek patches, or well spotted or striated plumages. Some laughing-thrushes sport red or yellow tails, or red wings. Many are more somber-colored, but often show pale eyes or bold bill colors.

Several of the other groups, including genus *Leiothrix*, minlas, mesias, barwings, and liocichlas, have fantastically colorful and intricate patterns on the folded wings. These include complex barred patterns (barwings); feather edging that changes from yellows at the base to oranges, and reds toward the tips; red-wax colored secondary tips; and various other gorgeous patterns.

The parrotbills are often cinnamon or rufous in color, with a yellow bill. The scimitar babblers show complex streaked patterns and often bold white striping on the face. Also boldly striped are species of the wren-babblers, some of which have

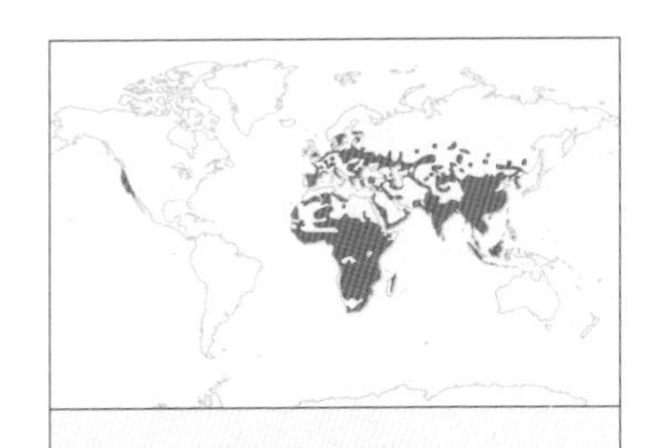

Number of genera 50
Number of species 273
Conservation Status
1 species Critically Endangered, 6 species Endangered, 22 Vulnerable, 47 species Near Threatened, 1 species Data Deficient
Distribution Europe, Asia, Africa; one representative in North America

a bright white throat, although others are brown and nondescript. The fulvettas range from nondescript, warm-colored, unstreaked species to others with contrasting black caps, similar to the related sibias, and recalling some of the Old World warblers in genus *Sylvia*. The yuhinas have a triangular crest. They range in color from dull olive species to warmer and brownish, with some bold wing patterning, and pale eye rings and dark throat stripes or face stripes. In fact, as far as plumage is concerned, babblers show almost all of the patterns seen in songbirds.

White-crested Laughing-thrush, found in the Himalayas and Southeast Asia, is highly social, and more conspicuous than other laughing-thrushes.

Behavior Babblers are generally social species, sometimes found in extended family groups, and even breeding cooperatively. Many species perform social bonding behavior such as allopreening, a reciprocal preening behavior that helps birds rid themselves of parasites, but also aids pair and group cohesion. Many babblers roost side by side, touching each other, rather than maintaining a wider spacing as is typical in most passerines. Cooperative breeders often feed each other. The birds are known to enter into disputes with each other, competing for the task of helping another babbler! This is not common behavior in birds, and has led to several babblers becoming focal species in long-term studies of sociality and the role of altruism in the evolution of bird behavior.

Arabian Babblers (*Turdoides squamiceps*) are cooperative breeders living in territorial groups, which may number as many as 22 individuals. The dominant male and female will breed, and the other group members help with incubating the eggs, feeding the young, and removing fecal sacs, as well as instructing and caring for the fledglings.

In Southern Pied Babbler (*T. bicolor*), groups range from three to 14 birds, although most have fewer than eight. Groups maintain a stable hierarchy, largely through displays and other interactions. Dominant individuals may become aggressive to assert dominance, while submissive displays, such as the use of chick begging calls, are given by lower-ranked individuals. In addition, there is a complex interaction between these birds and an unrelated species, Fork-tailed Drongo (*Dicrurus adsimilis*). Much of it is beneficial: the drongos guard groups of Pied Babblers and give alarm calls when predators are nearby. However, drongos will also give false alarm calls when the babblers have found a particularly good food item, allowing the drongo to steal the food from them.

While the *Turdoides* babblers are easy to observe in their open, arid habitats, the laughing-thrushes keep to forest and their study is more difficult. However, one species, White-crested Laughing-thrush (*Garrulax leucolophus*), is notable for the constant calling between flock members, and for the birds' communal habit of prancing around the forest floor with their wings extended and crest raised. Nest-helpers have also been observed for this species.

The North American Wren-Tit is monogamous, long-lived, and territorial year-round. The pairs remain in close contact vocally, perform a good deal of allopreening, and often roost together. Unusually for a passerine, males often incubate the eggs, and like females they have a well-developed brood patch; they also help in nest-building.

Voice As the name 'babbler' suggests, this group has some interesting vocalizations. Babblers tend to be loud and often noisy. Many species sing in duets, or as groups. Cooperative species, such as Pied Babbler, defend

A **Jungle Babbler** (*Turdoides striata*) preens its companion. This 'allopreening' is common behavior in flock members. India.

Wren-Tit (*Chamaea fasciata*) is a resident of coastal and mountain areas in western North America. Although shy, it has a loud, distinctive call. California, U.S.

their territory communally with songs and displays at territory boundaries. They also have a number of alarm calls to warn other group members of various dangers; these alarm calls are learned by young birds in a social context, allowing them to know when it is appropriate to use a specific alarm call.

Hwamei, or Melodious Laughing-Thrush (*Garrulax canorus*), and Red-billed Leiothrix, are widely renownd as singers, and as a result have become highly popular as cage birds in parts of Asia. Several species of laughing-thrush are also adept mimics, incorporating learned and copied sounds into their songs.

Habitat The *Turdoides* babblers are birds of arid zones ranging from desert margins to deciduous scrub and thickets. Other babbler groups, particularly the Asian species, are forest birds, with habitats ranging from open woodlands to lowland tropical forest and wet montane forest. Melodious Laughing-Thrush and Red-billed Leiothrix have both been introduced to Hawaii, possibly as a result of cage birds escaping.

Movements Most babblers are year-round residents of their territory. This lifestyle is reflected by their short, rounded wings, which do not serve well for long-distance flight.

Diet Babblers are omnivorous, but forage mainly on invertebrate prey. Some species also take fruits and berries. Certain of the larger species, particularly those in arid habitats, will also take small lizards or larger invertebrates such as scorpions. The parrotbills use their oddly shaped bill to extract and feed on seeds. The sibias (genus *Heterophasia*) are adept at feeding on nectar.

Breeding Many babblers have domed, covered nests, often rather coarse in nature, made of dead leaves, sticks, and other vegetation, and with a side entrance. Placement of the nest varies from on or near the ground to high up in a tree. Numerous other babbler species have open, cup-shaped nests, placed in bushes or trees away from the ground. Overall, nests are well concealed. This appears to be not only to thwart predation, but to decrease brood parasitism, as many babbler species are parasitized by cuckoos (family Cuculidae).

Clutch sizes vary from two to five eggs, depending on the species. Eggs range in color from white to blue or greenish, unspotted or well spotted, glossy to matte. Those of the large genus *Turdoides* are highly glossy, blue, and unspotted, while those of *Garrulax* are similar but often lightly spotted. In cooperative breeders, such as many *Turdoides* babblers, more than one female may lay in one nest. Pied Babbler may nest up to three times in one year. In most of the babbler groups, incubation takes 14–15 days, and nestlings fledge at 13–16 days.

Taxonomy Many taxonomists have found the limits of the babblers very difficult to define; in fact, the current classification is not yet settled, and many of the constituent groups will bounce in or out of the family in the near future. The fact that many groups' names make reference to other bird groups, such as thrushes and tits, is just one illustration of this problem.

Silver-eared Mesia (*Leiothrix argentauris*) is one of the smallest but most colorful babbler species, although females and juveniles are not as bright as the males. Malaysia.

WHITE-EYES *Zosteropidae*

Broad-ringed White-eye (*Zosterops poliogastrus*), as its name suggests, has a particularly obvious eye ring. Kenya.

A widely distributed Old World family, the white-eyes are named for the white ring of feathers around the eye in most species of *Zosterops*, the largest of the genera. As well as *Zosterops*, the family includes nine other genera of white-eyes, four species in genus *Speirops*, and the monotypic Bonin Honeyeater (*Apalopteron familiare*), Mountain Black-eye (*Chlorocharis emiliae*), and Cinnamon Ibon (*Hypocryptadius cinnamomeus*). Most species in each genus are very similar in size, structure, and plumage.

Structure These small, compact birds are 3–6 in. (7–16 cm.) in length and 0.3–1.1 oz. (8–31 g.) in weight. In some species, the females are smaller than the males. The bill is short; in most species it is pointed and slightly decurved, but in a few it is straight. The two species in genus *Woodfordia* are larger than the *Zosterops* species, with a prominent bill. The tongue is brush-tipped, to aid the collection of juice and pulp from fruit. The wings are short and rounded, with nine primaries. The tail in *Zosterops* species is short and generally square-tipped, while in some of the other genera it is slightly longer. The legs are mostly short and slender in the arboreal white-eyes, and thicker and stronger in the terrestrial species.

Plumage The sexes are generally alike, but in some species the males are slightly brighter than the females. Most of the *Zosterops* white-eyes are bright to dull green on the head and upperparts and yellow, grayish, green, or white on the underparts, although Japanese White-eye (*Z. japonicus*) and Chestnut-flanked White-eye (*Z. erythropleurus*) have prominent chestnut flanks on an otherwise white belly. The birds have a variably thin to broad, white eye ring that either encircles the eye or is partly broken. Some species have minor variations including a black forehead or crown, or an eye ring that merges with white lores. Some lack a white eye ring. Black-masked White-eye (*Lophozosterops goodfellowi*) and Mountain Black-eye have a broad eye ring that is black. *Woodfordia* species are mainly dull olive-brown; one has a poorly defined pale area around the eye, while the other has bare black face broadly surrounded by a band of white. One of the most brightly colored species, Golden White-eye (*Cleptornis marchei*), is unique in the family as it lacks a white eye ring and is almost entirely golden-yellow, with a bright orange bill and legs. Also lacking an eye ring is Cinnamon Ibon from the Philippines, which has cinnamon upperparts and warm buff underparts. *Speirops* species are mostly gray-brown and differ in having patches of white, a white eye ring, or a white head.

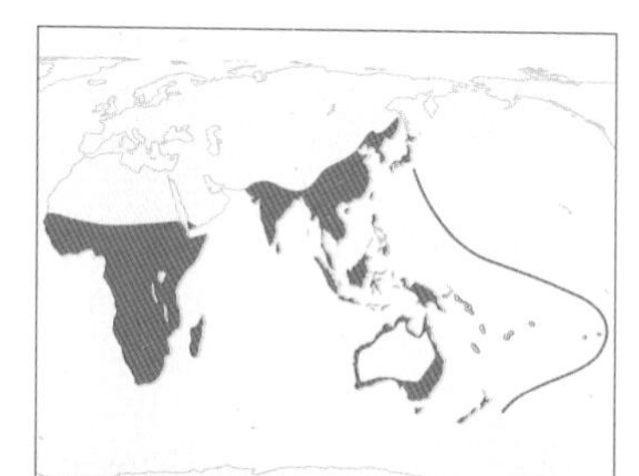

Number of genera 14
Number of species 95
Conservation Status 6 species Critically Endangered, 5 species Endangered, 10 species Vulnerable, 13 species Near Threatened
Distribution Africa, Asia, New Guinea, Australia, Pacific Ocean islands

Behavior Highly social birds, white-eyes are usually found in flocks of up to 20 birds; they may also join mixed-species foraging flocks with other insect-eating passerines. Most species actively and restlessly forage

Oriental White-eyes (*Z. palpebrosus*) form large flocks outside the breeding season. Panna National Park, India.

in foliage and along branches at all levels in trees and undergrowth. They eat fruit by piercing it with the bill, then extracting the juice and pulp with the brush-tipped tongue.
Voice Calls are often similar for all species in a genus. In *Zosterops*, the call is usually a single or disyllabic, nasal, thin or wispy, down-slurred *zeeee-er*. In other genera, calls are a shrill or squeaky, buzzy *seer*, harsh or rasping *chup*, chirps, or chattering notes, often followed by a short whistle. The songs are mostly rambling series of high- and low-pitched notes, often including call notes.
Habitat White-eyes are naturally found in Africa, Asia, New Guinea, Australia, New Zealand, and numerous islands in the Pacific Ocean, but have also been introduced to Hawaii and Tahiti. Most species are woodland or forest residents, but the birds also occur in coastal mangroves, the edges of montane forests, scrub, and gardens.

A disproportionately high number of species are threatened. Among the most critically endangered are two species with populations currently considered at less than 50 individuals, both of which occur on remote islands. Sangihe White-eye (*Z. nehrkorni*) is restricted to the Indonesian island of the same name, which has been almost entirely cleared for agriculture; the bird occurs in small areas of degraded and remnant broad-leaved forest. White-chested White-eye (*Z. albogularis*) is endemic to indigenous forest on Norfolk Island, but very few have been seen in recent years; the last sighting was in 2000. One subspecies of Slender-billed White-eye (*Z. tenuirostris*), Robust White-eye (*Z. t. strenuus*), was once common on Lord Howe Island but became extinct in the 1920s following the introduction of Black Rats (*Rattus rattus*) to the island.

Silver-eye (*Z. lateralis*) is one of the most migratory white-eyes; birds from Tasmania make a 1,000-mile (1,600-km.) journey north to winter in Queensland. Australia.

Movements Most species are sedentary, but Japanese White-eye and Chestnut-flanked White-eye are summer migrants into the north of their breeding ranges in central and northern China and eastern Russia. Other species disperse over wide areas. Silver-eyes (*Z. lateralis*) from southeast Australia colonized New Zealand, crossing more than 1,250 miles (2,000 km.) of open sea.
Diet A wide range of insects, fruit, nectar, and pollen.
Breeding White-eyes are territorial and monogamous; most species pair for life. The nest is a deep cup of plant fibers and down, bound with cobwebs and placed in a fork of a tree or bush branch. The clutch of one to three eggs is incubated by both parents for up to 13 days. The nestlings fledge at about 12 days, and are fed by the parents for several days after leaving the nest.
Taxonomy The relationships of various genera to each other and to *Zosterops* are not always clear. The *Speirops* birds, restricted to islands in the Gulf of Guinea, are fairly similar and clearly closely related to each other or to a common ancestor. The true affinities of Bonin Honeyeater remain uncertain; it was previously included with the honeyeaters (family Meliphagidae), but has similarities to the white-eyes.

Green-backed White-eyes (*Z. xanthochroa*) feeding on papaya. White-eyes help farmers by pollinating plants, but can cause problems by damaging fruit. New Caledonia.

FAIRY-BLUEBIRDS *Irenidae*

This tiny family contains only two species: Asian Fairy-bluebird (*Irena puella*) and Philippine Fairy-bluebird (*I. cyanogastra*). Both are strikingly colored birds that resemble orioles or thrushes in general appearance.

Structure Slender but stoutly built, fairy-bluebirds are 8–10 in. (21–27 cm.) in length; males are, on average, slightly larger than females. The birds have deep blood-red eyes and a fairly large, powerful bill. They also have short, pointed wings, a square-tipped tail, and fairly short legs.

Plumage In male fairy-bluebirds, the face, body, and flight and tail feathers are jet black, while the upperparts are glistening bright blue or black. In addition, the Asian species has blue tips to the inner greater wing coverts, while in the Philippine species, the wing coverts and tertials are mostly a glossy deep blue. The sexes of Philippine Fairy-bluebird are alike; in contrast, female Asian birds are dull bluish or green with blackish wings and tail. Two races of Asian Fairy-bluebird have extremely long upper- and undertail coverts that cover the tail or extend slightly beyond it.

A male **Asian Fairy-bluebird** eats a papaya. Fruit forms the majority of the diet for both species of fairy-bluebird. India.

Number of genera 1
Number of species 2
Conservation Status Least Concern
Distribution South Asia to Borneo and Philippines

Behavior Usually found in pairs or small groups, although single birds also occur with other species in fruiting trees. Fairy-bluebirds forage at all levels in forest trees, plucking fruit from foliage; they also fly-snatch fruit from distant branch tips and bushes, and pursue emerging termites in flight.

Voice The birds produce a variety of strident and far-carrying whistles, including a single or repeated *weet*, occasionally extended to a *tu-lip wae-waet-too*. The courtship song of the male Asian Fairy-bluebird includes notes like those of a starling; the contact note between pairs is a hoarse croaking. Philippine birds are very noisy when breeding: their song is like that of the Asian species, but they may also mimic other birds.

Habitat Tropical and subtropical lowland evergreen forests; also fruiting trees in woods and groves.

Movements Both species are sedentary, but may make local movements outside the breeding season.

Diet The birds mostly eat fruit, especially figs and berries. They also take nectar from blossoms, and insects including beetles, termites, and termite larvae. The young mainly eat insects until after fledging.

Breeding Asian Fairy-bluebirds breed between January and June in India and to August in Borneo. The male has an elaborate courtship display, which includes raising the crown feathers, stretching the neck, jerking the head, and tail-pumping to show off the glossy plumage. The nest is a small, shallow, flimsy cup of twigs and mosses situated up to 20 ft. (6 m.) high, on the fork of a branch or a frond in the understory. The two eggs are incubated by the female for two weeks, and the chicks fledge after a further two weeks.

KINGLETS *Regulidae*

A group of small songbirds belonging to the single genus *Regulus*, which are easily recognized by their calls. All the species are very similar in structure and plumage but show sexual dichromatism in the color of their hidden crown patches. In addition to the continental species, endemic forms occur on the Canary Islands, Madeira, and Taiwan.

Structure Among the smallest of all passerines, kinglets are 3.75–4 in. (8–10 cm.) in length. They have a small, thin bill, a rounded body, and a medium-length tail. The wings are relatively short and have 10 primaries, with the tenth reduced in size.

Plumage Most species have boldly marked heads, particularly with lateral black crown stripes. All but one species have colorful crown patches, often concealing a brighter or deeper crest. The body is grayish to bright olive-green, with bright olive-edged flight feathers. The wings have two white bars, although the upper wing bar is not always obvious.

Behavior Kinglets usually occur in pairs, but form small to moderate-sized flocks during migration and in winter. When foraging, they tend to quiver or flick their wings. They often feed by 'hover-gleaning,' as well as by searching in crevices in bark and inside flowers and buds. Males raise their crest to reveal their hidden crown patch during territorial disputes and courtship.

Voice Songs consist of high-pitched introductory notes followed by a warbling that is surprisingly loud and rich for such small birds. Calls typically comprise single, thin notes.

Habitat Most continental species breed in coniferous forests at high temperate latitudes in North America, Europe, and Asia. Madeira Firecrest (*R. ignicapilla madeirensis*) prefers laurel forest. High-latitude breeders winter in various habitats, including deciduous and second-growth woodlands and shrublands. Goldcrest (*R. regulus*) is highly tolerant of the cold winters and short daylight hours in these regions, but often suffers high mortality in prolonged periods of severe weather.

A male **Ruby-crowned Kinglet** (*Regulus calendula*) with its crest raised. This species is unlike most kinglets in that it does not have a striped head. U.S.

Movements The continental kinglet species are generally short-distance migrants, with many lingering in the southern portions of their breeding range. Migrating birds are frequently observed in feeding flocks, in low shrubbery or canopy. Island forms are non-migratory.

Diet Kinglets feed on insects year-round. In winter, they find their prey in buds and under bark; this allows them to winter farther north than other insectivores.

Breeding The breeding season is in the northern hemisphere summer. Kinglets often build their nests in conifers, at varying heights. Generally, they form monogamous pairs that defend relatively small territories. The birds are mainly double-brooded and produce large clutches, comprising 6–13 eggs. The incubation period is 14–17 days. The young are altricial, and fledge in 16–22 days.

Number of genera 1
Number of species 5
Conservation Status Least Concern
Distribution North America, Europe, Asia, North Africa

WRENS *Troglodytidae*

One of the most familiar of the bird families, wrens are smallish, compact birds with a tail that, in many species, is cocked upward. The Troglodytidae is primarily a New World family, but also includes Winter Wren (*Troglodytes troglodytes*), which is found in many other parts of the world.

Cactus Wren (*Campylorhynchus brunneicapillus*) is the largest species in North America, at 7 to 9 in. (18–22 cm.). It is found mainly in the desert areas of the southwest. Texas, U.S.

Structure Several species are less than 4 in. (9 cm.) long, but Giant Wren (*Campylorhynchus chiapensis*) is the largest, at 9 in. (22 cm.). The bill is slender and often slightly downcurved. In a minority of species, such as Sumichrast's Wren (*Hylorchilus sumichrasti*) and Nava's Wren (*H. navai*), it is relatively long. Wrens in genus *Troglodytes* have a very short tail, while those in genus *Microcerculus* are virtually tailless. In contrast, species in genus *Campylorhynchus* have a long tail.

Plumage Colors tend to be rather dull: mostly brown, chestnut, buff, black, and white, often with pale supercilia. *Troglodytes* wrens are various shades of russet and brown, generally paler on the underparts. Some *Campylorhynchus* species have barred upperparts; Boucard's Wren (*C. jocosus*) has a blackish-brown crown, eyestripes, and mustachial stripes, contrasting with a pale buff throat and supercilia. The *Henicorhina* species have prominent dark markings on the cheeks. In some species, for example those in genus *Thryothorus*, the plumage has striking patterns although the colors are not bright. Coraya Wren (*T. coraya*) has a mostly black head with a white throat. Bay Wren (*T. nigricapillus*) and Riverside Wren (*T. semibadius*) have fine blackish bands on the underparts. White-headed Wren (*C. albobrunneus*) is unlike any other member of the family, with unmarked dark brown upperparts and a white head and underparts. Sexual dichromatism exists in a minority of wren species; where it occurs, males may show more barring or better-defined coloration.

A **Bewick's Wren** (*Thryomanes bewickii*), in typical 'wren' posture, with the tail cocked. Oregon, U.S.

Behavior Wrens are generally very active birds. Many are noisy and conspicuous, although some give their presence away only when they sing. They forage in vegetation, on tree trunks, on the ground, or on cliffs. Many species turn over leaf-litter or pull moss from trees in search of invertebrates. Some species flycatch. Canyon Wren (*Catherpes mexicanus*) creeps across even vertical cliff faces and forages in crevices. Some species are usually seen only alone or in pairs, while others may forage in small groups. Wrens often join other birds to mob a predator. Many have a habit of quickly raising and lowering their tail, especially if excited.

Voice Songs are loud, powerful, often complex, and sometimes beautiful. Southern Nightingale Wren (*Microcerculus marginatus*) has one of the more elaborate songs, a sequence

Number of genera 17
Number of species 77
Conservation Status
4 species Critically Endangered, 2 species Endangered, 4 species Vulnerable, 4 species Near Threatened
Distribution Mostly New World, but one species also found across Old World

that may last for up to 150 seconds. Happy Wren (*Thryothorus felix*) has a series of gurgling whistles: *chee wee, chee wee, cheery cheery*; Giant Wren delivers a series of chortling phrases described as *kar-a-u-too, kar-rale-du-ow*; and Cactus Wren sings a succession of harsh, unmusical notes, repeated many times: *jar-jar-jar-jar-jar*. In many species, both sexes sing in unison; in some, the female's song differs from the male's. Songs are often delivered from a prominent perch, such as a snag or the top of a cactus. Others species sing from the ground. Winter Wren sometimes sings in flight and at night. (It has been estimated that the power output of a singing Winter Wren is, weight for weight, 10 times that of a cockerel.) In contrast, wrens' calls include a broad range of harsh churring and grating sounds.

Habitat Wrens occupy a wide range of habitats, including rain forest, temperate forest, arid scrub, grassland, *paramo*, reedbeds, cliffs, and urban areas. Some species, such as Winter Wren, are generalists, inhabiting a variety of forests, scrub, grasslands, and wetlands. Others are extreme specialists: for example, Zapata Wren (*Ferminia cerverai*) occurs only in sawgrass swamp in Cuba, Inca Wren (*Thryothorus eisenmanni*) in bamboo forest in northern South America, and Sumichrast's Wren only in humid forest on limestone in southern Mexico. Some wrens live at great altitude; for example, Santa Marta Wren (*Troglodytes monticola*) is found at up to 15,748 ft. (4,800 m.) in Colombia.

Movements Most are sedentary, although some practice altitudinal and other seasonal movements. A few are long-distance migrants, such as the *stellaris* subspecies of Sedge Wren (*Cistothorus platensis*), which breeds in the Great Lakes region, the Midwest, and central Canada, and winters in the southern and eastern U.S. and northern Mexico.

Black-capped Donacobius (*Donacobius atricapilla*) lives in the lowland marshes of tropical Brazil. Breeding pairs often duet; they sit together on a branch, facing in different directions, wagging their tails and exposing the orange patches on their neck. Brazil.

Diet Primarily invertebrates, with some very small vertebrates and some vegetable matter.

Breeding Most species are highly territorial and boldly repel intruders; for example, Giant Wren has been seen attacking squirrels and snakes.

Nests are often carefully constructed, domed structures. Those of Winter Wren are typically built with moss, grass, and leaves, and lined with feathers and hairs. The female incubates the clutch of three to nine eggs for about 16 days. Nestlings are fed by both parents, although only the female broods them. The young fledge at 14–19 days, and the parents care for them for a further 9–18 days. Pairs often have two or even three broods per season. Polygyny is frequent in some populations, and eggs from two females are sometimes laid in the same nest.

Other species are cavity nesters. One such is House Wren (*Troglodytes aedon*). Usually, the female incubates the clutch of four to eight eggs for 12–14 days. Both parents feed the young, which fledge in 16–18 days and are fed for a further 13 days. Females usually have two broods per season. Most males are monogamous.

The most complex breeding system is that of the South American species Stripe-backed Wren (*Campylorhynchus nuchalis*), which is a cooperative polyandrous breeder. Groups of up to 14 birds occupy a territory in which only the principal pair breeds. The other group members (the pair's previous offspring) help feed the latest brood and defend the territory against predators. Young birds remain in the group for about a year before dispersing.

Taxonomy Black-capped Donacobius (*Donacobius atricapilla*) has been included here in family Troglodytidae, but the species' placement is still disputed, and the bird is often classified as *incertae sedis*.

A **House Wren** sits on top of a fence post as it defends its territory with song. U.S.

GNATCATCHERS *Polioptilidae*

The small Polioptilidae family of passerines comprises the gnatcatchers and the gnatwrens. These birds look like the Old World warblers, but are found exclusively in the Americas. The gnatcatchers are all classified in genus *Polioptila*. The gnatwrens comprise two species in genus *Microbates*, and also Long-billed Gnatwren (*Ramphocaenus melanurus*).

Structure Gnatcatchers and gnatwrens are small, slender birds with a long, thin bill and a long, narrow tail. The gnatwrens have a particularly long, needlelike bill.

Plumage The gnatcatchers are very similar in plumage, most being gray above and white below, with varying amounts of black on the head. Males and females are usually different. The male Black-capped Gnatcatcher (*Polioptila nigriceps*) has an entirely black cap during the breeding season, while the recently described Iquitos Gnatcatcher (*P. clentsi*) has no black on the head. Gnatcatchers' tails are typically black, with varying amounts of white on the outer feathers. Several species have white eye rings. The gnatwrens are generally brown, although Collared Gnatwren (*Microbates collaris*) has white eyebrows and a black collar, Tawny-faced Gnatwren (*M. cinereiventris*) has rufous sides to the face, and Long-billed Gnatwren is paler on the throat and underparts.

A **Black-tailed Gnatcatcher** (*P. melanura*) sings to advertise its territory. Only the males sing. California, U.S.

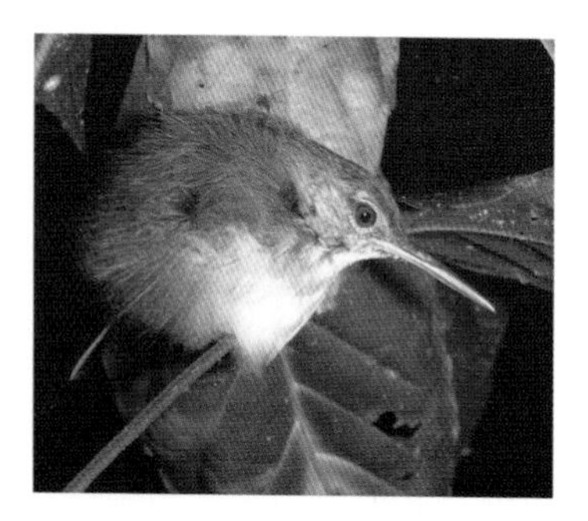

A tiny **Long-billed Gnatwren** roosts in dense tropical vegetation. These birds are more often heard than seen. Peru.

Behavior Gnatcatchers are usually seen in pairs. Some species are found in the canopy of tall rain forest, where they occasionally join mixed-species flocks. Males are territorial. The birds forage by gleaning small insects from leaves and branches; they often cock their tail while foraging. Gnatwrens typically join mixed-species flocks in the understory of tropical rain forest. Long-billed Gnatwren has a distinctive habit of wagging its tail.

Voice Only males sing. The songs of gnatcatchers and gnatwrens vary from a long series of sweet whistles to short, nasal notes. Calls are generally nasal or raspy *mew* notes.

Habitat Gnatcatchers and gnatwrens are often found in arid environments, and some species exist in tropical habitats: all three of the gnatwrens are restricted to the Neotropics. Iquitos Gnatcatcher, restricted to northeastern Peru, is listed by the IUCN as Critically Endangered, because it has a very small known range and a limited habitat.

Movements Most species are resident within their range and preferred habitat. The exception is Blue-gray Gnatcatcher (*P. caerulea*), which has subspecies that are highly migratory. In the fall, these birds vacate their northern breeding areas, in the eastern and central U.S., to winter in Florida, the Caribbean, and northern Mexico. Black-capped Gnatcatcher of northern Mexico has been known to disperse into Arizona outside the breeding season (and occasionally to breed there as well).

Diet All species are insectivorous.

Breeding The birds usually build a small cup nest out of grass, leaves, and other vegetation. They lay two to six eggs, and incubation lasts for 11–14 days. The young are fed for 10–16 days. Both males and females participate in all aspects of the breeding cycle. Young birds stay with the adults for some time after fledging, in small family groups.

Number of genera 3
Number of species 15
Conservation Status 1 species Critically Endangered
Distribution Southern and central North America; much of South America

NUTHATCHES AND WALLCREEPER *Sittidae*

A small family found across the Old and New Worlds, the Sittidae comprises two subfamilies: the Sittinae (nuthatches) and the Trichodromadinae (wallcreeper). The nuthatches are all in genus *Sitta,* and there is only one Wallcreeper (*Tichodroma muraria*). Nuthatches are named for their habit of wedging nuts in tree crevices and pecking them open; wallcreepers, as their name suggests, are specialist climbers that forage on rocks and cliff faces.

Structure Nuthatches are fairly small: 4–8 in. (9.5–20 cm.) in length.They are rather flat-headed, with a fairly short but robust, pointed, or chisel-shaped bill and a short neck. They have a plump body, usually held horizontally; rounded wings; a short, square-tipped tail; and short legs with long toes and claws. Wallcreepers are larger, at 6.5 in. (16.5 cm.). They have a more rounded head, with a long, slender, decurved bill; fairly long, broad, rounded wings; and a short, square-tipped tail.

Plumage In all nuthatches, the sexes are very similar but females often have paler underparts. Wallcreepers are also similar, but the sexes differ in breeding plumage and have a distinct nonbreeding plumage. Most nuthatches are characterized by blue-gray upperparts and tail and broad black eye stripes occasionally extending to the sides of the nape. The underparts vary from partly or entirely white to buffish, rufous, or chestnut. Some species differ in having a gray, brown, or black crown and broad white supercilium. Three have violet-blue upperparts with a velvet-black band across the forehead and fine yellowish eye rings, the underparts varying from white to pale lavender. Blue Nuthatch (*S. azurea*) differs distinctly from all others with black or blackish-blue on the head, upperparts, and belly to the undertail; a broad, pale blue eye ring; pale blue wing coverts, tertials, and flight feathers all outlined with black; and pure white throat and breast. Wallcreepers are mostly gray, and slightly darker on the wings and underparts. The male has a broad black chin to breast bib; the bib is smaller and narrower on females, with pale or whitish sides. The wings have crimson wing coverts with black flight feathers and white patches mid-way along the primaries. The tail is black, with white tips to the outer feathers. In winter they are generally paler gray on the body, with a whiter throat.

Nuthatches, like this **Eurasian Nuthatch** (*S. europaea*), are able to run down tree branches and trunks head-first as easily as they can climb them. U.K.

Behavior Nuthatches are usually found in pairs or small groups that regularly patrol their territory or home range year-round; outside the breeding season, several species often join mixed-species foraging flocks. Wallcreepers are also territorial throughout the year.

Nuthatches forage by climbing up and down tree-trunks, along branches, and occasionally among outer twigs, picking insects and spiders from the surface or probing under bark and into crevices. Wallcreeper and the rock nuthatches forage in the same manner but on rocks, vertical cliffs, and screes. Wallcreepers habitually flick their wings while foraging, and flit from one vertical face to another. The more northerly nuthatches collect seeds in the fall and store them in crevices in bark, rocks, or under moss for consumption during the winter.

Voice Nuthatches frequently advertise or defend their territory with a series of loud, sharp whistles and rising, piping or chattering trills. Wallcreepers are generally silent, but have a short, quiet chirping whistle and a soft, squeaky, whistling song.

Habitat Most species are fairly common in lowland tropical or subtropical forests, or in temperate forests and woodlands (both coniferous and broad-leaved). Wallcreeper is found on rocks or cliff faces.

The family is widespread through the Holarctic region, extending

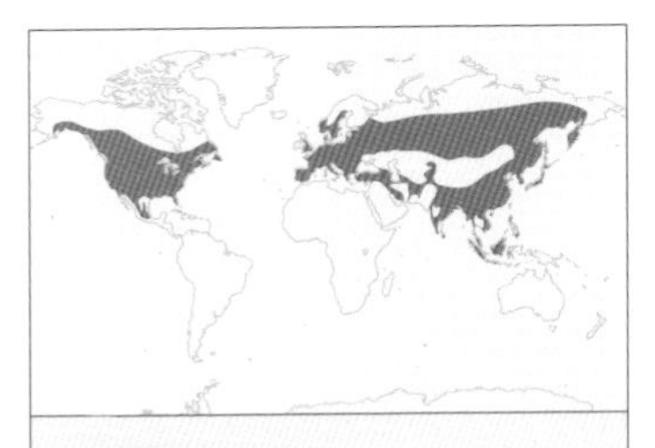

Number of genera 2
Number of species 25
Conservation Status 2 species Endangered
Distribution North America, Europe, Asia, North Africa

Pygmy Nuthatch (*S.pygmaea*), the smallest member of the family, is found in western North America. California.

north to southern Alaska and northern Siberia and south to Southeast Asia and parts of Indonesia. It reaches its greatest diversity in southern Asia, where half of the nuthatches are found. Four species of nuthatch are native to North America. Three occur in Europe; Eurasian Nuthatch (*S. europaea*) extends east through Asia to China and Japan and south into northern Africa. Wallcreepers are found from southern Europe east to Central Asia, the Himalayas, and China.

Several species have restricted distributions. Corsican Nuthatch (*S. whiteheadi*) is found only in montane pine forests on the Mediterranean island of Corsica, but is not thought to be at risk. However, two other species with equally small ranges, White-browed Nuthatch (*S. victoriae*) and Algerian Nuthatch (*S. ledanti*), are considered Endangered due to their very small ranges and declining areas of suitable habitat. Two other species from southern Asia are regarded as Vulnerable.

Movements Mainly sedentary, but northern-breeding populations are often irruptive and move south, particularly in years of poor spruce or pine seed crops. Red-breasted Nuthatch (*S. canadensis*) occasionally reaches northern Mexico, southern Texas, and northern Florida. The Siberian-breeding populations of Eurasian Nuthatch move west to Scandinavia. Wallcreepers are altitudinal migrants, wintering at lower levels.

The black supercilium on this **Velvet-fronted Nuthatch** (*S. frontalis*) shows it to be a male. This species lives in southern and eastern Asia. Goa, India.

Diet The birds eat mostly invertebrates in summer, and nuts and seeds in winter.

Breeding The birds are monogamous. They nest in holes in trees or rock crevices. Several species spread the noxious scent of certain insects or sticky resin around the nest entrance to repel predators. Certain species close up the entrance with mud, and rock nuthatches close the entrance and enter via a tube. Nests are mostly made from plant fibers, bark strips, moss, and feathers. The clutch of four to nine eggs is incubated by the female for 14–18 days. The young fledge after 20–25 days.

Taxonomy The Australian sitellas (family Neosittidae) were previously included in this family, due to similar shape and behavior, but are now considered to be unrelated and the result of convergent evolution. Wallcreeper's relationship to the nuthatches has been the subject of long debate, with some taxonomists placing it with the treecreepers (Certhiidae) or in a family of its own.

Wallcreeper forages even on vertical cliff faces, hopping across the rock and fluttering its wings for balance, or making short sallies after insects. Switzerland.

TREECREEPERS *Certhiidae*

As their name suggests, this family of small brown-and-white birds is known for their habit of creeping or scurrying up tree trunks. All are in genus *Certhia*, apart from Spotted Creeper (*Salpornis spilonotus*). Most species are similar in size and structure; they differ mainly in the degree of streaking on the upperparts and the color of the underparts.

Structure Treecreepers are 5–6 in. (12–15 cm.) in length, with a fine, decurved bill, slender body, and short, rounded wings with 10 primaries. The legs are short, with long toes and long, curved claws. In all species except Spotted Creeper the tail is long, pointed, and stiffened, with the vanes projecting at the tips of the feathers. This feature enables the bird to use the tail as a support when climbing. Spotted Creeper has a short, rounded tail.

Plumage The sexes are alike. Treecreepers have fine white supercilia, finely streaked or spotted ear-coverts, and intricate brown, black, and buff or white streaked markings on the upperparts. The rump and tail are usually brighter or rufous; in Bar-tailed Treecreeper (*C. himalayana*), the tail is finely barred. Spotted Creeper is almost entirely spotted and barred in brown, dark brown, and buff. Immatures in all species are very similar to adults.

A **Brown Creeper** (*C. americana*) climbs a tree, using its long claws and stiff-vaned tail for support. Connecticut, U.S.

The tail feathers are molted from the outermost ones inward, and the central pair replaced only once all the others have regrown. As a result, the bird can continue to use its tail throughout the molt. This feature is shared only by woodpeckers.

Behavior The birds may be found alone, in pairs, or in mixed-species foraging flocks. All the *Certhia* species forage for insects and spiders in a similar fashion, almost exclusively on trees, but occasionally on walls or more rarely on the ground. Typically, they fly to the base of a tree and work upward in a jerky motion either directly or going around the trunk in a spiral pattern, and often moving off to work along the undersides of branches before moving on to the base of the next tree. Spotted Creeper forages in a similar way but does not use the tail as a support in climbing trees. It may cling upside-down to the undersides of branches.

Voice Mostly thin or high-pitched whistles and trills. The songs are similarly high-pitched, sweet warbles or cadences.

Habitat Widely distributed throughout Holarctic forests and Asiatic woodlands. Spotted Creeper has a more southerly distribution, with separate populations in open *Acacia* woodlands in central India and sub-Saharan Africa. No species are currently considered at risk, but several of the restricted-range species may be vulnerable to continuing destruction of submontane forests.

A **Eurasian Treecreeper** (*C. familiaris*) brings food to its young, which are hidden in a cavity in a tree trunk. U.K.

Movements Mostly sedentary.

Diet Insects and other arthropods.

Breeding *Certhia* treecreepers build a loose nest, wedged in a crevice in a tree-trunk or placed behind loose bark. Spotted Creeper builds a cup-shaped nest on or in the fork of a branch, and decorated with cobwebs and spiders' egg sacs. The clutch of three to six eggs is incubated by the female for 14–15 days. The nestlings fledge after 15–16 days.

Number of genera 2
Number of species 8
Conservation Status Least Concern
Distribution North America, Eurasia, Southeast Asia, Africa

MOCKINGBIRDS AND THRASHERS *Mimidae*

A New World family, mockingbirds are named for some species' habit of mimicking the songs of other birds. The family Mimidae comprises four groups of birds: mockingbirds (16 species), thrashers (14 species), catbirds (two species), and tremblers (two species). Within each of the groups, all species are of similar size and structure. The family shares certain features and behavior, and most have loud, powerful voices; they differ mainly in principal habitat niches, plumage, and songs.

Structure The birds are 8–12 in. (20–30 cm.) in length, with the long, graduated tail making up at least a third of the overall length. The bill varies from strong but slightly decurved, in catbirds and some mockingbirds, to long and prominently arched in thrashers. The body is slender, with short, rounded wings and fairly long, strong legs and feet.

Plumage In all species, the sexes are alike or very similar, and there are no seasonal differences in plumage.

Northern Mockingbird is recognizable both by its loud song and by its white wing and tail patches, which are visible in flight and when the bird is displaying.

Most of the family are muted in color, ranging from black to gray above, with paler underparts. Several species have dark ear-coverts; two have broad, prominent spots or streaks, or rufous undertail-coverts. Northern Mockingbird (*Mimus polyglottos*) has broad white patches in the wing and prominent white outer tail feathers. Several of the thrashers have distinctive pale tips to the wing coverts, forming wing bars.

The two species in genus *Melanotis* are the most brightly colored of the mockingbirds: Blue Mockingbird (*M. caerulescens*) has entirely slaty-blue plumage, and Blue-and-white Mockingbird (*M. hypoleucus*) has slaty-blue head, upperparts, and tail. Brown Thrasher (*Toxostoma rufum*) is the most colorful of the thrashers, with rich rufous upperparts and tail. Black Catbird (*Melanoptila glabrirostris*) is entirely black.

Northern Mockingbird, the two tremblers, and some of the thrashers have white, pale, or bright yellow eyes giving a rather staring appearance; the *Melanotis* mockingbirds, and one thrasher, have red eyes.

The four species of mockingbirds found on the Galápagos islands are very similar, with brown upperparts, wings, and tail, and white underparts. They differ in the extent of black on the face, pale collar, and streaking or patches on underparts. All also have different eye colors.

Behavior Most species are mainly terrestrial, apart from the tremblers, which are entirely arboreal. The

Hood Mockingbirds suck blood from a young Masked Booby (*Sula dactylatra*). It has been suggested that this unusual behavior enables the mockingbirds to get sufficient moisture in a habitat lacking freshwater. Galápagos Islands, Ecuador.

Number of genera 12
Number of species 34
Conservation Status 2 species Critically Endangered, 3 species Endangered, 2 species Vulnerable, 1 species Near Threatened
Distribution The Americas

birds are usually solitary or found in pairs. Mockingbirds and thrashers spend long periods skulking or foraging low down in undergrowth or scrub, or on the ground. Galápagos Mockingbird (*Nesomimus parvulus*) forages along beaches, and Hood Mockingbird (*N. macdonaldi*) forages among breeding colonies of seabirds. Several thrashers, including California Thrasher (*T. redivivum*), Le Conte's Thrasher (*T. lecontei*), and Crissal Thrasher (*T. crissale*), use their long, curved bill for digging in soft earth and leaf-litter. The tremblers forage along branches and in clumps of leaves, tearing up and tossing aside vegetation in their search for invertebrates.

Mockingbirds are often very expressive in their actions: they often twitch the tail from side to side, raising or fanning it, as part of territorial or courtship displays. Tremblers are named for their habit of drooping and visibly trembling their wings while foraging or when alarmed. Several of the mockingbirds and both tremblers are noted for their inquisitive and confiding nature.

Voice Songs are typically loud, and often given from an open or exposed perch. Most are a varied repetition of musical warbles or slightly scratchy or gruff phrases; mockingbirds include a great deal of repetition and mimicry. Calls include whistles, mewing, scolding, and chacking notes.

Habitats Mockingbirds and thrashers are widely distributed throughout the Americas and on neighboring islands including the Galápagos; the highest number of species occur in Central America and on Caribbean islands. Habitats vary from rain forest to forest understory, plantations, parks and gardens, dry, open country scrub, and desert margins. Northern Mockingbird is one of the most widespread and widely known species of North American birds, and often occurs in parks and gardens. Several other species are fairly common and widespread, including those in restricted ranges, but three mockingbirds and two thrashers are at risk. Cozumel Thrasher (*T. guttatum*), restricted to Cozumel Island off the Yucatan peninsula, Mexico, and once fairly common, declined drastically following severe hurricanes in 1988 and 1995 and has not been seen since. If it survives at all it is regarded as Critically Endangered since, although large areas of suitable habitat remain intact, this is the part of Mexico most often hit by hurricanes. Charles Mockingbird (*Nesomimus trifasciatus*) is now found only on two tiny islands off Floreana in the Galápagos; the total population is about 150 individuals.

Gray Catbird, named for its mewing call, is found in thickets and gardens across North America. New Jersey, U.S.

Movements Most species are sedentary. The most migratory birds are the populations that breed in Canada and the northern U.S. Gray Catbird (*Dumetella carolinensis*) and Brown Thrasher move south to winter around the Gulf and in Mexico. Northern Mockingbirds also move south into southern U.S. states, but small numbers remain north into southern Canada if food is available.

Diet Mostly insects including spiders and centipedes, and some worms and insect larvae; some species also take fruit and berries. Galápagos Mockingbird feeds on small crabs and carrion. Hood Mockingbird steals and pecks open the eggs of various birds, from ground-doves to large albatrosses, and those of marine iguanas. It also regularly drinks blood from open wounds on iguanas, sealions, and nestling seabirds.

Breeding Most species are territorial and solitary. The nests are large, bulky cups of twigs, plant fibers, and grass, built by both sexes and placed either low down, on the ground, or up to 50 ft. (15 m.) high in a tree. The clutch comprises two to six eggs, with northern birds laying more. Incubation takes 12–13 days, and is done mostly by the female, except in the thrashers, where both sexes share the task. Both parents feed and care for the nestlings, which take up to 19 days to leave the nest. Most species begin nesting in spring, and may raise two or three broods; California Thrasher, unusually, nests during both spring and fall.

Taxonomy Black-capped Donacobius (*Donacobius atricapilla*) was formerly included in family Mimidae, but recent DNA evidence now indicates that this monotypic species has a closer relationship with the wrens (family Troglodytidae).

Le Conte's Thrasher forages for insects on the ground, and travels by running, typically with its tail cocked. California, U.S.

PHILIPPINE CREEPERS *Rhabdornithidae*

This tiny, little-known family is endemic to the Philippines. It comprises two species in a single genus: Stripe-headed Creeper (*Rhabdornis mystacalis*) and Stripe-breasted Creeper (*R. inornatus*). The two are very similar; differences are mostly in size, bill size, extent of streaking, and the color of the crown. The name 'creeper' is actually a misnomer: although the birds superficially resemble the true treecreepers in genus *Certhia*, they lack the morphological features of that family and rarely, if ever, creep along tree branches or trunks in the same way as treecreepers. The name 'rhabdornis' is often used instead.

A **Stripe-headed Creeper** (*R. mystacalis*). This is the more common of the two species, and is found in lowland forest.

Structure The birds are 4.5–7 in. (14–17 cm.) in length. They have a short neck and a short to medium-length bill, which is stout or strong, with a pronounced hook at the tip. The wings are short and pointed, the tail short to medium-length, and the legs and feet slender.

Plumage Both species have warm or dark brown upperparts, wings, and tail, and a broad black mask through the eye and ear coverts. The underparts vary from mostly white with heavily streaked flanks to duller or buffish with more extensive brown streaks. Stripe-headed Creeper is more heavily marked, with fine but extensive white streaks from the head to the mantle.

Behavior The birds occur alone or in pairs, in the middle to upper levels of trees. They are also often seen in mixed-species foraging flocks, often with bulbuls (genus *Hypsipetes*), tits (genus *Parus*), fantails (genus *Rhipidura*), or nuthatches (genus *Sitta*), which roam through the forests. Both species roost communally (occasionally up to several hundred together) in the canopy of a tree at the forest edge. They forage by hopping or jumping from branch to branch, gleaning insects from among the foliage or under the bark, and most often perch crosswise on branches. Flocks have also been recorded taking termites in flight.

Voice Not well known: mainly high-pitched or unmusical *zip* or *zeet* notes, given singly or as a series, and often run together as a faster trill.

Habitat Lowland to submontane forests and secondary woodlands. Although both species are uncommon, neither is currently considered to be threatened.

Movements Sedentary.

Diet Insects form the main diet, but small fruits and seeds are also taken. Long-billed Creeper (*R. i. grandis*), the Luzon subspecies of Stripe-breasted Creeper, has been known to eat small tree frogs.

Breeding Very little is known about the birds' breeding behavior. Nests of both species have been recorded in cavities in trees. Incubation and brooding habits are unknown.

Taxonomy The family shares similarities with some of the babblers in family Timaliidae, particularly the wren-babblers in genus *Ptilocichla*; however, they are considered sufficiently distinct to warrant their own family. Long-billed Creeper is sometimes considered a full species.

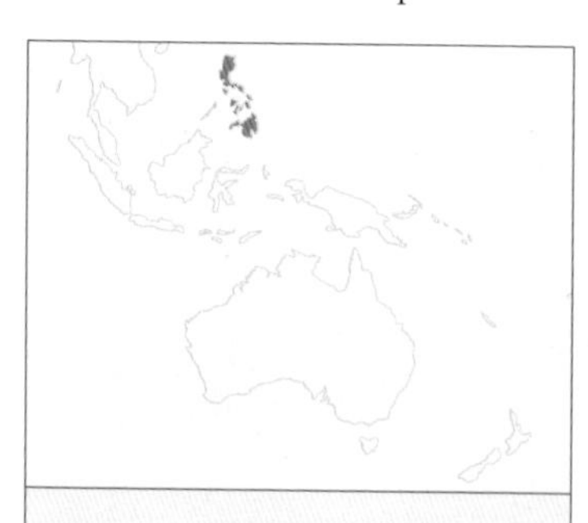

Number of genera 1
Number of species 2
Conservation Status Least Concern
Distribution Philippines

STARLINGS *Sturnidae*

A large family of Old World passerines, the starlings include some of the world's most familiar species, as well as some of the rarest and most endangered. The family includes the glossy starlings, mynas, and oxpeckers. Most species are fairly similar, differing only in plumages. Many of the African species have glossy plumage or are brightly colored; the oxpeckers differ mostly by being browner, with a slightly more elongated body and a larger, brightly colored bill.

Structure The birds range from 6 to 18 in. (16–45 cm.) in length, and may be slim or stocky. The bill is slim, narrow or pointed in starlings, larger and more broadly based in mynas, and downcurved in oxpeckers. The wings are short and rounded, or slightly longer and more pointed in several of the tropical species. The tail may be fairly short and square, but in several species it is longer and graduated or pointed. In White-eyed Starling (*Aplonis brunneicapillus*) the central feathers are elongated streamers.

Plumage The sexes are usually alike or similar, but some species are dichromatic, with the males being more brightly colored. Many species are black or blackish with a metallic sheen or spots. Most of the African starlings are highly glossed green or bluish, often with dark ear coverts and tips to the wing coverts. Several species have a pale orange or chestnut breast and belly, whereas others are more uniformly brown with a white vent and undertail. Starlings in genus *Onychognathus* are mostly glossy black with bright red patches on the primaries.

Noisy and sociable, **European Starlings** (*Sturnus vulgaris*) are a common sight in Europe and western Asia, and tame around people; this group is taking a bath in a garden. U.K.

The dazzling **Superb Starling** (*Lamprotornis superbus*) is most common in East Africa and also lives in Arabia. Dubai, U.A.E.

One of the most colorful of all species is Golden-breasted Starling (*Lamprotornis regius*), which has a glossy blue-green head and upperparts, bronze-green tail, purple breastband, and deep yellow underparts. Male Amethyst Starlings (*Cinnyricinclus leucogaster*) have deep metallic purple on the upperparts, head and breast, and white underparts, while the females are brown above and white below, with thrush-like spots.

Many of the *Aplonis* starlings are brown or only slightly glossy. Several of the Asian starlings are pale gray or whitish with darker wings and tail, but Daurian Starling (*Sturnus sturninus*) has a deep purple mantle and back, and Rosy Starling (*Sturnus roseus*) has a black head, wings, and tail and a deep pink body.

Mynas are mostly black with a golden head and face, or gold extending onto the neck, breast, and rump, as in Golden-breasted Myna (*Mino anais*); most also have white patches in the wing. However, both Bare-eyed Myna (*Streptocitta albertinae*) and White-necked Myna (*S. albicollis*) are largely black and white. The monotypic Coleto (*Sarcops calvus*) is mostly gray, with large,

Number of genera 25
Number of species 115
Conservation Status 2 species Critically Endangered, 2 species Endangered, 5 species Vulnerable
Distribution Much of Eurasia and Africa; introduced to other continents

Common Myna (*Acridotheres tristis*) is admired for its ability to mimic a range of sounds including human speech. Australia.

bare pink areas around the eyes. Both of the oxpeckers are mostly olive-brown with buff underparts, but have different bill colors; Red-billed Oxpecker (*Buphagus erythrorhynchus*) also has a yellow eye ring.

Several species, mainly among the mynas, have prominent areas of bare, brightly colored skin around the head and face; Wattled Starling (*Creatophora cinerea*) has a bald yellowish head and black wattles during the breeding season. Two species have floppy or filamentous crests, and three others have a prominent, rigid crest. A considerable number of the starlings, and several of the mynas, have pale or white eyes.

Behavior Most species are social, usually found in pairs, family groups, or larger flocks, especially outside the breeding season: communal roosts of European Starlings (*Sturnus vulgaris*) and Common Mynas (*Acridotheres tristis*) may total over a million birds. Conversely, some of the *Aplonis* starlings are often solitary. Some European Starlings roost in city or town centers, and as the birds gather they perform spectacular aerial turns and swooping maneuvers. Most starlings and mynas forage in a variety of habitats, from open fields, shorelines, margins of cultivated land, and gardens to woods and forest canopy. Oxpeckers forage on a wide range of animals (including domestic cattle), climbing over them while removing ticks and other parasitic insects from their hide; some are also known to take blood from their hosts.

Voice Calls include a wide range of whistles, squawks, dry rattles, and shrill phrases. Songs are often a series of call notes, run together and faster or more melodious, and often with phrases repeated. Eurasian Starling, when in full song, inflates the throat feathers and accompanies the song with wing-flapping. Most of the mynas have a series of sharp or high-pitched whistles and clicking notes. Several species (together with some starlings) are adept mimics and have an extensive repertoire of harsh, discordant noises. Common Hill Myna (*Gracula religiosa*) is especially good at imitating human voices, doing so as well as any parrot.

Habitat Spread across Europe, Africa and Asia to islands in the central Pacific Ocean. Several species have been introduced to North America, New Zealand, Australia, and islands in the Pacific, Indian, and Atlantic Oceans. Two species are Critically Endangered. Bali Myna (*Leucopsar rothschildi*), which is endemic to Bali, declined rapidly in the second half of the 20th century due to illegal trapping for the cage-bird trade; it is thought that fewer than 15 birds remain in the wild, although there are more than 1,000 in captivity. Pohnpei Starling (*Aplonis pelzelni*), endemic to the Micronesian island of Pohnpei, in the western Pacific, has been seen only once in the last 50 years; a tiny population is presumed to survive despite continuing threats from loss of habitat, hunting, and introduced predators. Two other species are considered as Endangered, and five others as Vulnerable.

Oxpeckers feed on insect parasites that live on animals; this **Red-billed Oxpecker** is foraging on an impala. Kenya.

Movements Most species are resident or make short-distance nomadic movements. Longer distances are covered by species such as Daurian Starlings, which breed north into eastern Siberia and winter south into Indonesia. European Starlings that breed over Scandinavia, eastern Europe, and western Russia move south to winter in Europe and around the Mediterranean.

Diet Starlings and mynas take mostly fruit and insects, as well as seeds, nectar, and pollen. Oxpeckers are specialized insectivores.

Breeding The birds are monogamous, and solitary or semicolonial when breeding. Their nests are usually untidy bundles of dry plant fibers, grass, and other suitable material, placed in holes in trees, buildings, ruins, or nest-boxes. Some species excavate their own holes; others utilize those of woodpeckers. Shining Starling (*Aplonis metallica*) is unique in building a hanging nest like that of a weaver (family Ploceidae), with a side entrance; often, a whole colony occupies a single tree. The clutch of one to six eggs is incubated by the female for 11–18 days, and the young fledge in 15–25 days.

THRUSHES *Turdidae*

Song Thrush (*Turdus philomelos*), one of the most familiar of all thrushes, is found extensively in the Old World. Spain.

Thrushes are known mainly for their rich variety of complex musical songs but also for their attractive plumages and, in several species, their close association with people in rural or urban areas. Previously thought to be closely related to the robins, chats, and babblers, they are now considered as a single family. The largest genus, the *Turdus* thrushes, includes American Robin (*T. migratorius*), which nests over much of North and Central America, and Blackbird (*T. merula*) and Song Thrush (*T. philomelos*) of Europe and Asia.

The family originated in the Old World, probably in central or southern Asia, where the greatest diversity of species is still found. Today, thrushes live on all continents except Antarctica, and on most of the larger islands; many small islands have their own endemic species or subspecies. In New Zealand, which had no endemic thrushes of its own, Blackbird and Song Thrush were introduced by colonists from Europe.

Structure Small to medium-sized passerines, thrushes range from 6 to 13 in. (15–33 cm.) in length. They have a rounded head and a slender, pointed bill with a few short rictal bristles at the base. Wing shapes vary from fairly short and rounded to longer and pointed (in migratory species). The tail is moderately long; it is usually square-tipped but in a few species is slightly rounded or graduated. The legs are usually fairly long and strong, but are shorter and slightly weaker in the South American solitaires (genus *Myadestes*).

Plumage In most thrushes, coloration comprises soft or subtle shades that blend in easily against a forest background. Brown predominates, with variations from gray to olive, and often with brighter orange or red. The eye is dark in most species, but bright red in some.

The members of genus *Turdus* have a wide variety of plumages.

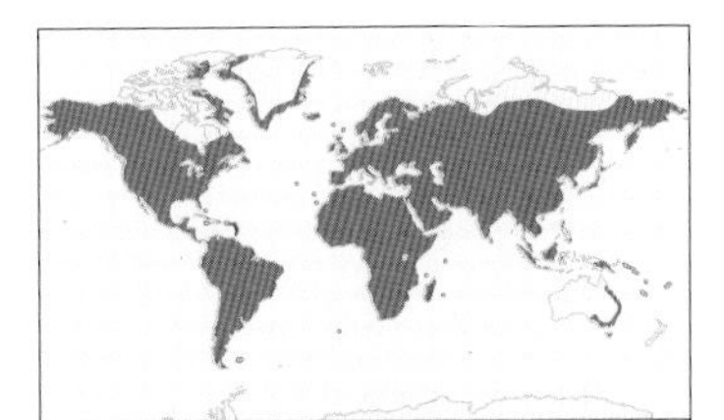

Number of genera 25
Number of species 167
Conservation Status 5 species Critically Endangered, 3 species Endangered, 9 species Vulnerable
Distribution Worldwide

Many species are colored in subtle shades of gray, brown, and orange. In contrast, some have striking features such as bright eye rings, glossy plumages and contrasting patterns. One example is Ring Ouzel (*T. torquatus*), which is mostly black but with a contrasting white crescent on the breast. Interestingly, the uniform Blackbird, which has all black plumage and a bright yellow

An **American Robin** takes a drink. Although they are native to forests, this species has adapted to forage in urban and suburban environments. New Jersey, U.S.

bill, has almost identical South American counterparts in Glossy-black Thrush (*T. serranus*) and one race of Yellow-legged Thrush (*Platycichla flavipes*).

Several species in other genera are largely black or deep blue, and a number have cryptically marked black and white face patches or lines. For example, the whistling thrushes (genus *Myophonus*) are mainly deep blue with bright, iridescent patches on the forehead and wing coverts; in the forest understory these birds look almost black, but in the sunlight their plumage gleams bright metallic blue.

The ground-thrushes in genus *Zoothera* are characterized by a prominent, black and white zig-zag underwing pattern. Several species are also very cryptically marked, with dull golden-brown plumage crossed with fine dark crescents; the handsome Pied Thrush (*Z. wardii*) has black and white plumage with a long, tapering, white supercilium.

The solitaires are mainly gray but also include species with rufous throats and bright chestnut upperparts, and black and white faces. The *Catharus* thrushes are mostly cryptically marked with brown or olive-brown, and are either white or partly spotted on the underparts. However, Veery (*C. fuscescens*) has distinctly more rufous upperparts and Hermit Thrush (*C. guttatus*) has a rich chestnut tail. Wood Thrush (*Hylocichla mustelina*) is similar to the *Catharus* species, but has more reddish-brown on the nape.

Behavior Most thrushes are solitary, or are found in pairs or occasionally in loose flocks. Some of the more northerly-breeding species gather to form large roosting or foraging flocks in winter. Some tropical-breeding species also roost communally, often engaging in noisy calling before they settle for the night.

All thrushes show similarities in their diet, but the various genera forage in different ways. The whistling thrushes forage close to water, usually along the edges of fast-flowing hill and mountain streams. The *Zoothera* or ground-thrushes are shy birds that live in dense forests, where they forage on the ground in the undergrowth and leaf litter. The smaller *Catharus* thrushes are similarly retiring forest birds, which feed in trees and bushes as well as foraging on the ground.

Mountain Bluebird (*Sialia currucoides*), found in western North America, is famed for the males' brilliant plumage. U.S.

The widespread *Turdus* thrushes are more generalist; they feed on the ground, running or hopping and then stopping to examine the area in front of them. It is thought that, by turning or tilting their head, they can detect the presence of worms or large insects below the surface of the ground and make darting pounces to grab the prey as it emerges. Ground-foraging thrushes turn over leaves and fallen vegetation in their search for prey. Some species dig into the leaf-litter and shallow soil, often forming holes into which they virtually disappear from view.

Song Thrush and Blue Whistling-thrush (*Myophonus caeruleus*) are among the few species that make use of a tool in obtaining their food. Snails are held in the bill and with a flick of the head beaten against a regularly used stone or 'anvil' until the shell breaks open; well-used anvils often accumulate large piles of shell fragments around them.

Flight is strong and undulating in most species, although many of those living in forest undergrowth rarely move long distances.

Olive Thrush (*Turdus olivaceus*) occurs throughout sub-Saharan Africa. It has 16 subspecies, which vary in coloration and brightness. The striking Northern Olive Thrush (*T. o. abyssinicus*), shown here, is sometimes considered to be a separate species. Ethiopia.

Despite its boldly scalloped plumage, **Bassian Thrush** (*Zoothera lunulata*) is rarely seen due to its shy nature and its habit of foraging in dense undergrowth. Tasmania, Australia.

Voice Thrushes are among the most accomplished and versatile songsters. They have a series of melodic, rich, fluting, whistled, or warbled songs, given either as a succession of pure notes or repeated in short phrases. Among the most exquisite are the songs of some solitaires, particularly Rufous-throated Solitaire (*Myadestes genibarbis*), and of the *Catharus* thrushes—in particular, Wood Thrush, Hermit Thrush, and the well-named nightingale-thrushes. Their repertoires of spiraling, flute-like phrases, ventriloquial cadences, modulated whistles, buzzes, and trills create an extremely evocative sense of true wilderness in the otherwise silent forest.

Some species are also very good mimics. For example, in Europe, Blackbirds and Song Thrushes have been heard imitating the ringing of telephones. Similarly, the South American Lawrence's Thrush (*T. lawrencii*) has little song of its own but is adept at mimicking the songs of more than 170 other bird species.

Most species have strident alarm notes or a harsh, chattering rattle; thrushes are often the first birds to raise the alarm whenever a predator is approaching.

Habitat Thrushes are mainly forest-loving birds, but have also moved into a wide range of other habitats, including upland moorlands and bare mountain-tops and, in contrast, human-created environments such as parks and gardens.

Movements Tropical species are sedentary, or make short-distance movements after the breeding season. Some species breeding farther north move to lower altitudes or warmer areas; for example, almost all American Robins breeding in Canada and Alaska move south to winter in the U.S. and central Mexico.

The species making the longest migrations are those breeding in the northern latitudes and wintering around the equator or farther south. Some Gray-cheeked Thrushes (*Catharus minimus*) breed in far eastern Siberia and Alaska and winter in the Amazon rain forest of South America. In Europe, Fieldfares (*Turdus pilaris*), Redwings (*T. iliacus*), Song Thrushes, Ring Ouzels, and Blackbirds that breed in northern Scandinavia and Russia move south to winter in central Europe, around the Mediterranean, and in North Africa. Farther east, several species, such as Naumann's Thrush (*T. naumanni*), Dusky Thrush (*T. eunomus*), Red-throated Thrush (*T. ruficollis*), and Black-throated Thrush (*T. atrogularis*), cross large areas of China and the Himalayas to winter in northern India and southeast Asia.

Diet Mostly insects and their larvae; also seeds, buds, and fruit. Insects form the main part of the diet in the spring and summer, and fruit from late summer. Some species, however, probably exist entirely on insects.

The types of insects taken include beetles, weevils, craneflies, caterpillars, bees, wasps, crickets, grasshoppers, bugs, ants, termites, sawflies, and flies. Fruit includes a wide variety of berries, such as those of hawthorn and juniper, and cherries, apples, pears, and plums. Thrushes also take earthworms, snails, slugs, and small mollusks; whistling-thrushes may take crabs, amphibians, frogs, geckos, and small lizards.

Breeding Most species are monogamous and breed in a well-defended territory, some pairs even remaining together during the winter. Fieldfares are almost unique among the thrushes in nesting semicolonially.

Bare-eyed Robin (*T. nudigenis*) is resident in tropical South America and the Caribbean. Trinidad.

The nest is typically an open cup made from plant stems, fibers, twigs, moss and feathers; some species line their nests with mud. Preferred nest-sites differ according to genus. Most thrushes place their nests on branches or forks in trees or low shrubs. Whistling thrushes, in contrast, nest in holes in banks, ledges in caves, in trees, under roots, or occasionally behind waterfalls.

The clutch of three to six eggs is incubated by the female for 12–15 days. The young are fed by both parents, often every three to four minutes, and fledge at 12–15 days.

Taxonomy Two southern African species, Rockrunner (*Chaetops pycnopygius*) and Rockjumper (*C. frenatus*), are sometimes included in the thrush family.

Fading Away: The Rarest Thrushes

Thrushes are a well-known and widely distributed family of birds: almost every continent has at least one species with which most people are familiar. For example, in North America, the best-known species is American Robin, and in Europe, Blackbird and Song Thrush. These birds are often seen in urban and suburban areas. Farther afield in woods and forests, there are other, less confiding species that are known for their various calls or evocative songs but rarely come close to human habitats. More distant still, in remote and pristine forests are a number of species about which we know very little.

Abyssinian Thrush (*Z. piaggiae*) is native to montane forests in northeast Africa. While not rare, the species is thinly spread. Ethiopia.

Some of these birds seem to be restricted to very small areas, often on small islands, or are found only in particular parts of tropical rain forests.

These little-known species appear, from what is known about them, to maintain small populations while remaining extremely scarce, and are often widely and very thinly distributed over large areas. A number of scarce or infrequently seen species occur in the tropical forests of Central Africa; the little that is known about them comes from collectors who made expeditions to these areas for scientific or personal collections; from local tribespeople; or from more recent and specific surveys in which researchers are attempting to map out the birds' exact distribution and determine what it is that makes them so enigmatic. Some of these thrushes are among the least known and most elusive of all passerines, as many seem either to have a very limited song period or, in some cases, to give no song at all.

For many of these little-known species, their entire population is unlikely to consist of more than a few thousand pairs, and in some cases it is probably fewer. Prime examples are the thrushes in genus *Zoothera*, one of the oldest members of the family and probably the most primitive, as it shows little evidence of being adaptable to changes in its existing or more recent habitats. Of the 32 species in this genus, about half are extremely rare or little-known, and of these, 15 are island endemics or have a restricted range. Three of the rarest forms—New Britain Thrush (*Z. talaseae talaseae*), San Cristobal Thrush (*Z. t. margaretae*), and Guadalcanal Thrush (*Z. t. turipavae*), subspecies of South Seas Thrush (*Z. talaseae*)—are known from only a handful of sightings. Moluccan Thrush (*Z. dumasi*), Ashy Thrush (*Z. cinerea*), and Amami Thrush (*Z. major*) are considered to be in decline due to the modification and degradation of their habitat.

Two species endemic to Hawaii, Oloma'o (*Myadestes lanaiensis*) and Kama'o (*M. myadestinus*), may already be extinct due to the introduction of mosquitoes carrying diseases to which they had no resistance, as well as to the loss of their habitat.

The underlying causes for the scarcity of these species is unclear but is unlikely to be attributable to a single cause. In most cases natural predation or poor breeding rates may be strongly limiting factor. Alternatively, it could be the birds' specialized habitat or feeding needs that controls their numbers and range, particularly in cases where that niche is similarly limited and fully occupied by these thrushes or by other species. For those species which are more numerous and widely distributed it is assumed that they have a broader choice of habitats in which to live and thrive than those which are ecologically or geographically restricted. ■

CHATS AND OLD WORLD FLYCATCHERS *Muscicapidae*

The huge, diverse family Muscicapidae comprises two groups: the true flycatchers (subfamily Muscicapinae) and the robins and chats (subfamily Saxicolinae). These birds are found across the Old World, with the greatest centers of diversity in Africa and southern Asia, and including several species in Indonesia and one on New Guinea; none, however, reach Australia or New Zealand.

European Robin (*Erithacus rubecula*) is one of the most widespread of the Old World chats, and is a familiar sight across Europe. Netherlands.

Structure Small or medium-sized, slender passerines, the species range from 4 to 11 in. (10–28 cm.) in length. The head may be large or broad, with large eyes in the case of niltavas and forest flycatchers. The bill is short, pointed in most chats but broad-based and hook-tipped in most flycatchers, and slightly longer or stouter in the forest flycatchers. The wings are short and rounded in most of the forest or sedentary species, longer and pointed in migratory species, and longest and almost sickle-shaped in the highly aerial Grandala (*Grandala coelicolor*). Tail shapes vary from fairly short to medium-long and square-tipped or slightly rounded; the tail is longest in scrub-robins, shamas, and forktails, the latter having a graduated and forked tail. The legs are short and fairly weak in some of the flycatchers, but longer and sturdier in ground-loving wheatears and forktails; the last two groups also have strong claws for gripping wet rocks.

Number of genera 48
Number of species 275
Conservation Status 2 species Critically Endangered, 8 species Endangered, 22 species Vulnerable
Distribution Europe, Africa, Asia

Plumage Within the entire family, sexual dichromatism is common; however, in a number of the flycatcher genera, including *Muscicapa*, *Myioparus*, *Bradornis*, and *Rhinomyias*, the sexes are alike. Many of the flycatchers are predominantly dull gray to brown or black with streaked, pale, or white underparts, some having broad white supercilia. In contrast, several East Asian-breeding species are bright yellow and black, or have underparts that are partly orange. In other species, the males vary from pale blue, as in Asian Verditer Flycatcher (*Eumyias thalassinus*), to deep blue in the *Cyanoptila* and *Cyornis* flycatchers, usually with black lores or faces, and some with light to deep orange underparts; females range from dull

Many flycatchers, like this **Spotted Flycatcher** (*Muscicapa striata*), have subtly colored, rather dull plumage. U.K.

gray to warm or light orange-brown. This coloration is also seen in the rock thrushes (genus *Monticola*). It is at its most intense in the *Niltava* flycatchers, in which the upperparts are mostly deep blue, often with vivid cobalt blue on the forehead, crown, sides of the neck, or wing coverts, and deep orange underparts.

The chats show a similar color range of gray to brown, black, or blue upperparts, often with paler underparts or occasional splashes of brighter colors, as in the red markings of European Robin (*Erithacus rubecula*), Siberian Rubythroat (*Luscinia calliope*), and Himalayan Rubythroat (*L. pectoralis*). The African akalats and robin-chats in genera *Sheppardia* and *Cossypha* are predominantly brown or black and orange; several genera have their own distinctive markings, such as the bright red tail in the redstarts (genus *Phoenicurus*); white or red side panels in the tails of the *Luscinia* chats; or a white or sandy rump and tail, usually with a black bar or an inverted 'T,' in the *Cercomela* chats and *Oenanthe* wheatears. The forktails are characteristically black and white, often with a broad white frontal band; Chestnut-naped Forktail (*Enicurus ruficapillus*) is more brightly marked, with deep chestnut on the crown and nape.

Rufous-bellied Niltava (*Niltava sundara*) appears mostly dark when in dense vegetation, but its plumage shows dazzling deep blues and rich orange in the light. Pakistan.

In male **Bluethroats** (*Luscinia svecica*), the coloration varies: the central spot can be either white or chestnut. Finland.

Behavior Chats and Old World flycatchers are usually found alone or in pairs, but several of the tropical forest species often occur as regular members of mixed-species foraging flocks. Most are very active and typically forage from exposed perches. Others, including most of the forest-living chats, and particularly the akalats and shortwings, are skulking and rarely seen. Some of the forest flycatchers spend long periods sitting quietly on a secluded perch. Rock thrushes, similarly, watch and wait on a high perch, before swooping to seize their prey. Forktails forage in the manner of dippers or wagtails, on rocks or along the sides of fast-moving streams and rivers.

Some species have distinctive wing- and tail-flicking movements when alarmed or disturbed. *Luscinia* and *Cercomela* chats, the scrub-robins, and the wheatears occasionally fan their tail, especially when displaying or as a threat. The scrub-robins and the *Cossypha* chats often hold their tail vertically.

Voice The chats include some of the finest songsters in the bird world: several have achieved widespread celebration in poetry and songs for the richness and beauty of their vocal capabilities. In some species, vocalizations also include accurate mimicry of other species. In Europe, Common Nightingale (*Luscinia megarhynchos*) is considered to be the finest of all singers, with a rich series of ventriloquial, liquid, rising and falling, bubbling phrases combined with shorter fluted notes and whip-like whistles and flourishes. In Asia, the shamas have powerful songs combining a series of rich fluting notes and melodious phrases. In

Buff-streaked Chat (*Campicoloides bifasciatus*) is a resident of dry tropical or subtropical parts of Africa. South Africa.

A **Common Nightingale** in full voice. This species often sings by day as well as throughout the night. Hungary.

Africa, the scrub-robins and robin-chats are the finest singers. The songs of several other species are equally notable but lack the versatility and repertoire of their relatives.

The flycatchers, by comparison, are poor singers with less well developed or complex songs, most consisting of a series of high-pitched whistles interspersed with scratchy notes. Blue-and-white Flycatcher (*Cyanoptila cyanomelaena*) is noted as one of the best, with a series of rich, fluting notes ending with a distinctive *si-si-si*, given at short intervals in the early dawn and at dusk.

Habitat Most of the family occupies a wide range of habitats, including forest, clearings, light woodland, scrubby areas, grasslands, and desert margins. The monotypic Grandala lives on high Himalayan mountain peaks and in ravines, and the forktails are birds of streams and rivers.

Certain species, such as European Robin, are very common, but some island species have small or even endangered populations. Of the two species that are critically threatened, one, Ruck's Blue Flycatcher (*Cyornis ruecki*) of northern Sumatra, is probably extinct, and no suitable habitat remains for it. Another species, Seychelles Magpie-Robin (*Copsychus sechellarum*), has suffered from introduced cats and rats and human persecution. In the mid-1960s, the total population was about 15 individuals, but it has since, through translocation and predator eradication programs, increased to more than 140 on three islands.

Movements Most tropical species are sedentary or make only short movements, including several that breed in the Himalayas and move to lower elevations in winter. In southern Asia, Brown-breasted Flycatcher (*Muscicapa muttui*) is notable for moving west to winter in southern India and Sri Lanka, and in Africa a proportion of the Fairy Flycatchers (*Stenostira scita*) that breed in South Africa move north into Namibia and Botswana.

Species breeding at higher latitudes migrate for longer distances. Of those breeding in Siberia and far eastern Russia, several move long distances south to winter in the Philippines, Malaysia, Indonesia, and northern New Guinea. In Europe and western Russia, a small number of species move south or southwest to winter in sub-Saharan Africa. The Rufous-tailed Rock Thrushes (*Monticola saxatilis*) that breed in eastern Asia also migrate, via the Middle East, to winter in sub-Saharan Africa: a journey, for some birds, of more than 5,000 miles (8,000 km.). The longest-distance movements are made by Spotted Flycatcher (*Muscicapa striata*), which breeds as far north as northern Scandinavia and Russia and east to Lake Baikal, and winters in

Blue-and-white Flycatcher is admired for its melodious song as well as for its dazzling plumage. Japan.

A **Desert Wheatear** (*Oenanthe deserti*) gives a threat display, crouching with its tail fanned, probably to protect its territory or young. Location unknown.

A breeding pair of **Common Redstarts** (*Phoenicurus phoenicurus*) with food; the female is on the left and the male on the right. Both parents feed the young. Wales, U.K.

southern Africa. Perhaps the most remarkable migrations are made by Northern Wheatear (*Oenanthe oenanthe*). Some of the birds breed in Alaska and move westward through Russia, then turn southwest to cross into Africa; these meet with birds that have crossed the Atlantic (often direct) from Greenland to western Europe and then moved south across the Mediterranean.

Diet Primarily insects. In the flycatchers, various flies make up the majority of the diet, including large groups such as caddis flies, stone flies, and dragonflies, as well as beetles, bees, wasps, crickets, grasshoppers, moths, and spiders. The chats, and some flycatchers, take woodlice, millipedes, snails, and earthworms. Some of the larger species also take cockroaches, tree-frogs, and small lizards; Large Niltava (*Niltava grandis*) may take small snakes. The forktails eat mainly aquatic insects, crustaceans, and larvae. The birds may also take seasonally available fruit: mostly various berries, including those of Black Nightshade (*Solanum nigrum*), which the bird may pick from the shrub or take while hovering over the plant. Fiscal Flycatchers (*Malaenornis silens*) are known to take nectar from aloe plants.

Breeding Most species are monogamous, but some are polygamous and several have cooperative helpers at the nest. All species establish and defend territories for courtship, nesting, and care of nestlings. The nest is a woven dome, cup, or loose platform, made of grasses, plant fibers, twigs, leaves, fern fronds, moss, and lichens. Many species place their nest in a secluded site on a branch, in a fork, or in a hole in a tree. Several African species use abandoned weavers' nests, while in Europe, Pied Flycatchers (*Ficedula hypoleuca*) regularly take to nestboxes specifically erected for them. Many of the chats nest on the ground under tussocks or stones, or occasionally in walls, while rock thrushes use cavities in rocks. Some wheatears in semi-desert areas utilize and often share the holes of pikas and sousliks.

Clutches comprise two to six eggs (generally fewer in southern-breeding species). In most of the flycatchers, both parents incubate the eggs; however, in several of the *Muscicapa* and *Ficedula* species, together with most of the chats, the female alone incubates, for up to 15 days. The young fledge at 10–18 days.

Taxonomy Until recently, the two subfamilies were considered separate, the flycatchers within an enlarged grouping of Muscicapidae and the chats grouped with the thrushes in the Turdidae. There are good reasons to maintain these former arrangements, as many of the robins and chats merge in structure and behavior through the larger chats to the rock-thrushes and true thrushes. However, studies of genetic and anatomical features have shown that the flycatchers have closer affinities with the redstarts and robins than with any other group. This classification has not received unanimous support, and the debate over the true position of both still continues, supported in part by similarities in the groups' behavior.

The New World flycatchers in family Tyrannidae, although similar in structure and behavior, are not closely related, and nor are the fantails (Rhipiduridae) or the monarchs (Monarchidae).

A female **Pied Flycatcher** at the nest. These birds naturally nest in tree trunks, but will also occupy artificial nest boxes. U.K.

DIPPERS *Cinclidae*

A remarkable family, the dippers are the only aquatic passerines adapted to walk and swim underwater.

Structure The birds are 6–8 in. (15–20 cm.) in length; males are larger than females. Dumpy and wrenlike in shape, they have a short neck, short, pointed wings, and a short tail. They also have a straight, slender bill, and medium-long legs with strong feet and claws. The eye has a protective whitish membrane, which covers the eyeball when the bird is underwater. There is a large oil-gland at the base of the tail, which the bird uses in preening sessions to keep its feathers waterproof.

Plumage The feathers are very soft and waterproof, with thick down. The plumage may be entirely dark brown or chocolate, as in Brown Dipper (*Cinclus pallasii*), or gray-brown, as in American Dipper (*C. mexicanus*). Otherwise, the birds may be mostly dark, with white from the chin to the breast, as in White-throated Dipper (*C. cinclus*); have a white crown, nape, chin, and throat, as in White-capped Dipper (*C. leucocephalus*); or have a rufous throat and upper breast, as in Rufous-throated Dipper (*C. schulzi*). The eyes are dark but the eyelids are covered with fine white feathers.

An **American Dipper** by an icy stream. Dippers inhabit some of the coldest and most torrential rivers and streams year-round, even when the water is partly frozen. Alaska, U.S.

American Dipper feeding. Dippers partially or entirely submerge themselves to pursue prey. Alaska, U.S.

Behavior Dippers are usually seen in or by streams. To forage for food, they watch while occasionally bobbing up and down, flicking the tail, and blinking the white eyelid before wading in after their prey. Three of the species dive up to 3ft. (90 cm.) in often torrential water; they stay in position underwater by rapidly beating the wings or running.

Voice Calls are a high-pitched *zink* or *zeet* and penetrate above the noise of the water; the birds often call several times when alarmed. The songs are also high-pitched, consisting of a musical, bubbling, rising and falling series of phrases.

Habitat Mostly areas by fast-flowing hill or montane streams, and rivers with rocks, eddies, and shoals; occasionally, edges of lakes and seashore.

Movements Almost entirely sedentary. Some dippers breeding in Scandinavia and northern Russia winter farther south, in central Europe.

Diet Mainly aquatic invertebrates and larvae, including stonefly, dragonfly, and caddis-fly larvae. Dippers may also take mayflies, beetles, tadpoles, small fish, and fish eggs. They beat large insects and fish against rocks before swallowing them.

Breeding Some pairs maintain and defend a territory year-round. The domed nest of leaves, plant fibers, and grass is located near running water, often in a hole or bank, or among rocks. The clutch of two to six eggs is incubated by the female, for 14–20 days. The young leave the nest after a further 20–24 days.

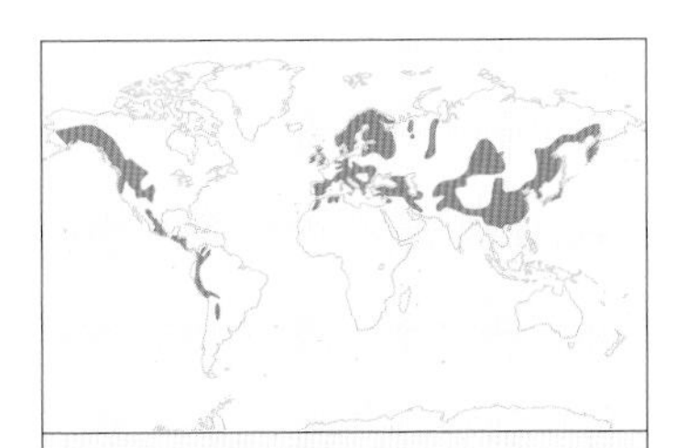

Number of genera 1
Number of species 5
Conservation Status 1 species Vulnerable
Distribution North, Central, and South America; Palearctic; Asia

LEAFBIRDS *Chloropseidae*

A small family of Asian passerines. Formerly considered as part of a larger grouping that included the ioras (family Aegithinidae) and fairy bluebirds (Irenidae), the leafbirds are now regarded as sufficiently distinct to be treated separately. The species are very similar in size, shape, and plumage, differing mainly in the patterns on the face and underparts.

Structure The birds are 6–10 in. (15.5–25 cm.) in length. They have a slender, tapering, slightly decurved bill; a brush-tipped tongue; a slim body with short, rounded or slightly pointed wings; a square-tipped tail; and short, robust legs and feet.

An immature **Blue-winged Leafbird** (*Chloropsis cochinchinensis*). This species is widespread across Southeast Asia, occurring from India and Sri Lanka to Borneo. Thailand.

Plumage Leafbirds are mainly bright green (hence their name); this coloration provides excellent camouflage among foliage. Males have colorful patterns from face to breast, which vary in extent and shape between species. Some leafbirds have turquoise blue in the wings and tail, and/or orange on the head and underparts. Several share a broad but indistinct, metallic blue tear-drop malar line. Females are generally plain green and lack the prominent features, although some species have a bluish malar line.

Behavior All species are arboreal, spending all their time high in the middle to canopy levels of forest trees. They may be found alone, in pairs, or frequently in mixed-species foraging flocks. The birds forage by actively and thoroughly searching foliage and flowers, gleaning insects from leaves or branches or taking food as they hover briefly around the outermost foliage.

This male **Golden-fronted Leafbird** (*Chloropsis aurifrons*) shows the typical teardrop-shaped malar streak. Location unknown.

Voice Most species have rich rising and falling, melodious songs, which include whistles, chuckles, and chattering notes, together with perfect mimicry of other birds' calls and songs. Calls include short, liquid notes often repeated monotonously.

Habitat Most leafbirds are common and widespread in evergreen and monsoon forests and dry scrub woodland within their range. Philippine Leafbird (*Chloropsis flavipennis*), is considered Vulnerable. This bird is more secretive than other species and has always been rare. It has disappeared from several islands within its former range, and is now known from only two localities.

Movements Sedentary.

Diet Insects, as well as seeds, berries, and small fruit, including figs; also nectar and the pulp of larger fruit such as ripe guavas.

Breeding Nests, built largely by the females, are neat, shallow platforms of plant fibers, grass, and lichens, bound together with large amounts of cobwebs. They are suspended from a fork in a branch or fine leaf stems, usually hidden high up in the foliage of tall trees. The clutch of two or three eggs is incubated by the female, but both parents feed and care for the nestlings.

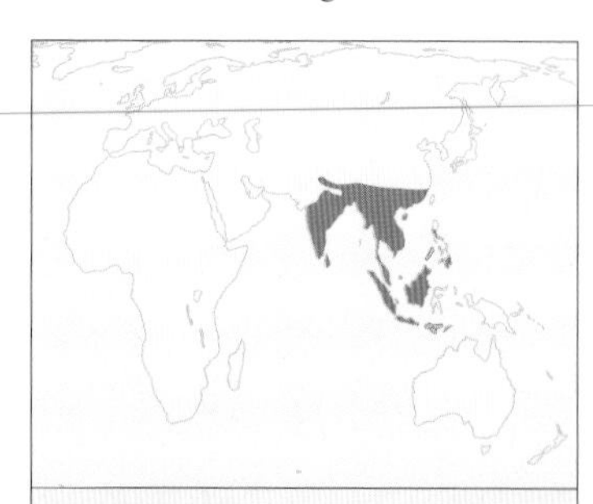

Number of genera 1
Number of species 8
Conservation Status 1 species Vulnerable
Distribution South and Southeast Asia

FLOWERPECKERS *Dicaeidae*

Flowerpeckers, like this **Crimson-breasted Flowerpecker** (*Prionochilus percussus*), are named for their typical feeding behavior, in which they peck at the flowers, fruits, or seeds of trees and bushes. Malaysia.

An Old World family concentrated in Southeast Asia, flowerpeckers are small, rather dumpy birds frequently associating with flowers, fruiting trees, or berries such as mistletoe. All species are very similar in appearance; they also closely resemble the sunbirds (family Nectariniidae), but are smaller, and have a shorter and less curved bill. The sexes are dichromatic in most species, with the males often being very brightly colored.

Structure The birds are small, measuring 3–4.5 in. (7.5–11 cm.) in length, and with a weight range of 0.2–0.7 oz. (5–20 g). Most species have a short, slightly decurved, pointed bill, but in a few species the bill is very short and stubby, with finely serrated edges. Flowerpeckers have a tongue with a double-grooved tip, specially adapted for eating pollen and nectar. The gizzard and stomach are similarly well adapted for separating the consumption and digestion of insects and fruit. The body is stout and the wings are short, with 10 primaries and rounded tips. The two genera, *Prionochilus* and *Dicaeum*, differ in wing structure, with the outermost primary being well developed in *Prionochilus* species but vestigial in most of the *Dicaeum* species. The tail is short and stumpy. The legs and feet are fairly short and slender, used for clinging to the flowers and stems of plants rather than climbing through vegetation.

Plumage Coloration varies between species, with some having brightly colored males and drab females, while in others both sexes are generally all dull or dingy.

Brightly colored males have glossy bluish-black to grayish-blue upperparts with black wings and tail. Several species are variously marked broad bright crimson from forehead to rump, with a black face. One has a scarlet head. Another has a black head; yellowish body, wings, and tail; and a bright red breast enclosed by a broad black band. Many species have a bright red to pinkish throat and breast, or white or yellow underparts. The males of Fire-breasted Flowerpecker (*D. ignipectum*), Black-fronted Flowerpecker (*D. igniferum*), and Red-chested Flowerpecker (*D. maugei*) have a

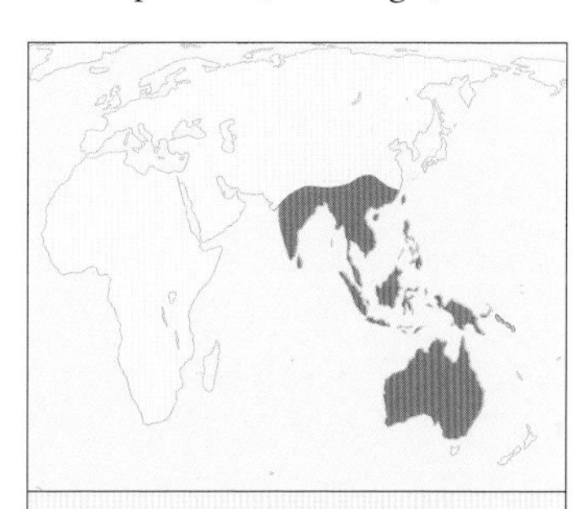

Number of genera 2
Number of species 44
Conservation Status 1 species Critically Endangered, 2 species Vulnerable, 5 species Near Threatened
Distribution Southeast Asia to Australia

pale breast with a bright red patch at the center. Most females and drab species have dull green upperparts with pale yellow or off-white underparts, but one or two species have brighter yellow undertail-coverts and are streaked broadly darker.

Behavior The birds are usually found alone or in pairs, although several individuals may congregate at a fruiting tree. Most species forage in the foliage of flowering and fruiting trees and bushes. Many are active foragers, and visit a number of trees in rapid succession, while others spend long periods in only a few trees before moving on. Flowerpeckers are highly instrumental in pollinating blossoms and dispersing seeds from fruit. Seeds pass rapidly through their digestion system; within a short time of being eaten they are excreted whole, often several together, in a glutinous thread, often within the area where the fruit was taken. In Australia, Mistletoebird (*D. hirundinaceum*) specializes on mistletoe berries, which the birds eat whole; excreted seeds stick to the tree in which the bird is feeding, and germinates. It is notable that this species is widespread over most of Australia, wherever mistletoes are found, and is absent from Tasmania, where mistletoes also do not occur.

Voice Most species are recognizable by their short, sharp, often high-pitched single contact call notes, which are given either while perched or in flight. The songs are similarly short series of call notes, or are mingled with a twittered series of rapid rising and falling phrases.

Habitat Flowerpeckers occupy a range of evergreen and deciduous forests, woodlands, and thickets. They occur in southern Asia (north to central China and Taiwan), Indonesia, the Philippines, New Guinea, the Solomon Islands, and Australia. Most species are fairly common or widespread. In contrast, Cebu Flowerpecker (*D. quadricolor*), from the island of the same name, is listed as Critically Endangered. It was thought to have become extinct in the 20th century, but in the 1990s it was rediscovered at three separate sites; however, it remains severely threatened due to illegal logging, deforestation, and hunting. Two further species are considered Vulnerable, also largely due to increased habitat destruction, and five others are classed as Near Threatened.

A male **Orange-bellied Flowerpecker** (*D. trigonostigma*). This species occurs in clearings and margins of evergreen tropical forest, and in mangroves. Singapore.

A male **Mistletoebird** feeding. The birds are well known for eating mistletoe berries; they excrete the seeds less than 30 minutes after eating them. Northern Territory, Australia.

Movements Most species are entirely sedentary, but some species make seasonal altitudinal movements. Mistletoebirds are nomadic outside the breeding season.

Diet Nectar and pollen taken directly from flowers, as well as berries, small fruits, and fruit pulp. Several species also take small insects.

Breeding Monogamous and solitary, but most pairs remain together within small family groups outside the breeding season. The nests, built mostly by the females, are elaborate bag- or purse-like structures made of plant fibers, down, and large amounts of cobwebs, decorated with cocoons, seeds, insect pellets, and dried excreta, and hung from a fine branch or leafy twig. The clutch of one to three eggs is incubated for about 12 days, and the nestlings fledge after a further 15 days.

Taxonomy The flowerpeckers are closely related to the sunbirds. The family was previously thought to include the berrypeckers (family Melanocharitidae) of New Guinea and the pardalotes (Pardalotidae) of Australia, but both are now considered sufficiently distinct to warrant separation into their own families.

SUNBIRDS *Nectariniidae*

The sunbirds are small, often brightly colored birds that feed on nectar and insects from flowers. Found in the tropics and subtropical areas of the Old World, they are the equivalents of the hummingbirds in the Americas, but are entirely unrelated, having evolved separately. The sunbirds' closest relatives are probably the flowerpeckers (family Dicaeidae) of Southeast Asia and Australasia. The greatest diversity, with over half of the species in the family, is found in Africa, with a secondary concentration of species between Malaysia and Australia. The family also includes the spiderhunters of southeast Asia, which, despite their name, feed mainly on nectar.

A male **Collared Sunbird** (*Hedydipna collaris*). The iridescent coloring is due to the scattering of light by tiny structures within the feathers. Kenya.

Structure Sunbirds range from 4 to 12 in. (10–30 cm.) in length. Bill shapes vary, but are usually short or medium-length, slender, decurved, and pointed or serrated at the tip; spiderhunters' bills are longer and more heavily decurved. In some species, the bill is adapted for specialization on certain flowers. The tongue is long, tubular, and divided at the tip. Wings vary from short to rounded or pointed, and all have 10 primaries. Tail shapes vary according to sex: the males of several species have long, pointed, needle-like, or graduated tails, and one species has a forked tip to the tail, but the females have a shorter, square-tipped tail. The legs are slim, and the feet short with fine, sharp claws.

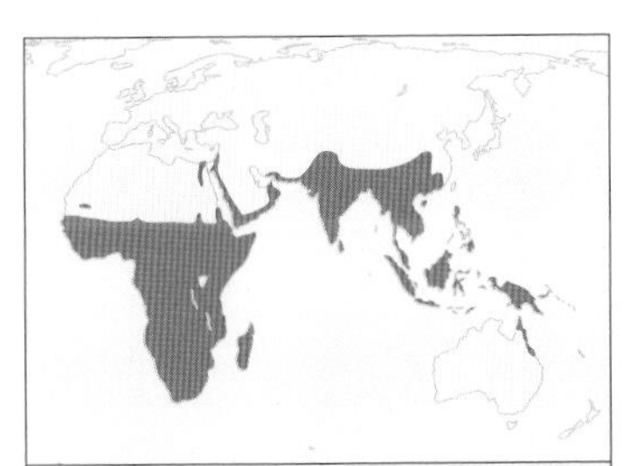

Number of genera 16
Number of species 127
Conservation Status 2 species Endangered, 4 species Vulnerable, 8 species Near Threatened
Distribution Africa, Middle East, southern Asia, Indonesia, Philippines, and Australia

Plumage Sunbirds are highly dichromatic. Males show a great variety of vivid and often iridescent colors including deep purple, brilliant reds, orange, yellow, and green, sometimes with variable intergrading shades; females are generally duller green to olive or gray, some species with brighter patches.

The brilliant colors of male sunbirds are usually seen only in sunlight: a seemingly dark bird will suddenly show bright blues, purple, or red as the angle of the light changes. One example is Scarlet-chested Sunbird (*Chalcomitra senegalensis*) which has a bright scarlet breast and is otherwise mostly deep velvet-black, but when it turns its head, deep, iridescent bottle green appears on the crown and sides of the chin. The larger or longer-tailed *Nectarinia* species, such as the brilliant green Yellow-tufted Malachite Sunbird (*N. famosa*), the glossy brownish Bronzy Sunbird (*N. kilimensis*), and the green and purple Tacazze Sunbird (*N. tacazze*), are especially spectacular. Outside the breeding season, males of several species molt into a duller plumage similar to that of the females.

In the spiderhunters, the sexes are similar; the birds are characterized by their dull leaf-green or brown plumage, often streaked darker and with a paler face and underparts; some, however, may have a bright or prominent eye ring.

Behavior Both sunbirds and spiderhunters spend long periods of each day defending territories that include flowering trees or shrubs, or searching for plants about to come into blossom; they have a regular circuit of flowers that they visit. Some sunbird species are extremely aggressive and defend their territories against all intruders, including other, and frequently larger, species. When feeding, sunbirds perch either on a flowerhead or adjacent to it, and reach in with the bill to sip nectar from the base of the flower. To extract nectar from longer-petaled flowers, the bird may pierce the flower near the base from the outside. When not foraging for nectar, sunbirds and spiderhunters feed on insects, which may be gleaned in and around flowers, pursued and taken in flight, or frequently taken from spiders' webs. Flight is often dashing and direct, but the birds occasionally hover at the outer edge of foliage as they feed.

Voice Most species have a poorly developed repertoire of calls, limited to thin, sharp, or squeaky metallic notes. These are used in contact between pairs, to signal alarm, or as increasingly excitable, higher-pitched, and persistent calls to drive intruders away from feeding or nesting areas. The songs of most species are high-pitched, sibilant, warbling series of rising and falling notes; one species, Amethyst Sunbird (*Chalcomitra amethystina*) from southern Africa, has a more varied, rapid, bubbling song interspersed with twitters, chirps, and musical phrases.

Habitat The family occupies a wide range of habitats, from dense rain forests to mangroves, dry *Acacia* scrub, and the edges of cultivated land: wherever flowering trees and shrubs are found.

Movements Largely sedentary. Some birds, however, travel considerable areas in search of food: certain African species are known to cover 70–75 miles (130–140 km.) a day to reach feeding areas. Several African species that breed at high elevations move to lower levels in winter.

Diet Mostly nectar, and small insects, mainly flies and larvae. Sunbirds also take the fleshy pulp of fruits and berries. The large African species, including the two malachite sunbirds, may take small lizards.

Breeding All species are monogamous. Sunbirds build a long, oval or purse-shaped nest of leaves, plant fibers, grasses, and feathers, bound together with spiders' webs and with a side entrance. The nest is slung from a branch in a tree or bush, usually less than 12 ft. (3.7 m.) from the ground. Spiderhunters attach their round or cup-shaped nest to the underside of large, broad leaves, by sewing long plant fibers or cobwebs through the leaf with their sharply pointed bill. One species, Long-billed Spiderhunter (*Arachnothera robusta*), has a long entrance tunnel to the nest; this is similarly bound to the underside of the leaf.

The normal clutch in all species is two or three eggs, which are incubated by the female for 14–17 days. The chicks' bills are straight on hatching, and only begin to curve as the bill lengthens, from 9–10 days. The young leave the nest at about 17 days old.

Socotra Sunbird (*Chalcomitra balfouri*) is endemic to Socotra Island, living in the wadis and wooded, shrubby hillsides. The sexes are alike. Socotra Island, Yemen.

A male **Green-tailed Sunbird** (*Aethopyga nipalensis*) inserts its bill into a flower. As in all sunbirds, the bill and tongue are well adapted for feeding on nectar. Thailand.

SUGARBIRDS *Promeropidae*

The sugarbirds of southern Africa are specialized to feed on *Protea* shrubs and their nectar. It seems likely that they evolved with these flowering shrubs, on which they depend for food, nest material, nest sites, and shelter. The relationship between bird and plant is partly symbiotic, as sugarbirds are the main pollinators of protea.

Structure Sugarbirds are 14–17 in. (37–44 cm.) in length, two-thirds of which is the long, ribbonlike tail. They have a longish, decurved, and sharply pointed bill, short wings, and slender legs. The sexes are alike, but females have shorter tails.

Plumage Both species have olive-brown upperparts, streaked darker; streaked flanks; a whitish belly; dark brown or blackish wings and tail; and a bright yellow undertail. Cape Sugarbird (*Promerops cafer*) is yellowish from forehead to crown, dark from lores to eyes, with a buffish submoustachial stripe and white chin and throat divided by a darker malar stripe, and an olive-green rump. Gurney's Sugarbird (*P. gurneyi*) has a rufous-brown crown, less well marked face pattern, a rufous breast, and a yellowish rump.

Behavior Sugarbirds often occur in pairs, or are loosely colonial when breeding. They are often seen sitting on top of bushes such as *Protea*. They forage by perching on a flower and inserting the bill into the flowerhead, fluttering their wings for balance; they may disappear into the flower so that only their tail shows.

Voice The songs comprise jumbled harsh, liquid, and scratchy notes with occasional deeper notes or phrases, or a rambling medley of twangy and squeaky notes. Gurney's Sugarbird has a slightly faster, higher-pitched song with fewer harsh notes.

A **Cape Sugarbird** perches on a protea. The birds are particularly reliant on proteas in the breeding season: the plants provide nectar for adults and insects for the young. South Africa.

Habitat Grasslands and scrubby hillsides with *Protea*; also riverine bush, parks, and large gardens.

Movements Sedentary, but makes short-distance movements outside the breeding season to find food.

Diet In the breeding season, sugarbirds principally take nectar from a wide variety of proteas, and occasionally take small beetles, flies, moths, and spiders, many of which are fed to nestlings. In winter, they feed on flowering agaves, aloes, spider-gum trees, and other eucalypts, and catch flying insects.

Breeding Territorial when breeding, males perform elaborate, looping courtship display flights. The nest is a deep cup of heather, plant fibers, twigs, grasses, and seed down, usually in a small, dense *Protea* bush. The clutch of two or three eggs is incubated by the female for 16–18 days. The young remain in the nest for a further three weeks.

Taxonomy Sugarbirds closely resemble the sunbirds (family Nectariniidae) in structure, but have also been classed with starlings (Sturnidae), thrushes (Turdidae), and Australian honeyeaters (Meliphagidae).

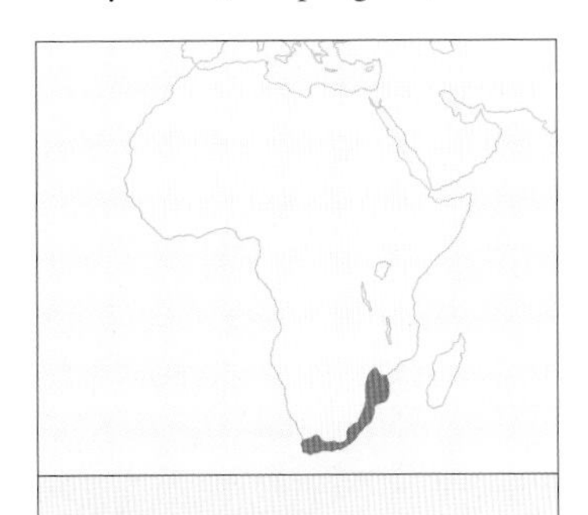

Number of genera 1
Number of species 2
Conservation Status Least Concern
Distribution Southern Africa

OLD WORLD SPARROWS *Passeridae*

Sparrows are one of the best-known of all bird families: several species associate closely with humans, and nest in or around buildings in villages, towns, and cities. Snowfinches are their montane relatives, only rarely venturing below the snowline, and are specially adapted to living at high altitudes and low temperatures.

The family originated in the Old World, but many species have now spread worldwide. House Sparrows (*Passer domesticus*) probably have the widest distribution of any land bird. This species, and Eurasian Tree Sparrow (*P. montanus*), have been introduced into the Americas, Australia, and islands in the Atlantic, Pacific, and Indian Oceans. None of the family are considered to be at risk, but the most common species have undergone several population declines in parts of their range, the causes of which are not clear.

A **Eurasian Tree Sparrow**. This species can be distinguished from male House Sparrows by the crown, which is chestnut instead of gray. Poland.

Structure These small birds range from 4 to 8 in. (10–20 cm.) in length. They have a rounded head with a short, pointed, or conical bill and a short neck. The body is stocky, slightly larger or plumper in the snowfinches. The wings are mostly blunt-tipped in sparrows and more sharply pointed in snowfinches, and have 10 primaries. The tail is medium-length (or slightly shorter in the petronias), square-ended or slightly notched, and has 12 feathers. The legs and feet are fairly short and stout.

White-rumped Snowfinch (*Onychostruthus taczanowskii*) is the palest of the snowfinches; the white rump is clearly seen when the bird is in flight. China.

Plumage Sexual dichromatism is common; only in the petronias, social weavers, gray-headed sparrows, and Tree Sparrow are the sexes alike. Colors range from brown to rufous or paler buffish, gray, and yellowish.

Sparrows are usually darker brown, with distinctive head and face patterns. Many have a pale face and prominent black chin and throat patches, while the rock sparrows and petronias have a pale or yellow throat patch. Female sparrows of all species are very similar, generally duller brown with darker streaks; the species differ slightly in size and markings, with some having prominent pale supercilia and broad, pale stripes down the back. Three of the sparrows are predominantly bright or golden yellow or deep chestnut, and have no head pattern. The closely related group of African gray-headed sparrows have entirely gray heads and bodies, brown upperparts and tails, and white patches in the wings. The petronias are more plainly colored than the sparrows, except for the more heavily striped Rock Sparrow (*Petronia petronia*). The sparrow weavers are larger versions of the

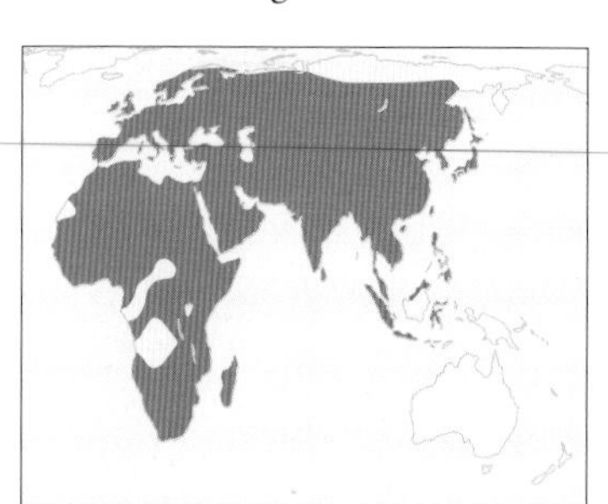

Number of genera 11
Number of species 40
Conservation Status Least Concern
Distribution Europe, Africa, and Asia to Indonesia; also introduced to rest of world

Males of the **Sudan Golden Sparrow** (*Passer luteus*) are the brightest of the sparrows; the gold color of their head becomes even more intense in the breeding season. Sudan.

sparrows, and often have prominent patterns on the head.

The snowfinches are generally slightly paler versions of the sparrows, most having black on the face or chins. Rufous-necked Snowfinch (*Pyrgilauda ruficollis*) is more brightly colored cinnamon-rufous on the nape and the sides of the breast; two other species have black and white wings and tails, and have plumage patterns very similar to that shown by male Snow Buntings (*Plectrophenax nivalis*, family Emberizidae).

Behavior Most sparrows, sparrow weavers, and snowfinches are highly social, usually seen in family groups or small flocks. The majority of species are terrestrial, especially the snowfinches. They mainly forage in a variety of low-growing vegetation, although sparrows often feed in trees and bushes. Petronias are more arboreal and, apart from the two species of rock sparrows, rarely forage on the ground. All species fly mostly in short, swift, undulating bounds, seldom for long distances.

The desert-living species, such as Desert Sparrow (*Passer simplex*), Golden Sparrow (*P. luteus*), and Saxaul Sparrow (*P. ammodendri*), can withstand midday temperatures of around 104°F (40°C), although they rarely sit in the open for long.

Some House and Eurasian Tree Sparrows spend their entire lives around human habitation, often in the middle of densely populated cities, and rarely encounter natural habitats. In Europe, House Sparrows are one of the main species that benefit from having food put out on garden birdtables. White-winged Snowfinches (*Montifringilla nivalis*), in parts of their range, have learned to forage around skiing areas, where they readily take scraps left by skiers.

Voice Calls and songs are mostly a series of loud chirps and twitters, given with varying emphasis and tone, and occasionally interspersed with higher-pitched notes and harsh churrs. Alarm calls are usually high-pitched chirps or squealing notes.

Habitat Overall, the family occupies a wide range of habitats, including lowland desert edges and oases, dry scrublands, villages and city suburbs, woodlands, edges of forests, agricultural areas, montane pastures, and open alpine plateaux.

Movements Almost all sparrows are sedentary; some of those breeding in towns and cities rarely move more than a mile (1.5 km.) from the place where they were hatched. In contrast, populations of Spanish Sparrow (*Passer hispaniolensis*) that breed north of the Mediterranean eastward into Central Asia move nomadically, or migrate southward to winter along the edge of the Sahara, down the Nile Valley, and into Arabia and northwest India. Similarly, Pale Rock Sparrow (*Carpospiza brachydactyla*), which breeds in the southern Caucasus and throughout Iran, winters in Arabia and central

Rock Sparrows naturally live and breed in rocky areas, but this bird has found a niche for itself in the stones of a building. Armenia.

A male **Spanish Sparrow** (*Passer hispaniolensis*) feeds on the seedhead of a low-growing plant. Lesvos, Greece.

Egypt. Most snowfinches are altitudinal migrants, moving to lower levels to escape severe weather, although many remain at high altitudes or make only local movements even in conditions of deep snow.

Diet All species are mainly seedeaters, either feeding on low seedheads or taking fallen or windblown seeds. Sparrows, sparrow weavers, and petronias also forage in rice and cereal fields, occasionally perching on the ears of standing crops, and may eat the seeds, buds, and shoots of trees, including fruiting bushes and trees. Desert Sparrows take windblown seeds, together with millet, and are especially active in feeding on the plant-growth that rapidly springs up and blooms after rains, having lain dormant up to that time. Snowfinches forage for seeds in alpine areas, including at the edges of glaciers and melting snow patches. Most species also take some insects, especially aphids, flies, and ants.

Breeding Some sparrows occur alone or in pairs, but several species, and some of the petronias, nest in loose colonies of up to 100 pairs. Usually, each pair builds their own nest, but Black-capped Social Weaver (*Pseudonigrita cabanisi*) occasionally builds a compound nest containing up to five individual nests. Most of the sparrows have noisy, active territorial displays involving chases and aggressive fights between rival males; some of these encounters have been known to end in the death of one of the contestants.

Nests are mostly ball-shaped and made of grasses, plant fibers, and some moss, occasionally with animal hair and pieces of string or paper as well. They may be placed at the end of a tree branch, in a crevice in a cliff, in a hole in a wall or building, or occasionally in the nest of another species that they may have evicted. House Sparrows have been known to nest 2,000 ft. (600 m.) down a coal-mine shaft in England. Snowfinches also nest in rocky crevices, or may utilize the narrow, deep holes made by the hamster-like pikas (genus *Ochontona*), and often share the burrows with the animals.

Clutches comprise three to six eggs; in some species, the eggs are incubated by the female alone, while in others, they are incubated by both parents. The young fledge at around 15 days.

White-browed Sparrow Weaver (*Plocepasser mahali*) is a cooperative breeder, with other family members helping a breeding pair rear their young. Kenya.

Taxonomy The family is most closely related to the weavers (Ploceidae), through similarities in feeding behavior, courtship displays, and nest structures. This relationship is best seen in the sparrow weavers and social weavers, and suggests that they originated in Africa but are sufficiently distinct to be recognized as a closely allied family. The Passeridae are not closely related to the New World sparrows, which are part of the Emberizidae family.

A female **White-winged Snowfinch** carries nesting material into her nest burrow. This species typically shares the burrows of Plateau Pika (*Ochotona curzoniae*). Turkey.

WEAVERS *Ploceidae*

A large, diverse family of Old World passerines, weavers are named for their extraordinary nest-building abilities. Some males weave intricate nests that are among the most striking and enduring constructions made by any animal; others build large communal structures requiring a high degree of social organization and cooperation.

There are three major groups: the buffalo weavers (genera *Bubalornis* and *Dinemellia*); bishops and widowbirds (genus *Euplectes*); and true weavers (*Sporopipes*, *Amblyospiza*, *Ploceus*, *Malimbus*, *Anaplectes*, *Quelea*, *Foudia*, and *Brachycope*). Most species are similar in structure, differing mainly in size, bill shape, and plumage. In some species, the sexes may differ in plumage and size.

The majority of species are distributed across sub-Saharan Africa, São Tomé, Madagascar, and several of the islands in the western Indian Ocean. One species, Rüppell's Weaver (*Ploceus galbula*), occurs in Arabia, and five others between southern and southeast Asia to southeast China and Indonesia.

A pair of **African Masked Weavers** (*Ploceus velatus*). Although male weavers build the nests, females contribute by lining the nest chamber with soft materials. South Africa.

Structure Species vary in length from 4 to 10 in. (11–26 cm.), although *Euplectes* species that have longer tails when in breeding plumage can reach up to 26 in. (65 cm.). They have a smallish head and short neck. The bill is usually short and straight, but varies between species from slender and sharply pointed to thick and blunt-tipped, or massive and conical, as in Grosbeak-Weaver (*Amblyospiza albifrons*). The body is slender or slightly plump. The wings are fairly long and rounded, with 10 primaries. The tail is usually short and square or partly rounded at the tip. The legs are fairly short but strong, with powerful toes.

Plumage In dichromatic species, males have different breeding and nonbreeding plumages, the latter very similar or identical to that of the females. Plumage also varies greatly between subfamilies and genera.

In the buffalo weavers, the *Bubalornis* species are dark brown or black. White-headed Buffalo Weaver (*Dinemellia dinemelli*) is mostly white, with brown back, wings, and tail, and rufous rump and undertail.

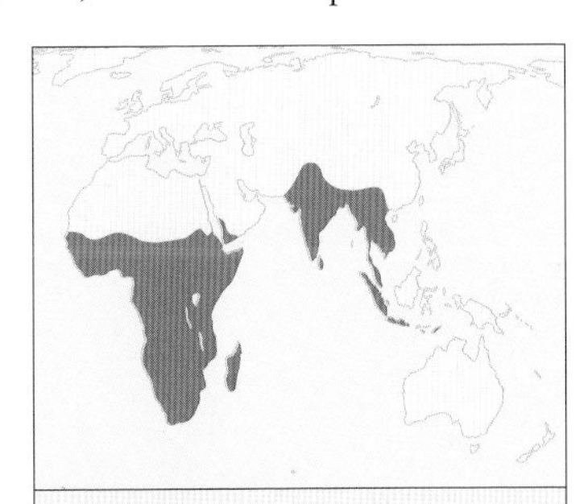

Number of genera 11
Number of species 108
Conservation Status 1 species Critically Endangered, 6 species Endangered, 7 species Vulnerable
Distribution Sub-Saharan Africa, south and Southeast Asia

White-headed Buffalo Weaver, with the characteristic bright rufous area visible on the rump. Kenya.

Among the true weavers in genus *Ploceus*, the sexes are alike in some species, while in others, males have a distinct breeding plumage. This may be bright or golden yellow with brown or black on the head, face, chin to throat, or entire upperparts. In some cases it is entirely black; in others, black with a yellow head; and in certain species, the upperparts are greenish. Females and non-breeding males are usually dull olive-yellow to greenish and streaked darker; it is often hard to determine the species. Several species have white rather than yellow on the underparts and back, while others are predominantly chestnut or cinnamon with a black head and yellow undertail. Malimbes (genus *Malimbus*) are predominantly black, with crimson or yellow on the head and breast or across the entire underparts. Red-headed Weaver (*Anaplectes melanotis*) males are mostly red on the head and breast, white on the belly and undertail, and olive-brown upperparts; females resemble female *Ploceus* weavers but with whiter underparts. Grosbeak-Weaver is mainly dark brown with a white patch on the forehead and a smaller white patch in the wing; females are slightly paler and more streaked below.

A male **Southern Red Bishop** (*Euplectes orix*) performs a display by fluffing out his loose, shaggy feathers. South Africa.

In the three species of queleas, breeding males have bright red on the head or face in two species, and a black face, with a golden-buff or pinkish surround from the crown to the breast, on Red-billed Quelea (*Quelea quelea*). In the six species of fody, males in breeding plumage have red on the head and breast or on the rump, or are almost entirely red. The upperparts are usually heavily streaked, as are those of non-breeding birds and females. The species also differ in the color of the underparts. In one, they are yellow; males of the other species have bright yellow head and underparts, or a yellowish face, with olive on the rest of the head and underparts.

Male bishops and widowbirds in breeding plumage are mostly black, with bright or golden yellow, orange, or red areas on the crown,

A male **Red-collared Widowbird** (*Euplectes ardens*). Both the extravagant tail and the bright red crown and scapulars are essential elements of courtship displays for this highly polygamous species. Kenya.

shoulders, back to rump, uppertail, and wings. Several of the widowbirds have long, flowing tails of glossy black, with patches of red, brown, or yellow on the head, breast, shoulders, or mantle. Female bishops and widowbirds have dowdy brown, streaky plumage.

One species, Crested Malimbe (*Malimbus malimbicus*), has a prominent crest. Several of the bishops (*Euplectes*) and malimbes have loose or fluffy plumage on the crown, upperparts, or underparts.

Behavior Many species are gregarious and are usually found in small to large flocks. Others are solitary for much of the year, or may join foraging flocks with woodpeckers, bulbuls, sunbirds, and tits, or occur in pairs that maintain a territory year-round. The most highly social are the queleas, which form huge flocks of hundreds or thousands together. The most gregarious of all are the Red-billed Queleas, which often travel across large areas of the African savanna in their millions, in search of seeds and insects; these birds, also known as 'locust birds,' have become an agricultural pest in millet- and rice-growing areas.

Weavers forage at all levels in trees and shrubs, in dense undergrowth, and on the ground. Olive-headed Weaver (*Ploceus olivaceiceps*) and Preuss's Weaver (*P. preussi*) have adapted to climbing tree-trunks and epiphyte-covered branches in search of insects, in a similar way to nuthatches (family Sittidae), including hanging upside-down on occasions.

Flight is usually straight or direct, but in highly social species, flocks move swiftly and with frequent turns and maneuvers.

Voice Breeding males are vocal, and some species have a wider repertoire than others. Songs usually comprise a series of rising and falling, harsh and soft chatters or chuckles, interspersed with rattling churrs, trills, and clicking notes, occasionally given rapidly. The contact calls are single or repeated thin whistles, hisses, and *chack* or *chek* notes; females also have sharp contact calls.

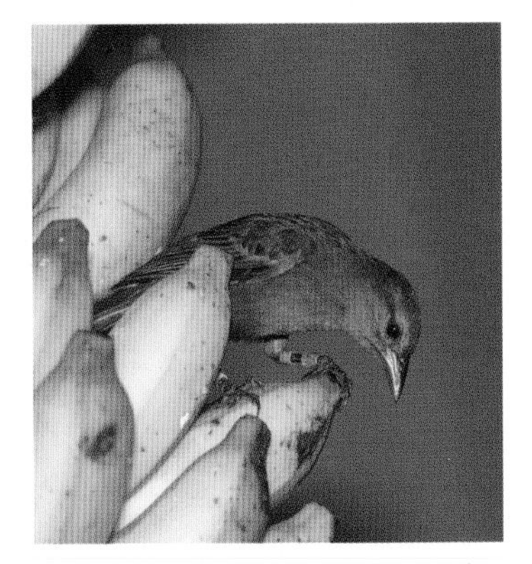

Seychelles Fody (*Foudia sechellarum*). The bird is banded as part of a conservation program, which has succeeded in increasing the population. Seychelles.

A pair of **Red-headed Weavers** with a skeleton nest. In many species, the female visits the nest while the male is building it; only if she selects it will he go on to finish it. Zimbabwe.

Habitats The Ploceidae occupy a wide range of habitats, including mountains, tropical rain forest, lowland secondary woodland, edges of cultivated land, scrub, savanna, marshes, and desert edges. Most species are common and widely distributed. However, Mauritius Fody (*Foudia rubra*), confined to a small area of southern Mauritius, is Critically Endangered due to its small and declining population, as well as from habitat clearance and introduced predators, mostly Black Rats (*Rattus rattus*) and Crab-eating Macaques (*Macaca fascicularis*). These factors led to almost total breeding failure in parts of the range; by 2000 the total population was slightly in excess of 125 pairs. Six further species are regarded as Endangered, and another seven are listed as Vulnerable.

Movements None of the weavers are migratory, although several species wander nomadically outside the breeding season, and some Himalayan species move altitudinally to lower levels in winter. Within Africa, several weavers and queleas make local movements following the rains, in search of food and water.

Diet Weavers take a wide variety of seeds, including grasses, millet, cereals and small herbaceous plants, as well as buds, flowers, nectar, and fruit. They also take insects, including adult and larval grasshoppers, locusts, ants, termites, weevils, beetles, small moths, and butterflies, as well as spiders. Some larger species, such as buffalo weavers, take small frogs. Nestlings and young are fed on green grass seeds and insects.

Breeding Varies between species from monogamous pairs to polygamous males in scattered or dense colonies, or cooperative breeding in family groups. Most males perform a loud display of singing and wing-flapping; in several *Ploceus* species, this is done while hanging upside-down from the nest. Most of the bishops have courtship displays that involve puffing out the brightest areas of the plumage while singing a buzzing song, and defend their territories aggressively against rival males. Jackson's Widowbird (*Euplectes jacksoni*) performs a dancing and aerial display, arching its long tail, at a lekking area of flattened grass.

The nests (see box, below) are highly elaborate, and placed in trees or grass. Clutches usually comprise three to five eggs, incubated by the female for 9–17 days. The young fledge after 11–24 days.

Nest Construction: Weaving a Masterpiece

Weaver nests are very distinctive and show wide variation in size, design, structure, materials and in the skill or expertise of the builder(s). Most are built to a design that is specific to one species or to several closely related species. Styles include individual ball- or dome-shaped structures, hung from the outer edges of branches; communal nests comprising huge bundles of straw placed in the canopy of trees; and large, ball-like structures in grasslands and marshes. In some cases, weavers' nests are so well made that they last for more than a year before wear and the weather begin to take effect.

Baya Weaver (*Ploceus philippinus*) builds a nest with a tunnel opening downward, which may stop snakes entering. India.

In the *Ploceus* weavers, the largest and most widespread genus, nests are built by individual males, although several often utilize the same tree. The birds weave thin strips of grass or palm fronds into a ball-shaped container with the entrance underneath, suspended from the end of a thin branch. Some, including the *Anaplectes* and some of the *Malimbus* species, add a narrow, sleevelike entrance tunnel, which hangs down at one side of the nest.

The bird begins to build by tying a few strips to the selected branch tip. Then, by hanging upside-down and using the bill as a shuttle, he adds more strips by weaving them into a vertical ring, which forms the walls of the nest chamber. Once the ring is complete, he hangs from it to complete the remainder of the construction. When the nest is finished, the male displays from it to attract a female. She will try out the chamber and, if she chooses it, will complete the building by adding a lining of fine material, including soft flowerheads and feathers. After mating, she incubates the eggs and rears the chicks alone, while the male begins nest-building anew in the hope of attracting more females.

Another accomplished nest-builder is Sociable Weaver (*Philetairus socius*), which, despite its name, is an Old World sparrow (family Passeridae). These nests are a common sight across southern Africa. They are huge structures like haystacks, placed in a tree, with the entrances to individual nests on the underside. The nests may house as many as 100 breeding pairs, with new sections added as the colony grows. They are also often adopted as homes by other species; some, such as Pygmy Falcon (*Polihierax semitorquatus*), owls, and chats, nest inside with the weavers, while eagles nest on top.

The craftsmanship needed to build these nests appears to be learned, as most nests are not built by birds in their first years of life and those of flimsy construction are usually by first-time builders. The skill seems to be unrelated to the size of the bill; neatly woven nests may be made by large- or stout-billed species such as Grosbeak-Weaver, which uses thin shreds of grass to build a compact, ball-shaped nest between reed stems, with a side entrance. ■

WAXBILLS *Estrildidae*

A large Old World family of small passerines, the waxbills, or estrildid finches, are found from sub-Saharan Africa to Australia. The greatest diversity of species exists in Africa; the family probably originated here, and has subsequently spread eastward to Arabia, south and Southeast Asia, Australia, and some Pacific islands.

In addition to the waxbills in genus *Estrilda*, mostly found in Africa, the family contains a variety of other small finches. Other African groups include the antpeckers (genus *Parmoptila*), nigritas (*Nigrita*), pytilias, twinspots and crimsonwings (*Amadina* and *Cryptospiza*), bluebills (*Spermophaga*), seedcrackers (*Pyrenestes*), firefinches (*Lagonosticta*), quailfinches (*Ortygospiza*), and cordon-bleus and grenadiers (*Uraeginthus*). The munias and mannikins (*Lonchura*), and most of the avadavats (*Amandava*), are found in Asia as well as Africa, while the parrotfinches (*Erythrura*) occur in Asia and Australia. There are also seven genera of Australian finches.

Black-faced Waxbill (*Estrilda erythronotos*), native to eastern and southern Africa, has the stubby, conical, shiny bill typical of the waxbill family. South Africa.

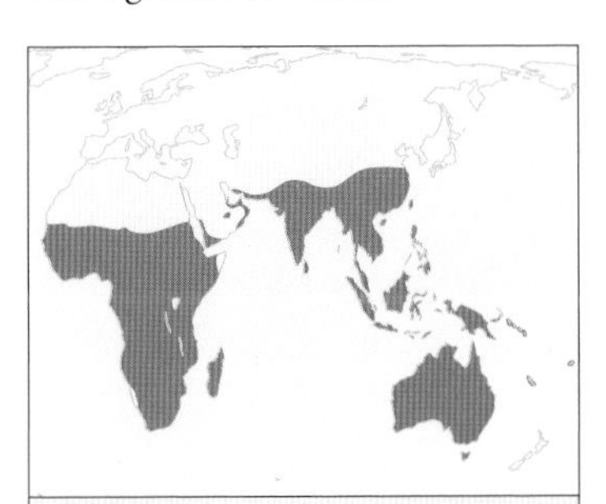

Number of genera 26
Number of species 130
Conservation Status 2 species Endangered, 8 species Vulnerable, 8 species Near Threatened
Distribution Sub-Saharan Africa, Arabia, south and Southeast Asia, Australia, Pacific Ocean islands

The latter include the widely known Zebra Finch (*Taeniopygia guttata*), commonly kept and bred as a cagebird. Most species are characterized by small size and patches of pink or red on the face, rump, tail, or underparts. Several species have a silver or bright waxy-red bill.

Structure The species range from 3.5 to 6.5 in (9–17 cm.) in length. Bill sizes and shapes differ between genera. In most species, the bill is short, stout, and conical, but it is large to proportionately massive in the bluebills and seedcrackers (both of which also have large or domed heads), small in the nigritas, and fine or slender in the antpeckers. The wings are usually short and rounded, with 10 primaries. The tail shape also varies. In many species, it is short and slightly rounded or graduated at the tip. In contrast, genera *Uraeginthus* and *Neochmia* have a long tail with pointed central feathers, and in two Australian species, Gouldian Finch (*Erythrura gouldiae*) and Long-tailed Finch (*Poephila acuticauda*), the elongated central feathers are thin and wispy. The legs and feet are fairly short; in some species they are stronger, and used for clinging to grass stems and seedheads, but in more terrestrial species the legs are weaker.

Plumage In most genera, the sexes are alike or very similar. In other species, the females lack some of the male's bright areas of plumage. Certain waxbills are truly dichromatic, such as several of the firefinches, quailfinches, parrotfinches, and mannikins, and some species of munia and antpecker.

Although popular as a captive bird, **Zebra Finch** is still commonly seen in the wild. Queensland, Australia.

Gouldian Finch has two color morphs: in most birds the head is black, but some, like this one, have a red head. Australia.

The family shows a wide range of plumages, from generally somber brown to extremely colorful or with patches or flashes of color. Most waxbills have a bright red bill and a fine band of red or black through the eyes. Several species are finely barred gray or pale brown, often with a pinkish or reddish tinge on the rump and underparts; some, however, are plainer brown with a gray head or orange face, and two species are almost entirely gray.

Nigritas are black and gray in two species, chestnut and gray-brown in a third, and brown and white in the fourth. Olivebacks (genus *Nesocharis*) are mostly gray with olive-green upperparts. Pytilias are characterized by a bright red face and bill, gray or deep green upperparts, barred underparts, and bright orange or red in the wing. The twinspots range from gray to green or brown above, but the underparts have bold white spots arranged in pairs on a brown or black base color.

The seedcrackers are boldly patterned red and black or brown, as are the bluebills, which are mostly black and red with a metallic blue bill and white spots across the underparts. Firefinches are delicate shades of plum red to pink, brown, or gray, often with fine white spots on the breast or belly, and several species have a black face. The cordon-bleus are pale brown with light blue from the head or face to the underparts, and the two grenadier species are colored with light purple on the face and deep blue on the rump and underparts. Cut-throat Finch (*Amadina fasciata*) is well named for its bright slash of red across the cheeks and throat. The quailfinches, and Locust Finch (*Ortygospiza locustella*), are black or gray and deep orange, and finely spotted or barred.

The munias and mannikins range from pale brown to chestnut and black-and-white. The parrotfinches are mostly bright leaf-green, with shades of deep blue to bright red on the head, face, and underparts.

Several of the Australian species recall the plumage patterns of the African species, with a mixture of browns, black and red, plus fine barring and white spots. Several of genus *Emblema* have pale blue eye rings. Two of the most colorful are Crimson Finch (*Neochmia phaeton*) and Gouldian Finch. In the latter species, the head is variably colored from red to orange and black; it also has green upperparts, a blue rump, a lavender to purple breast, and yellow on the rest of the underparts.

Behavior Most waxbills are gregarious, and are found in small family groups to large flocks. Some, however—particularly the antpeckers, some of the parrotfinches, and the Australian species—occur alone or in pairs. Certain species join mixed-species foraging flocks outside the breeding season.

Most species forage on the ground or in low vegetation, searching for seeds. Several of the smaller waxbills and munias feed by clinging to ripe seedheads of grass, cereals, or rice, while forest-living species such as the seedcrackers, bluebills, and crimson-wings feed on larger seeds taken either from ripening trees and bushes or on the ground. Due to the dry nature of the seeds they eat, most open-country species make regular visits to water to drink.

Several species, including bluebills, pytilias and nigritas, also eat insects, bluebills often digging into termite tunnels. The antpeckers forage for ants in trees and on the ground, and may also feed on nectar

A **Scaly-breasted Munia** (*Lonchura punctulata*), also known as Nutmeg Mannikin, perches on grass stems to eat the seeds. The birds also feed on the ground. India.

A group of **Arabian Waxbills** (*Estrilda rufibarba*). These highly sociable birds are found in areas with good water supplies, such as wadis and irrigated land. Yemen.

and pollen. The olivebacks are more arboreal. Most forage on insects in the forest canopy; however, White-collared Oliveback (*Nesocharis ansorgei*) has been reported to feed only on the seeds of the plant *Melanthera scandens*.

Voice Waxbills are highly vocal, with a great deal of calling between pairs and flock members. Forest-living species give frequent contact-calls. The calls are usually sharp, high-pitched whistles, chirps, and chirrups; several species, including the waxbills and pytilias, give a querulous or rising *tsway*. Songs, usually given at the start of the breeding season, are mostly a series of short, soft, musical warbling phrases.

Habitat The family occurs in a broad range of habitats, from desert margins to open grasslands, reedbeds, scrub, open woodlands, and edges of forests. Most species are fairly common but two are considered to be Endangered and eight are regarded as Vulnerable, mostly due to habitat loss and fragmentation. One of the least well known and most restricted in range, Anambra Waxbill (*Estrilda poliopareia*), occurs in a small area of southern Nigeria; the habitat is plentiful but the population is very small, and is possibly being limited by as yet unknown factors. Another species, Java Sparrow (*Padda oryzivora*), has been extensively trapped for the cage-bird trade, and the population in the wild has declined dramatically as a consequence.

Movements None are truly migratory, although many species, particularly those dependent on ripening seeds (such as waxbills, quailfinches, and Locust Finch), are nomadic when not breeding, moving around over a wide area in search of food.

Diet Mostly grass seeds; also insects such as termites and ants, and nectar and pollen.

Breeding The birds may be social, semicolonial, or solitary breeders. They are monogamous; pair bonds are strong and are usually maintained outside the breeding season. The nests, in most cases, are untidy domes or balls, with a side entrance or occasionally an entrance tunnel; they are made of grasses, leaves, feathers, and bark strips, collected by the male and built by the female. Some species, however, nest in holes in trees, or utilize the disused or abandoned hanging nests of weavers (family Ploceidae). The clutch of three to eight eggs is incubated by both parents for 11–18 days, and the chicks remain in the nest for up to 25 days. The nestlings of all species have distinctive, highly patterned bill and tongue markings called tubercles or papillae; in Gouldian Finch and the parrotfinches, these reflect light, enabling the young to be seen in the dark nest. The nests of several African waxbills are parasitized by whydahs and indigobirds (family Viduidae).

Taxonomy Waxbills were formerly considered to be part of a larger family, Viduidae, but are now separated into their own grouping. They are also distantly related to the weavers and finches. Some taxonomists consider the Australian species to be more distantly related, and their true relationships have yet to be determined.

A **Blue-breasted Cordon-bleu** (*Uraeginthus angolensis*). These birds are known for breeding close to wasps' nests; the wasps effectively protect the nest against predators. South Africa.

INDIGOBIRDS *Viduidae*

An Old World family, the indigobirds and whydahs are restricted to sub-Saharan Africa. They are closely related to the waxbills (family Estrildidae). Many species are specialized brood-parasites on waxbills. Most of the indigobirds are very similar and often difficult to identify unless they are seen with their brood-host species. In contrast, male whydahs in breeding plumage are instantly recognizable by their streamerlike tail.

Structure The birds are usually 4–5 in. (10–13 cm.) in length; however, male whydahs in breeding plumage reach 12–17 in. (30–43 cm.) long, including the tail. The bill is typically short, stout, and conical, especially so in Parasitic Weaver (*Anomalospiza imberbis*). The wings are short and rounded. The tail is fairly short and square-tipped, except in breeding male whydahs, in which it is extremely long, with rounded or pointed tips to the central feathers.

Plumage All species are sexually dichromatic, with male indigobirds and whydahs having a distinct breeding plumage; males' nonbreeding plumage resembles that of females. Male indigobirds in breeding plumage are mostly black, variably glossed black, purple, or greenish, and have a red, white, or pale blue bill and legs. Females (and nonbreeding males) are more somber, streaked brown above and off-white to buff below, with broad stripes on the head. Male whydahs in breeding plumage vary from all black to black-and-white or straw-colored; the paradise-whydahs have an orange collar and chestnut breastband and a long, broad tail. Females and nonbreeding males resemble the indigobirds. The nestlings of indigobirds and whydahs have the same mouth and bill markings as the nestlings of their brood-host.

In their breeding plumage, male **Purple Indigobirds** (*Vidua purpurascens*) are black with a sheen to the feathers. Kenya.

The male **Pin-tailed Whydah** (*Vidua macroura*) performs courtship displays in which he flies slowly or hovers while his tail wafts in the air. South Africa.

Behavior The birds may occur alone, in pairs, or in small family groups, often in close company with waxbills, especially those that they parasitize; most whydahs occur in large flocks outside the breeding season. They forage on the ground or clinging to tall grasses and sunflowers.

Voice The songs are mostly a series of imitations of the host species' song, together with high-pitched rattles and trills. The calls are mostly dry chirps or repeated *chick* notes.

Habitat Mainly grasslands with bushes; also meadows, open woodlands and the margins of cultivated land, especially rice-fields, often in close proximity to their brood-host species. Several species have small or restricted ranges, although none are considered to be threatened.

Movements Mostly sedentary.

Diet Mostly seeds and small insects such as ants and termites. Female Parasitic Weavers eat the eggs of the host species—mainly cisticolas or prinias (family Cisticolidae).

Breeding All species are parasitic on waxbills or small Old World warblers, mainly using a single host species. Males of several species, mainly the paradise-whydahs, have slow-flight courtship displays. Females lay single eggs in a series of host species' nests. The nestlings imitate the begging behavior of the host species' young.

Taxonomy Parasitic Weaver closely resembles a small canary (genus *Serinus*) or a weaver (*Ploceus*), and has previously been classed together with both of these genera, but recent DNA research shows that it is most closely related to the indigobirds.

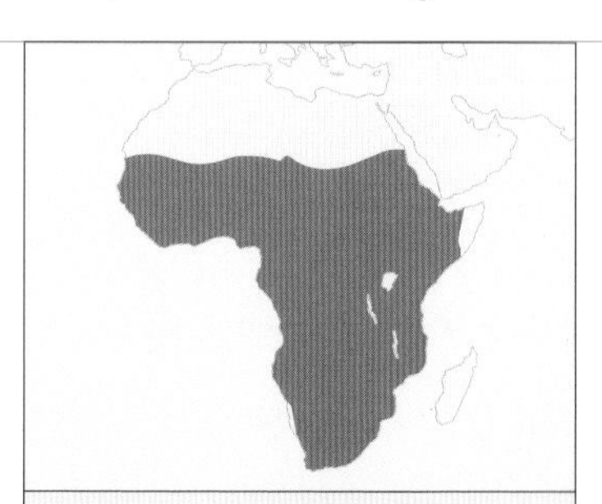

Number of genera 2
Number of species 20
Conservation Status Least Concern
Distribution Sub-Saharan Africa

ACCENTORS *Prunellidae*

Dunnock (*Prunella modularis*) is a familiar bird in Europe. Unlike most accentors, which live at high elevations, it occurs in various lowland habitats, including backyards. U.K.

The accentors are a small but well-defined Old World family found in the temperate parts of the northern hemisphere. Most of the species are not especially well known, but the common European representative, Dunnock (*Prunella modularis*), is among the best studied of European species. Accentors are generally nondescript in appearance; this fact is even reflected in the name Dunnock, which is derived from an Old English word meaning 'little brown bird.' They can be shy and retiring birds that mainly hide in dense vegetation close to the ground, but when singing they become much more prominent and easy to see.

The strikingly marked **Radde's Accentor** (*P. ocularis*) occurs in upland areas of central Asia, the Middle East, and India. Armenia.

Structure Species range from 5.5 to 7 in. (14–18 cm.) in length. Accentors are small passerines, similar in size and shape to buntings (Emberizidae), although accentors can be distinguished by their thin, warbler-like bills. In detail, the bill is thick at the base and thins out to a sharp tip. In general, accentors are stocky birds with no outstanding, unique structural features. The tail is of medium length and is usually notched at the tip. Although the birds are terrestrial, the legs are not particularly strong nor long.

Plumage Males and females are similar in plumage, and juveniles resemble adults. Accentors are often brownish above, sometimes with rufous tones, and typically streaked. Most also show strongly streaked flanks. Some species are largely grayish below, while others are buffish or cinnamon, some showing a white or black throat. Many have a contrasting face pattern with a noticeably paler or warm-colored supercilium, while on others the supercilium and ear coverts are grayish.

Behavior Accentors are ground foragers. They tend to hop or shuffle around, typically on bent legs, with the belly close to the ground. Displays are usually restricted to singing, but Dunnock also performs a display on the ground, in which two or more birds hop about and raise one or both wings to each other.

Voice The songs are complex and varied, made up of high-pitched trills and warbles. Their complex nature suggests the song of a Winter Wren (*Troglodytes troglodytes*). Many species' songs are not well known, but those that are known tend to be similar to Dunnock's song, differing in how melodious, dry, or complex they are. The calls tend to be thin *seep* notes, or a multiple-syllable series, *tsee-tee-tee*.

Habitat Accentors occur across the Palearctic, from the British Isles to Japan, and are generally confined to

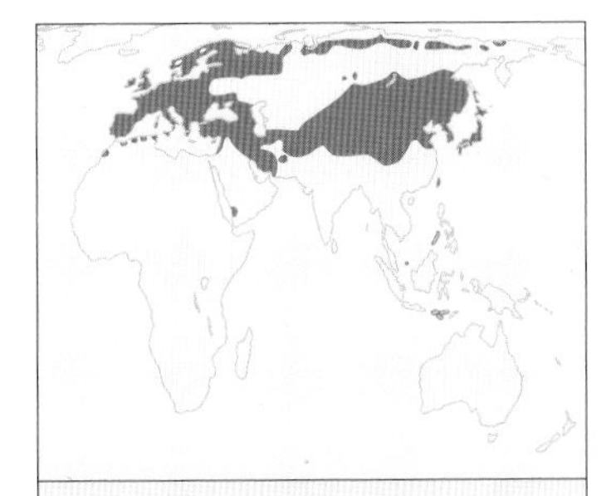

Number of genera 1
Number of species 13
Conservation Status 1 species Near Threatened
Distribution Palearctic, from British Isles to Japan: south to North Africa and southern Asia

An **Altai Accentor** (*P. himalayana*) feeding. Accentors forage on the ground and in grass, taking insects, spiders, and worms in summer, or seeds and berries in winter. Kazakhstan.

temperate latitudes, reaching south to North Africa and southern Asia. Dunnock has also been introduced to New Zealand. They are birds of open country and forest margins rather than forest interior, and some are drawn to thickly vegetated habitats; for example, the other English name for Dunnock, 'Hedge-Sparrow,' alludes to its preference for hedges and dense vegetation in suburban or rural areas. On the other hand, highland species, such as Alpine Accentor (*Prunella collaris*) and Radde's Accentor, are found above the treeline, in truly alpine conditions, at elevations as high as 20,000 ft. (7,000 m.).

Maroon-backed Accentor (*P. immaculata*), found in Himalayan forests, is unusually colorful for an accentor. However, its sharp bill is typical of the family. China.

Movements Most accentors are residents or short-distance migrants. These are hardy birds, and winter at temperate latitudes rather than in the tropics. However, most species that breed in far northern areas such as Siberia, or at high elevations, are migratory, and in more widely ranging species the northern populations migrate while the southern ones are residents. Siberian Accentor (*P. montanella*) is a rare vagrant to western North America in the fall (Alaska) or winter (Pacific Northwest).

Diet Omnivorous, foraging on insects and other invertebrates mainly in the summer, and taking seeds and berries during the winter.

Breeding The habits of most accentors are little known, although Dunnock has been closely studied. While drab in appearance, Dunnock is highly colorful in its breeding behavior. Some pairs are monogamous, but many females mate and have young with more than one male (polyandry), and sometimes several females and several males are involved (polygynandry). The birds also practice the curious behavior of cloacal pecking. The female raises the tail and quivers the wings while distending the cloaca and offering it to the male. The male pecks at it, before copulating with the female. It is thought that cloacal pecking enables the male to remove sperm deposited by other males before he himself mates. Males help at the nest roughly in proportion to their mating success in that nest, so sometimes two males will feed the young at a single nest. Males may also assist in incubating the eggs.

The female builds the nest, often close to the ground, and lays three to six eggs, which are incubated for 11–15 days. The young leave the nest at 12–14 days. Dunnocks suffer nest parasitism by Common Cuckoo (*Cuculus canorus*), but do not reject foreign eggs in their nest, as do many of the cuckoo's other hosts; they will accept any egg morph of the cuckoo.

Taxonomy The family was initially thought to be related to the thrushes (family Turdidae), partly because of their terrestrial habits and thin bill, but more recent genetic data suggests that they are closer to the true sparrows (Passeridae) and the weaver finches (Estrildidae).

Alpine Accentor is notable for being polygynandrous, with groups of several males and females interbreeding. Spain.

OLIVE WARBLER *Peucedramidae*

An adult male **Olive Warbler**, showing the characteristic bright tawny orange head and upper breast, black lores and ear patch, gray back, two white wing bars, and dirty white underparts. New Mexico, U.S.

The single species in this family, a small arboreal bird, was long thought to be an aberrant wood-warbler (family Parulidae), but recent genetic studies have determined that Olive Warbler (*Peucedramus taeniatus*) should be placed in its own family.

The adult female is duller than the male, with a yellowish head and a mottled gray-black ear patch. Arizona, U.S.

Structure Olive Warbler is 5.2 in. (13 cm.) in length. It is similar to wood-warblers in genus *Dendroica*, but with a slightly longer bill and a decurved culmen, a distinctly notched tail, and, usually, 10 primaries (the tenth is much reduced or absent).

Plumage The male has a tawny orange head and breast, a black ear patch, gray upperparts, and off-white underparts. The wings have white bars and a white patch at the base of the primaries. The northernmost subspecies (*P. t. arizonae*) is larger and paler than the four southern ones, and males show a unique molt pattern in which breeding plumage is not attained until the second year.

Behavior Forages for insects high in conifers. Kinglet-like wing flicking is a distinctive habit.

Voice The song consists of loud whistled notes, *peeta peeta peeta*, similar to that of Tufted Titmouse (*Parus bicolor*). Females also sing. The call is a soft note like that of Western Bluebird (*Sialia mexicana*).

Habitat Montane coniferous and mixed pine-oak forest in Arizona, New Mexico, and Mexico. Currently fairly common within its range, but could be negatively affected by logging activities.

Movements Northernmost subspecies is partially migratory, with some birds moving south from Arizona to Mexico and returning to breeding grounds in early April. Southern subspecies are sedentary.

Diet Insects.

Breeding Apparently monogamous. The nest is sited high in a conifer, far from the trunk; the birds let it become soiled with droppings from the nestlings, a behavior unknown in wood-warblers. Three or four eggs are laid in June. Incubation and fledging periods are unknown. The young are altricial.

Taxonomy The basihyal bone and jaw musculature are similar to those of the Old World warblers (family Sylviidae), but some behavioral differences, and recent genetic evidence, suggest that Olive Warbler is more closely related to the finches (family Fringillidae). It is generally considered to be more primitive than the wood-warblers.

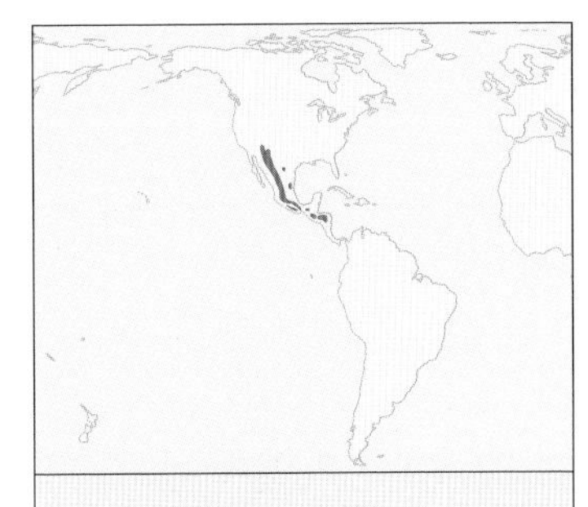

Number of genera 1
Number of species 1
Conservation Status Least concern
Distribution Southwestern U.S. and northern Mexico

WAGTAILS AND PIPITS *Motacillidae*

A **Blue-headed Wagtail** singing. This is a subspecies of Yellow Wagtail (*Motacilla flava*) in which breeding males have a blue-gray head. Spain.

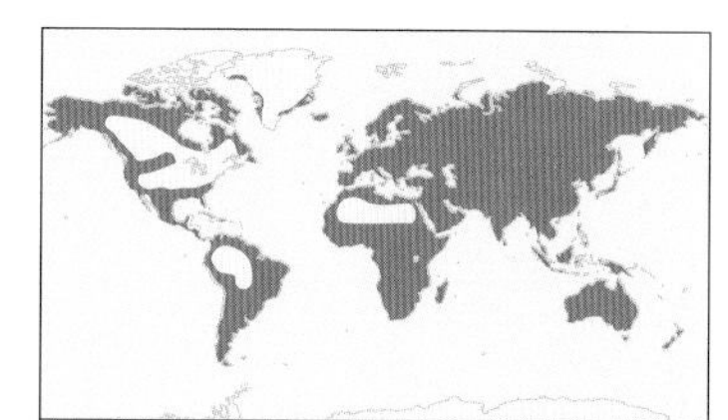

Number of genera 5
Number of species 65
Conservation Status 2 species Endangered, 3 species Vulnerable
Distribution Mostly North America, Europe, Asia, and Africa, but representatives worldwide

This family of small passerines has two main groups: wagtails (genus *Motacilla*) and pipits (*Anthus*). It also includes longclaws (*Macronyx*). Pipits and wagtails are slender birds, and have a long tail with white outer feathers. They share the characteristic behavior of 'pumping' their tail as they walk: wagtails usually more vigorously, but pipits more regularly.

Structure The birds are generally small to medium-sized, 5–8.75 in. (12.5–22 cm.) in length. Pipits are slender birds with a long, pointed bill, rather long tail, and long legs and toes adapted for walking on the ground. Wagtails are similar in shape, but with a longer tail. The longclaws of Africa have an elongated hind claw (as do many pipits).

Plumage In pipits the sexes are alike, but in wagtails the sexes differ greatly. Pipits are generally brown or buff-colored, with streaking above and below, and are well camouflaged in the grasslands that most of them occupy. Most have varying amounts of white on the outer tail feathers. Red-throated Pipit (*Anthus cervinus*) is unique in having a bright salmon-red throat. Yellow-breasted Pipit (*A. chloris*) and Golden Pipit (*Tmetothylacus tenellus*) are notable for pipits in having bright yellow underparts. Wagtails are generally either black or gray above and white below, or green or gray above and yellow below, and different species have varying degrees of black on the throat or head. All have a very long tail with white outer tail feathers. Forest Wagtail (*Dendronanthus indicus*) is unique, having ploverlike black breast bands. The longclaws, while shaped like pipits, have plumage more like the meadowlarks (family Icteridae) of North and South America.

Behavior These birds are often found near water. Both pipits and wagtails forage mainly for small insects by walking on the ground, usually bobbing their tail continuously. Both have undulating flights. Some species have a unique flight behavior: one example is Sprague's Pipit (*A. spragueii*) which, when flushed, rises high above the ground and circles with a fluttering flight, before plummeting back to the earth. During the breeding season, both pipits

A **Tawny Pipit** (*A. campestris*) bathing. This bird is found in dry areas of Europe, but winters in India and tropical Africa. Kenya.

and wagtails are generally found singly or in pairs, but migratory species often form species-specific flocks during the winter.

Voice The pipit gets its name from the calls of some of the better-known species. Calls generally consist of single or double notes, often squeaky, and sometimes explosive. Pipits' primary songs vary from simple, monotonous songs to a complex jumble of high-pitched notes. Wagtail songs are quite different: generally a short series of high-pitched notes strung together. Calls are also given in flight: they are typically metallic, loud, one or two-noted, and repeated often. In some cases, different species have similar calls.

Habitat Most species are found in temperate zone habitats: pipits and wagtails often breed in tundra or grasslands, or near streams and rivers. Some species winter in agricultural areas, often in flocks. The longclaws inhabit grasslands and arid areas in Africa.

Movements The family includes both migratory and nonmigratory species. While some, such as Long-billed Pipit (*A. similis*) and Japanese Pied Wagtail (*Motacilla grandis*) are resident, most are highly migratory, typically breeding in northern latitudes and wintering in southern regions. Some species migrate a limited distance, such as Buff-bellied (American) Pipit (*A. rubescens*), which breeds across the North American tundra and winters mainly in the southern U.S. and northern Mexico. Others migrate much longer distances: for example, White Wagtail (*M. alba*) breeds as far north as northern Siberia and winters in southeast Asia and northern Africa. Migratory birds are often seen in species-specific flocks, following coastlines and rivers.

Diet During the breeding season, pipits and wagtails feed primarily on insects, spiders, and other small invertebrates, and sometimes even small snails. They are also insectivorous during migration and in winter, but may supplement their diet with seeds from grass and weeds.

Breeding During the breeding season, male pipits give display songs, generally in flight, high in the air. Some species sing from the ground, and a few species, such as Olive-backed Pipit (*A. hodgsoni*), from trees. The flight displays often last from a few seconds to as long as 20 minutes. Male wagtails sing from exposed perches in trees, on rocks, or on buildings, with songs much shorter than those of pipits. Pipits and wagtails are monogamous and typically build cup nests on or near the ground. The nests are usually constructed out of grasses, leaves, mosses, sometimes lined with hair or feathers. The clutch size is generally three to seven eggs, with an incubation period of 10–14 days, but likely varies with latitude. Both parents participate in incubating the eggs and feeding the young, which are altricial.

A **Sharpe's Longclaw** (*Macronyx sharpei*) with its prey. The elongated claws visible at the back of the feet are thought to help the bird walk through long grass. Kenya.

FINCHES *Fringillidae*

The Fringillidae is an extensive family of small birds, whose members are found across the world. The family derives its name from a group of small finches in genus *Fringilla,* found mainly in the Old World temperate zone, perhaps the best-known of which is Chaffinch (*F. coelebs*). The most familiar groups include the finches, siskins, and grosbeaks. In contrast, the most atypical are the Hawaiian honeycreepers—a highly diverse collection of birds endemic to the Hawaiian islands.

Structure Fringillids vary in length from 5 to 10 in. (12–25 cm.), with most being medium-sized. Most have a conical bill, but bill size varies across the family, from very tiny in Hoary Redpoll (*Carduelis hornemanni*) to massive and triangular in Hawfinch (*Cocothraustes cocothraustes*) and some east Asian grosbeaks (genus *Eophona*). Typical fringillid bills are adapted for cracking open seed husks. More unusual are the crossbills (genus *Loxia*), whose uniquely-shaped bills cross at the tip, and are adapted for opening cones from conifers. The bills of Hawaiian honeycreepers are very varied in size and shape, from conical, like those of other fringillids, to long, thin, and decurved, as in I'iwi (*Vestiaria coccinea*).

A male **Chaffinch** in breeding plumage. Chaffinch is the most common species of finch in western Europe, and is often seen in gardens and hedgerows. U.K.

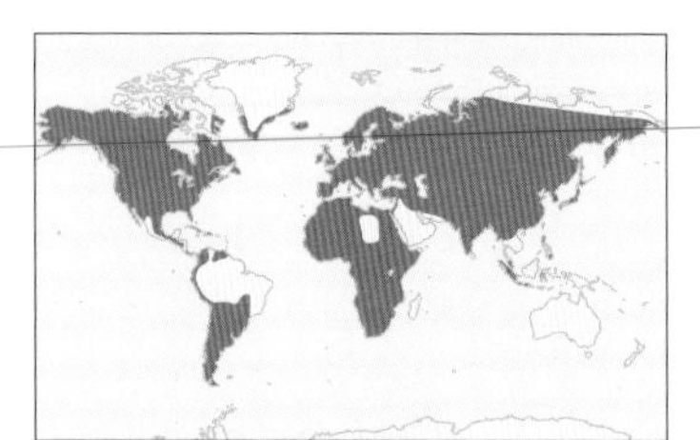

Number of genera 42
Number of species 168
Conservation Status 10 species Critically Endangered, 10 species Endangered, 9 species Vulnerable
Distribution Eurasia, Americas, continental Africa, Asia; introduced to Australia and New Zealand

Plumage Most of the fringillids are highly sexually dichromatic, and there is a huge variety of coloration across the family. The goldfinches and siskins are closely related (all in genus *Carduelis*) but range from bright yellow and black, as in American Goldfinch (*C. tristis*) to brown and white, with a red face, as in European Goldfinch (*C. carduelis*); brown and streaked, as in Pine Siskin (*C. pinus*); and even bright scarlet and black, as in Red Siskin (*C. cucullata*). All, however, have varying degrees of bright yellow (or red) in the wings, usually visible when the bird is perched. The male *Carpodacus* finches are mostly red or pink, with differing degrees of streaking or spotting on the head and body. The serins and canaries are also varied: many are bright yellow, while others are dull in color.

The family includes a number of species with especially vivid plumage. Males of genus *Pyrrhula* have bright underparts, as in Eurasian Bullfinch (*P. pyrrhula*), which is a stunning pink below. The monotypic Scarlet Finch (*Haematospiza sipahi*) from the Himalayas is entirely scarlet. The North American

A male **Red Crossbill** (*Loxia curvirostra*) opens a conifer cone to take the nuts. The crossed tip of the bill gives leverage so the bird can prise the scales apart. Germany.

Grosbeaks, like this **Golden-winged Grosbeak** (*Rhynchostruthus socotranus*), are finches with an especially large, triangular bill. Socotra Island, Yemen.

Evening Grosbeak (*Hesperiphona vespertina*) is golden yellow with black and white wings. The unique Gold-naped Finch (*Pyrrhoplectes epauletta*), which is found in the Himalayas, is entirely black apart from a golden-orange crown.

Hawaiian honeycreepers also vary greatly in plumage. Akepa (*Loxops coccineus*) is bright red all over. Apapane (*Himatione sanguinea*), or Crested Honeycreeper, is mostly brown with streaks of black and white and a white crest. Other species are a rather plain yellowish-green, such as Akekee (*Loxops caeruleirostris*).

Behavior Although members of the family pair off during the breeding season, fringillids are well known for forming large single-species flocks during the winter; for example, it is not uncommon to see flocks of hundreds of House Finches (*Carpodacus mexicanus*) or Hooded Siskins (*Carduelis magellanica*).

There is a wide diversity of foraging styles. Some finches forage on the ground, while many cling to plant stalks or branches in search of seeds. Certain species visit seed feeders put up by people in winter. Crossbills typically cling to a conifer cone, insert their crossed bill, and pry open the cone to get at the seeds inside. Hawaiian honeycreepers feed on nectar from flowers.

Voice Song and calls are very important in the Fringillidae. Outside the breeding season, finches that gather into flocks rely on contact calls to keep together. During the breeding season, males defend their territory by singing elaborate songs while perched, and occasionally in flight; females also sometimes give soft songs. Several species make particular calls while in flight. In North America, finches, goldfinches, and grosbeaks are usually first detected by their flight calls. Brambling (*Fringilla montifringilla*) and Chaffinch both give distinctive nasal flight notes. The crossbills' flight calls vary between and even within species. Red Crossbill has been detected giving seven to nine different types of call, which have been correlated with different bill sizes and shapes of birds in different locations and habitats. Hawaiian honeycreepers produce a variety of sweet songs and nasal calls.

Yellow-crowned Canary, or Cape Canary (*Serinus canicollis*), is a resident of eastern and southern Africa. Kenya.

Hawaiian honeycreepers such as this **I'iwi** are genetically related to finches, despite their strange appearance. Hawaii, U.S.

Habitat The family is widely distributed around the world. The rosefinches in genus *Carpodacus*, and goldfinches and siskins in genus *Carduelis*, are found in temperate areas across Europe, Asia, and North America. Several of the siskins and goldfinches are also found south into South America. The Old World serins and canaries in genus *Serinus* are, with the exception of a handful of species, widely distributed in Africa, the ancestral home of the genus. Some members of the family are under severe threat, and many Hawaiian honeycreepers have gone extinct in recent decades due to habitat loss and predator pressure.

Movements Most of the fringillids breeding in the tropics and subtropics are resident within their range and habitat. In contrast, most of the temperate-breeding birds are migratory to some extent.

The *Carpodacus* finches move short distances. Other species vary in their movements. For example, Common Redpoll (*Carduelis flammea*) breeds at the northern edges of

North America and Eurasia, moving southward during the winter, but in some years may irrupt in great numbers to areas even farther south when food supply in their normal winter range fails. This phenomenon occurs in a number of northern finches, including Pine Siskin (*Carduelis pinus*), Evening Grosbeak, and Pine Grosbeak (*Pinicola enucleator*). Certain finches are highly migratory. Common Rosefinches (*Carpodacus erythrinus*) breeding in Siberia have been found as vagrants on islands in the Bering Sea in western Alaska, while those breeding across northern Europe and Russia to Siberia move south to winter in India and southeast Asia. Other long-distance migrants are Brambling, which breeds in Scandinavia and western Siberia but winters in the U.K. and southern Europe, and the siskins.

The rosy finches (genus *Leucosticte*), mainly found high in the Rocky Mountains of North America, and in the Himalayas and Siberia in Asia, are known to form large flocks and 'invade' lower elevations when the high ground is covered in snow.

A male **Brambling** in winter plumage feeds on a flowerhead. In the breeding season, the male has a brighter chest, darker upperparts, and an entirely black head. U.K.

Brown-capped Rosy Finch (*Leucosticte australis*) nests in rocks at high elevations, but winters on lower, open ground. U.S.

Diet The Fringillidae are best known as seed-eaters, but some will take fruit or insects. The typical finches, goldfinches, and siskins feed mainly on the seeds of annual plants. Crossbills mainly take seeds from conifer cones. Serins and canaries feed on both insects and seeds. Hawaiian honeycreepers show more diversity in their diet, with some feeding on seeds, some on insects, and others on nectar from flowers.

Breeding Fringillids are monogamous. Nests are typically open cups constructed out of grasses, leaves, twigs, and other plant matter. Females do most of the nest construction, and egg incubation, while males actively defend territories. Clutches comprise two to six eggs, incubated for about 14 days. The young are fed by both parents until they fledge, at 11–17 days.

Taxonomy Recent genetic studies suggest that the euphonias and chlorophonias belong in family Fringillidae, but this reclassification is not generally accepted and has not been followed here; these groups are more usually placed with the tanagers. Hawaiian honeycreepers have also been added recently to Fringillidae, having previously been placed in their own family, the Drepanididae; however, this arrangement is also still under discussion. These groups, while showing genetic affinities with more typical finches, are so different in structure, behavior, and habitat that some authorities argue they should be given their own taxonomic grouping.

A pair of **Twites** (*Carduelis flavirostris*). These birds typically breed in moorlands and nest in bushes. Armenia.

WOOD-WARBLERS *Parulidae*

An exclusively New World family of small songbirds, also known as New World warblers. Most North American species of wood-warblers are migrants, wintering in the Neotropics. Their return in spring, in their finest, most colorful breeding plumage and full song, is eagerly anticipated by birdwatchers and ornithologists.

More than 60 percent of all the species in this large family are classified in just four genera. The dominant temperate-breeding genera are *Dendroica* (29 species) and *Vermivora* (9 species), while the dominant tropical and subtropical genera are *Basileuterus* (25 species) and *Myioborus* (12 species).

Structure Wood-warblers range from 4.5 to 7.5 in. (11.5–19 cm.) in length. The different species show a variety of physical characteristics, often related to the features of their environment. Yellow-breasted Chat (*Icteria virens*) is unusually large and thick-billed, and Wren-thrush (*Zeledonia coronata*) is unusually dull, long-legged, and short-tailed. Some species have rictal bristles, an adaptation to foraging aerially for insect prey. Species that forage on tree bark may have slightly longer and more decurved bills. Birds that perch in trees and bushes generally have slender legs, while some terrestrial species have thicker, sturdier legs. A few species have tiny fringes on their tongue to help them feed on nectar.

A male **Lucy's Warbler** (*Vermivora luciae*). Although plainer than many other wood-warbler species, these birds are distinctive due to the rufous rump and, in males, the reddish crown patch. New Mexico, U.S.

Plumage There is a wide variation between species. Many are brightly colored, while others are rather dull. Sexual dichromatism is common in temperate species and rare in tropical species, with a few exceptions in both regions. Most North American species molt into a special breeding (alternate) plumage, which is more colorful than the nonbreeding (basic) plumage worn in winter and by juveniles. In contrast, most tropical and some temperate species have similar breeding and nonbreeding plumages.

In many species, the main body colors are yellow, green, and olive,

A male and female **Bay-breasted Warbler**, showing sexual dimorphism; the male is in front.

Number of genera 24
Number of species 112
Conservation Status 1 species Critically Endangered, 8 species Vulnerable
Distribution North, Central, and tropical South America

and a number of species are boldly marked with white and black on the head and underparts. Birds of genus *Ergaticus* are pinkish to bright red.

Wood-warblers have nine visible primaries, with the tenth greatly reduced or absent. Some species show prominent wing bars. Many species have white spots on the outer tail feathers. As in other passerines, the juveniles of many species show more pointed rectrix tips than do adults.

Behavior Wood-warblers use a wide range of foraging techniques, especially to find insects and other invertebrates. Some species glean and probe crevices in bark, others sally like tyrant flycatchers (family Tyrannidae), and some forage on the ground. Resident tropical species, as well as some migrants from North America, join mixed-species feeding flocks that move through forests in search of fruiting trees.

Voice The males of many species have two song types: a primary song that is often accented at the end, and is used for defending territory, and an alternate song, often unaccented, that is used in a variety of situations, including near the nest when the bird is paired with a female. The songs are quite diverse, from a dull, monotonous chipping to sweet and varied notes, or may even be loud and ringing. In a few species, females are known to sing. Call notes are generally very short; they include harsh scold notes (infrequent), and contact calls ('chip notes'), which are often heard during the day as the birds are migrating. Most North American species also give short, typically high-pitched flight calls during their nocturnal migrations.

Habitat Many eastern North American species breed in mixed and coniferous boreal forest, with a few 'southern' species breeding in deciduous woodland and swamps. Many western North American species breed in montane coniferous and deciduous forest, while a few prefer second growth and shrublands.

Tropical species breed in habitats including montane cloud forests, lowland rain forests, and riverine woodlands. The nine species of yellowthroats (genus *Geothlypis*) favor freshwater marshes. Several of the species that breed in the Neotropics have small, remote ranges and are poorly known.

Movements Most North American wood-warblers are migratory, and use various migration strategies (see feature box). Caribbean, Central American, and South American species are mainly sedentary, with some altitudinal migration among

A female **Painted Redstart** (*Myioborus pictus*) flashes her black-and-white tail. This display is thought to startle insect prey, which the bird then pursues and captures. Arizona, U.S.

The striking **Rufous-capped Warbler** (*Basileuterus rufifrons*) is normally a resident of tropical Central America, but small numbers of birds are occasionally found in the southernmost states of the U.S. Panama.

A female **Black-and-white Warbler** (*Mniotilta varia*) searches for food in tree bark, in a manner similar to the foraging behavior of nuthatches and treecreepers. California, U.S.

the montane species. Southward migration has not been documented, as few species occur in southern hemisphere temperate zones.

Diet On the breeding grounds, northern species feed mainly on insects, larvae, caterpillars, and small spiders, most of which they obtain from vegetation. In winter, eastern Yellow-rumped Warbler (*Dendroica coronata*) feeds extensively on fruit, especially wax-myrtle (*Myrica cerifera*) and poison ivy (*Toxicodendron radicans*); this change in diet allows the birds to winter farther north than other wood-warblers. Some species alter their diet and feed on nectar and berries when in the tropics. The diets of tropical species are little known, but many of these birds are also largely insectivorous.

Breeding The northern temperate wood-warblers have been well studied, but the breeding habits of many tropical species are almost completely unknown.

Northern wood-warblers breed in the summer months (May to August), while tropical species are less seasonal, sometimes breeding year-round. Typically, males form and defend single territories (although some species are loosely colonial). To do so, they use song; they also employ song, together with their brightly colored breeding plumage, in courtship displays to attract mates. Males of northern species form new monogamous pair bonds with females each spring, but there is no evidence that the birds pair for life. The males of some species may mate with several females in adjacent territories.

Typically, the females build the nest, while both sexes share in incubating the eggs and feeding the young. Many species build open cup nests well hidden either in high trees, in shrubs, or on the ground in undergrowth. Some build domed nests, usually located on the ground or among roots and tangled vegetation. Only two species, Prothonotary Warbler (*Protonotaria citrea*) and Lucy's Warbler, use cavities for nesting. As with many bird families that include both temperate and Neotropical species, temperate wood-warblers have larger clutches (two to nine eggs), mainly have single broods, and have shorter incubation periods (10–14 days) and fledging periods (8–14 days). Tropical species generally lay fewer eggs (one to four), and have more extended fledging periods. The young are altricial in all species.

Wood-warblers are among the most common targets of parasitism by cowbirds. Some species have adapted to this problem by removing the cowbirds' eggs, or in the extreme case of Yellow Warbler, by building a new nest on top of the old one.

Hybridization is not uncommon among wood-warblers. In particular, two species—Blue-winged Warbler (*Vermivora pinus*) and Golden-winged Warbler (*V. chrysoptera*)—breed in similar habitats and tend to hybridize extensively where their ranges come into contact, despite significant differences in plumage and pattern. The most common form of hybrid is called Brewster's Warbler. The songs are similar, and both species share an alternate song.

NEW WORLD BLACKBIRDS *Icteridae*

The male **Red-winged Blackbird** (*Agelaius phoeniceus*) is well known for its vivid red shoulder flashes. U.S.

This New World family includes the American blackbirds, as well as the New World orioles, meadowlarks, grackles, caciques, oropendolas, and cowbirds. The family is named after the Old World Blackbird (*Turdus merula*), but other than being black, the Icteridae and Blackbird have no relation to each other.

While many species are black, most have highly colorful plumage. In addition to their dazzling appearance, the Icteridae are remarkable for their elaborate displays and vocalizations. In fact, this family is renowned in the Americas for fancy displays of flash colors.

A male **Baltimore Oriole** (*Icterus galbula*). The birds do not show full adult plumage until their second year. Texas, U.S.

Structure New World blackbirds are sturdy, fairly large songbirds, ranging from 6 to 21 in. (15–55 cm.) in length. They all have strong bills, varying in shape from the short, more finchlike bill of Bobolink (*Dolichonyx oryzivorus*) to the sharper bill of orioles, the stout, heavy bill of the large grackles, and the impressive beak of the oropendolas. The culmen is flattened and straight; this feature enables the birds to perform their characteristic 'gaping' behavior (see below). Overall, the birds are stocky, but some orioles are slim, while the meadowlarks are rotund. In most species, the tail is of medium length, extreme variations being the short tail of the meadowlarks and the long, keel-shaped tail of the grackles. All have strong legs, particularly those species that feed on the ground.

Plumage Most New World blackbirds show black, yellow, orange, or red in the plumage. Some species are sexually dichromatic, with females being brownish and streaked.

Many of the American blackbirds and grackles have largely or fully black plumage, often with blue or green iridescence. Flash colors may be present as red on the shoulders or yellow on the head. The black species often have bright yellow eyes.

The meadowlarks show upperparts that are streaked and patterned for camouflage. However, the underparts are bright yellow in northern species and red in southern species.

Body plumages in the orioles are a striking mix of yellow, orange, or chestnut with black. Female orioles vary in appearance; most migratory species are dull and yellowish, while resident tropical birds tend to resemble the males in plumage.

The Neotropical blackbirds include some very showy species. The presence of bright yellow, orange, or red flash colors is typical. Most caciques and oropendolas have bright yellow sides to the tail, and red or yellow flash colors on the rump or shoulders. Some oropendolas are mainly green in color, but even these birds generally have striking plumage, with a colorful bill and bright yellow on the tail.

Behavior All members of the family share a special musculature that enables them to open their bill with great force—a behavior called 'gaping.' This feature allows them to forage by inserting the bill tip in the ground or a crevice and then powerfully opening it, to create a hole from which to extract food. Another common behavior is 'bill-tilting,' in which a bird points its bill up at the

Number of genera 26
Number of species 98
Conservation Status 1 species Extinct, 1 species Critically Endangered, 6 species Endangered, 6 species Vulnerable, 1 species Near Threatened
Distribution North, Central, and South America

Great-tailed Grackle (*Quiscalus mexicanus*), as its name suggests, is distinctive for its large, keel-shaped tail. This bird is often seen in urban and suburban areas. Texas, U.S.

sky. This is an aggressive signal, and often is given to another bird nearby during feeding and territorial disputes. When singing to advertise their territory, the birds make use of their flash colors, flaring red shoulders, exposing yellow breasts, or twisting yellow heads. In displays, grackles deeply keel their tail, ruffle their plumage, and drop their wings. Black species often show off their yellow eyes. Oropendolas' displays can be extreme, with males bowing in an exaggerated manner, with the tail elevated, while singing, and hanging almost upside-down by the end of their song. Many territorial species are highly social outside the breeding season, forming large mixed-species flocks.

A male **Yellow-headed Blackbird** (*Xanthocephalus xanthocephalus*) defends his marshland territory with distinctive song and posture. U.S.

Voice Generally loud and obvious. The songs of American blackbirds and grackles can be rather unmusical, while those of orioles and meadowlarks are sweet and melodious. Some of the orioles and caciques are fine mimics. Certain oropendola species, when displaying, produce two different sounds at the same time, one descending and one ascending; this effect creates a distinctive, complex, cascading song.

Habitat The family is found from Alaska to Tierra del Fuego. They occupy a variety of habitats, from lowland tropical forest in the case of oropendolas and caciques, to grassland and farmland in the meadowlarks, Bobolink, various American blackbirds, and some grackles, and freshwater marshes in other American blackbirds and grackles.

Movements The most migratory species breed in North America. Bobolink shows the longest migration, wintering as far south as Argentina. The orioles are also strong migrants, but most reach only to Central America, few getting as far as South America. Tropical blackbirds tend to be more sedentary.

Diet Variable, from seeds (wintering flocks) and fruit (orioles, caciques, oropendolas) to invertebrates and small vertebrates. Grackles and some blackbirds have been known to eat songbirds at times of food stress.

Breeding The family shows some of its most interesting behavior in its mating systems. Complicated mating strategies, and brood parasitism, are common in this group.

Orioles tend to be largely monogamous. In contrast, the caciques, oropendolas, larger grackles, and marsh-nesting American blackbirds are polygamous. In the large grackles and many oropendolas and caciques, the females form a breeding colony in a tree, and the largest and most aggressive male defends this tree and mates with the 'harem' of females. The practice of defending a harem is common in mammals but almost unknown in birds.

Most species build open cup nests of twigs, leaves, and lichens, cemented with mud or rotting vegetation; some are placed over water. Clutches comprise three to six eggs, incubated for 12–15 days. The young are altricial, but fledge in 9–35 days. The cowbirds are brood parasites, laying their eggs in other species' nests.

Chestnut-headed Oropendola (*Psarocolius wagleri*) often has its nests parasitized by the related Giant Cowbird (*Molothrus oryzivorus*), but its young apparently suffer no ill effects. Panama.

BANANAQUIT *Coerebidae*

The small, active Bananaquit (*Coereba flaveola*) belongs to a Neotropical family that contains only one species, divided into 41 subspecies. It is common throughout its range, and is often tame and confiding with humans.

Structure A short-tailed, short-legged bird, Bananaquit is only 4.5 in. (11 cm.) in length. It has a relatively long, thin, distinctly decurved bill with small, red, fleshy corners at the gape. Its fringed, brushlike tongue is specialized for feeding on nectar.

Plumage The sexes are alike. The plumage is generally black, white, and yellow, with a white eye stripe, a white spot at the base of the primaries, and a yellow rump. Most subspecies have a gray or a white throat. There is a variable amount of yellow on the underparts. A black color-morph exists on two Caribbean islands. Immature birds are duller and paler, with a yellowish eye stripe.

A **Bananaquit** sticks its bill and tongue into a banana blossom to extract nectar. Puerto Rico, U.S.

The **Bananaquit's** sharp bill enables it to pierce the base of flower petals to take nectar. Trinidad.

Behavior An active bird, Bananaquit is fairly social, though territorial, at food sources. It often joins mixed flocks of tanagers, warblers, and honeycreepers. The bird forages by piercing the bases of larger flowers to gain access to nectar, and often hangs upside down or creeps along branches when feeding. It builds both dormitory and breeding nests, using dormitory nests for communal roosting at night. In addition, Bananaquit often comes to bird tables and feeders if people put out fruit or nectar for it.

Voice The song varies throughout the bird's range, but is generally a high-pitched or rather shrill and buzzy twitter. Calls include short *tsip* notes and an ascending *zhree*.

Habitat Bananaquit is generally found in lowlands, in a variety of open and semi-open habitats in both humid and arid regions.

Movements Sometimes individuals wander to the mainland U.S. (Florida), but the bird is generally sedentary throughout its range.

Diet Primarily nectar, but the bird also takes fruits and berries, in addition to small insects and spiders.

Breeding The bird builds an untidy globular nest, quite unlike those of dacnises, honeycreepers, warblers, and most tanagers. Nests are constructed by both sexes, although males do most of the work. Breeding is seasonal in areas where rainfall is seasonal; otherwise, it occurs year-round. The bird often produces several broods each year. It usually lays two eggs (although clutches of up to six have been reported), which are incubated for 12–13 days. The young are altricial, but fledge after about 17–19 days.

Taxonomy Of uncertain taxonomic relationships, Bananaquit is often placed near the tanagers (family Thraupidae) and wood-warblers (Parulidae). Formerly, the exclusively Neotropical family Coerebidae included the honeycreepers, conebills, dacnises, and flowerpiercers, but all of these have been assigned to other families (mainly tanagers). Bananaquit is the sole remaining species in the family, and recent DNA evidence places it with certain island species currently classified with buntings and American sparrows (Emberizidae). As a result, the American Ornithologists' Union (2005) proposed that the family Coerebidae is likely no longer valid.

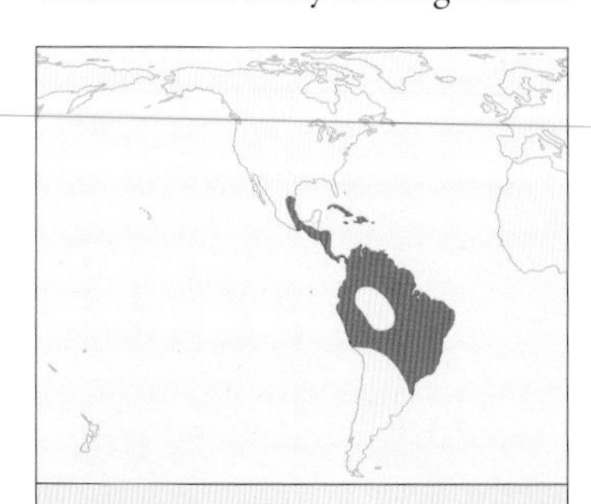

Number of genera: 1
Number of species 1
Conservation Status Least Concern
Distribution Caribbean, Central and South America

BUNTINGS AND AMERICAN SPARROWS *Emberizidae*

Most of the well-known species in this family are called 'sparrows,' after the 'true' sparrows that originated in the Old World (family Passeridae), but the two families are only distantly related. This pattern has been repeated through history, with largely New World (or Australian) groups being named by comparison with unrelated, but similar-looking, groups from Europe. In Europe, members of the family Emberizidae are known as buntings; however, this name has been used in the Americas for another, entirely unrelated group in the cardinal-grosbeak family (Cardinalidae). As such, there is no good English name for this group that is not somehow misleading. Here, they are called the buntings and American sparrows, or 'emberizids' for short.

The emberizids include some of the best known of all birds, such as Song Sparrow (*Melospiza melodia*) and White-crowned Sparrow (*Zonotrichia leucophrys*) in North America,

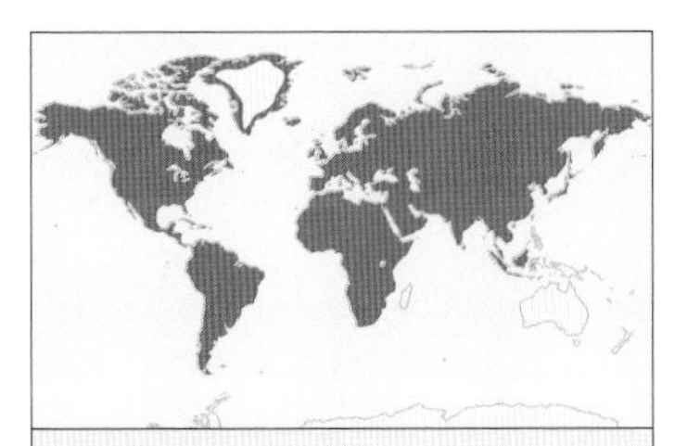

Number of genera 73
Number of species 308
Conservation Status 6 species Critically Endangered, 11 species Endangered, 19 species Vulnerable, 26 species Near Threatened
Distribution Worldwide, apart from Australia and Antarctica; most species are found in the Americas

Seaside Sparrow (*Ammodramus maritimus*) lives in salt marsh on the Atlantic and Gulf coasts of North America. New Jersey, U.S.

and Yellowhammer (*Emberiza citrinella*) in Europe. Even in temperate Latin America, one of the most widespread and familiar species belongs to this group: Rufous-collared Sparrow (*Zonotrichia capensis*).

Structure Most members of this family are aptly described as 'sparrow-sized,' although the seedeaters in genus *Sporophila* are smaller, and the towhees (genus *Pipilo*) are larger and longer-tailed than a sparrow. The species range from 4 to 9.5 in. (10–24 cm.) in length.

All emberizids have a conical bill, adapted for eating seeds. Within this general scheme, however, shapes vary from the short, thick, parrotlike bills of *Sporophila* seedeaters to the massive bills of seed-finches (genus *Oryzoborus*). In addition, there are the longer bills of Gough Bunting (*Rowettia goughensis*) and Slender-billed Finch (*Xenospingus concolor*). In contrast, Warbler-Finch (*Certhidea olivacea*) has a thin, warblerlike bill, as its name suggests. The full range of emberizid bill shapes can be found within one group, on one set of islands: the Darwin's finches of the Galápagos Islands. Even within a single genus, *Geospiza*, shapes range all the way from massive, in some of the Darwin's finches, to slender, in the cactus-finches. It is no mystery why this group helped confirm Charles Darwin's ideas concerning evolution and plasticity in species.

In body shape, emberizids range from stocky, as in buntings and American sparrows, to elongated in birds such as towhees and brush finches (genus *Buarremon*).

The wings in all species have nine primaries, but wing shapes differ dramatically depending on migratory tendencies. Tropical resident species have short, rounded wings, while highly migratory species, such as many buntings (*Emberiza*), longspurs (*Calcarius*), and Snow Bunting (*Plectrophenax nivalis*), have notably long, relatively pointed wings.

The tail is comprised of 12 rectrices. It varies from short, in many of the Darwin's finches, to long, in towhees, grass finches (genus *Emberizoides*), and Long-tailed Reed-Finch (*Donacospiza albifrons*).

As emberizids are terrestrial birds, they have strong legs and feet. Species that forage by scratching, such as towhees and Fox Sparrow (*Passerella iliaca*), have relatively

Lincoln's Sparrow (*Melospiza lincolnii*) can be recognized by its striped crown and its finely streaked, buffy breast. U.S.

A male **Snow Bunting** in breeding plumage. In winter, the birds have a rusty-colored crown and nape, and much of the black coloration is not visible. Alaska, U.S.

large feet with strong claws. The longspurs (genus *Calcarius*) are named for their long hind claw, or calcar.

Plumage On the whole there is no specific plumage pattern that unites all emberizids. However, most species are streaked somewhere on their body. The typical American sparrows, and the *Emberiza* buntings, are usually streaked on the upperparts to provide some camouflage in their terrestrial habitats, and some species are also well streaked on the underparts. Those that are not streaked, like the towhees, and some tropical groups, have streaked juvenal plumages. While the birds are generally camouflaged when seen from above, several groups have contrasting white areas on the tail, which they use to signal other species, or perhaps to deflect predators.

Although this is not a hard and fast rule, many buntings and American sparrows have stripes somewhere on the face, be they dark lateral throat stripes, pale malar stripes, bold eye stripes, pale supercilia, or crown stripes. Some species have all of these.

Many *Emberiza* buntings are boldly patterned on the face and head, or have colorful underparts. An extreme example is the male Snow Bunting, which, in breeding plumage, is white with a black back, as well as black-and-white wings and tail. In contrast, the male Lark Bunting (*Calamospiza melanocorys*), in breeding plumage, is entirely black apart from a large white wing patch. The striking Crested Bunting (*Melophus lathami*) is another black species, with a noticeable crest and contrasting rufous wings and tail.

Although considered merely 'little brown birds' by many people, a good proportion of North American sparrows are, in fact, superbly patterned and colored when observed at close quarters. One example is LeConte's Sparrow (*Ammodramus leconteii*), with its rich coloration and complex streaked patterning.

The most colorful genus is the red-crested cardinals of South America (*Paroaria*), which have a bright red head contrasting with white underparts and blackish or gray upperparts. Some of these birds are kept in captivity on account of their fine song and striking plumage.

The brush finches are also attractively colored, often yellow or white below and olive or gray above, with varying patterns of white, yellow, or rufous on the crown.

The ground sparrows (genus *Melozone*) are a small group that show unusual head patterns, including white patches on the face, or white ear patches that look rather like headphones.

The male *Sporophila* seedeaters vary from black-and-white to rusty and gray. Many species have contrasting darker caps, earning them the colloquial name 'cappuccinos' in Argentina. Female seedeaters, on the other hand, are among the most nondescript of all passerines, being pale dun, lacking streaks or patterns of any type. In fact, many females in this group cannot be readily identified as to species in the field.

Behavior Many emberizids flock outside the breeding season, although they are strictly territorial while breeding. Mixed-sparrow flocks are a highlight of birding in the fall and winter in parts of North America. When breeding, emberizids can usually be located by their songs, given from an exposed perch.

The birds usually forage on the ground, stripping seed heads and removing the seed husks with their strong bill. The towhees have a singular habit of 'double-scratching' the ground with both feet at once, to unearth food.

Voice The singing ability of emberizids varies between species. Grassland species have weak, insectlike songs. On the other hand, Song Sparrow and White- throated Sparrow (*Zonotrichia albicollis*) are known for their pleasant voice. Some open-country buntings, as well as Lark Bunting and the longspurs, have longer, more complicated songs, which they give in flight during aerial displays.

Habitat Buntings and American sparrows occupy a wide range of habitats. Northern-hemisphere species

Yellowhammer (*Emberiza citrinella*) is found across Europe, but has suffered a decline in population due to modern intensive agricultural practices. U.K.

Lapland Longspur (*Calcarius lapponicus*), which breeds in Arctic Eurasia and North America, spends most of its time on the ground, foraging for seeds or insects. Norway.

are often associated with open country, including tundra, grassland, and marshes, although many are also margin and shrub birds. The deserts of Mexico and the United States hold a good assortment of emberizids, and some Old World species are also arid habitat specialists. The brush finches are closely associated with montane forests, often cloud forest. Few emberizids occur in lowland tropical forest; those that do tend to live along rivers, on river islands, or at wetland edges. In contrast, many species occupy dry forest and scrubland in South America. Some, such as the sierra finches (genus *Phrygilus*), yellow finches (*Sicalis*), or *Melanodera* finches, occur in alpine habitats, often at higher levels than almost all other passerines.

Movements Most tropical species are sedentary, but migration is marked in temperate species in both the northern and southern hemispheres. Emberizids are hardy: although they will retreat from far northern breeding sites in the northern hemisphere, they may winter at relatively high latitudes. The most highly migratory species include some of the Old World *Emberiza* buntings, as well as the Arctic-breeding *Plectrophenax* buntings and longspurs.

Tricolored Brush Finch (*Atlapetes tricolor*) is resident in moist montane forests in northwest South America. Ecuador.

Diet As the birds' bill shape suggests, seeds are an important part of the diet. The short bills of the seedeaters are specially adapted to foraging on grass seeds. Species with large bills take larger, harder-shelled seed types. Forest species, such as brush finches, also take a good proportion of invertebrates as food. In fact, the diet of many emberizids includes a higher proportion of insects and other invertebrates during the breeding season, and the young are usually provisioned with insect prey.

Breeding In general, members of this group are monogamous. However, some species do not form pair bonds at all and are polygynous.

Most emberizids nest on or close to the ground. The nests are cup-shaped, and in grassland species they are well concealed from above by overhanging blades of grass. Some of these species may also create a grass tunnel leading to the nest itself. Clutch sizes range from three to five eggs, with tropical species having smaller clutches than high-latitude emberizids; some Arctic species may lay as many as six to seven cream or greenish, spotted eggs. All incubation is done by the female. Males in most species help feed the young at the nest, particularly in tropical species, but less so in some Old World buntings. Many species rear more than one brood in a season, particularly those from the mid-latitudes.

Taxonomy Recent genetic work has complicated the issue of defining which birds are actually emberizids. Several largely South American emberizids may in fact be tanagers (family Thraupidae) or cardinal-grosbeaks (Cardinalidae), while the bush-tanagers (genus *Chlorospingus*), which are treated here as tanagers, may be emberizids. This is the problematic legacy of our original classifications, which relied heavily on bill shape; it is now known that bill shape is very plastic and can change quite quickly, giving us false impressions of which groups are related. The closest relatives of the Emberizidae appear to be the New World blackbirds (family Icteridae).

Large Ground Finch (*Geospiza magnirostris*) is one of the so-called 'Darwin's finches,' which have been seen to evolve rapidly in response to environmental pressures. Galápagos Islands, Ecuador.

TANAGERS *Thraupidae*

The tanagers and their allies are remarkably varied in terms of coloration, habitat preference, and behavior. Many species are brightly colored or even gaudy. This New World family reaches its greatest diversity in Central and northern South America.

Tanagers and their allies make up one of the largest of all bird families, although there is much debate about the inclusion of several large groups, including most North American tanagers (genus *Piranga*), ant tanagers (*Habia*), bush tanagers (*Chlorospingus*), euphonias (*Euphonia*), and chlorophonias (*Chlorophonia*).

Summer Tanager (*Piranga rubra*) is well known in southern North America for the male's scarlet breeding plumage. However, it is sometimes seen as a pest for eating bees. U.S.

Structure Tanagers are small to medium-sized birds, ranging in length from Short-billed Honeycreeper (*Cyanerpes nitidus*) and several of the euphonias, at 3.5 in. (9 cm.), to Magpie-Tanager (*Cissopis leverianus*), at 11.5 in. (29 cm.).

There is a wide variation in morphology, especially in bill structure and tail length. Ant tanagers are relatively bulky, with strong legs and feet; mountain tanagers (*Buthraupis* and *Anisognathus*) have a large body and a fairly long tail; Magpie-Tanager has a very long tail; and *Bangsia* and *Wetmorethraupis* tanagers are stocky, medium-sized birds with a short tail. At the other end of the scale, conebills in genus *Conirostrum*, honeycreepers (*Chlorophanes*, *Cyanerpes*, and *Iridophanes*), and dacnis (*Dacnis*) are small and delicate, while euphonias are very small, short-tailed, and often rotund.

Bills differ greatly in length and shape. Dacnis have a short, pointed bill, whereas conebills have a longer, finely pointed bill, honeycreepers have a slender, decurved bill, and ant tanagers have a relatively long, conical bill. In contrast, *Buthraupis* have a strong, relatively chunky, bill; *Ramphocelus* birds have a swollen, silvery-white base to the mandible; and Magpie-Tanager has a very powerful bill. These structures tend to reflect the birds' feeding specializations. For example, *Diglossa* flowerpiercers have a slender, upturned bill with a hooked tip, which enables them to pierce flowers at the base of the corolla and feed on the nectar.

Plumage As befits such a large family, there is a wide assortment of plumages. Some species show the most brilliant colors seen in any birds, and/or complex patterns. Others are bright but simply patterned. At the opposite end of the spectrum, there are many duller species. Sexual dichromatism is a feature of some genera, but not others.

Multicolored Tanager (*Chlorochrysa nitidissima*) of western Colombia has one of the most complex of all color patterns. The male has a yellow face and throat; a bright green nape; black patches on the sides of the neck, bordered by chestnut below; a yellow back; a greenish-blue rump; an emerald-green tail and wings; and bright blue underparts with a black median patch. Females have similar, though duller, plumage, lacking the black on the breast and belly, and with a green

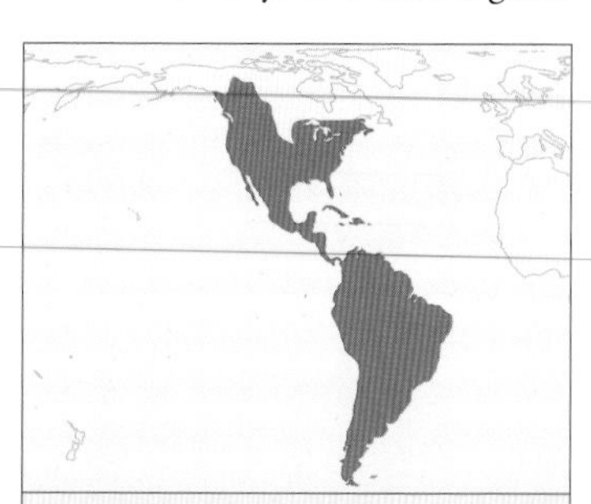

Number of genera 62
Number of species 271
Conservation Status 2 species Critically Endangered, 6 species Endangered, 14 species Vulnerable, 16 species Near Threatened
Distribution North and South America

back. *Tangara* species are often similarly bright. The northern Brazil race of Paradise Tanager (*Tangara chilensis paradisea*) has a highly complex plumage, with a bright green head; a violet throat; lustrous blue underparts; a black nape, mantle, wing-coverts, and tail; and a crimson upper and bright yellow lower rump. The euphonias and chlorophonias are also brightly colored, with shades of blue, yellow, and green prominent.

Species in genus *Dacnis* are colorful, but rather simpler in pattern. For example, male Blue Dacnis (*D. cayana*) are mostly bright, lustrous blue and black, while the females are mostly bright leaf-green. Male honeycreepers are usually bright blue or purplish, while females are greenish above and streaked below; both sexes often have red or yellow legs. *Buthraupis* mountain tanagers have yellow underparts with a black throat, and blue, green, or gold upperparts. *Iridosornis* species have iridescent, bright yellow on the head, either on the throat or the nape, and orange undertail-coverts. *Ramphocelus* species are lustrous red and black; and *Bangsia* and *Wetmorethraupis* tanagers also exhibit bright, contrasting patterns.

Not all tanagers are so bright. *Thlypopsis* species are mostly gray, with touches of orange and rufous. The hemispinguses (*Hemispingus*) are mostly dull olivaceous with touches of black and gray; and *Chlorospingus* are mostly yellowish-olive above and gray below, with some white and yellowish. Male ant tanagers are usually reddish, with a crest, and females are mostly yellowish below and olive above, or umber all over. The flowerpiercers are mostly black or very dark blue, some with russet on the underparts.

With its vivid turquoise underparts, chestnut back, and almost fluorescent green head, **Paradise Tanager** is the most spectacular of the tanagers. Location unknown.

Behavior Some of the family are solitary or live only in pairs. Other species are highly gregarious, regularly joining mixed-species foraging parties; they include those in genera *Hemispingus*, *Chlorospingus*, *Tangara*, and *Euphonia*. Swallow-Tanager (*Tersina viridis*) sometimes forms single-species parties of up to a dozen birds; this species forages high in the canopy in forest, but closer to the ground in more open habitats. *Thraupis* species usually form pairs or family parties and tend to be highly animated, often bickering. In contrast, *Bangsia* and *Wetmorethraupis* birds are relatively inactive.

Some tanagers have distinctive feeding behaviors. Giant Conebill (*Oreomanes fraseri*) creeps along trunks and branches, nuthatch-like, as it searches for insects. Plush-capped Finch (*Catamblyrhynchus diadema*) is a restless forager, clinging upright or upside-down on bamboo stalks, and pressing its bill into leaf whorls. Flowerpiercers spend much of their time piercing flowers at the base of the corolla to reach the nectar. Magpie-Tanager is noisy and conspicuous, wagging its long tail and fanning it when agitated.

Voice Most tanagers are not renowned songsters. Many species' songs comprise simple series of repeated high notes. However, some birds have more memorable songs. For example, Rosy Thrush-Tanager (*Rhodinocichla rosea*) repeats loud, rich, wrenlike phrases: *wheeo, chee-oh, chweeo*; often, both members of a pair duet. Some ant tanagers deliver sweet songs; as with other tanagers, these are mostly given around dawn, but certain species have a second daytime song. In some species

In **Western Tanager** (*Piranga ludoviciana*), the red coloration on the male's face is actually acquired from the bird's diet—possibly from pigment in insects. U.S.

of *Piranga* and *Thraupis*, the female sings a similar, but weaker, song to the male. Contact calls are generally thin and high-pitched; alarm calls are usually sharper.

Habitats Most tanagers are birds of forest and woodland. Many live in wet, mossy rain forests, while others occupy forest margins, open country, suburban backyards, and even semiarid scrub. Within forest interiors, some forage at ground level or in the understory, but more feed at higher levels, especially in the canopy. Depending on the dietary preferences of the species, they may favor areas with flowers or fruit. Plush-capped Finch is a *Chusquea* bamboo specialist, although it is not entirely dependent on these plants. Giant Conebill generally lives only in high Andean *Polylepis* forest.

The birds are roughly evenly divided between those species whose range is centered below 3,300 ft. (1,000 m.) and those above. Many species live at sea level, and others range in altitude to just below the treeline. Giant Conebill lives at elevations up to 14,750 ft. (4,500 m.), and several other tanagers reach close to this altitude.

Movements Most species are sedentary. However, those breeding at the highest latitudes in the temperate northern and southern hemispheres migrate seasonally from those regions to avoid winter food shortages. The species that breeds farthest north, Western Tanager (*Piranga ludoviciana*), ranges as far north as the southern Northwest Territories of Canada. In the fall, it vacates Canada and the western U.S., to winter south to Costa Rica.

In the tropics, some species are nomadic, roaming in response to cycles of rainfall and food supply. For example, in Peru, Black-and-white Tanager (*Conothraupis speculigera*) will leave areas during dry periods, relocating to regions where rainfall has been plentiful.

Rosy Thrush-Tanager (*Rhodinocichla rosea*) is a striking but shy bird that forages on the ground. It is more often heard than seen, and is admired for its beautiful song.

Diet Most tanagers take both fruit and insects, and a sizeable minority are also known to eat flowers or nectar; the flowerpiercers and honeycreepers feed on nectar for a good part of the year.

When eating small fruits, tanagers typically stretch to reach them while perched, then mash the fruit in their bill before swallowing it. Usually, they swallow the entire fruit, including the seeds. Arils, *Cecropia* fruits, and mistletoe are particular favorites; the euphonias and chlorophonias are specialized mistletoe feeders.

Common invertebrate food includes larvae and caterpillars, beetles, spiders, grasshoppers, and crickets. Different groups employ a

Blue-winged Mountain Tanager (*Anisognathus somptuosus*) is found in the Andean cloud forest, at altitudes of about 6,560–8,200 ft. (2,000–2,500 m.). Ecuador.

Red-crowned Ant Tanager (*Habia rubica*) forages in the forest understory, in pairs or family groups. Males are recognizable by their bright red, erectile crest. Panama.

variety of methods to find invertebrate food. Some inspect the undersides of branches while moving along them; others hang over the edge of their perches to check the undersides of leaves; and others concentrate their attentions on epiphytes or masses of moss growing on trees. Giant Conebill pulls fragments of bark from the trunks and branches of *Polylepis* trees to access the invertebrates beneath. Some tanagers take insects in flight: for example, *Piranga* species when bees are plentiful. Others follow army ant swarms, feeding on invertebrates disturbed by the ants; ant tanagers and Gray-headed Tanager (*Eucometis penicillata*) are examples. Many species forage on the ground.

Breeding Tanagers usually breed at times when food is most plentiful. For temperate-breeding species such as Summer Tanager (*Piranga rubra*) and Scarlet Tanager (*P. olivacea*), that means during the northern-hemisphere summer, when insects are most abundant. For the tropical species, breeding takes place when fruit, insects, or—for the nectivorous tanagers—plants in flower are most readily available.

Some species, at least, sing to defend a territory. However, much remains unknown about this behavior as well as other aspects of the breeding ecology in many species.

Most tanagers build an open-cup nest, usually in a tree but sometimes on the ground. Euphonias and chlorophonias build a dome-shaped nest with a side entrance. Uniquely within the family, Swallow-Tanager nests in a cavity in an earth bank, a cliff, or even the wall of a building. In most species, the female alone builds the nest.

Usually, two or three eggs are laid, though some species have clutches of up to five. In most tanagers, only the female incubates, although Rosy Thrush-Tanager is an example of a species in which this duty is shared between the parents. Incubation is typically 12–18 days. Both parents feed the chicks. Cooperative feeding has been noted in Golden-hooded Tanager (*Tangara larvata*), with subadults and other adults helping parents feed young. Fledging takes place at 11–24 days; the time taken to fledge is longest in the euphonias and chlorophonias.

Taxonomy Much debate concerns the inclusion of several genera within the Thraupidae. DNA evidence now suggests that birds in genera *Habia* and *Piranga* should be included in family Cardinalidae, and that *Euphonia* and *Chlorophonia* should be in family Fringillidae. Plush-capped Finch is treated as *incertae sedis* by some authorities.

The different races of **Orange-bellied Euphonia** (*Euphonia xanthogaster*) vary in crown color. Some have a yellow crown, as shown here; in others, the crown is rufous. Peru.

GROSBEAKS AND ALLIES *Cardinalidae*

A male **Northern Cardinal** (*Cardinalis cardinalis*). The reds, oranges, and yellows seen in colorful species like this are derived from carotenoid pigments in the birds' diet. U.S.

Colorful New World songbirds, the Cardinalidae are noted for their conical bill and, in many species, the males' bright plumage. The main groups are the grosbeaks, sometimes called cardinal-grosbeaks, which are named for their stout bill; the saltators of tropical South America, named for their jumping gait; and the bright buntings of genus *Passerina*.

A male **Rose-breasted Grosbeak** (*Pheucticus ludovicianus*). The stout, conical bill is typical of birds in family Cardinalidae. U.S.

Structure The species range from 5 to 8 in. (12–21 cm.) in length; males typically are slightly larger than females. The most notable feature is the conical bill, which is specialized for crushing seeds. The bill is proportionately largest in grosbeaks of genus *Pheucticus*, and smallest in the *Passerina* buntings; the culmen may be straight or curved. The body is short and somewhat stocky, and the head proportionately large. The wings vary from short and rounded, in more sedentary species, to long and pointed, in highly migratory species. The tail is generally of medium length, with a slightly notched to rounded tip.

Plumage In most species the sexes are strongly dichromatic, but in the tropical species in genus *Caryothraustes* they are alike. In addition, some northern hemisphere-breeding species (such as some in genera *Passerina* and *Pheucticus*) show marked seasonal variation.

Coloration ranges from relatively drab to gaudy, with male *Passerina* buntings being especially dazzling. In some species, the males (and, to a lesser extent, the females) show bright reds, oranges, and yellows. The male plumage of genera *Cyanocompsa*, *Cyanoloxia*, and *Passerina* includes various shades of blue, while the females exhibit shades of brown or olive. Many species have distinctive mask, head, or breast markings, contrasting wing bars or spots, or tail spots. Genus *Cardinalis* has a prominent crest.

Iris colors range from dark brown to red. The bill varies from uniform to strongly bicolored, in black or shades of brown, gray, ivory, red, or orange. The legs and feet are usually gray, but browner in some species.

Juveniles have a distinct, though briefly held, plumage. The first postjuvenal plumage is duller than that of adults, resembling that of the adult female in sexually dichromatic species. Birds may not attain adult plumage until their second year (for example, in *Pheucticus* grosbeaks and *Passerina* buntings). Adults have a single complete postbreeding molt. Some also have a partial prebreeding molt, and *Passerina* buntings have a partial third or supplemental molt.

Behavior Most species are solitary, but a few join flocks or form loose aggregations outside the breeding season. Black-faced Grosbeak (*Caryothraustes poliogaster*) is the most gregarious, and can occur in groups of

Number of genera 11
Number of species 42
Conservation Status Least Concern
Distribution New World

50 or more. Yellow-shouldered Grosbeak (*Parkerthraustes humeralis*) and Crimson-collared Grosbeak (*Rhodothraupis celaeno*) join mixed-species foraging flocks.

The birds forage on the ground or in vegetation; grosbeaks and saltators tend to be the most arboreal. Many species are difficult to observe as they skulk in vegetation.

Voice Songs are usually given by males, and are usually complex, consisting of repetitive, buzzy notes, slurred whistles, or warbling phrases; the females of some species countersing. Call notes, used during flight, or for location or alarm, may be single or double, and in many species are squeaky or metallic in quality.

Habitat A variety of arid to humid habitats, including grasslands, scrub, secondary growth, forest margins, and forest interior, at elevations from sea level to 11,480 ft. (3,500 m.).

Movements Tropical species are generally sedentary, but northern hemisphere breeders include sedentary species and short- to long-distance migrants. Rose-breasted Grosbeak (*Pheucticus ludovicianus*) is the family's longest-distance migrant, breeding as far north as Canada and wintering as far south as Peru.

Diet A wide variety of seeds, fruit, and flowers. All grosbeaks initially feed insects to their young, switching to a typical adult diet once the young have fledged. Yellow-shouldered Grosbeak, however, is generally insectivorous year-round.

Breeding Seasonal in temperate species, but often year-round in tropical birds; breeding is usually synchronized with peaks in insect abundance. Pairs are monogamous.

The nest, usually built by both members of the pair, is an open cup woven with grasses and other plant materials and lined with finer materials, and placed off the ground in a bush or tree. The female incubates the eggs and broods the young, while the male provides food for her. Clutches comprise two to four eggs, with tropical species laying the fewest eggs. The birds produce one to three broods per season, depending on species, and will lay a replacement clutch if their nest is destroyed before the young have fledged.

Buff-throated Saltator (*Saltator maximus*), a Neotropical species, is unusual among the Cardinalidae in that the sexes are alike in plumage coloration. Panama.

The chicks hatch naked, or with sparse down (as in *Passerina* buntings and Dickcissel), with eyes closed, and with colorful mouth or gape flanges. They fledge in 7–15 days (tropical species are poorly known), typically before they are capable of flight. The young are cared for by both parents, or by the male (if the female starts another brood), for two to three weeks, until fully grown. Some may be fed for much longer: for example, those of Northern Cardinal (*Cardinalis cardinalis*) are fed up to 56 days. Many species are parasitized by cowbirds (genus *Molothrus*).

A male **Painted Bunting** (*Passerina ciris*), North America's most colorful bird. The population has declined, possibly due to habitat loss and capture for the pet trade.

Taxonomy Most members of family Cardinalidae known as 'cardinal-grosbeaks' are typified by their stout, conical bill, but no morphological or behavioral characters clearly delineate this group from its presumed closest relatives. Ongoing genetic studies may clarify relationships that are currently clouded by convergent morphology and behavior.

The genera here recognized as Cardinalidae are *Spiza*, *Pheucticus*, *Cardinalis*, *Caryothraustes*, *Parkerthraustes*, *Rhodothraupis*, *Periporphyrus*, *Saltator*, *Cyanoloxia*, *Cyanocompsa*, and *Passerina*. A recent genetic study supports absorbing the blue seedeaters (genus *Amaurospiza*) from family Emberizidae, as well as three genera (*Piranga*, *Chlorothraupis*, and *Habia*) currently placed with the tanagers, and an odd 'warbler' genus (*Granatellus*, which has an atypically heavy bill). This study also suggests that genus *Saltator* and Yellow-shouldered Grosbeak are not as closely allied to the cardinal-grosbeaks as previously believed.

ADDITIONAL READING

Whether you wish to observe birds in the field or study particular species in greater depth, the information provided below will supplement the articles in this book.

General Ornithology

del Hoyo, Josep.; Elliott, Andrew.; Sargatal, Jordi (eds.); Christie, David (eds), *Handbook of the Birds of the World*, Volumes 1–13, Lynx Edicions, Barcelona 1992–2008
A multi-volume work of international scholarship which, when completed in 2011, will discuss and illustrate every living species of bird. There are 13 volumes to date, with three further volumes still to come.

Dickinson, Edward C. (ed.), *The Howard and Moore Complete Checklist of the Birds of the World*, 3rd edition, Christopher Helm 2003
A checklist of every bird species in the world, with subspecies and ranges.

Gill, Frank B., *Ornithology*, 3rd edition, W. H. Freeman 2006
An authoritative, encyclopedic treatment of bird anatomy and behavior.

Proctor, Noble S.; Lynch, Patrick J., *Manual of Ornithology*, Yale University Press 1993
An in-depth guide to the anatomy and body systems of birds.

Regions of the World

This section is divided geographically and lists a small sample of the most respected and up-to-date field guides and regional handbooks.

North America

Alderfer, Jonathan (ed.), *Complete Birds of North America*, National Geographic Society 2006
A comprehensive guide to the characteristics, distribution, and populations of all wild bird species reliably recorded in North America.

Floyd, Ted, *Smithsonian Field Guide to the Birds of North America*, Collins 2008
Descriptions and high-quality photographs of more than 750 species.

Howell, Steve N. G.; Webb, Sophie, *A Guide to the Birds of Mexico and Northern Central America*, Oxford University Press 1995

Peterson, Roger Tory; Chalif, Edward L., *A Field Guide to Mexican Birds*, Houghton Mifflin Co. 1999
A field guide covering Mexico, Belize, Guatemala, and El Salvador.

South America

de la Pena, Martin; Rumboll, Maurice, *Birds of Southern South America and Antarctica*, Princeton University Press 1998
A field guide for regions from the jungles of the Amazon to the Pantanal of Brazil, the islands of the Southern Ocean, and Antarctica.

Hilty, Steven, *Birds of Tropical America*, University of Texas Press 2005

Restall, Robin; Rodner, Clemencia; Lentino, Miguel, *Birds of Northern South America*, Volumes 1 and 2, Yale University Press 2007
Full descriptions of more than 2,300 species and their confirmed distributions, covering the area from Ecuador to French Guyana. Volume 1 comprises species accounts; Volume 2 is a field guide.

Stutchbury, Bridget J.M.; Morton, Eugene S., *Behavioral Ecology of Tropical Birds*, Academic Press 2001

Europe, Middle East, Africa

Cramp, Stanley; Simmonds, K.E.L (eds.), *Handbook of the Birds of Europe, the Middle East and North Africa: The Birds of the Western Palearctic*, Volumes I–IX, Oxford University Press 1977–1994
Standard reference to European birds; contains comprehensive descriptions of physical characteristics and behavior for each species, as well as a field guide. Now out of print, but an updated version is available on CD-ROM.

Fry, C. Hilary; Keith, Stuart; Urban, Emil K. (eds.), *Birds of Africa*, Volumes 1–7, Princeton University Press 1982–2004

Mullarney, Killian; Svensson, Lars; Zetterstrom, Dan; Grant, Peter J., *Collins Bird Guide*, HarperCollins 1999

Sinclair, I.; Ryan, P., *Birds of Africa South of the Sahara*, Princeton University Press 2004
The first book to describe and illustrate all of the birds found in sub-Saharan Africa, and covers more than 2,100 species.

Sinclair, I.; Langrand, O., *Birds of the Indian Ocean Islands*, Struik 2003
Covers Madagascar, Mauritius, Réunion, Rodrigues, Seychelles and the Comoros Islands. Comprehensive, richly illustrated species accounts.

Asia

Rasmussen, Pamela C. ; Anderton, John C., *Birds of South Asia: The Ripley Guide*, Volumes 1 and 2, Lynx Edicions/Smithsonian Institution 2005
Volume 1 is a field guide. Volume 2 gives information on the physical characteristics, distribution, and taxonomy of about 1,500 species.

Australia, New Zealand, Antarctica

Marchant, S.; Higgins, P. J., *The Handbook of Australian, New Zealand and Antarctic Birds*, Volumes 1–6, Oxford University Press (USA) 1990–2003

Pizzey, G.; Knight, F., *Birds of Australia* (Harper Collins 1997)
Definitive guide to Australian birdlife, with information on nearly 800 bird species.

Robertson, Hugh; Heather, Barrie, *The Hand Guide to the Birds of New Zealand*, Oxford University Press (USA) 2001
The standard field guide for birding in New Zealand.

Worldwide

Elphick, Jonathan, *Atlas of Bird Migration*, Firefly Books 2007

Enticott, Jim; Tipling, David, *Seabirds of the World*, Stackpole Books 1997

Taylor, Don, *Shorebirds of North America, Europe, and Asia*, Princeton University Press 2006

Web Sites

www.bsc-eoc.org/avibase/avibase.jsp
A database of every bird species and subspecies in the world, with information including names in different languages, taxonomy, and distribution.

www.iucnredlist.org
A database of every species in the world, with information on distribution and conservation status.

animals.jrank.org
A free Web encyclopedia designed for students; the Birds Scientific References comprise several volumes of information on a wide range of bird families, including physical characteristics, geographic range, diet, behavior, and reproduction, together with accounts of selected species.

www.birds.cornell.edu/AllAboutBirds/
Authoritative, accessible Web site about birds in North America, including extensive species list and information on observation and conservation.

www.americanbirding.org
American Birding Association: gives information and education about birds and birding.

ADDITIONAL READING

www.rspb.org.uk
Web site for the UK's premier bird protection organization, offering information on major UK species and conservation issues.

www.birdsinbackyards.net
Factsheets on Australian birds, produced in association with the Australian Museum.

Books on Bird Families

For some families without specific books, we have given a general reference that includes detailed information. In some entries, families are grouped together and not all families are listed. In addition, the volumes of the *Handbook of the Birds of the World* (see General ornithology) feature 'book-length' monographs on every family.

TINAMOUS, OSTRICHES, RHEAS, CASSOWARIES, EMU

Davies, Stephen, *Bird Families of the World: Ratites and Tinamous*, Oxford University Press 2002
Pizzey, G.; Knight, F., *Birds of Australia*, Harper Collins 1997
Simpson, K.; Day, N., *Field Guide to the Birds of Australia*, 3rd edition, A&C Black / Christopher Helm Publishers 1998
Sinclair, I.; Ryan, P., *Birds of Africa South of the Sahara*, Princeton University Press 2004

KIWIS

Feduccia, A., *The Origin and Evolution of Birds*, Yale University Press 1996
Heather, B. and Robertson, H., *Field Guide to the Birds of New Zealand*, Oxford University Press 1997

MEGAPODES

Jones, Darryl N.; Dekker, Rene W. R. J.; Roselaar, Cees S., *Bird Families of the World: The Megapodes*, Oxford University Press (USA), 1995

CURASSOWS, GUANS

Delacour, Jean; Amadon, Dean, *Curassows and Related Birds*, Lynx Edicions 2004

GUINEAFOWL

Sinclair, I.; Ryan, P., *Birds of Africa South of the Sahara*, Princeton University Press 2004

GROUSE, PHEASANTS

Johnsgard, Paul A., *The Grouse of the World*, University of Nebraska Press 1983
Johnsgard, Paul A., *The Pheasants of the World*, Smithsonian Institution Press 1999
Madge, Steve; McGowan, Phil, *Pheasants, Partridges, & Grouse*, Princeton University Press 2002

SCREAMERS, MAGPIE-GOOSE, DUCKS, GEESE, SWANS

Kear, Janet, *Bird Families of the World: Ducks, Geese and Swans*, Volumes 1 and 2, Oxford University Press (USA) 2005
Madge, Steve; Burn, Hilary, *Waterfowl*, Houghton Mifflin 1988

PENGUINS

Lynch, Wayne, *Penguins of the World*, Lynx Edicions 2007
Williams, Tony D.; Davies, J. N., *Bird Families of the World: The Penguins*, Oxford University Press (USA) 1995

ALBATROSSES, PETRELS, SHEARWATERS

Brooke, Michael, *Bird Families of the World: Albatrosses and Petrels across the World*, Oxford University Press (USA) 2004
Onley, Derek; Schofield, Paul, *Albatrosses, Petrels and Shearwaters of the World*, Princeton University Press 2007

GREBES

Fjeldsa, John, *Bird Families of the World: The Grebes*, Oxford University Press (USA) 2005
Ogilvie, Malcolm; Rose, Chris, *Grebes of the World*, Bruce Coleman 2002

FLAMINGOS, STORKS, IBISES, SPOONBILLS, HAMMERKOP, SHOEBILL

Hancock, J. A.; Kushlan, J. A.; Kahl, M. P., *Storks, Ibises and Spoonbills of the World*, Christopher Helm Publishers 1992
Mackenzie, John P. S., *Birds of the World: Wading Birds*, Key Porter Books 1997

HERONS, BITTERNS

Hancock, James; Kushlan, James, *The Herons Handbook*, Christopher Helm Publishers 1990
Hancock, James; Kushlan, James, *Bird Families of the World: The Herons*, Oxford University Press (USA) 2005

TROPICBIRDS, FRIGATEBIRDS, PELICANS, GANNETS, BOOBIES, CORMORANTS, ANHINGAS

Johnsgard, Paul A., *Cormorants, Darters, and Pelicans of the World*, Smithsonian Institution Press 1993
Nelson, J. Bryan, *Bird Families of the World: Pelicans, Cormorants, and their Relatives*, Oxford University Press (USA) 2006

NEW WORLD VULTURES

Palmer, R.S., *Handbook of North American Birds*, Volume 4, Yale University Press 1988

FALCONS, CARACARAS, OSPREY, KITES, HAWKS, EAGLES

Brown, Leslie; Amadon, Dean, *Eagles, Hawks & Falcons of the World*, The Wellfleet Press 1989
Ferguson-Lees, James; Christie, David A. (eds.), *Raptors of the World*, Princeton University Press 2006

BUSTARDS

Johnsgard, Paul A., *Bustards, Hemipodes, and Sandgrouse: Birds of Dry Places*, Oxford University Press (USA) 1991

RAILS, COOTS

Taylor, Barry; van Perlo, Ber, *Rails: A Guide to the Rails, Crakes, Gallinules and Coots of the World*, Yale University Press 1998

CRANES

Johnsgard, Paul A., *Cranes of the World*, Indiana University Press 1983

THICK-KNEES, SHEATHBILLS, OYSTERCATCHERS, CRAB PLOVER, STILTS, AVOCETS, PLOVERS, JACANAS, SANDPIPERS, SNIPE, COURSERS, PRATINCOLES

Hayman, Peter; Marchant, John; Prater, Tony, *Shorebirds: An Identification Guide*, Houghton Mifflin 1991

GULLS, TERNS, SKIMMERS

Grant, Peter J, *Gulls: A Guide to Identification*, Princeton University Press 1997
Howell, Steve N. G.; Dunn, Jon, *Gulls of the Americas*, Houghton Mifflin 2007
Olsen, Klaus Manning; Larsson, Hans, *Terns of Europe and North America*, Princeton University Press 1995

JAEGERS, SKUAS

Olsen, Klaus Manning; Larsson, Hans, *Skuas and Jaegers: A Guide to the Skuas and Jaegers of the World*, Yale University Press 1997

AUKS

Gaston, Anthony J.; Jones, Ian L., *Bird Families of the World: The Auks*, Oxford University Press (USA) 1998
Snow, D. W. and Perrins, C.M., *The Birds of the Western Palearctic (Concise Edition)*, Volume 1: Non-passerines, Oxford University Press 1998

SANDGROUSE

Johnsgard, Paul A., *Bustards, Hemipodes, and Sandgrouse: Birds of Dry Places*, Oxford University Press (USA) 1991

ADDITIONAL READING

PIGEONS, DOVES
Gibbs, David; Barnes, Eustace; Cox, John, *Pigeons and Doves*, Yale University Press 2001

PARROTS, COCKATOOS
Forshaw, Joseph M., *Parrots of the World: An Identification Guide*, Princeton University Press 2006
Juniper, Tony; Parr, Mike, *Parrots: A Guide to Parrots of the World*, Yale University Press 1998

TURACOS
Forshaw, Joseph M., *Turacos: A Natural History of the Musophagidae*, Nokomis Editions 2002

CUCKOOS
Payne, Robert B., *Bird Families of the World: The Cuckoos*, Oxford University Press (USA) 2005

BARN OWLS, OWLS
Duncan, James R., *Owls of the World: Their Lives, Behavior and Survival*, Firefly Books 2003
König, Claus; Weick, Friedhelm, *Owls of the World*, Christopher Helm Publishers 2008

FROGMOUTHS, OILBIRD, POTOOS,NIGHTJARS, NIGHTHAWKS, OWLET-NIGHTJARS
Cleere, Nigel, *Nightjars: A Guide to Nightjars and Related Nightbirds*, Pica/Christopher Helm Publishers 2000
Holyoak, David, *Bird Families of the World: Nightjars and their Allies*, Oxford University Press (USA) 2001

SWIFTS, TREESWIFTS
Chantler, Phil; Driessens, Gerald, *Swifts: A Guide to the Swifts and Treeswifts of the World*, 2nd edition, Yale University Press 2000

HUMMINGBIRDS
Fogden, Michael & Patricia, *Hummingbirds of Costa Rica*, Firefly Books 2006
Howell, Steve N. G., *Hummingbirds of North America: The Photographic Guide*, Academic Press 2002
Williamson, Sheri L., *A Field Guide to Hummingbirds of North America*, Houghton Mifflin Co. 2002

TROGONS
Johnsgard, Paul A., *Trogons and Quetzals of the World*, Smithsonian Institution Press 2000

ROLLERS, GROUND ROLLERS, CUCKOO-ROLLER, KINGFISHERS, TODIES, MOTMOTS, BEE-EATERS
Fry, C. Hilary; Fry, Kathie; Harris, Alan, *Kingfishers, Bee-eaters and Rollers*, Princeton University Press 2000

HORNBILLS, GROUND-HORNBILLS
Kemp, Alan, *Bird Families of the World: The Hornbills*, Oxford University Press (USA) 1995

BARBETS, TOUCANS, HONEYGUIDES
Short, Lester; Horne, Jennifer, *Bird Families of the World: Toucans, Barbets and Honeyguides*, Oxford University Press (USA) 2002

WOODPECKERS
Short, Lester L., *Woodpeckers of the World*, Delaware Museum of Natural History 1982
Winkler, Hans; Christie, David A. (eds.); Nurney, David, *Woodpeckers: An Identification Guide to the Woodpeckers of the World*, Houghton Mifflin Co. 1995

NEW ZEALAND WRENS
Robertson, Hugh; Heather, Barrie, *The Hand Guide to the Birds of New Zealand*, Oxford University Press 2001

BROADBILLS, ASITIES, SAPAYOA, PITTAS
Lambert, Frank; Woodcock, Martin, *Pittas, Broadbills and Asities*, Pica/Christopher Helm Publishers 2000

COTINGAS
Snow, David, *The Cotingas, Bellbirds, Umbrellabirds, and Other Species*, Cornell University 1982

ANTBIRDS, GNATEATERS, TAPACULOS, ANTTHRUSHES, ANTPITTAS, OVENBIRDS, WOODCREEPERS
Skutch, Alexander F., *Antbirds and Ovenbirds: Their Lives and Homes*, University of Texas Press 1996

LYREBIRDS, SCRUB-BIRDS
Higgins, Peter J.; Peter, John M.; Steele, W. K. (eds), *The Handbook of Australian, New Zealand and Antarctic Birds*, Volume 5: Tyrant-flycatchers to Chats, Oxford University Press (USA) 2001
Smith, L. H., *The Life of the Lyrebird*, W. Heinemann (Australia) 1988

BOWERBIRDS
Frith, Clifford B. and Dawn W., *Bird Families of the World: The Bowerbirds*, Oxford University Press (USA) 2004
Rowland, Peter, *Bowerbirds*, CSIRO Publishing 2008

AUSTRALASIAN TREECREEPERS, BRISTLEBIRDS, THORNBIRDS
Higgins, Peter J.; Peter, John M.; Steele, W. K. (eds), *The Handbook of Australian, New Zealand and Antarctic Birds*, Volume 5: Tyrant-flycatchers to Chats, Oxford University Press (USA) 2001

AUSTRALASIAN WRENS
Higgins, Peter J.; Peter, John M.; Steele, W. K. (eds), *The Handbook of Australian, New Zealand and Antarctic Birds*, Volume 5: Tyrant-flycatchers to Chats, Oxford University Press (USA) 2001
Rowley, Ian; Russell, Eleanor, *Bird Families of the World: Fairy-Wrens and Grasswrens*, Oxford University Press (UK) 1997

HONEYEATERS, PARDALOTES, BERRYPECKERS
Higgins, Peter J.; Peter, John M.; Steele, W. K. (eds), *The Handbook of Australian, New Zealand and Antarctic Birds*, Volume 5: Tyrant-flycatchers to Chats, Oxford University Press (USA) 2001
Higgins, Peter J.; Peter, John M. (eds), *The Handbook of Australian, New Zealand and Antarctic Birds*, Volume 6: Pardalotes to Shrike-Thrushes, Oxford University Press (USA) 2003

AUSTRALASIAN BABBLERS, LOGRUNNERS
Higgins, Peter J.; Peter, John M. (eds), *The Handbook of Australian, New Zealand and Antarctic Birds*, Volume 6: Pardalotes to Shrike-Thrushes, Oxford University Press (USA) 2003

WATTLED CROWS
Robertson, Hugh; Heather, Barrie, *The Hand Guide to the Birds of New Zealand*, Oxford University Press 2001

WHIPBIRDS, WEDGEBILLS, JEWEL-THRUSHES, QUAIL-THRUSHES
Higgins, Peter J.; Peter, John M. (eds), *The Handbook of Australian, New Zealand and Antarctic Birds*, Volume 6: Pardalotes to Shrike-Thrushes, Oxford University Press (USA) 2003

WATTLE-EYES, BATISES
Fry, C. Hilary; Keith, Stuart; Urban, Emil K. (eds.), *The Birds of Africa*, Volume V: Thrushes to Puffback Flycatchers, Princeton University Press 1997

HELMET-SHRIKES, PUFFBACKS
Fry, C. Hilary; Keith, Stuart; Urban, Emil K. (eds.), *The Birds of Africa*, Volume V: Thrushes to Puffback Flycatchers, Princeton University Press 1997
Fry, C. Hilary; Keith, Stuart; Urban, Emil K. (eds.), *The Birds of Africa*, Volume VI: Picathartes to Oxpeckers, Princeton University Press 2000
Harris, Tony, *Shrikes & Bush-Shrikes*, Princeton University Press 2000

ADDITIONAL READING

VANGAS
Sinclair, I.; Langrand, O., *Birds of the Indian Ocean Islands*, Struik 2003

BOATBILL, BUTCHERBIRDS, WOODSWALLOWS, SITTELLAS, SHRIKE-TITS
Higgins, Peter J.; Peter, John M.; Cowling, Sid J. (eds), *Handbook of Australian, New Zealand and Antarctic Birds*, Volume 7: Boatbills to Starlings, Oxford University Press (USA) 2006

IORAS, CUCKOO-SHRIKES, WHISTLERS
Marchant, Stephen; Higgins, Peter J. (eds), *The Handbook of Australian, New Zealand and Antarctic Birds*, Volume 6: Pardalotes to Shrike-Thrushes, Oxford University Press (USA) 2002

SHRIKES
Harris, Tony, *Shrikes & Bush-Shrikes*, Princeton University Press 2000
Lefranc, Norbert, *Shrikes: A Guide to the Shrikes of the World*, Yale University Press 1997

VIREOS
Alderfer, Jonathan (ed.), *Complete Birds of North America*, National Geographic Society 2006

SHRIKE-THRUSHES, AUSTRALASIAN ROBINS
Marchant, Stephen; Higgins, Peter J. (eds), *The Handbook of Australian, New Zealand and Antarctic Birds*, Volume 6: Pardalotes to Shrike-Thrushes, Oxford University Press (USA) 2002

MONARCHS
Fry, C. Hilary; Keith, Stuart; Urban, Emil K. (eds.), *The Birds of Africa*, Volume V: Thrushes to Puffback Flycatchers, Princeton University Press 1997

CROWS, JAYS
Goodwin, Derek, *Crows of the World*, 2nd edition, University of Washington Press 1987
Madge, Steve; Burn, Hilary, *Crows & Jays*, Princeton University Press 2001

BIRDS OF PARADISE
Cooper, W. T.; Forshaw, J. M., *The Birds of Paradise and Bower Birds*, Collins (Sydney) 1977
Frith, Cliffort B.; Beehler, Bruce M., *Bird Families of the World: The Birds of Paradise*, Oxford University Press (USA) 1998

BALD CROWS
Fry, C. Hilary; Keith, Stuart; Urban, Emil K. (eds.), *The Birds of Africa*, Volume VI: Picathartes to Oxpeckers, Princeton University Press 2000

CHICKADEES, TITS, PENDULINE TITS, LONG-TAILED TITS
Harrap, Simon; Quinn, David, *Chickadees, Tits, Nuthatches & Treecreepers*, Princeton University Press 1996

SWALLOWS, MARTINS, LARKS
Turner, Angela K., *A Handbook to the Swallows and Martins of the World*, Houghton Mifflin 1989

OLD WORLD WARBLERS
Baker, Kevin; Baker, Jeff, *Warblers of Europe, Asia, and North Africa*, Princeton University Press 1997
Shirihai, Hadoram; Gargallo, Gabriel; Helbig, Andreas J., *Sylvia Warblers*, Princeton University Press 2001

BABBLERS, PARROTBILLS
Fry, C. Hilary; Keith, Stuart; Urban, Emil K. (eds.), *The Birds of Africa*, Volume Vi: Picathartes to Oxpeckers, Princeton University Press 2000

WRENS, MOCKINGBIRDS, THRASHERS, DIPPERS
Brewer, David, *Wrens, Dippers, and Thrashers*, Yale University Press 2001

NUTHATCHES, TREECREEPERS, WALLCREEPER
Harrap, Simon; Quinn, David, *Chickadees, Tits, Nuthatches & Treecreepers*, Princeton University Press 1996

STARLINGS
Feare, Chris; Craig, Adrian, *Starlings and Mynas*, Princeton University Press 1999

THRUSHES
Clement, Peter, *Thrushes*, Princeton University Press 2001

CHATS, OLD WORLD FLYCATCHERS
Urquhart, Ewan, *Stonechats: A Guide to the Genus Saxicola*, Yale University Press 2003

FLOWERPECKERS, BERRYPECKERS, SUNBIRDS, SUGARBIRDS
Cheke, Robert A.; Mann, Clive F., *Sunbirds: A Guide to the Sunbirds, Spiderhunters, Sugarbirds and Flowerpeckers of the World*, Yale University Press 2001

OLD WORLD SPARROWS
Clement, Peter, *Finches & Sparrows: An Identification Guide*, Princeton University Press 1994

WEAVERS, WAXBILLS, INDIGOBIRDS
Fry, Hilary; Keith, Stuart; Urban, Emil K. (eds.), *The Birds of Africa*, Volume VII, Christopher Helm Publishers 2004

WAGTAILS, PIPITS
Alstrom, Per; Mild, Krister, *Pipits & Wagtails*, Princeton University Press 2003

FINCHES
Clement, Peter, *Finches & Sparrows: An Identification Guide*, Princeton University Press 1994
Pratt, H.D., *Bird Families of the World: The Hawaiian Honeycreepers*, Oxford University Press (USA) 2005

OLIVE WARBLER, WOOD WARBLERS
Curson, Jon; Quinn, David; Beadle, David, *Warblers of the Americas*, Houghton Mifflin 1994
Dunn, Jon N.; Garrett, Kimball, *A Field Guide to Warblers of North America*, Houghton Mifflin 1997

NEW WORLD BLACKBIRDS
Jaramillo, Alvaro; Burke, Peter, *New World Blackbirds: The Icterids*, Princeton University Press 1999

BUNTINGS, AMERICAN SPARROWS
Beadle, David; Rising, James D., *Sparrows of the United States and Canada: The Photographic Guide*, Princeton University Press 2001
Byers, Clive; Olsson, Urban, *Sparrows and Buntings: A Guide to the Sparrows and Buntings of North America and the World*, Houghton Mifflin 1995

TANAGERS, CARDINALS, SALTATORS, GROSBEAKS
Beadle, David; Rising, James D., *Tanagers, Cardinals, and Finches of the United States and Canada: The Photographic Guide*, Princeton University Press 2006
Isler, Morton L.; Isler, Phyllis R., *The Tanagers: Natural History, Distribution, and Identification*, Smithsonian Institution Press 1987

GLOSSARY

A

aftershaft A small supplementary feather growing from the calamus of a larger one; mainly seen in primitive birds such as screamers.

air sac Air-filled structure in a bird's body, which supplies air to the lungs.

allofeeding Feeding another bird, such as a mate or a member of a social group.

allopreening Preening another bird, such as a mate or member of a social group.

altricial A word used of young that stay in the nest until they are almost fully grown.

anisodactylous A type of toe arrangement in which the first toe points backward and the other three face forward.

arboreal Living entirely or mainly in trees.

austral Relating to the southern hemisphere.

axillars The feathers of a bird's 'armpit,' where the underside of the wing joins the body.

B

barb One of the soft, horizontal strands that project from the *shaft* of a feather.

barbules Tiny hooks that join the *barbs* of feathers together.

bare parts Those areas of a bird's body that are completely without feathers.

bill A bird's beak: the two hard parts that project from the face.

boreal Relating to the northern hemisphere: in particular, the area just south of the Arctic Circle.

breeding The set of activities that birds perform when raising young, including courtship, nest-building, mating, and brooding.

breeding plumage A colorful or ornate plumage that some species grow for the *breeding season*.

breeding season The time of year when *breeding* takes place.

brood A group of chicks that all come from the same *clutch*.

brood parasite A bird that lays its eggs in another species' nest, leaving the young for the other species to rear.

C

calamus See *quill*.

canopy The topmost continuous layer of a forest (usually *rain forest*).

carpal joint The joint halfway down a bird's wing, which forms the 'angle' in the wing; equivalent to the wrist joint in a human arm.

caruncle A natural lump of bare, fleshy skin, usually occurring on the face or neck.

casque A bony growth on the top of the *bill*, seen in species such as hornbills and cassowaries.

cere A bare, fleshy area of skin just above the *bill*, which contains the nostrils; seen in species such as *raptors* and budgerigars.

character A distinctive physical feature.

clutch A batch of eggs produced together at the same time.

colony A group of birds gathered together for *breeding*.

congeners Birds belonging to the same *genus*.

conspecifics Birds belonging to the same *species*.

convergent evolution A process in which birds from unrelated *families* or *genera* develop similar *characters* as a result of adapting to similar environments; examples include swifts and swallows, or hummingbirds and sunbirds.

cooperative breeding A form of *breeding* behavior in which a pair of birds raises young with the help of other adult or juvenile birds.

coverts Smaller feathers that overlay larger ones on the wings and tail, or grow over a bird's ears.

covey A group of birds; usually *game birds*, such as partridges.

crèche A group of young birds within a flock; the crèche is usually tended by one or more adult birds, which keep the young birds safe and teach them skills such as foraging.

crest A tuft or plume of feathers on a bird's crown.

crop A pouch, connected to a bird's gullet, that is used to store and carry food.

crown The top of a bird's head.

cryptic A word used of *plumage* that has subtle, camouflage-type patterns.

culmen The ridge along the top of a bird's *bill*.

D

dichromatism A form of *dimorphism* in which male and female birds have different *plumage*.

dimorphism A difference in size, shape, or color between males and females.

dispersal A movement away from a *breeding* area after the *breeding season*; it may involve moving to an adjacent area, or could be a prelude to migration.

display A behavior used to convey a specific message to other birds or animals; examples include courtship displays and threat displays.

double-brooded Producing two *clutches* or *broods* in a single *breeding season*.

down A very soft, fluffy type of feather; the first form of *plumage* in chicks, or an insulating layer under the main feathers in adults.

E

eclipse plumage A *plumage* seen on males of certain species, outside the *breeding season*; it is usually much duller than breeding plumage, and often similar to females' plumage.

endemic Confined to one specific area or habitat, such as an island or a country.

F

family A group of closely related *genera*; each article in this book covers one family, such as owls.

fledging The process in which a young bird grows its first full set of feathers.

fledgling A young bird that has recently completed *fledging*.

flight feathers The long feathers on the rear (trailing) edge of each wing; they comprise the *primaries* and the *secondaries*.

frugivorous Fruit-eating.

G

game bird A species traditionally hunted for food or sport; examples include grouse and pheasants.

gape The opening of the mouth, comprising the *bill* and the jaws; certain birds, such as nightjars, have a tiny bill but a huge gape.

generalist A bird that eats a wide variety of foods.

genus A group of closely related *species*; in *taxonomy*, indicated by a scientific name in italics, with a capital letter, e.g. *Bubo*.

gizzard The part of the digestive system where food is ground up; some species swallow stones or grit, which enter the gizzard and help break up food.

gonys A bulge near the tip of the lower *mandible*, most commonly seen in gulls.

gonydeal angle The angle where the *gonys* begins.

granivorous Seed-eating.

gular To do with the throat.

H

hallux A bird's 'big toe;' most birds have four toes, with the hallux pointing backward and the other three toes facing forward.

hypertrophied Enlarged; used of an organ or other body structure.

I

immature Not fully adult, either in some specific area of development, such as plumage, or as a whole.

incubation The process of keeping eggs at the correct temperature until they hatch; most birds sit on the eggs to keep them warm, but megapodes incubate their eggs by burying them in mounds of earth and rotting vegetation.

insectivorous Insect-eating.

irruption A sudden or unusual movement of birds outside their normal range, usually as a result of food shortages.

J

juvenal plumage *Plumage* worn by *juvenile* birds; some species go through a sequence of successive immature plumages before attaining adult plumage.

juvenile A bird wearing its first set of feathers after molting out of its natal down.

K

keel A projection that runs along the breastbone, to which the flight muscles are attached; much reduced or absent in flightless birds.

L

lamellae Comblike structures on the inner edges of the *bill*, in birds such as flamingos and some ducks, which are used to filter food items out of water.

lek A gathering of male birds at which they perform courtship *displays* to win females.

lobate A word used for the toes of some aquatic birds, such as grebes; lobes are flat, rounded growths of skin that aid swimming.

lores The area between the eye and the *bill*.

M

mandible One half of the *bill*; sometimes used specifically of the lower mandible.

mantle The *plumage* on a bird's upper back and shoulders.

maxilla The upper half of the *bill*; also called the upper *mandible*.

migration The movement of birds

GLOSSARY

from one part of the world to another, to breed or to spend winter in warmer conditions.

molt The orderly, seasonal replacement of old feathers with new feathers.

monogamous Taking only one mate at a time.

monophyletic A word used for a group of *species* that are all descended from a single ancestral species.

monotypic A word used of a *species* that is the sole member of a *genus* or even of a *family*.

morph One of two or more types of *plumage* coloration seen within the same *species*; sometimes known as a phase.

morphology The physical structure of a bird.

N

nestling A young bird still living in the nest.

nidicolous A word used of a young bird that remains in the nest until it is able to fly.

nidifugous A word for a young bird that leaves the nest very shortly after hatching.

nomadic Traveling constantly around an area following no established routes and with no fixed home territory.

nominate race In *species* with two or more *races*, the nominate race is the one first described by science; often indicated by a three-part scientific name with the second part repeated, e.g. *Phoenicopterus ruber ruber*.

O

obligate A word used of behavior that a bird is obliged to perform; for example, certain antbirds are obligate followers of army ant swarms, as this is the only way they obtain their food.

omnivorous Eating both animal and plant food.

opportunistic Taking advantage of temporary beneficial conditions, such as the abundance of a particular food.

order A taxonomic grouping that contains several *families*.

oscine A subgroup of *passerines*, comprising about 4,000 species, which have a *syrinx* adapted for singing; often called songbirds.

P

pair bond An association between a *breeding* pair of birds; pair bonds may last for part or all of a single breeding season, or for life, and are often reinforced by behaviors such as *allopreening* or *allofeeding*.

passerine A bird belonging to the *order* Passeriformes, with feet adapted for perching; divided into *oscines* and *suboscines*.

pectinate Comblike; some birds, such as herons, have a pectinate edge on their middle toe, used for grooming themselves.

plumage The collective word for all of a bird's feathers.

pneumatized A word used of a bird's body that is extensively supplied with air sacs, as in screamers, which have air sacs in their bones and skin as well as among their internal organs.

polyandrous A word for a female that takes several mates in the course of a breeding season.

polygamous Taking two or more mates in a breeding season.

polygynous A word for a male that mates with several females in a *breeding season* (either serially or at the same time).

polymorphic Having two or more *morphs*.

polytypic A word for a taxonomic category containing two or more representatives of the category immediately below it. For example, a polytypic *species* contains two or more *races*.

precocial A word used of young birds that are capable of independent activity, such as walking and feeding, soon after hatching.

preen gland A gland on the rump, which secretes oil that the bird uses in *preening*; also called the uropygial gland.

preening The act of cleaning and maintaining the feathers; a bird uses its bill to neaten individual feathers, stroke them down into place, and to apply preen oil.

preen oil Oil from the *preen gland*, used to waterproof feathers and keep them in good condition.

primaries The large *flight feathers* on the outer wing; birds typically have 9 or 10 primaries.

Q

quill The bare base of a feather, where it emerges from the skin.

R

race A distinct group within a *species*, also called a subspecies, and indicated by three scientific names; for example, Greater Flamingo comprises the races *Phoenicopterus ruber ruber* and *P. r. roseus*.

rachis See *shaft*.

ramus See *barb*.

range The geographical area over which a species is distributed.

raptor A bird of prey, typically with strong feet and long *talons*.

ratites Primitive, flightless birds such as ostriches, emus, and kiwis; typified by the absence of a *keel*.

rectrices (singular: rectrix) Another word for tail feathers.

relict A surviving *species* from an otherwise extinct group.

remiges (singular: remex) Another name for *flight feathers*.

rictal bristles Bristle-like feathers at the corners of the *gape*; seen particularly in insectivorous birds, possibly to protect the birds' eyes from the flailing legs and wings of their prey.

S

sally A sudden, darting flight, usually in pursuit of prey.

scapulars The feathers of a bird's shoulder, where the top of the wing joins the body.

secondaries The *flight feathers* on the inner wing, between the *carpal joint* and the body.

sedentary A word used of birds that occupy one specific territory year-round.

shaft The long, thin, stiff part at the center of a feather, from which the *barbs* grow.

specialist A bird that eats only particular plants or animals.

species A distinct group within a *genus*; it has its own English name as well as a two-part scientific name, e.g. Mute Swan (*Cygnus olor*).

speculum A colorful or iridescent patch seen on the *secondaries* in some species (e.g. Mallard).

suboscine A subgroup of *passerines*, comprising about 1,100 *species*, which are not adapted for singing; they typically have a simpler *syrinx* than the *oscines*.

substrate A surface on which a bird walks or forages.

supercilium A line of color running just above a bird's eye, resembling an eyebrow.

syndactylous A type of toe arrangement in which two of the toes are fused together along part or all of their length, as in kingfishers.

syrinx A bird's vocal organ, or 'voice box,' located at the lower end of the windpipe.

T

talons Sharp, curved claws, adapted for seizing and killing prey.

tarsal scutellation Scales on part or all of the *tarsus*.

tarsus The bare lower leg of a bird.

taxon (plural: taxa) Any distinct groups defined in *taxonomy*, such as a *family*, *genus*, *species*, or *race*.

taxonomy The scientific classification of organisms: assigning names and relationships.

terrestrial Living entirely or mainly on the ground.

tertials Prominent feathers, usually three in number, that overlay the *secondaries* of the folded wing; often referred to as the innermost secondaries.

tomial Referring to the tomium: the cutting edge of a bird's *bill*.

U

underparts The breast and belly of a bird.

undertail coverts The feathers under a bird's tail that cover the bases of the *rectrices*.

underwing coverts The feathers under a bird's wing that cover the bases of the *flight feathers*; also known as the wing linings.

upperparts The shoulders, back, rump, and wings of a bird.

uppertail coverts The feathers on top of a bird's tail that cover the bases of the *rectrices*.

upperwing coverts The feathers on top of a bird's wing that cover the bases of the *flight feathers*.

V

ventriloquial A word used for a bird call that does not seem to come from the bird itself, thus helping disguise a bird's location.

vermiculation A threadlike pattern of fine lines on *plumage*.

W

wattle A loose, wrinkled natural growth of skin that hangs from a bird's lower face or throat.

wing bars One or two contrasting bars of color running across a bird's wing, formed by the pale tips of the greater and/or median secondary *coverts*.

Z

zygodactylous A type of toe arrangement in which two toes face forward and two face backward; seen in birds such as parrots.

INDEX

Bold page numbers indicate illustrations.

INDEX

INDEX

INDEX

INDEX

INDEX

INDEX

INDEX

INDEX

INDEX

INDEX

INDEX

INDEX

PHOTOGRAPHY AND ILLUSTRATION CREDITS

Photography Credits

Alamy: Natural History Museum, London 153b, Blickwinkel 272t; **Ardea:** Dennis Avon 16b, Francois Gohier 17t, Rolf Kopfle 43b, Jean Paul Ferrero 84b, 272b, Pat Morris 160tr, Alan Greensmith 210t, Ellis McNamara 220, 253, Don Hadden 274, 301b, Duncan Usher 283t, Joanna Van Gruisen 300b; **Arthur Grosset:** 209; **Bob Steele Photography:** 15t, 223, 228b, 243, 259b, 266t, 266b, 306t, 351; **Brian E. Small:** 358, 359b, 363b; **Dennis Buurman:** 51; **FLPA:** Michael & Patricia Fogden 17b, 198r, 200t, 200b, Konrad Wothe 20, 71, 145, 197b, Tui De Roy 25t, 63t, 185, Hugh Lansdown 25b, Jurgen & Christine Sohns 26r, 129b, 134t, 241t, 340t, Andrew Forsyth 28t, Tom Vezo 31b, 147, David Hosking 33t, 49, 92, 237b, 240b, 291b, 318, 320t, 335t, R. & M. Van Nostrand 33b, Malcolm Schuyl 47, 115b, 127t, Flip Nicklin 48t, FLPA 61b, Flip De Nooyer 68t, 132, Dickie Duckett 72, 135b, Robin Chittenden 74, 187b, Pete Oxford 83t, 89, 146t, 160b, 179b, Frank W. Lane 83b, 195b, 324b, Winfried Wisniewski 91b, Neil Bowman 93, 178t, 224, 250, 261t, 265t, 290b, 316b, 326b, 332tr, Bill Baston 94, 294b, S., D. & K. Maslowski 100t, 346b, Roger Tidman 111b, 151b, 152t, 294t, S. Jonasson 119b, Tony Hamblin 120, Wendy Dennis 122t, Gerard Lacz 133b, Gerry Ellis 144b, 254b, Michael Gore 148b, 242, Frans Lanting 152b, 191t, 347tr, Martin B. Withers 158t, 241b, 262, 289t, 314b, 334tl, Terry Andrewartha 160tl, ZSSD 176t, SA Team 178b, John Holmes 179tr, 264b, 281b, 297b, 308tr, 325, 330b, 342t, 342bl, Claus Meyer 196t, Tom & Pam Gardner 222, 238t, 254t, 302, Geoff Moon 237t, Len Robinson 265b, Cyril Ruoso 267b, Mike Lane 268b, John Watkins 270t, Stefan Huwiler 280, John Hawkins 284t, Roger Wilmshurst 288b, Danny Ellinger 296t, Francois Merlet 331t, Richard Brooks 332tl, John Karmali 334b, 340b; **Gary Rosenberg:** 37b, 76, 107, 108, 119t, 129t, 141, 155b, 159, 168, 177b, 182, 187t, 196b, 199, 201, 202, 203t, 203b, 212t, 212b, 213t, 213b, 214, 279, 295t, 348bl, 361t; **Jim Zipp:** 14, 204, 256t, 283b, 308tl, 309b, 323t, 323b, 352t, 362t, 362b; **John Jackson:** 48b; **Kevin T. Karlson:** 355t, 355b; **Marc Anderson Photography:** 248; **Mike Danzenbaker:** 63b, 192b, 197t, 206, 207t, 207b, 211t, 211b, 216, 217, 259t, 269t, 311t, 311b, 315b, 317b, 350t, 350b, 353br, 357bl, 357br, 360t, 361b, 363t; **Natural Visions:** Richard Coomber 240t; **NaturePL:** Gabriel Rojo 16t, 35, 87t, 95l, Tony Heald 18, 28b, 56t, Rolf Nussbaumer 30, 53b, 75, 91t, 117, 165t, 167, 181b, 278, 352b, 353t, 354b, John Cancalosi 31t, 62l, 127b, 137, 157, 329, 335b, David Tipling 34b, 43t, 205, Hugh Pearson 44t, Pete Oxford 44b, 50, 56b, 65t, 69tr, 82b, 95r, 130t, 163t, 163b, 188, 198l, 360b, Michael Pitts 45t, 97b, Todd Pusser 45b, Tom Vezo 61t, Mike Wilkes 64, 112t, 181t, Cindy Buxton 65b, Peter Reese 69tl, 189, Vincent Munier 81t, 142t, Andrew Parkinson 81b, Nick Garbutt 82t, 162b, 192t, 244b, Hermann Brehm 85, Nature Production 86, 175, 292b, 321tr, Laurent Geslin 87b, Gary K. Smith 100b, Roger Powell 105t, 249, 307, Andy Sands 110b, Bernard Castlelein 112b, 116, 169, Hanne & Jens Eriksen 118, 121t, 123, 261b, 289b, 291t, 313b, 320br, 328t, 339t, 347tl, Christophe Courteau 130b, Mike Read 139, Dietmar Nill 143, Mark Carwardine 146b, Martha Holmes 148t, Kim Taylor 156, Charlie Hamilton James 164, Aflo 165b, Richard Du Toit 166t, Dave Watts 166b, 230t, 230b, 246t, 317t, T. J. Rich 170b, Jose B. Ruiz 171, Luiz Claudio Marigo 215, William Osborn 226t, Artur Tabor 260, Steven David Miller 275t, Paul Hobson 286, Nigel Bean 299t, 354t, Nick Upton 310, David Kjaer 319b, 322b, Markus Varesvuo 321tl, Juan Manuel Borrero 342br; **Neil Bowman:** 2–3, 21, 36, 98, 102t, 102b, 115t, 124, 125, 126t, 126b, 128, 131, 136, 138t, 138b, 161t, 170t, 174b, 184, 221, 225, 226b, 227, 228t, 231, 232, 233t, 233b, 234, 245, 256b, 257t, 257b, 263t, 263b, 264t, 267t, 268t, 270b, 271t, 271b, 275b, 276l, 276r, 285b, 287t, 288tl, 288tr, 290t, 292t, 300t, 301t, 313t, 314t, 315t, 319t, 320bl, 321b, 327, 328b, 330t, 331b, 332b, 337b, 338b, 341t, 341b, 344, 345t, 345b, 346t, 347b, 348t, 348br, 356b, 357t; **Pete Morris:** 106; **Photolibrary Group:** Juniors Bildarchiv 11, 57t, 58b, 60, 134b, 194, 282b, 324t, Konrad Wothe 19, Chris McLennan 23, Patricio Robles Gil 26l, Richard & Julia Kemp 27, Mark Hamblin 52, 96, Barbara Von Hoffmann 53t, Gerard Soury 69b, Fritz Polking 70t, Carlos Sanchez Alonso 80, James Hager 103, J. & C. Sohns 133t, 177t, Alan G. Nelson 151t, Arthur Gloor 172t, Tony Tilford 273t, Adam Jones 303; **Photoshot:** A.N.T. Photolibrary 15b, 150b, 239, 246b, 251, 252, 255, 296b, Nigel J. Dennis 29, 180t, 180b, 244t, Bill Coster 46, 67t, Mike Lane 58t, 322t, Henry Ausloos 68b, Daniel Heuclin 84t, Haroldo Palo Jr 135t, 183t, 186, 305t, Bruce Beehler 144t, 150t, 235, 236, 238b, Gerald Cubitt 149, Morten Strange 153t, 247, 273b, 299b, 326t, Tony Crocetta 158b, 179tl, Nick Garbutt 162t, 190, Cede Prudente 173, 191b, Kevin Schafer 195t, Dave Watts 218, 219, Joe Blossom 269b, 298, Jari Peltomaki 282t, Roger Tidman 295b, Melvin Grey 297t, 309t, Mark Bowler 306b, Eric Soder 308b, E. Hanumantha Rao 336; **Rob Curtis:** 287b, 353bl, 356t, 359t; **Robert Royse:** 258, 316t, 343t, 343b, 349; **Shutterstock:** Muriel Lasure 32, Marek Cech 34t, Ying 37t, Eric Gevaert 38, Malewitch 39b, Lynsey Allan 41t, Ifstewart 41b, Armin Rose 42t, 42b, Kristian Sekulic 54t, 99t, Brad Thompson 54b, Calin Tatu 55, Liv Falvey 57b, Chris Fourie 59, Franck Camhi 62r, Henk Bentlage 66t, Zimmytws 66b, Javarman 67b, Bob Blanchard 70b, EcoPrint 77, 140t, 334tr, Ismael Montero Verdu 78b, FloridaStock 79, 99b, Steffen Foerster Photography 97t, Johan Swanepoel 104, 105b, 121b, 122b, 161b, 174t, 176b, Gerrit De Vries 111t, 337t, 339b, Mollypix 114t, Dennis Donohue 114b, Ronnie Howard 154, Teekaygee 183b, Bruce MacQueen 281t, Andre Klopper 284b, Karel Broz 285t, Paul S. Wolf 304t, Michael J. Thompson 304b, Michael Woodruff 305b, Jan Erasmus 333, Joanne Harris & Daniel Bubnich 338t; **Stefan Oscarsson:** 39t, 73, 78t, 90, 109, 110t, 140b, 142b, 293t, 293b; **Still Pictures:** K. Hennig 172b; **Vireo:** C. H. Greenewalt 155t, S. Elsom 208, Doug Wechsler 210b.

Illustration Credits

BRG: 19, 21, 22, 24, 50, 52, 88, 89, 90, 118l, 118r, 167, 168, 186, 188l, 188r, 197, 199, 208, 209, 269, 274, 298; **Michael Woods:** 98, 193, 229, 277, 312; **NGS:** Thomas R. Schultz 13t, 13b, 113, 117, Cynthia J. House 40, Diane Pierce 60, N. John Schmitt 74, 310, Kent Pendleton 77, Jonathan Alderfer 101, John C. Pitcher 109, Donald L. Malick 143, H. Douglas Pratt 258, 349.

NATIONAL GEOGRAPHIC COMPLETE BIRDS OF THE WORLD
Edited by Tim Harris

PUBLISHED BY THE NATIONAL GEOGRAPHIC SOCIETY
John M. Fahey, Jr. *President and Chief Executive Officer*
Gilbert M. Grosvenor *Chairman of the Board*
Tim T. Kelly *President, Global Media Group*
John Q. Griffin *President, Publishing*
Nina D. Hoffman *Executive Vice President; President, Book Publishing Group*

PREPARED BY THE BOOK DIVISION
Kevin Mulroy *Senior Vice President and Publisher*
Leah Bendavid-Val *Director of Photography Publishing and Illustrations*
Marianne R. Koszorus *Director of Design*
Barbara Brownell Grogan *Executive Editor*
Elizabeth Newhouse *Director of Travel Publishing*
Carl Mehler *Director of Maps*

STAFF FOR THIS BOOK:
THE BROWN REFERENCE GROUP LTD
Katie John *Project Editor*
David Poole *Design Manager*
Tony Truscott *Designer*
Sophie Mortimer *Picture Manager*
Sean Hannaway *Picture Researcher*
Encompass Graphics Ltd *Cartography*
Lindsey Lowe *Editorial Director*
Alastair Gourlay *Production Director*
Tim Harris *Managing Editor*
Chris Perrins *Consultant*
Peter Clement *Consultant*

NATIONAL GEOGRAPHIC SOCIETY
Jonathan Alderfer *Chief Consultant of Birding Program*
Barbara Levitt *Editor*
Carol Norton *Art Director of Birding Program*
Sanaa Akkach *Designer*
Meredith Wilcox *Illustrations Specialist*
Cameron Zotter *Design Assistant*
Jennifer Thornton *Managing Editor*
R. Gary Colbert *Production Director*
Connie D. Binder *Indexer*

MANUFACTURING AND QUALITY MANAGEMENT
Christopher A. Liedel *Chief Financial Officer*
Phillip L. Schlosser *Vice President*
Chris Brown *Technical Director*
Nicole Elliott *Manager*
Monika D. Lynde *Manager*
Rachel Faulise *Manager*

Founded in 1888, the National Geographic Society is one of the largest nonprofit scientific and educational organizations in the world. It reaches more than 285 million people worldwide each month through its official journal, *National Geographic*, and its four other magazines; the National Geographic Channel; television documentaries; radio programs; films; books; videos and DVDs; maps; and interactive media. National Geographic has funded more than 8,000 scientific research projects and supports an education program combating geographic illiteracy.

For more information, please call 1-800-NGS LINE (647-5463) or write to the following address:

National Geographic Society
1145 17th Street N.W.
Washington, D.C. 20036-4688 U.S.A.

Visit us online at www.nationalgeographic.com

For information about special discounts for bulk purchases, please contact National Geographic Books Special Sales: ngspecsales@ngs.org

For rights or permissions inquiries, please contact National Geographic Books Subsidiary Rights: ngbookrights@ngs.org

ISBN: 978-1-4262-0403-6 (Trade)

Printed in China

ACKNOWLEDGMENTS

The editors wish to thank especially Jonathan Alderfer, the Chief Consultant of the National Geographic Birding Program, for his thorough checking of text, captions, and photographs, and for his unstinting support throughout.

FREE PUBLIC LIBRARY UNION, NEW JERSEY
3 9549 00436 7927